Genome editing for precision crop breeding

It is widely recognised that agriculture is a significant contributor to global warming and climate change. Agriculture needs to reduce its environmental impact and adapt to current climate change whilst still feeding a growing population, i.e. become more 'climate-smart'. Burleigh Dodds Science Publishing is playing its part in achieving this by bringing together key research on making the production of the world's most important crops and livestock products more sustainable.

Based on extensive research, our publications specifically target the challenge of climate-smart agriculture. In this way we are using 'smart publishing' to help achieve climate-smart agriculture.

Burleigh Dodds Science Publishing is an independent and innovative publisher delivering high quality customer-focused agricultural science content in both print and online formats for the academic and research communities. Our aim is to build a foundation of knowledge on which researchers can build to meet the challenge of climate-smart agriculture.

For more information about Burleigh Dodds Science Publishing simply call us on +44 (0) 1223 839365, email info@bdspublishing.com or alternatively please visit our website at www.bdspublishing.com.

Related titles:

Advances in breeding techniques for cereal crops
Print (ISBN 978-1-78676-244-3); Online (ISBN 978-1-78676-246-7, 978-1-78676-247-4)

Plant genetic resources: A review of current research and future needs
Print (ISBN 978-1-78676-451-5); Online (ISBN 978-1-78676-453-9, 978-1-78676-454-6)

Chapters are available individually from our online bookshop:
https://shop.bdspublishing.com

climate•SMART•publishing
IN AGRICULTURAL SCIENCE

BURLEIGH DODDS SERIES IN AGRICULTURAL SCIENCE
NUMBER 97

Genome editing for precision crop breeding

Edited by Dr Matthew R. Willmann, Cornell University, USA

burleigh dodds
SCIENCE PUBLISHING

Published by Burleigh Dodds Science Publishing Limited
82 High Street, Sawston, Cambridge CB22 3HJ, UK
www.bdspublishing.com

Burleigh Dodds Science Publishing, 1518 Walnut Street, Suite 900, Philadelphia, PA 19102-3406, USA

First published 2021 by Burleigh Dodds Science Publishing Limited

Library of Congress Control Number: 2021931537

British Library Cataloguing in Publication Data
A catalogue record for this book is available from the British Library

ISBN 978-1-78676-447-8 (Print)
ISBN 978-1-78676-450-8 (PDF)
ISBN 978-1-78676-449-2 (ePub)
ISSN 2059-6936 (print)
ISSN 2059-6944 (online)

DOI 10.19103/AS.2020.0082

Typeset by Deanta Global Publishing Services, Dublin, Ireland

Contents

Part 2 Applications

Series list

Title	Series number
Achieving sustainable cultivation of maize - Vol 1 From improved varieties to local applications Edited by: Dr Dave Watson, CGIAR Maize Research Program Manager, CIMMYT, Mexico	001
Achieving sustainable cultivation of maize - Vol 2 Cultivation techniques, pest and disease control Edited by: Dr Dave Watson, CGIAR Maize Research Program Manager, CIMMYT, Mexico	002
Achieving sustainable cultivation of rice - Vol 1 Breeding for higher yield and quality Edited by: Prof. Takuji Sasaki, Tokyo University of Agriculture, Japan	003
Achieving sustainable cultivation of rice - Vol 2 Cultivation, pest and disease management Edited by: Prof. Takuji Sasaki, Tokyo University of Agriculture, Japan	004
Achieving sustainable cultivation of wheat - Vol 1 Breeding, quality traits, pests and diseases Edited by: Prof. Peter Langridge, The University of Adelaide, Australia	005
Achieving sustainable cultivation of wheat - Vol 2 Cultivation techniques Edited by: Prof. Peter Langridge, The University of Adelaide, Australia	006
Achieving sustainable cultivation of tomatoes Edited by: Dr Autar Mattoo, USDA-ARS, USA & Prof. Avtar Handa, Purdue University, USA	007
Achieving sustainable production of milk - Vol 1 Milk composition, genetics and breeding Edited by: Dr Nico van Belzen, International Dairy Federation (IDF), Belgium	008
Achieving sustainable production of milk - Vol 2 Safety, quality and sustainability Edited by: Dr Nico van Belzen, International Dairy Federation (IDF), Belgium	009
Achieving sustainable production of milk - Vol 3 Dairy herd management and welfare Edited by: Prof. John Webster, University of Bristol, UK	010
Ensuring safety and quality in the production of beef - Vol 1 Safety Edited by: Prof. Gary Acuff, Texas A&M University, USA & Prof. James Dickson, Iowa State University, USA	011
Ensuring safety and quality in the production of beef - Vol 2 Quality Edited by: Prof. Michael Dikeman, Kansas State University, USA	012
Achieving sustainable production of poultry meat - Vol 1 Safety, quality and sustainability Edited by: Prof. Steven C. Ricke, University of Arkansas, USA	013
Achieving sustainable production of poultry meat - Vol 2 Breeding and nutrition Edited by: Prof. Todd Applegate, University of Georgia, USA	014

Acknowledgements

We wish to acknowledge the following for their help in reviewing particular chapters:

- Chapter 1: Dr Colby Starker and Professor Feng Zhang, University of Minnesota, USA
- Chapter 2: Dr Sylvia de Pater, University of Leiden, The Netherlands
- Chapter 3: Professor Kan Wang, Dr Alan Eggenberger, Dr Jacob Zobrist and Dr Minjeong Kang, Iowa State University, USA
- Chapter 4: Dr Magdy Mahfouz, King Abdullah University of Science and Technology, Saudi Arabia
- Chapter 5: Dr Florian Veiilet, INRAE, France; and Dr Sachin Rustgi, Clemson University, USA
- Chapter 8: Professor Yaoguang Liu, South China Agricultural University, China; and Dr Jimmy Botella, The University of Queensland, Australia
- Chapter 9: Dr Sakiko Okumoto, Texas A&M University, USA
- Chapter 10: Dr Jay Shockey, USDA-ARS, USA
- Chapter 11: Professor Jim Dunwell, University of Reading, UK; and Professor Piet van der Meer, Ghent University, Belgium
- Chapter 12: Dr Sebastian Gasparis, IHAR, Poland
- Chapter 13: Professor Dave Jackson, Cold Spring Harbor Laboratory, USA
- Chapter 16: Dr Ning Zhang, Cornell University, USA
- Chapter 17: Dr Chida Kanchiswamy, Plantedit, Ireland

Introduction

Genome editing is rapidly transforming plant research. The technique offers unparalleled precision in breeding without the need to introduce foreign DNA into plants. CRISPR/Cas systems have established themselves as the leading technique in genome editing. This collection of chapters takes stock of the wealth of research on these techniques and their potential in crop breeding in improving traits such as yield, disease resistance, drought tolerance and nutrient use efficiency.

Part 1 of this volume reviews advances in techniques such as TALENS and zinc finger nucleases, double-strand break repair techniques, insertion-based genome edits, base editing, guide RNAs and gRNA/Cas9 constructs. Chapters in this part of the book also discuss advances in screening plants as well as regulatory issues. Part 2 examines applications of gene editing in key cereal crops, specifically focusing on barley, maize and sorghum. Other chapters cover crops such as brassicas, tomatoes and perennials.

Part 1 Genome editing techniques

Part 1 begins with a chapter that examines using transcription activator-like effector nucleases (TALENs) for genome editing in plants. TALENs are powerful tools for precise and efficient locus-specific editing. Among the various genome editing tools that were developed prior to the CRISPR/Cas system, TALENs have been the most widely used, and they remain worthy of consideration due to their unique targeting properties and the current intellectual property landscape, which give them distinct advantages over CRISPR/Cas for some applications. Chapter 1 briefly reviews the use of TALENs in plants including their important distinguishing features, as well as design principles and tools, methods for construct assembly, and other available resources to assist researchers using this technology.

Chapter 2 reviews double strand break (DSB) repair pathways in plants and their application in genome engineering. In genome engineering, after targeted induction of DSBs researchers take advantage of the organisms' own repair mechanisms to induce different kinds of sequence changes into the genome. Therefore, understanding of the underlying mechanisms is essential. The chapter describes in detail the two main pathways of DSB repair in plant cells, non-homologous end joining (NHEJ) and homologous recombination (HR) and sums up what we have learned over the last few decades about them. The chapter also summarises the different models that have been proposed and set these into relation with the molecular outcomes of different classes of

DSB repair. Moreover, it describes the factors that have been identified to be involved in these pathways. Applying this knowledge of DSB repair should help us to improve the efficiency of different types of genome engineering in plants.

The next chapter assesses advances in the generation of insertion-based genome edits in plants. Tremendous progress has been achieved in the field of gene editing in plants, such as with the use of zinc finger nucleases (ZFNs), transcription activator-like effector nucleases (TALENs), and clustered regularly interspaced short palindromic repeats (CRISPR). Because of the potential advantages associated with mutant creation and crop germplasm innovation, genome editing technology has been rapidly developed and widely used in crop improvement in recent years. Chapter 3 documents some of the important recent developments and applications of genome-editing tools, especially with respect to gene knock-ins. It introduces the mechanism underlying knock-ins and different outcomes of insertion. The chapter also discusses genome editing tools and methods developed to improve insertion efficiencies. Additionally, the chapter reviews the recent trends in genetic editing biotechnologies; several strategies are being developed to further improve the efficiency of plant gene knock-ins. Undoubtedly, CRISPR/Cas technology will boost the development of new plant breeding techniques tremendously.

The subject of Chapter 4 is using viruses as vectors for the delivery of gene editing reagents. A significant challenge for plant gene editing is the delivery of editing reagents to germline or regenerable cells to recover heritable genetic modifications. Reagent delivery using biolistics or Agrobacterium is only possible with a limited range of species and genotypes, and inefficient editing or lengthy tissue culture steps further limit throughput. Viruses are natural vectors for nucleic acids, and both DNA and RNA plant viruses have been engineered to extend or replace conventional vectors for delivery of gene editing reagents. The chapter reviews aspects of viral biology essential for engineering vectors, highlights landmark studies using viruses to overcome traditional limitations in gene editing, and outlines important considerations for the use of viral vectors in new systems or for new targets. Motivated by fundamental differences in both their infection modes and utility as vectors, DNA and RNA viruses are discussed separately.

Moving on from Chapter 4, Chapter 5 addresses the progress that has been made in precise and predictable genome editing in plants with base editing. Base editors are gene editing tools that allow targeted nucleic acid conversions, most commonly C>T and A>G, through pairing of deamination domains with impaired nucleases. Multiple deaminase domains and architectures have been demonstrated in planta across a wide array of species, with both cytosine and adenine base editing frequencies being observed at over 80%. The chapter first focuses on progress that has been made in mammalian systems, then moves on to discuss cytosine and adenine base editing in plants. A section on protospacer adjacent motif (PAM) accessibility is also included, followed by a discussion on sequence diversification in crops, off-target base editing and current applications of base editors in crops.

The next chapter analyses the advances in guide RNA design for editing plant genomes using CRISPR-Cas systems. To date, many bioinformatic tools have been developed to facilitate the design of sgRNAs for CRISPR-based genome editing in eukaryotes. Chapter 6 focuses specifically on the general rules for selecting target sites for genome editing using CRISPR-Cas technology and summarises the bioinformatic tools that can be used to design sgRNA sequences. It also discusses predicting the outcome of CRISPR-mediated genome editing, before concluding with an overview of why designing a sgRNA is a crucial step for CRISPR-based genome editing.

Chapter 7 focuses on the advances in assembling gRNA/Cas9 constructs in genome editing of plants. It reviews the principles of CRISPR cloning in binary vectors and the different methods and elements employed, including the nucleases alternative to Cas9. It pays special attention to modular cloning strategies and multiplexing tools as well as the engineering of expanded Cas activities. Finally, the chapter includes a case study of the cloning of a nine-gRNA multiplex construct and the analysis of its transformants in tobacco plants.

The subject of Chapter 8 is strategies for CRISPR/Cas9-mediated genome editing: from delivery to production of modified plants. Despite the conspicuous and rapid development of genome editing tools, implementing this technology in plants often remains constrained by our inability to regenerate fertile genome-modified plants. Finding a viable combination of delivery method, genome editing reagents, and plant regeneration system has often been both crop-specific and crop-limited. Recent progress using morphogenic genes such as Wuschel2 (Wus2) and Babyboom (Bbm) has alleviated some of these crop-specific challenges, and along with other improvements, continues to expand the range of plants that can be edited. The chapter focuses specifically on delivery of genome editing components into plant cells, then moves on to examine the delivery methods for genome editing reagents in terms of delivery into single cells and intact tissues. A section on alternatives to DNA delivery is also provided. This is followed by discussions on morphogenic genes and how they can increase transformation efficiency, extend genotype range and permit transformation in non-traditional explants.

Chapter 9 considers the advances in screening plants for edits and off targets. Genome editing technology greatly accelerates the development of plant molecular research and crop breeding. However, accurately detecting mutations induced by sequence-specific nucleases between large-scale plants produced from transformation become a challenge. The chapter introduces some frequently used molecular methods of identify genome editing induced mutation or off-targets, which developed in recent years. It also discusses the advantages and disadvantages of these methods and the scope of application, providing a useful reference for researchers who will identify genome-edited plants.

The next chapter assesses targeted modification of promoters. Although most genome editing efforts focus on modifications to gene coding regions, this chapter emphasizes genome editing of the upstream regulatory regions. Thoughtful editing of the promoter region will ultimately lead to improved plants, modified for more precise control of the intensity and specificity of native gene expression. Chapter 10 presents an overview of the promoter or upstream regulatory region of a gene and describes how this sequence is defined and studied. It then describes how the composition and arrangements of cis-regulatory elements within the promoter and the leading intron associated with the promoter region have been studied using classical transgenic approaches to reveal what regulatory components might be suitable for genome editing approaches. Finally, the chapter offers some suggestions for pursuit of promoter editing and gene expression modulation, which will eventually lead to modified plants with an altered regulation of native gene expression.

The final chapter of Part 1 reviews the regulation of genome edited crops. To explore the regulation of genome edited crops, Chapter 11 describes the regulatory systems adopted by different countries as well as how the Cartagena Biosafety Protocol might apply. It finds there is a continuum of oversight, with some countries applying their GMO biosafety laws to all genome edited crops and other countries exempting certain genome edited techniques and/or products from any special oversight. For countries with what on the surface seem like similar regulatory policies, when their regulations are applied to identical products, they often produce different results, which will make international harmonization difficult. In addition, how much oversight to impose on different crops tends not to be based on scientific evidence. Finally, the regulatory treatment for multi-edited products is unclear. It will take time and additional the development of additional genome edited crops before there is sufficient clarity on how all types of genome edited crops will be regulated.

Part 2 Applications

Part 2 opens with a discussion of genome editing of barley. Although barley is of great importance for the brewing and animal feed industries and is regarded as a model for small grain cereals, only a few results on targeted gene modification using CRISPR/Cas endonuclease technology have been published to date. Chapter 12 examines the frontiers and achievements of the currently used techniques in barley genome modification. It focuses specifically on the expression systems used for targeted mutagenesis, enhancing plant development and performance and also examines the molecular understanding of resistance against plant pathogens. A section on targeted mutagenesis for molecular farming purposes is also included. The chapter concludes by emphasising the importance of developing genome editing techniques in barley.

Moving on from Chapter 12, Chapter 13 focuses on genome editing of maize. Developed over thousands of years largely through human intervention, the modern maize genome can now be precisely modified for agricultural improvement and scientific research. The chapter focuses on the recent progress that has been made in utilizing site-specific nuclease (SSN) technologies in maize genome engineering. Many SSNs, such as meganucleases, zinc finger nucleases (ZFNs), transcription activator-like effector nucleases (TALENs) and clustered regularly interspaced short palindromic repeat (CRISPR)/CRISPR-associated proteins (Cas) have been used in maize for both functional analysis and trait improvement. The chapter summarizes the recent innovations related to maize genome editing using SSN technologies, the type of approaches, target genes and traits, and reagent delivery methods. It also discusses the current challenges as well as potential improvements for maize genome engineering protocols.

Chapter 14 addresses genome editing in sorghum. Genetic engineering plays a key role in plant functional research and genetic improvement. A novel and powerful gene editing technique, CRISPR/Cas9, which was developed from a type II bacterial immune system, opened up a new era in precision genetic engineering in plants. This technique is based on a non-permanent transgene system and is starting to be adopted for precise gene editing in major cereal crops. It offers tremendous potential to accelerate crop improvement in a way that potentially reduces or eliminates the cumbersome and expensive regulatory processes associated with traditional transgenic crops. The chapter describes the advance of gene editing applied to sorghum, a drought tolerant C4 crop, and a successful strategy of CRISPR/Cas9 mediated gene family editing to improve sorghum digestibility and protein quality. It also discusses future prospects of CRISPR/Cas9 gene editing for sorghum genetic improvement.

The next chapter examines CRISPR/Cas9-mediated genome editing in Brassica. CRISPR/Cas9 is a valuable tool for both basic and applied research that has been widely applied to different plant species. Chapter 15 reviews the application of the CRISPR/Cas9 genome editing toolkit in various *Brassica* crops. The chapter also provides a case study in *Brassica napus*, specifically describing how mutants are generated using the CRISPR/Cas9 toolbox. Collectively, the results demonstrate that CRISPR/Cas9 is an efficient tool for creating targeted genome modifications at multiple loci in B. napus. These findings open many doors for biotechnological applications in oilseed crops.

Chapter 16 focuses on genome editing of tomatoes and other Solanaceae. Application and effectiveness of different editing approaches based on CRISPR/Cas of solanaceous species are outlined in this chapter. The chapter begins by discussing gene editing proof-of-concept investigations, which is then followed by a review of DNA sequence insertion. The chapter moves on to examine base editing and modification of gene function for improved disease tolerance, specifically focusing on Solanaceae species. Gene editing to affect fruit and tuber quality is also discussed and a section on plant growth and development modifications is provided. The chapter also provides a case study on application of gene editing for de novo domestication.

The final chapter of the book examines genome editing of woody perennial trees. Chapter 17 discusses the challenges facing woody perennial trees, namely the heterozygosity of outcrossing species and limited genomic resources. Gene knockouts (KO) represent the predominant applications of the CRISPR technology in woody perennial crops to date and will be the focus of this chapter. The issues considered for gene KO are generally applicable to other CRISPR applications targeting regulatory sequences or non-coding genes. Case studies from the Populus 4-coumarate:CoA ligase (4CL) family are presented to demonstrate the power of CRISPR in elucidating functional redundancy as well as specificity of duplicated genes in a perennial woody species.

Part 1

Genome editing techniques

Chapter 1

Using TALENs for genome editing in plants

Hilal Betul Kaya, Manisa Celal Bayar University, Turkey; Rhitu Rai, ICAR-National Institute for Plant Biotechnology, India; and Adam J. Bogdanove, Cornell University, USA

1 Introduction

1.1 Why consider TALENs?

Given the ease of use, proven efficacy, and focused attention on CRISPR/Cas systems ('CRISPR') for genome editing, it is reasonable to ask why anyone should consider using transcription activator-like (TAL) effector nucleases (TALENs). Of the four main site-directed nucleases used to date for genome editing–zinc finger nucleases, meganucleases, TALENs, and CRISPR (reviewed in Carroll, 2014; Bogdanove et al., 2018)–CRISPR, is by far the most widely used. In contrast to the other reagents, targeting CRISPR to a sequence of choice is a simple matter of generating an RNA construct that hybridizes to that sequence. The specificity of the other nucleases is determined by their direct interactions with the DNA and thus requires protein engineering to customize. This is most challenging for meganucleases, which recognize long sequences (typically 12–40 bases and up to 60 bases (Stoddard, 2014)) through DNA contacts that are structurally complex. It is less so but still arduous for zinc finger nucleases, in which individual structural units ('fingers') specify triplet nucleotides in the bound DNA sequence but are influenced by the neighboring fingers. TAL effectors emerged as the first truly modular DNA-binding proteins, their specificity being determined by polymorphic structural repeats that individually and independently specify single nucleotides in the

http://dx.doi.org/10.19103/AS.2020.0082.02

bound DNA sequence, in a contiguous fashion. TALENs are customized simply by assembling an array of repeats in the TAL effector domains that specify the desired sequence. Several efficient protocols using publicly available libraries of repeat modules have been developed for low-cost assembly in as little as three days using standard molecular biology techniques. Along with the ease of engineering TALENs, there are three important features of TALENs distinct from CRISPR that make these reagents worth considering, described below.

1.2 TALENs target long sequences

TALENs are described in detail in the next section of this chapter. Relevant here, they function as two monomers that bind distinct sequences on the DNA separated by a short spacer and on opposite strands. Generally, TAL effector-binding sites are no shorter than 12 bases and can measure as long as 30 or more bases (Boch and Bonas, 2010). Thus, the two monomers together specify a binding site length rivaling that of meganucleases, likely to be unique even within a large genome. This contrasts with the 20 base recognition sequence common for CRISPR. Theoretically, TALEN monomers could be designed with super-long repeat arrays. However, extending a repeat array past about 18 repeats has been observed to decrease overall specificity by increasing the non-specific affinity for DNA, while leaving affinity for the targeted DNA sequence relatively unchanged (Rinaldi et al., 2017).

1.3 TALENS offer the potential for specificity tuning

Different repeat types in TAL effectors differ not only in their base preference but in the stringency of that preference. Indeed, there are highly specific repeat types for each of the four bases, there are repeat types with dual specificity, and there are 'wildcard' repeat types that are able to accommodate any of the four bases (Miller et al., 2015; Yang et al., 2014). Thus, a repeat array could be assembled to judiciously relax specificity at particular positions, enabling targeting of multiple, moderately conserved sequences that match a consensus. With CRISPR, in which target recognition is governed by Watson and Crick base pairing, this can be achieved by introducing multiple gRNAs. But where the number of gRNAs needed becomes a limiting factor, TALENs could be a useful alternative, taking advantage of different repeat types to tune the specificity to capture the range of sequences desired (discussed further in Bogdanove et al., 2018).

1.4 The TALEN intellectual property landscape is settled

In contrast to CRISPR, for which (at the time of this writing) an intellectual property (IP) dispute continues, ownership of the dominant IP for TALENs is uncontested. The TAL effector-DNA binding 'code' was patented by members of one of the groups that co-discovered it (Boch et al., 2009), and rights to that intellectual property are managed by the 2Blades Foundation, a non-profit organization dedicated to advancing technologies to improve crop disease resistance. For uses outside of plants, 2Blades licensed the code exclusively to ThermoFisher Scientific. For uses in plants, however, 2Blades retains the rights to grant additional licenses. For the TALEN technology itself, the University of Minnesota and Iowa State University own the rights but granted an exclusive license to the biopharmaceutical company Cellectis. Cellectis established the subsidiary Calyxt, an independent company now that uses TALENs to develop healthier food crops. Commercial use of TALENs requires licenses both from 2Blades and from Calyxt. Depending on the specific architecture and delivery methods used, additional licenses from these or other parties may be required.

The remainder of this chapter describes how TALENs work, discusses how they have been used in plant genome editing, and provides a brief overview of the workflow for a TALEN-mediated genome editing experiment.

2 How TALENs work

TALENs create targeted double-strand breaks in DNA for genome editing via non-homologous end-joining (NHEJ) or homology-dependent repair (HDR) (Carroll, 2014). Each TALEN monomer is a fusion of a customized TAL-effector DNA recognition and binding domain to the N-terminus of the non-specific catalytic domain of the Type IIS restriction endonuclease *Fok*I. *Fok*I functions as a dimer. Thus, TALEN monomers function in pairs, binding the DNA on opposite strands and on opposite sides of a short-spacer sequence, in a tail-to-tail orientation, so that the *Fok*I domains come together across that spacer to make the break (Fig. 1a). TAL effectors in their native context are produced by members of the plant pathogenic bacterial genus *Xanthomonas*, which, upon interacting with the host plant, secrete them into the host nucleus, where the proteins mimic host transcription factors by binding to and activating specific sequences in promoters of host genes that may contribute to disease development, or trigger defense (Christian et al., 2010). A TAL effector is structurally composed of an N-terminal secretion signal, a central repeat region (CRR) with nearly identical direct repeats of 33–35 amino acid residues followed by a truncated half repeat with 20 amino acids, and a C-terminal region with nuclear localization signals and an acidic activation domain. TAL effectors, including the CRR, are highly conserved, except at positions 12 and 13 in each repeat, which are together known as the repeat-variable-diresidue (RVD). The RVD of each repeat determines its base specificity, such that the sequence of RVDs across the CRR determines the overall DNA sequence specificity of the protein (Boch et al., 2009; Moscou and Bogdanove, 2009). Importantly,

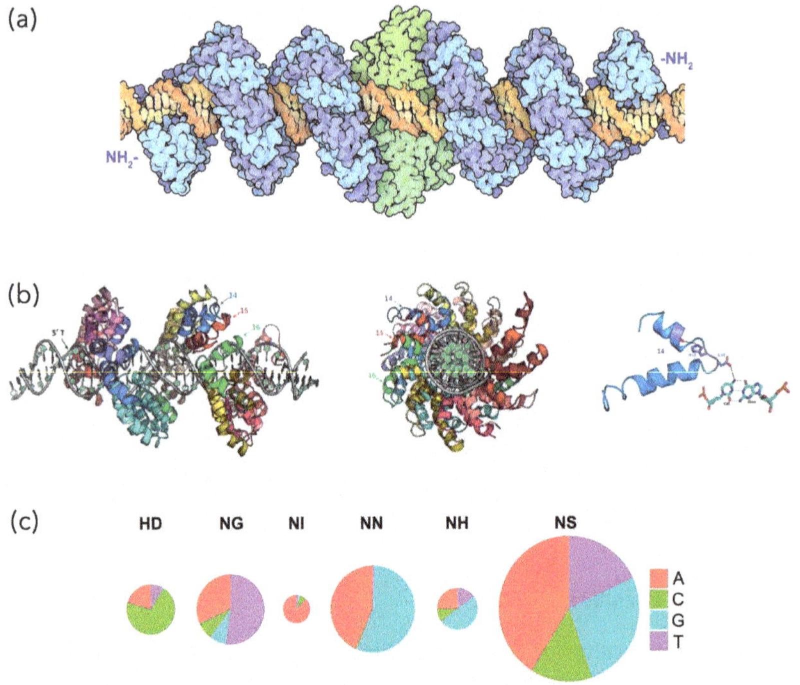

Figure 1 TALEN structure and mechanism of DNA recognition: (a) A TALEN drawn by David Goodsell from PDB files 1FOK (nuclease/DNA) and 3UGM (TALE/DNA); CC license 3.0. The paired TAL effector-*Fok*I fusion proteins bind the DNA in tail-to-tail orientation. The TAL effector domains are blue, the *Fok*I domains are green, and the DNA is orange; (b) Ribbon cartoon views of TAL effector PthXo1 bound to DNA. Left to right: the full complex viewed along the long axis of the DNA, an end view of the complex, and a zoomed view of a single repeat, with the RVD 'HD' interacting with cytosine in the plus strand of the DNA. Images reprinted with permission from Bogdanove et al. (2018); (c) Base specificities and relative binding contributions of commonly used RVDs, determined by Miller et al. (2011). Data were generated by measuring relative binding in an ELISA of a test TAL effector in which one repeat in the middle of the CRR was substituted with a repeat harboring the indicated RVD, on DNA targets with an A, G, C, or T at the position corresponding to that repeat. The slices of each pie represent the ELISA signal for the TAL effector with that RVD on the target with that base, and the overall size of the pie represents the sum of those relative binding affinities.

TAL effectors bind efficiently only when the CRR-defined DNA sequence is preceded by a 'T'. A 'C' is sometimes tolerated, but is thought to reduce binding affinity (Yu et al., 2011; Miller et al., 2011; Sun et al., 2012). Immediately N-terminal to the CRR, a 'cryptic' repeat that does not contain RVDs appears to contribute to the specificity for this initial base, in a structurally complex way (Doyle et al., 2013). Each repeat in a TAL effector comprises two alpha helices, with the RVD at the interhelical loop, and the overall protein wraps the DNA as a super helix, tracking the major groove, with the RVDs of each repeat oriented inward for base-specific contacts (Mak et al., 2012; Deng et al., 2012) (Fig. 1b). The most common RVDs in nature, HD, NI, NG, NN, NS, and N* (missing the second amino acid of the RVD), respectively specify C, A, T, G/A, A/T/G/C and C/T (Moscou and Bogdanove, 2009; Boch et al., 2009), and the less common RVD NH specifies G (Streubel et al., 2012). For engineering, this small set thus provides RVDs specific for each of the four bases, two dual specificity RVDs, and a wildcard (Fig. 1c). As further detailed in this chapter later, generating a TALEN with the desired DNA sequence specificity is a simple matter of assembling a sequence of repeats with the necessary RVDs selected from among this set into the CRRs of the respective monomer constructs. Specificity and affinity contributions for all 400 possible RVDs have been characterized (Miller et al., 2015; Yang et al., 2014) and offer even greater potential for specificity tuning.

3 How have TALENs been used?

The precision, efficiency, and ease of use of TALENs offer great potential for a wide range of applications to accelerate crop improvement. TALEN-based genome editing is capable of yielding different types of mutations in target sequences, including targeted gene knockout, replacement, insertion, and large deletions and chromosomal rearrangements (Gurushidze et al., 2014; Sawai et al., 2014; Forner et al., 2015; Kelliher et al., 2017; Li et al., 2016b; Butler et al., 2016; Budhagatapalli et al., 2015; Wendt et al., 2013; Christian et al., 2013), and since the first two proof of principle reports of TALEN-based editing in *Arabidopsis* protoplasts and *Nicotiana benthamiana* leaves in 2011 (Cermak et al., 2011; Mahfouz et al., 2011), TALENs have been used in a wide variety of plant species including *Arabidopsis*, barley, *Brachypodium*, Brassica, maize, *N. benthamiana*, potato, rice, sugarcane, soybean, tobacco, tomato, and wheat (Table 1). TALENs have been applied in plants both to investigate gene function and to create new varieties with desirable traits such as enhanced nutritional content and biotic or abiotic stress tolerance (Table 1; reviewed also in (Zhang et al., 2017; Malzahn et al., 2017; Zhang et al., 2018)).

Most TALEN-mediated genome editing studies have been NHEJ-based (see Table 1). However, precise, HDR-based genome editing using TALENs has been successfully implemented in *N. benthamiana* and tobacco (Zhang et al., 2013; Mahfouz et al., 2011), barley (Budhagatapalli et al., 2015), potato (Forsyth et al., 2016; Butler et al., 2016), rice (Li et al., 2016b), and tomato (Cermak et al., 2015). To date, approximately 60 genes including protein-coding and non-protein-coding genes have been targeted for editing using TALENs in plants. Gene-regulatory regions have also been targeted by TALENs in some studies (Li et al., 2012; Holme et al., 2017). As shown in Table 1, a large portion of published studies is proof of concept and method development studies, which helped to optimize TALENs in plants (e.g. Zhang et al., 2013).

Table 1 Uses of TALENs in plants

Plant species	Targeted gene(s)	TALEN construction platform	Modification type	DNA repair pathway[a]	Inheritance tested	Purpose[b]	Tissue	Delivery technique	References
Arabidopsis	*ADH1*	Golden gate assembly	Random indel	NHEJ	No	MD	Protoplast	PEG-mediated transformation	(Cermak et al., 2011)
Arabidopsis	*ADH1, TT4, MAPKKK1, DSK2B, NATA2*	Golden gate assembly	Random indel	NHEJ	Yes	PP	Floral tissues	*Agrobacterium*-mediated transformation	(Christian et al., 2013)
Arabidopsis	*CLV3*	Unspecified	Precise deletion	NHEJ (2 DSBs)	Yes	PP, GA	Floral tissues	*Agrobacterium*-mediated transformation	(Forner et al., 2015)
Arabidopsis	*CRU3*	Golden gate assembly	Random indel	NHEJ	No	MD, CS (T|Z)	Leaf explants	*Agrobacterium*-mediated transformation	(Johnson et al., 2013)
Barley	*HvPAPhy_a*	Golden gate assembly	Precise deletion	NHEJ (2 DSB)	Yes	GA	Callus	*Agrobacterium*-mediated transformation	(Wendt et al., 2013)
Barley	*GFP (transgene)*	Unspecified	Random indel	NHEJ	Yes	PP, GA	Pollen	*Agrobacterium*-mediated transformation	(Gurushidze et al., 2014)
Barley	*GFP (transgene)*	Unspecified	Gene replacement	HR	No	PP	Leaf cells, callus	Particle bombardment	(Budhagatapalli et al., 2015)
Barley	*HvPAPhy_a*	Golden gate assembly	Random indel	NHEJ	Yes	GA, CS (T|C)	Callus	*Agrobacterium*-mediated transformation	(Holme et al., 2017)
Brachypodium	*ABA1, CKX2, SMC6, SPL, SBP, COII, RHT, HTA1*	Golden gate assembly	Random indel	NHEJ	No	PP	Protoplast/ callus	PEG-mediated transformation/*Agrobacterium*-mediated transformation	(Shan et al., 2013)
Brassica oleracea	*FRIGIDA*	Unit assembly	Random indel	NHEJ	No	PP, TD	Hypocotyl explants	*Agrobacterium*-mediated transformation	(Sun et al., 2013)
Maize	*GL2*	Golden gate assembly	Random indel	NHEJ	Yes	TD, GA	Callus	*Agrobacterium*-mediated transformation, particle bombardment	(Char et al., 2015)

Maize	*ZmPDS, ZmIPK1A, ZmIPK, ZmMRP4,*	Golden gate assembly	Random indel	NHEJ	No	PP, CS (T|C)	Protoplast/ unspecified	PEG-mediated transformation/*Agrobacterium*-mediated transformation	(Liang et al., 2014)
Maize	*MTL*		Random indel	NHEJ	Yes	GA	Unspecified	*Agrobacterium*-mediated transformation	(Kelliher et al., 2017)
Nicotiana benthamiana	*EBE of Hax3*	Unspecified	Precise deletion	HR, NHEJ (2 DSBs)	No	PP	Leaf explants	*Agrobacterium*-mediated transformation	(Mahfouz et al., 2011)
Nicotiana benthamiana	*ALS*	Unspecified	Random indel	NHEJ	No	MD	Protoplast	PEG-mediated transformation	(Stoddard et al., 2016)
Nicotiana benthamiana	*FucT, XylT*	Unspecified	Random indel	NHEJ	Yes	TD	Protoplast	PEG-mediated transformation	(Li et al., 2016a)
Potato	*SSR2*	Platinum gate system	Random indel	Unspecified	No	MD, TD	Internodal explants	*Agrobacterium*-mediated transformation	(Yasumoto et al., 2019)
Potato	*SSR2*	Golden gate assembly	Random indel	Unspecified	No	GA	Unspecified	*Agrobacterium*-mediated transformation	(Sawai et al., 2014)
Potato	*StALS1*	Conventional RE cloning	Insertion	HR	No	MD	Unspecified	*Agrobacterium*-mediated transformation	(Forsyth et al., 2016)
Potato	*StALS1*	Golden gate assembly	Gene replacement	HR	No	PP, TD, CS (T|C|Z)	Leaf explants	*Agrobacterium*-mediated transformation	(Butler et al., 2016)
Potato	*StGBSS*	Platinum gate system	Random indel	NHEJ	No	MD	Callus	*Agrobacterium*-mediated transformation	(Kusano et al., 2016)
Potato	*VInv*	Unspecified	Random indel	NHEJ	No	TD	Protoplast	PEG-mediated transformation	(Clasen et al., 2016)
Potato	*ALS*	Golden gate assembly	Random indel	NHEJ	No	PP	Protoplast	PEG-mediated transformation	(Nicolia et al., 2015)
Potato	*SBE1 and StvacINV2*	Golden gate assembly	Precise deletion	Unspecified	No	PP	Leaf explants	*Agrobacterium*-mediated transformation	(Ma et al., 2017)

(*Continued*)

Table 1 Uses of TALENs in plants (*Continued*)

Plant species	Targeted gene(s)	TALEN construction platform	Modification type	DNA repair pathway[a]	Inheritance tested	Purpose[b]	Tissue	Delivery technique	References
Rice	*SWEET14*	Golden gate assembly	Random indel	NHEJ	Yes	TD	Unspecified	*Agrobacterium*-mediated transformation	(Li et al., 2012)
Rice	*EPSPS*	TA cloning	Gene replacement	HR, NHEJ	Yes	PP	Callus	Particle bombardment	(Wang et al., 2015)
Rice	*MST7, MST8, PMS3, CSA, DERF1*	Golden gate assembly	Random indel	NHEJ	Yes	PP	Callus	*Agrobacterium*-mediated transformation	(Zhang et al., 2016)
Rice	*LOX3*	Golden gate assembly	Random indel	NHEJ	Yes	TD	Protoplast/ unspecified	PEG-mediated transformation/*Agrobacterium*-mediated transformation	(Ma et al., 2015)
Rice	*ALS*	Golden gate assembly	Gene replacement	HR	Yes	TD	Unspecified	Agrobacterium-mediated transformation, particle bombardment	(Li et al., 2016b)
Rice	*SWEET14*	GoldenTal method	Random indel	NHEJ	Yes	PP, TD	Unspecified	*Agrobacterium*-mediated transformation	(Blanvillain-Baufumé et al., 2017)
Rice	*WAXY*	Golden gate assembly	Random indel	NHEJ, MMEJ	Yes	MD	Callus	*Agrobacterium*-mediated transformation	(Nishizawa-Yokoi et al., 2016)
Rice	*DEP1, BADH2, CKX2, and SD1*	Golden gate assembly	Random indel	NHEJ	No	PP	Protoplast/ unspecified	PEG-mediated transformation/*Agrobacterium*-mediated transformation	(Shan et al., 2013)
Rice	*BADH2, CKX2 and DEP1*	Golden gate assembly	Random indel	NHEJ	Yes	TD	Callus	*Agrobacterium*-mediated transformation	(Shan et al., 2015)
Sugarcane	*COMT*	Unspecified	Random indel	NHEJ	No	PP, TD	Leaf whorl/ callus	Agrobacterium-mediated transformation/particle bombardment	(Kannan et al., 2018)

Sugarcane	*COMT*	Unspecified	Random indel	NHEJ	No	TD	Leaf whorl/ callus	Agrobacterium-mediated transformation/particle bombardment	(Jung and Altpeter, 2016)
Soybean	*PDS11, PDS18*	FastTALE™ TALEN Assembly Kit (SIDANSAI Biotechnology (Shanghai) CO., LTD)	Random indel	NHEJ	No	TD, CS (T|C)	Hairy root	*Agrobacterium*-mediated transformation	(Du et al., 2016)
Soybean	*FAD2-1A, FAD2-1B, FAD3A*	Unspecified	Random indel	NHEJ	Yes	TD	Protoplast	*Agrobacterium*-mediated transformation	(Demorest et al., 2016)
Soybean	*FAD2-1A, FAD2-1B*	Unspecified	Random indel	NHEJ	Yes	TD	Hairy root	*Agrobacterium*-mediated transformation	(Haun et al., 2014)
Tobacco	*ALS*	Golden gate assembly	Random indel, insertion, gene replacement	HR, NHEJ	No	MD	Protoplast	PEG-mediated transformation	(Zhang et al., 2013)
Tomato	*ANT1*	Golden gate assembly	Insertion	HR, NHEJ	Yes	MD, CS (T|C)	Protoplast	PEG-mediated transformation	(Cermak et al., 2015)
Tomato	*PROCERA*	Golden gate assembly	Random indel	NHEJ	Yes	PP	Unspecified	*Agrobacterium*-mediated transformation	(Lor et al., 2014)
Wheat	*MLO*	Golden gate assembly	Random indel, insertion	NHEJ	Yes	PP, TD	Protoplast/ unspecified	PEG-mediated transformation/ particle bombardment	(Wang et al., 2014)
Wheat	*uidA, lr21Ψ*	Unspecified	Precise deletion	Unspecified	Yes	TD	Protoplast	*Agrobacterium*-mediated transformation	(Luo et al., 2019)

[a] NHEJ, Non-homologous end-joining; HDR, homology-dependent repair; DSB, double-strand break.
[b] CS, comparative study; C, CRISPR; GA, gene functional analysis; MD, method development; PP, proof of principle; TD, trait development; T, TALEN; Z, ZFN.

However, several studies have focused more directly on crop improvement. TALENs have been used to knockout the potato *VInv* gene, encoding vacuolar invertase, to produce potatoes that convert less starch to sugar when damaged during transport and storage, resulting in lower levels of the neurotoxin and carcinogen acrylamide on frying (Clasen et al., 2016). To improve oil composition of soybean for longer shelf life, genes encoding fatty acid desaturase (*FAD*) were disrupted using TALENs (Haun et al., 2014). Shan and colleagues (Shan et al., 2015) used TALENs to knockout *BADH2* to produce fragrant rice. TALENs were also used in rice for seed storage improvement by knocking out the *LOX3* gene (Ma et al., 2015). Knockout of *ZmIPK* to reduce content of the anti-nutrient phytate in maize grain was also reported (Liang et al., 2014). In another study in maize, epicuticular wax in leaves was reduced by knocking out the *GL2* gene (Char et al., 2015). To improve cell wall composition and saccharification efficiency in sugarcane, a knockout of *COMT* was achieved with TALEN-mediated mutagenesis (Jung and Altpeter, 2016; Kannan et al., 2018). Rice genotypes resistant to bacterial leaf blight were obtained by TALEN-mediated mutagenesis of the promoter region of the *SWEET14* sucrose transporter gene (Li et al., 2012; Blanvillain-Baufumé et al., 2017). Wang and colleagues (Wang et al., 2014) successfully used TALENs to obtain powdery mildew-resistant wheat by knocking out three homologs of the mildew resistance locus (MLO). In addition, using TALENs, haploid induction in maize was achieved by creating frameshift mutations in the *MTL* gene. In *Brassica*, early flowering mutants were obtained by targeting *FRIGIDA* using TALENs. These examples, all achieved in the past several six or seven years, demonstrate the scope of potential uses of TALEN-based genome editing in plant breeding for crop improvement.

4 How to use TALENs

4.1 Architectures available

The specific architecture of TALENs is an important parameter for efficiency of targeted modifications and affects the optimum spacer length (Miller et al., 2011; Cermak et al., 2011). Multiple TALEN architectures have been developed by different research groups, with different N- and C-terminal truncations of the TAL effector component. In the very first TALENs that were reported, slightly truncated or full-length, naturally occurring TAL effectors were fused either to the N-terminal (Christian et al., 2010; Li et al., 2011) or to the C-terminal end (Li et al., 2011) of the non-specific *Fok*I nuclease domain. The N-terminal fusions worked well. The TALEN architecture created by Christian and colleagues (Christian et al., 2010) includes 287 amino acids before and 231 amino acids after the CRR, missing just the initiator methionine at the N-terminus and 56 amino acids at the C-terminus, and thus retaining the native nuclear localization signals but lacking the acidic activation domain. An additional, SV40-based nuclear localization signal was incorporated at the N-terminus. The optimal spacer length range for this architecture is 13–30 bp. The architecture created by Li and colleagues (Li et al., 2011) used the full-length TAL effector protein, and has an optimal spacer length range of 16–31 bp. Motivated both to optimize efficiency, and to minimize size for viral delivery systems used in human and other animal cells, Miller and colleagues (Miller et al., 2011) tested different N- and C- terminal truncations of the TAL effector component in two human cell lines. They observed that different truncations resulted in different levels of activity and in different optimal spacer length ranges. The observations were corroborated by similar, subsequent studies (Mussolino et al., 2011; Sun et al., 2012). Among the architectures tested by Miller and colleagues, a variant missing the N-terminal 152 amino acids and retaining just 63 amino acids after the CRR (the 'Δ152/+63' architecture) was found to be highly active and has since been widely adopted (e.g. Bedell et al., 2012; Ma et al., 2013; Carlson et al., 2012), including in plants (e.g. Cermak et al., 2017). This architecture has an optimum spacer length range of 15–20 bp.

In addition to determining the optimal TAL effector domain for efficient genome editing, researchers have tested TALENs with alternative nuclease domains. A so-called 'compact TALEN' (cTALEN) architecture was developed by replacing the *Fok*I domain with an I-TevI nuclease domain, which functions as a monomer (Beurdeley et al., 2013). The cTALEN architecture, which is simpler than a conventional TALEN to integrate into an expression construct, has been used to knock out mitochondrial genes in rice and rapeseed to cure cytoplasmic male sterility (Kazama et al., 2019). However, the I-TevI DNA cleavage domain in this single-chain TALEN architecture has its own, inherent DNA recognition specificity, which limits the targetable sequence. Another single-chain TALEN architecture was developed by replacing the single *Fok*I domain of the conventional TALEN architecture with two *Fok*I domains connected by a linker such that they can self-dimerize to cleave downstream of the TAL-effector-determined binding site in the DNA (Sun et al., 2014). In contrast to cTALENs, this single-chain TALEN architecture has no additional targeting constraints relative to a conventional TALEN, because the *Fok*I cleavage domain has no inherent DNA recognition specificity.

4.2 Design software

Irrespective of the architecture, determining the best sequence of RVDs for highly specific and effective targeting of the locus of interest is inevitably important for genome editing. Because of the varying stringency of the specificity of different RVDs for different nucleotides, predicting DNA sequences to which a given TAL effector will bind is a probabilistic exercise

(Moscou and Bogdanove, 2009). Thus, while an RVD sequence can be readily designed that matches a sequence of interest, determining the specificity of that TAL effector for that sequence within a complex genome requires a degree of computation.

To date, numerous web-based computational tools have been developed that can help design TALENs to minimize off-target activity. These tools vary in their off-target prediction approaches, features, off-target search capabilities, and computational speeds, and accessibility for offline use. Most of these tools accept DNA sequence in FASTA or plain text format, or common database gene IDs, as an input file. RVD sequences, intended target sites, or genomic coordinates are also accepted input files for some tools. Some of the tools offer stand-alone versions, such as TALENoffer and PROGNOS, which allow greater flexibility for implementation. Various incorporated parameters of these tools can be used to customize the process depending on the user's purpose, such as restriction-enzyme site analysis and setting the minimum and maximum repeat array and spacer lengths. Some tools also assist the user by providing primer design features for PCR-based screening for edits at the target site. Table 2 provides a quick overview of the readily available TALEN design tools to aid researchers in selection of the most suitable software for their purposes. A more comprehensive review of the various computational tools and databases, and comparisons among them, have been presented previously (Booher and Bogdanove, 2014; Bogdanove and Booher, 2016; Khan et al., 2017; Periwal, 2017).

4.3 Assembly principles and reagents

Once a TALEN is designed, a genetic construct for it must be assembled. The construction of customized TAL effector DNA-recognition domains is complicated by the necessity of assembling highly similar TAL-effector repeat units. To overcome this limitation, several approaches and platforms have been developed, including Golden Gate assembly (Cermak et al., 2011; Mussolino et al., 2011), REAL assembly (Reyon et al., 2012), Fast Ligation-based Automatable Solid-phase High-throughput (FLASH) assembly (Reyon et al., 2012), Iterative Capped Assembly (Briggs et al., 2012), Integrated Chip (Wang et al., 2012), Unit Assembly (Huang et al., 2011), USER-Based Ligation Mediated Assembly of TAL Effector (ULtiMATE) (Yang et al., 2013), and Ligation-Independent Cloning Assembly (Schmid-Burgk et al., 2013). These approaches and platforms are based variously on restriction enzyme digestion and ligation, ligation-independent assembly, and solid-state assembly.

Platforms based on standard serial digestion and ligation include Unit assembly (Huang et al., 2011), REAL (Restriction Enzyme and Ligation) (Sander et al., 2011), and REAL-Fast Assembly (Reyon et al., 2012). Although these methods can accurately assemble TAL effector repeats, they are relatively time-consuming, and not suitable for high throughput.

Golden Gate assembly is also based on restriction enzyme digestion and ligation, but relies on Type IIS restriction enzymes, which cut at a distance from their recognition site. This property is exploited to release DNA fragments with unique overhangs, allowing, in a cycled reaction with ligase in a single tube, ordered and oriented assembly of multiple fragments. To make TALEN constructs by Golden Gate assembly, a library of repeat modules staggered to generate a sequence of unique overhangs each matching the next is used, with variants at each position that encode different RVDs. Several research groups have developed and implemented variations of the protocol (Cermak et al., Mussolino et al., 2011; Li et al., 2011; Geissler et al., 2011; Weber et al., 2011), and Golden Gate assembly has become the most widely adopted approach for constructing TALENs. However, though it is straightforward and fast (a proficient molecular biologist can assemble a few dozen TALEN constructs in about a week), the approach is still not as amenable to high throughput as the remaining approaches discussed below.

For large-scale synthesis of TAL-effector repeat arrays, a few high-throughput, automatable assembly methods have been developed. These enable users to generate from dozens to hundreds of TALEN constructs in a single day. FLASH assembly (for Fast Ligation-based Automatable Solid-phase High-throughput), the first reported high-throughput assembly protocol, assembles the repeats iteratively on streptavidin-coated magnetic beads, using an immobilized biotinylated DNA double-strand adaptor with a restriction site. This allows automation of restriction digest, purification, and ligation steps using a liquid-handling robot (Reyon et al., 2012). FLASH assembles TALEs using oligomers from a large, pre-assembled combinatorial plasmid library, which limits its adaptivity to some extent. Iterative Capped Assembly and Integrated Chip are two other solid-phase, high-throughput approaches, but use individual monomers to build constructs, thus preserving the flexibility to introduce new repeat types (Briggs et al., 2012; Wang et al., 2012).

Ligation-Independent Cloning Assembly is another high-throughput protocol. It uses T4 DNA polymerase rather than ligation to assemble RVDs. The major advantage of this ligation-independent protocol is that it eliminates the need for many steps common to other protocols, including PCR, ligation, sequencing, and gel electrophoresis (Schmid-Burgk et al., 2013).

TALEN plasmids and kits for many of these different assembly platforms can be obtained from the non-profit repository Addgene (http://www.addgene.org/talen/). There are also several commercial sources for TALEN kits based on different assembly platforms and architectures (reviewed in Khan et al., 2017). It should be noted that when choosing an assembly platform, TALEN architecture should be considered.

Table 2 Computational tools for TALEN design and off-target prediction

TALEN tools	TALE-NT[a]	TALENoffer[b]	E-TALEN[c]	Mojo Hands[d]	SAPTA[e]	ChopChop[f]	PROGNOS[g]	ZiFiT[h]
Year	2012	2013	2013	2013	2014	2014	2014	2010
Current version	2	1	2.5	2	2014	3	2014	4.2
Purpose	TALEN design and target prediction	Target prediction	TALEN design and target prediction	TALEN design	TALEN design	TALEN design and target prediction	Target prediction	TALEN design and target prediction
Input	FASTA file	FASTA-formatted query sequences, an optional annotation file, and RVD sequences for each TALEN monomer	FASTA file, or RVD sequences, gene symbols or common database gene IDs	NCBI ID, FASTA file	DNA sequence in plain-text format	Gene ID, genomic coordinates, DNA sequences or organism selection option	RVD sequences or intended target sites	FASTA file or raw text
Automated primer design	No	No	No	No	No	Yes	Yes	No

Useful feature(s)	Ability to assess the specificity of designed TALENs by predicting all potential target sites	Multi-threaded to allow for simultaneously loading, parsing and scanning input data	The ability to check for off-targets simultaneously in the target genome and any repair, replacement, or insertion sequences to be introduced as part of the experiment. A unique feature of E-TALEN is the search for TALENs in up to 50 sequences/genes simultaneously, enabling a parallel design of multiple TALENs for larger experiments	Provides a guide to facilitate TALEN assembly by Golden Gate method and TALEN DNA kit from Addgene	Quantitative evaluation of target sites by assigning scores that reflect predicted TALEN activities	Accepts a wide range of inputs, provides a dynamic graphical output display that includes an interactive visualization of the gene	User-friendly, rapid, relatively low false positive ratios	Generates intuitive and easy-to-follow graphical overviews that guide users through an assembly process for TALEN-encoding plasmids that requires only standard restriction, digestion and ligation to practice
Implementation	Web application	Open-source Java library Jstacs, command line program or Galaxy web application	Web application	Web application	Web application	Web application	Web application	Web application
On- or offline	Online/Offline	Online/Offline	Online	Online/Offline	Online	Online	Online	Online

(Continued)

Table 2 *(Continued)*

TALEN tools	TALE-NT[a]	TALENoffer[b]	E-TALEN[c]	Mojo Hands[d]	SAPTA[e]	ChopChop[f]	PROGNOS[g]	ZiFiT[h]
Genome search options	Dropdown list or any in NCBI	Dropdown list	Dropdown list	Only zebrafish	Not available	Dropdown list	Dropdown list	Dropdown list
Programming language	Python	Java	Perl	Perl	JavaScript	Phyton	Perl	Unspecified
Standalone version	Yes	Yes	No	No	No	No	Yes	No
Reference	(Doyle et al., 2012)	(Grau et al., 2013)	(Heigwer et al., 2013)	(Neff et al., 2013)	(Lin et al., 2014)	(Montague et al., 2014)	(Lin et al., 2014)	(Reyon et al., 2012)

[a] https://tale-nt.cac.cornell.edu
[b] http://www.jstacs.de/index.php/TALENoffer
[c] http://www.e-talen.org/E-TALEN/
[d] http://www.talendesign.org
[e] http://bao.rice.edu/Research/BioinformaticTools/TAL_targeter.html
[f] https://chopchop.cbu.uib.no
[g] http://bao.rice.edu/Research/BioinformaticTools/prognos.html
[h] http://zifit.partners.org/ZiFiT/

4.4 Workflow for plant genome editing with TALENs

A generalized workflow for plant genome editing with TALENs is presented in Fig. 2. The main steps include selection of the target and TALEN design,

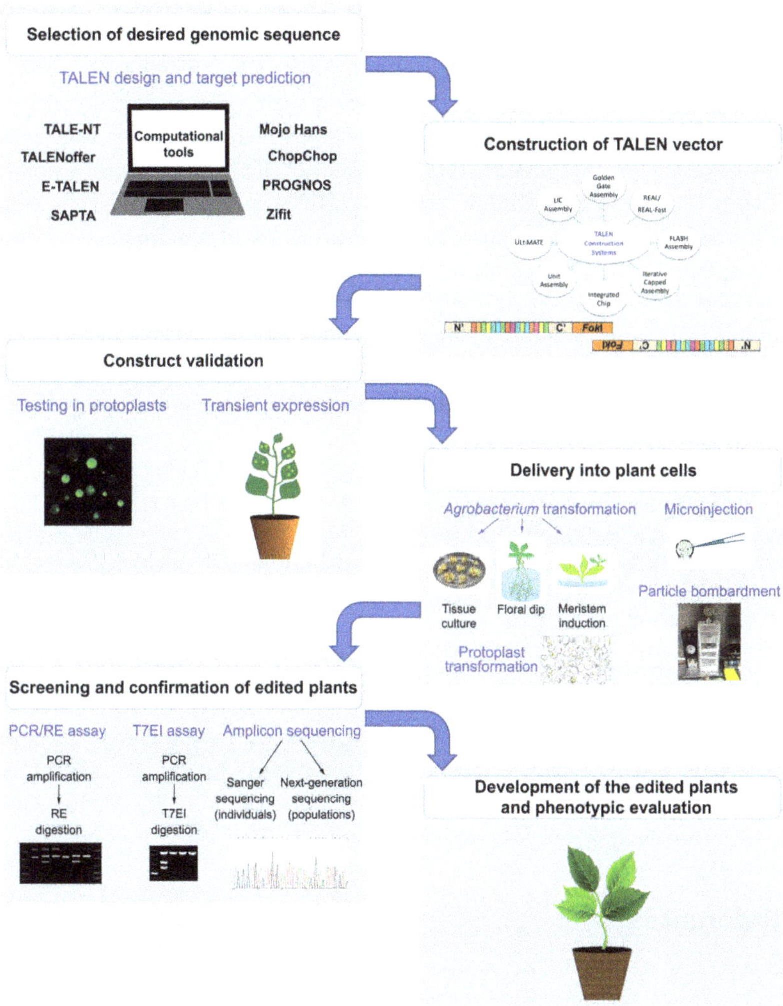

Figure 2 Example workflow for TALEN-mediated plant genome editing. RE, restriction enzyme; T7EI, T7 endonuclease I. T7EI is an enzyme that recognizes and cleaves DNA mismatches in heteroduplex DNA. The T7EI assay consists of initial amplification of nuclease target site, denaturation of amplicons, re-annealing and digestion with T7EI. Heteroduplexes formed by annealing of edited with wild-type strands are cleaved. Editing efficiency can be estimated by the percentage of cleaved amplicon.

assembly of the TALEN construct, validation of activity, delivery, screening for the desired edit, and development and testing of the edited plants.

5 Conclusion

The advent of CRISPR and its further development since has made earlier tools all but obsolete. Its ease and low cost, along with its multiplexing ability, and the rapidly expanding set of Cas nuclease variants that improve efficiencies and expand targeting capacity, make CRISPR the go-to tool for most plant genome editing applications for research. Even more so with the recent establishment of CRISPR prime editing, which for shorter sequences overcomes the limitations of the poor efficiency of HDR-based genome editing in plants (Anzalone et al., 2019; Lin et al., 2020). In the authors' own work, CRISPR is in fact our tool of choice. Nonetheless, TALENs merit consideration when specificity is paramount, when it is desirable to target large numbers of related sequences with a single reagent, or when freedom to operate commercially is important. As detailed in this chapter, TALENs have proven to be effective editing tools in a variety of plant species, for various purposes, and in various delivery contexts. And, as presented above and more generally elsewhere in this volume, there are many tools and approaches available for each of the various steps for plant genome editing with TALENs outlined in Fig. 2, providing users numerous options to choose from, to best suit their editing objectives and capabilities.

6 Acknowledgements

Recent work in plant genome editing in the laboratory of AJB has been funded by the Plant Genome Research Program of the U.S. National Science Foundation (IOS 1444511). RR was supported by a Fellowship under the Genome Editing Initiative of the Indo-U.S. Science and Technology Forum, Department of Biotechnology, Government of India. HBK was supported by Manisa Celal Bayar University Scientific Research Projects funds (2017-113).

7 References

Anzalone, A. V., Randolph, P. B., Davis, J. R., Sousa, A. A., Koblan, L. W., Levy, J. M., Chen, P. J., Wilson, C., Newby, G. A., Raguram, A. and Liu, D. R. 2019 Search-and-replace genome editing without double-strand breaks or donor DNA. *Nature* 576(7785), 149–157.

Bedell, V. M., Wang, Y., Campbell, J. M., Poshusta, T. L., Starker, C. G., Krug, R. G., Tan, W., Penheiter, S. G., Ma, A. C., Leung, A. Y., Fahrenkrug, S. C., Carlson, D. F., Voytas, D. F., Clark, K. J., Essner, J. J. and Ekker, S. C. 2012. *In vivo* genome editing using a high-efficiency TALEN system. *Nature* 491(7422), 114–118.

Beurdeley, M., Bietz, F., Li, J., Thomas, S., Stoddard, T., Juillerat, A., Zhang, F., Voytas, D. F., Duchateau, P. and Silva, G. H. 2013. Compact designer TALENs for efficient genome engineering. *Nat. Commun.* 4, 1762.

Blanvillain-Baufumé, S., Reschke, M., Solé, M., Auguy, F., Doucoure, H., Szurek, B., Meynard, D., Portefaix, M., Cunnac, S., Guiderdoni, E., Boch, J. and Koebnik, R. 2017. Targeted promoter editing for rice resistance to *Xanthomonas oryzae* pv. oryzae reveals differential activities for *SWEET14*-inducing TAL effectors. *Plant Biotechnol. J.* 15(3), 306–317.

Boch, J. and Bonas, U. 2010. *Xanthomonas* AvrBs3 family-type III effectors: discovery and function. *Annu. Rev. Phytopathol.* 48, 419–436.

Boch, J., Scholze, H., Schornack, S., Landgraf, A., Hahn, S., Kay, S., Lahaye, T., Nickstadt, A. and Bonas, U. 2009. Breaking the code of DNA binding specificity of TAL-type III effectors. *Science* 326(5959), 1509–1512.

Bogdanove, A. J., Bohm, A., Miller, J. C., Morgan, R. D. and Stoddard, B. L. 2018. Engineering altered protein-DNA recognition specificity. *Nucleic Acids Res.* 46(10), 4845–4871.

Bogdanove, A. J. and Booher, N. J. 2016. Online tools for TALEN design. In: Kühn, R., Wurst, W. and Wefers, B. (Eds), *TALENs: Methods and Protocols*. New York, Heidelberg, Dordrecht, London: Humana Press (Springer), pp. 43–47.

Booher, N. J. and Bogdanove, A. J. 2014. Tools for TAL effector design and target prediction. *Methods* 69(2), 121–127.

Briggs, A. W., Rios, X., Chari, R., Yang, L., Zhang, F., Mali, P. and Church, G. M. 2012. Iterative capped assembly: rapid and scalable synthesis of repeat-module DNA such as TAL effectors from individual monomers. *Nucleic Acids Res.* 40(15), e117.

Budhagatapalli, N., Rutten, T., Gurushidze, M., Kumlehn, J. and Hensel, G. 2015. Targeted modification of gene function exploiting homology-directed repair of TALEN-mediated double-strand breaks in barley. *G3 (Bethesda)* 5(9), 1857–1863.

Butler, N. M., Baltes, N. J., Voytas, D. F. and Douches, D. S. 2016. Geminivirus-mediated genome editing in potato (*Solanum tuberosum* L.) using sequence-specific nucleases. *Front. Plant Sci.* 7, 1045.

Carlson, D. F., Tan, W., Lillico, S. G., Stverakova, D., Proudfoot, C., Christian, M., Voytas, D. F., Long, C. R., Whitelaw, C. B. A. and Fahrenkrug, S. C. 2012. Efficient TALEN-mediated gene knockout in livestock. *Proc. Natl. Acad. Sci. U.S.A.* 109(43), 17382–17387.

Carroll, D. 2014. Genome engineering with targetable nucleases. *Annu. Rev. Biochem.* 83, 409–439.

Cermak, T., Baltes, N. J., Cegan, R., Zhang, Y. and Voytas, D. F. 2015. High-frequency, precise modification of the tomato genome. *Genome Biol.* 16, 232.

Cermak, T., Curtin, S. J., Gil-Humanes, J., Cegan, R., Kono, T. J. Y., Konecna, E., Belanto, J. J., Starker, C. G., Mathre, J. W., Greenstein, R. L. and Voytas, D. F. 2017. A multipurpose toolkit to enable advanced genome engineering in plants. *Plant Cell* 29(6), 1196–1217.

Cermak, T., Doyle, E. L., Christian, M., Wang, L., Zhang, Y., Schmidt, C., Baller, J. A., Somia, N. V., Bogdanove, A. J. and Voytas, D. F. 2011. Efficient design and assembly of custom TALEN and other TAL effector-based constructs for DNA targeting. *Nucleic Acids Res.* 39(12), e82.

Char, S. N., Unger, W. E., Frame, B., Briggs, S. A., Main, M., Spalding, M. H., Vollbrecht, E., Wang, K. and Yang, B. 2015. Heritable site-specific mutagenesis using TALENs in maize. *Plant Biotechnol. J.* 13(7), 1002–1010.

Christian, M., Cermak, T., Doyle, E. L., Schmidt, C., Zhang, F., Hummel, A., Bogdanove, A. J. and Voytas, D. F. 2010. Targeting DNA double-strand breaks with TAL effector nucleases. *Genetics* 186(2), 757–761.

Christian, M., Qi, Y., Zhang, Y. and Voytas, D. F. 2013. Targeted mutagenesis of *Arabidopsis thaliana* using engineered TAL effector nucleases. *G3 (Bethesda)* 3(10), 1697–1705.

Clasen, B. M., Stoddard, T. J., Luo, S., Demorest, Z. L., Li, J., Cedrone, F., Tibebu, R., Davison, S., Ray, E. E., Daulhac, A., Coffman, A., Yabandith, A., Retterath, A., Haun, W., Baltes, N. J., Mathis, L., Voytas, D. F. and Zhang, F. 2016. Improving cold storage and processing traits in potato through targeted gene knockout. *Plant Biotechnol. J.* 14(1), 169–176.

Demorest, Z. L., Coffman, A., Baltes, N. J., Stoddard, T. J., Clasen, B. M., Luo, S., Retterath, A., Yabandith, A., Gamo, M. E., Bissen, J., Mathis, L., Voytas, D. F. and Zhang, F. 2016. Direct stacking of sequence-specific nuclease-induced mutations to produce high oleic and low linolenic soybean oil. *BMC Plant Biol.* 16(1), 225.

Deng, D., Yan, C., Pan, X., Mahfouz, M., Wang, J., Zhu, J. K., Shi, Y. and Yan, N. 2012. Structural basis for sequence-specific recognition of DNA by TAL effectors. *Science* 335(6069), 720–723.

Doyle, E. L., Booher, N. J., Standage, D. S., Voytas, D. F., Brendel, V. P., Vandyk, J. K. and Bogdanove, A. J. 2012. TAL Effector-Nucleotide Targeter (TALE-NT) 2.0: tools for TAL effector design and target prediction. *Nucleic Acids Res.* 40, W117–W122.

Doyle, E. L., Hummel, A. W., Demorest, Z. L., Starker, C. G., Voytas, D. F., Bradley, P. and Bogdanove, A. J. 2013. TAL effector specificity for base 0 of the DNA target is altered in a complex, effector- and assay-dependent manner by substitutions for the tryptophan in cryptic repeat -1. *PLoS ONE* 8(12), e82120.

Du, H., Zeng, X., Zhao, M., Cui, X., Wang, Q., Yang, H., Cheng, H. and Yu, D. 2016. Efficient targeted mutagenesis in soybean by TALENs and CRISPR/Cas9. *J. Biotechnol.* 217, 90–97.

Forner, J., Pfeiffer, A., Langenecker, T., Manavella, P. A. and Lohmann, J. U. 2015. Germline-transmitted genome editing in *Arabidopsis thaliana* using TAL-effector-nucleases. *PLoS ONE* 10(3), e0121056.

Forsyth, A., Weeks, T., Richael, C. and Duan, H. 2016. Transcription activator-like effector nucleases (TALEN)-mediated targeted DNA insertion in potato plants. *Front. Plant Sci.* 7, 1572.

Geissler, R., Scholze, H., Hahn, S., Streubel, J., Bonas, U., Behrens, S. E. and Boch, J. 2011. Transcriptional activators of human genes with programmable DNA-specificity. *PLoS ONE* 6(5), e19509.

Grau, J., Boch, J. and Posch, S. 2013. TALENoffer: genome-wide TALEN off-target prediction. *Bioinformatics* 29(22), 2931–2932.

Gurushidze, M., Hensel, G., Hiekel, S., Schedel, S., Valkov, V. and Kumlehn, J. 2014. True-breeding targeted gene knock-out in barley using designer TALE-nuclease in haploid cells. *PLoS ONE* 9(3), e92046.

Haun, W., Coffman, A., Clasen, B. M., Demorest, Z. L., Lowy, A., Ray, E., Retterath, A., Stoddard, T., Juillerat, A., Cedrone, F., Mathis, L., Voytas, D. F. and Zhang, F. 2014. Improved soybean oil quality by targeted mutagenesis of the fatty acid desaturase 2 gene family. *Plant Biotechnol. J.* 12(7), 934–940.

Heigwer, F., Kerr, G., Walther, N., Glaeser, K., Pelz, O., Breinig, M. and Boutros, M. 2013. E-TALEN: a web tool to design TALENs for genome engineering. *Nucleic Acids Res.* 41(20), e190–e190.

Holme, I. B., Wendt, T., Gil-Humanes, J., Deleuran, L. C., Starker, C. G., Voytas, D. F. and Brinch-Pedersen, H. 2017. Evaluation of the mature grain phytase candidate *HvPAPhy_a* gene in barley (*Hordeum vulgare* L.) using CRISPR/Cas9 and TALENs. *Plant Mol. Biol.* 95(1–2), 111–121.

Huang, P., Xiao, A., Zhou, M., Zhu, Z., Lin, S. and Zhang, B. 2011. Heritable gene targeting in zebrafish using customized TALENs. *Nat. Biotech.* 29(8), 699–700.

Johnson, R. A., Gurevich, V. and Levy, A. A. 2013. A rapid assay to quantify the cleavage efficiency of custom-designed nucleases in planta. *Plant Mol. Biol.* 82(3), 207–221.

Jung, J. H. and Altpeter, F. 2016. TALEN mediated targeted mutagenesis of the caffeic acid O-methyltransferase in highly polyploid sugarcane improves cell wall composition for production of bioethanol. *Plant Mol. Biol.* 92(1–2), 131–142.

Kannan, B., Jung, J. H., Moxley, G. W., Lee, S. M. and Altpeter, F. 2018. TALEN-mediated targeted mutagenesis of more than 100 COMT copies/alleles in highly polyploid sugarcane improves saccharification efficiency without compromising biomass yield. *Plant Biotechnol. J.* 16(4), 856–866.

Kazama, T., Okuno, M., Watari, Y., Yanase, S., Koizuka, C., Tsuruta, Y., Sugaya, H., Toyoda, A., Itoh, T., Tsutsumi, N., Toriyama, K., Koizuka, N. and Arimura, S. I. 2019. Curing cytoplasmic male sterility via TALEN-mediated mitochondrial genome editing. *Nat. Plants* 5(7), 722–730.

Kelliher, T., Starr, D., Richbourg, L., Chintamanani, S., Delzer, B., Nuccio, M. L., Green, J., Chen, Z., McCuiston, J., Wang, W., Liebler, T., Bullock, P. and Martin, B. 2017. MATRILINEAL, a sperm-specific phospholipase, triggers maize haploid induction. *Nature* 542(7639), 105–109.

Khan, Z., Khan, S. H., Mubarik, M. S., Sadia, B. and Ahmad, A. 2017. Use of TALEs and TALEN technology for genetic improvement of plants. *Plant Mol. Biol. Rep.* 35(1), 1–19.

Kusano, H., Onodera, H., Kihira, M., Aoki, H., Matsuzaki, H. and Shimada, H. 2016. A simple Gateway-assisted construction system of TALEN genes for plant genome editing. *Sci. Rep.* 6, 30234.

Li, J., Stoddard, T. J., Demorest, Z. L., Lavoie, P. O., Luo, S., Clasen, B. M., Cedrone, F., Ray, E. E., Coffman, A. P., Daulhac, A., Yabandith, A., Retterath, A. J., Mathis, L., Voytas, D. F., D'aoust, M. A. and Zhang, F. 2016a. Multiplexed, targeted gene editing in *Nicotiana benthamiana* for glyco-engineering and monoclonal antibody production. *Plant Biotechnol. J.* 14(2), 533–542.

Li, T., Liu, B., Chen, C. Y. and Yang, B. 2016b. TALEN-mediated homologous recombination produces site-directed DNA base change and herbicide-resistant rice. *J. Genet. Genom.* 43(5), 297–305.

Li, T., Huang, S., Zhao, X., Wright, D. A., Carpenter, S., Spalding, M. H., Weeks, D. P. and Yang, B. 2011. Modularly assembled designer TAL effector nucleases for targeted gene knockout and gene replacement in eukaryotes. *Nucleic Acids Res.* 39(14), 6315–6325.

Li, T., Liu, B., Spalding, M. H., Weeks, D. P. and Yang, B. 2012. High-efficiency TALEN-based gene editing produces disease-resistant rice. *Nat. Biotechnol.* 30(5), 390–392.

Liang, Z., Zhang, K., Chen, K. and Gao, C. 2014. Targeted mutagenesis in *Zea mays* using TALENs and the CRISPR/Cas system. *J. Genet. Genom.* 41(2), 63–68.

Lin, Q., Zong, Y., Xue, C., Wang, S., Jin, S., Zhu, Z., Wang, Y., Anzalone, A. V., Raguram, A., Doman, J. L., Liu, D. R. and Gao, C. 2020. Prime genome editing in rice and wheat. *Nat. Biotechnol.* 38(5), 582–585.

Lin, Y., Fine, E. J., Zheng, Z., Antico, C. J., Voit, R. A., Porteus, M. H., Cradick, T. J. and Bao, G. 2014. SAPTA: a new design tool for improving TALE nuclease activity. *Nucleic Acids Res.* 42(6), e47.

Lor, V. S., Starker, C. G., Voytas, D. F., Weiss, D. and Olszewski, N. E. 2014. Targeted mutagenesis of the tomato PROCERA gene using transcription activator-like effector nucleases. *Plant Physiol.* 166(3), 1288–1291.

Luo, M., Li, H., Chakraborty, S., Morbitzer, R., Rinaldo, A., Upadhyaya, N., Bhatt, D., Louis, S., Richardson, T., Lahaye, T. and Ayliffe, M. 2019. Efficient TALEN-mediated gene editing in wheat. *Plant Biotechnol. J.* 17(11), 2026–2028.

Ma, A. C., Lee, H. B., Clark, K. J. and Ekker, S. C. 2013. High efficiency *in vivo* genome engineering with a simplified 15-RVD GoldyTALEN design. *PLoS ONE* 8(5), e65259.

Ma, J., Xiang, H., Donnelly, D. J., Meng, F.-R., Xu, H., Durnford, D. and Li, X.-Q. 2017. Genome editing in potato plants by agrobacterium-mediated transient expression of transcription activator-like effector nucleases. *Plant Biotechnol. Rep.* 11(5), 249–258.

Ma, L., Zhu, F., Li, Z., Zhang, J., Li, X., Dong, J. and Wang, T. 2015. TALEN-based mutagenesis of lipoxygenase LOX3 enhances the storage tolerance of rice (*Oryza sativa*) seeds. *PLoS ONE* 10(12), e0143877.

Mahfouz, M. M., Li, L., Shamimuzzaman, M., Wibowo, A., Fang, X. and Zhu, J. K. 2011. De novo-engineered transcription activator-like effector (TALE) hybrid nuclease with novel DNA binding specificity creates double-strand breaks. *Proc. Natl. Acad. Sci. U.S.A.* 108(6), 2623–2628.

Mak, A. N., Bradley, P., Cernadas, R. A., Bogdanove, A. J. and Stoddard, B. L. 2012. The crystal structure of TAL effector PthXo1 bound to its DNA target. *Science* 335(6069), 716–719.

Malzahn, A., Lowder, L. and Qi, Y. 2017. Plant genome editing with TALEN and CRISPR. *Cell Biosci.* 7, 21.

Miller, J. C., Tan, S., Qiao, G., Barlow, K. A., Wang, J., Xia, D. F., Meng, X., Paschon, D. E., Leung, E., Hinkley, S. J., Dulay, G. P., Hua, K. L., Ankoudinova, I., Cost, G. J., Urnov, F. D., Zhang, H. S., Holmes, M. C., Zhang, L., Gregory, P. D. and Rebar, E. J. 2011. A TALE nuclease architecture for efficient genome editing. *Nat. Biotechnol.* 29(2), 143–148.

Miller, J. C., Zhang, L., Xia, D. F., Campo, J. J., Ankoudinova, I. V., Guschin, D. Y., Babiarz, J. E., Meng, X., Hinkley, S. J., Lam, S. C., Paschon, D. E., Vincent, A. I., Dulay, G. P., Barlow, K. A., Shivak, D. A., Leung, E., Kim, J. D., Amora, R., Urnov, F. D., Gregory, P. D. and Rebar, E. J. 2015. Improved specificity of TALE-based genome editing using an expanded RVD repertoire. *Nat. Methods* 12(5), 465–471.

Montague, T. G., Cruz, J. M., Gagnon, J. A., Church, G. M. and Valen, E. 2014. CHOPCHOP: a CRISPR/Cas9 and TALEN web tool for genome editing. *Nucleic Acids Res.* 42, W401–W407.

Moscou, M. J. and Bogdanove, A. J. 2009. A simple cipher governs DNA recognition by TAL effectors. *Science* 326(5959), 1501–1501.

Mussolino, C., Morbitzer, R., Lutge, F., Dannemann, N., Lahaye, T. and Cathomen, T. 2011. A novel TALE nuclease scaffold enables high genome editing activity in combination with low toxicity. *Nucleic Acids Res.* 39(21), 9283–9293.

Neff, K. L., Argue, D. P., Ma, A. C., Lee, H. B., Clark, K. J. and Ekker, S. C. 2013. Mojo Hand, a TALEN design tool for genome editing applications. *BMC Bioinformatics* 14, 1.

Nicolia, A., Proux-Wéra, E., Åhman, I., Onkokesung, N., Andersson, M., Andreasson, E. and Zhu, L. H. 2015. Targeted gene mutation in tetraploid potato through transient TALEN expression in protoplasts. *J. Biotechnol.* 204, 17-24.

Nishizawa-Yokoi, A., Cermak, T., Hoshino, T., Sugimoto, K., Saika, H., Mori, A., Osakabe, K., Hamada, M., Katayose, Y., Starker, C., Voytas, D. F. and Toki, S. 2016. A defect in DNA Ligase4 enhances the frequency of TALEN-mediated targeted mutagenesis in rice. *Plant Physiol.* 170(2), 653-666.

Periwal, V. 2017. A comprehensive overview of computational resources to aid in precision genome editing with engineered nucleases. *Brief. Bioinform* 18(4), 698-711.

Reyon, D., Tsai, S. Q., Khayter, C., Foden, J. A., Sander, J. D. and Joung, J. K. 2012. FLASH assembly of TALENs for high-throughput genome editing. *Nat. Biotechnol.* 30(5), 460-465.

Rinaldi, F. C., Doyle, L. A., Stoddard, B. L. and Bogdanove, A. J. 2017. The effect of increasing numbers of repeats on TAL effector DNA binding specificity. *Nucleic Acids Res.* 45(11), 6960-6970.

Sander, J. D., Cade, L., Khayter, C., Reyon, D., Peterson, R. T., Joung, J. K. and Yeh, J. R. 2011. Targeted gene disruption in somatic zebrafish cells using engineered TALENs. *Nat. Biotechnol.* 29(8), 697-698.

Sawai, S., Ohyama, K., Yasumoto, S., Seki, H., Sakuma, T., Yamamoto, T., Takebayashi, Y., Kojima, M., Sakakibara, H., Aoki, T., Muranaka, T., Saito, K. and Umemoto, N. 2014. Sterol side chain reductase 2 is a key enzyme in the biosynthesis of cholesterol, the common precursor of toxic steroidal glycoalkaloids in potato. *Plant Cell* 26(9), 3763-3774.

Schmid-Burgk, J. L., Schmidt, T., Kaiser, V., Honing, K. and Hornung, V. 2013. A ligation-independent cloning technique for high-throughput assembly of transcription activator-like effector genes. *Nat. Biotechnol.* 31(1), 76-81.

Shan, Q., Wang, Y., Chen, K., Liang, Z., Li, J., Zhang, Y., Zhang, K., Liu, J., Voytas, D. F., Zheng, X., Zhang, Y. and Gao, C. 2013. Rapid and efficient gene modification in rice and *Brachypodium* using TALENs. *Mol. Plant* 6(4), 1365-1368.

Shan, Q., Zhang, Y., Chen, K., Zhang, K. and Gao, C. 2015. Creation of fragrant rice by targeted knockout of the *OsBADH2* gene using TALEN technology. *Plant Biotechnol. J.* 13(6), 791-800.

Stoddard, B. L. 2014. Homing endonucleases from mobile group I introns: discovery to genome engineering. *Mob. DNA* 5(1), 7.

Stoddard, T. J., Clasen, B. M., Baltes, N. J., Demorest, Z. L., Voytas, D. F., Zhang, F. and Luo, S. 2016. Targeted mutagenesis in plant cells through transformation of sequence-specific nuclease mRNA. *PLoS ONE* 11(5), e0154634.

Streubel, J., Blücher, C., Landgraf, A. and Boch, J. 2012. TAL effector RVD specificities and efficiencies. *Nat. Biotechnol.* 30(7), 593-595.

Sun, N., Bao, Z., Xiong, X. and Zhao, H. 2014. SunnyTALEN: a second-generation TALEN system for human genome editing. *Biotechnol. Bioeng.* 111(4), 683-691.

Sun, N., Liang, J., Abil, Z. and Zhao, H. 2012. Optimized TAL effector nucleases (TALENs) for use in treatment of sickle cell disease. *Mol. Biosyst.* 8(4), 1255-1263.

Sun, Z., Li, N., Huang, G., Xu, J., Pan, Y., Wang, Z., Tang, Q., Song, M. and Wang, X. 2013. Site-specific gene targeting using transcription activator-like effector (TALE)-based nuclease in *Brassica oleracea*. *J. Integr. Plant Biol.* 55(11), 1092-1103.

Wang, M., Liu, Y., Zhang, C., Liu, J., Liu, X., Wang, L., Wang, W., Chen, H., Wei, C., Ye, X., Li, X. and Tu, J. 2015. Gene editing by co-transformation of TALEN and chimeric RNA/

DNA oligonucleotides on the rice *OsEPSPS* gene and the inheritance of mutations. *PLoS ONE* 10(4), e0122755.

Wang, Y., Cheng, X., Shan, Q., Zhang, Y., Liu, J., Gao, C. and Qiu, J. L. 2014. Simultaneous editing of three homoeoalleles in hexaploid bread wheat confers heritable resistance to powdery mildew. *Nat. Biotechnol.* 32(9), 947-951.

Wang, Z., Li, J., Huang, H., Wang, G., Jiang, M., Yin, S., Sun, C., Zhang, H., Zhuang, F. and Xi, J. J. 2012. An integrated chip for the high-throughput synthesis of transcription activator-like effectors. *Angew. Chem. Int. Ed.* 51(34), 8505-8508.

Weber, E., Gruetzner, R., Werner, S., Engler, C. and Marillonnet, S. 2011. Assembly of designer TAL effectors by Golden Gate cloning. *PLoS ONE* 6(5), e19722.

Wendt, T., Holm, P. B., Starker, C. G., Christian, M., Voytas, D. F., Brinch-Pedersen, H. and Holme, I. B. 2013. TAL effector nucleases induce mutations at a pre-selected location in the genome of primary barley transformants. *Plant Mol. Biol.* 83(3), 279-285.

Yang, J., Yuan, P., Wen, D., Sheng, Y., Zhu, S., Yu, Y., Gao, X. and Wei, W. 2013. ULtiMATE system for rapid assembly of customized TAL effectors. *PLoS ONE* 8(9), e75649.

Yang, J., Zhang, Y., Yuan, P., Zhou, Y., Cai, C., Ren, Q., Wen, D., Chu, C., Qi, H. and Wei, W. 2014. Complete decoding of TAL effectors for DNA recognition. *Cell Res.* 24(5), 628-631.

Yasumoto, S., Umemoto, N., Lee, H. J., Nakayasu, M., Sawai, S., Sakuma, T., Yamamoto, T., Mizutani, M., Saito, K. and Muranaka, T. 2019. Efficient genome engineering using Platinum TALEN in potato. *Plant Biotechnol.* 36(3), 167-173.

Yu, Y., Streubel, J., Balzergue, S., Champion, A., Boch, J., Koebnik, R., Feng, J., Verdier, V. and Szurek, B. 2011. Colonization of rice leaf blades by an African strain of *Xanthomonas* oryzae pv. oryzae depends on a new TAL effector that induces the rice nodulin-3 *Os11N3* gene. *Mol. Plant Microbe Interact.* 24(9), 1102-1113.

Zhang, H., Gou, F., Zhang, J., Liu, W., Li, Q., Mao, Y., Botella, J. R. and Zhu, J. K. 2016. TALEN-mediated targeted mutagenesis produces a large variety of heritable mutations in rice. *Plant Biotechnol. J.* 14(1), 186-194.

Zhang, H., Zhang, J., Lang, Z., Botella, J. R. and Zhu, J.-K. 2017. Genome editing - principles and applications for functional genomics research and crop improvement. *Crit. Rev. Plant Sci.* 36(4), 291-309.

Zhang, Y., Massel, K., Godwin, I. D. and Gao, C. 2018. Applications and potential of genome editing in crop improvement. *Genome Biol.* 19(1), 210.

Zhang, Y., Zhang, F., Li, X., Baller, J. A., Qi, Y., Starker, C. G., Bogdanove, A. J. and Voytas, D. F. 2013. Transcription activator-like effector nucleases enable efficient plant genome engineering. *Plant Physiol.* 161(1), 20-27.

Chapter 2

Double strand break (DSB) repair pathways in plants and their application in genome engineering

Natalja Beying, Carla Schmidt and Holger Puchta, Karlsruhe Institute of Technology, Germany

1 Introduction

All organisms have to cope with DNA-damaging factors, with plants especially being exposed to many biological and environmental conditions causing DNA damage. Due to their sessile and phototrophic lifestyle, plants in particular have to face increased exposure to a wide variety of these factors such as UV radiation, site-specific biotic and abiotic environmental factors or reactive oxygen species. In the worst case, failure to repair the induced damage can lead to cell death. In order to ensure survival of cells or organisms, repair pathways have evolved as defense mechanisms to protect the integrity of the genetic information while still maintaining its malleability for further evolution. This means that mutations that are beneficial for survival are accumulated under selection pressure. These mutations result from error-prone repair processes and often have a direct impact on quantitative and qualitative traits of the organism.

For thousands of years, humans used this natural, dynamic variability in classical breeding and selected plants with characteristics that are interesting for agriculture. To further enhance genetic diversity, radiation such as X-rays or chemicals such as ethyl methanesulfonate (EMS) were used to increase the mutation rate. Both breeding processes of selection and mutagenesis are very complex and time-consuming, since the desired trait changes arise randomly and undirected, undesired mutations can occur elsewhere in the genome

http://dx.doi.org/10.19103/AS.2020.0082.04

without being noticed. Furthermore, the efficiency of classical breeding approaches depends on the amount of available functional diversity, which is limited in many elite varieties that have passed through genetic bottlenecks during domestication (Shi and Lai, 2015). Thus, the reliance on natural or randomly induced diversity is a limiting factor slowing down the breeding process (Watson et al., 2018) and contributing to an unpredictable breeding outcome (Scheben and Edwards, 2018). Consequently, in green biotechnology new breeding approaches were sought that result in precise and predictable outcomes.

Through site-specific induction of double-strand breaks (DSBs) in the genome, specific changes can be introduced. In early experiments, rare cutting endonucleases were used as tools. In plants, the homing endonuclease I-*Sce*I from *Saccharomyces cerevisiae* was applied for the induction of DSBs (Jacquier and Dujon, 1985; Perrin et al., 1993). As a meganuclease, I-*Sce*I recognizes an 18 nt sequence, which is often not present in the genome of the target organism. Thus, the recognition site must first be inserted into the genome, which leaves undesirable traces following DSB induction and repair, although no off-target effects are usually detected. Nevertheless, I-*Sce*I was successfully used in plants for DSB induction (Puchta et al., 1993) and for the investigation of basic mechanisms in DNA repair (Puchta et al., 1996; Puchta, 1998; Salomon and Puchta, 1998). With the introduction of synthetic nucleases, the targeted modification of DNA was brought to the next level. The first sequence-specific nucleases developed were genetically modified meganucleases (Arnould et al., 2007; Smith et al., 2006), followed by zinc finger nucleases (ZFNs) (Kim et al., 1996) and transcription activator-like effector nucleases (TALENs) (Boch et al., 2009; Voytas, 2013). With the discovery of the clustered regularly interspaced short palindromic repeats and CRISPR-associated (CRISPR/Cas) system and its adaption as a biotechnological tool (Jinek et al., 2012), molecular biology was revolutionized. Compared to ZFNs and TALENs, the CRISPR/Cas system is characterized by its simplicity, efficiency and low costs, as well as its ability to cut multiple genes at the same time. As these synthetic sequence-specific nucleases and their use in fast and high-precision genome editing are described in detail in other chapters, we will not discuss the characteristics of these different classes of enzymes here but will refer the reader to the respective chapters in this issue and to some recent reviews (Chen et al., 2019; Schindele et al., 2020).

With the availability of artificial site-specific nucleases, it is now possible to induce breaks at almost any site in the genome at will triggering the hosts own repair mechanisms to introduce genomic changes. For effective and targeted genome editing, it is therefore essential to understand the mechanisms behind DSB-induced modifications (Puchta, 2005).

As a DSB poses a major threat to genome integrity, plants have developed a complex repair network to cope with this kind of lesion. In each eukaryotic cell within the meristematic tissue, about 10 DSBs are generated per day (Martin et al., 1985). These DSBs can be caused by spontaneous oxidative damage, ionizing radiation or replication errors, but also two adjacent single-strand breaks (SSB) can lead to a DSB (Britt, 1996). They can occur during DNA repair and replication, but are also introduced specifically during meiosis or V(D)J recombination (Bleuyard et al., 2006). Since both DNA single strands are affected and the phosphate backbone of the DNA helix is broken, there is no intact strand as a repair template available. Therefore, the repair of DSBs is especially challenging for the organism.

DSB repair can be classified into two different main repair pathways - non-homologous end joining (NHEJ) and homologous recombination (HR) (Jasin and Haber, 2016; Puchta, 2005), which can be further divided into several sub-pathways. The DNA repair pathways are highly conserved between mammals and plants. All DSB repair pathways seem to compete with each other. The choice between these routes depends on the phase of the cell cycle and the initiation of the DNA resection (Ceccaldi et al., 2016; Grabarz et al., 2012; Kakarougkas and Jeggo, 2014). NHEJ is active throughout the cell cycle with the exception of mitosis, while HR is restricted to certain phases. In addition, the direct end joining can be completed faster than recombination and additionally NHEJ is thought to suppress HR (Tomimatsu et al., 2007). Taken together this makes NHEJ the predominant repair route in somatic cells in animals as well as plants. The second determining factor for the choice of the repair pathway is the processing of DSB ends to 3′ overhangs by means of 5′-3′ resection depending on which repair factors bind to the break site first (Chang et al., 2017; Ranjha et al., 2018). Unprocessed DSBs are repaired via NHEJ, whereas after resection homologies become accessible for strand exchange proteins to initiate HR. The two main repair routes of the DSB repair, NHEJ and HR, are described in more detail below.

2 DSB repair via non-homologous end joining

NHEJ is the main repair pathway of DSBs in somatic plant cells and is responsible for random integration of DNA into plant genomes. In principle, in NHEJ the two ends of a DSB are re-joined directly without using longer homologies. This can result in small insertions, deletions or a combination of both (Indels) (Salomon and Puchta, 1998). Therefore, this repair mechanism is usually described as inaccurate and error-prone (Charbonnel et al., 2010), but analyses in mammals and plants indicate that a significant part of NHEJ proceeds without loss of sequence information (Geisinger et al., 2016; Guo et al., 2018; Schmidt et al., 2019; Shou et al., 2018).

For NHEJ at least two mechanisms are known to operate in plant cells that are present in from other eukaryotes as well. These two mechanisms can be distinguished by the repair factors that are involved as well as the resulting sequence pattern at the re-joined break ends. The classic or canonical pathway (c-NHEJ) is characterized by the direct ligation of the break ends, while the alternative pathway (a-NHEJ) uses microhomologies (MHs) near the break site for re-ligation (Fig. 1). This is why this repair pathway is also referred to as microhomology-mediated end joining (MMEJ).

2.1 Basic mechanisms of NHEJ

c-NHEJ is also described as the KU-dependent pathway as the key player in this pathway is the KU70/KU80 heterodimer. In the first step, the DSB ends are recognized and bound by the protein complex of KU70 and KU80. This fast binding protects the broken ends from nucleolytic degradation. The KU heterodimer also acts as a platform for the recruitment of other NHEJ factors (Chang et al., 2017). In vertebrates, this includes DNA-PKcs and the

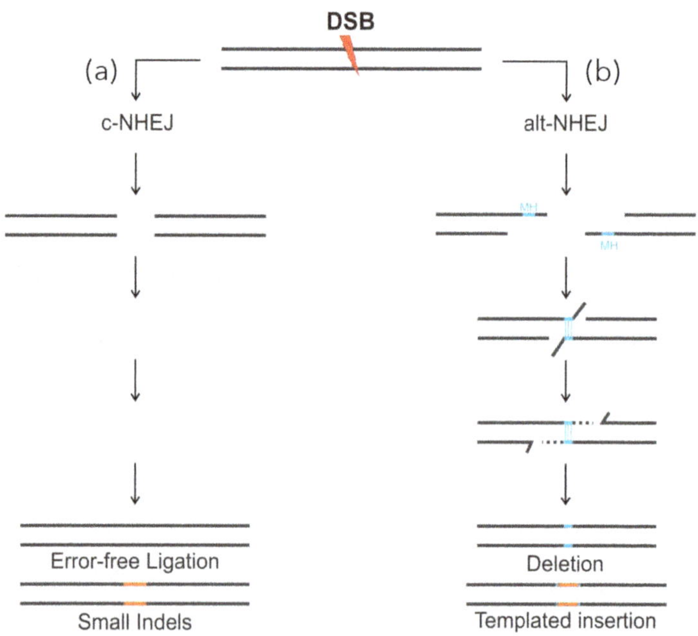

Figure 1 Non-homologous end joining (NHEJ)-mediated repair of DNA double strand breaks (DSB). In plant cells, two major NHEJ pathways are present: (a) the most abundant pathway is classical NHEJ (c-NHEJ). After binding of the KU70/KU80 heterodimer to the broken ends, they can be ligated directly leading either to error-free repair or to the induction of small insertions or deletions (Indels, orange), (b) in the absence of the KU protein, alternative pathways (a-NHEJ) take over. In contrast to c-NHEJ, more processing of the broken ends occurs and they are re-ligated via microhomologies (MH, blue) close to the break site. Therefore, a-NHEJ is error-prone and mainly leads to the deletion of the sequence between the MHs.

endonuclease Artemis as well as X-family polymerases, which act if processing of the break ends is required. KU is also involved in the final processing due to its own enzymatic activity (Roberts et al., 2010; Strande et al., 2012). For final ligation of the DSB, the KU protein recruits the ligase 4 complex comprised of LIG4, XRCC4 and XLF (Ma and Lieber, 2002). Thus, very little or no sequence information is lost. In plants, homologs of the main actors in c-NHEJ – KU70, KU80, LIG4 and XRCC4 – could be identified (Friesner and Britt, 2003; Tamura et al., 2002; West et al., 2000; West et al., 2002). However, no homologs have been detected for the other proteins, as yet. Mutant lines of the important c-NHEJ components in *Arabidopsis* show increased sensitivity to DNA-damaging agents, but do not show any phenotypic differences to the wild type (Bleuyard et al., 2006).

Due to the ubiquitous occurrence of the KU heterodimer and its rapid binding to DSBs, the repair of a DSB is usually carried out via c-NHEJ. However, if this pathway is compromised, the alternative end-joining mechanisms can take over. This KU-independent route to DSB repair is referred to in the literature with many different synonyms. These include backup-NHEJ, alternative NHEJ, microhomology-mediated end joining (MMEJ) or pol theta-mediated EJ/TMEJ. It is still not clear if those are synonyms for only one or if they describe several independent repair pathways. All alternative pathways are characterized by short MHs close to the break site.

a-NHEJ proceeds in five steps: (1) DSB recognition and end resection, (2) stabilization of the ends by annealing of MHs, (3) processing of flap structures, (4) fill-in synthesis and (5) ligation of the DSB (Wang and Xu, 2017). First, the break can be recognized via poly(ADP-ribose) polymerase 1 (PARP1) (Audebert et al., 2004; Robert et al., 2009). PARP1 is in direct competition for the binding of the break ends with KU70/KU80 (Wang et al., 2006). In addition to the detection of the DSB, PARP1 is also necessary for the subsequent recruitment of the other factors of a-NHEJ to initiate resection of the 3′ ends of the DSB, which creates short single-strand overhangs. This resection takes place via the same proteins which mediate the resection of the break ends in HR, CtIP and the MRN complex (Truong et al., 2013). By processing the ends, MHs are exposed on both sides of the damage, which are used for re-ligation. In a-NHEJ, MH between 2 bp and 20 bp are sufficient to hybridize the DNA strands, while further resection of the ends of 20–100 bps is necessary for the repair via single-strand annealing (SSA) and HR (chapter 3.1). If this extended processing occurs, an inhibitory effect on a-NHEJ was found (Deng et al., 2014). The pairing of the MH serves to stabilize the two ends of the DSB. Although the underlying mechanism has not yet been clarified in detail, in mammals it is assumed that polymerase theta (Pol θ) plays an important role in this step (Black et al., 2016; Seol et al., 2018; Wyatt et al., 2016; Zahn et al., 2015). The 3′-overhangs can then be removed by nucleases such as XPF-ERCC1 and MRE11 in mammals (Bennardo et al., 2008).

The plant homologs of this nuclease RAD1/RAD10 have not yet been reported to be involved in a-NHEJ. The 3′-hydroxyl ends generated allow POLθ to bind, catalyze the elongation of the DNA and fill in the gaps (Ahmad et al., 2008; Hogg et al., 2012). In the last step, the end ligation is carried out by the Xrcc1/ Ligase III complex or Ligase I (Liang et al., 2008; Masani et al., 2016).

Polymerase theta (Pol θ, encoded by the POLQ gene) was identified as the most important factor for a-NHEJ by several studies (Chan et al., 2010; Kent et al., 2015; Koole et al., 2014; Mateos-Gomez et al., 2015; Wood and Doublié, 2016; Yu and McVey, 2010). With more than 250 kDa, Pol θ is a comparatively large protein that, in addition to a polymerase domain at the C-terminus, has a helicase domain at the N-terminus (Black et al., 2016; Wood and Doublié, 2016). Pol θ can elongate both single and double-stranded DNA without a template via transferase activity. By adding individual nucleotides to the end of the break, new MHs can be created that serve a-NHEJ (Hogg et al., 2011, 2012; Kent et al., 2016). Furthermore, suppression of HR by binding of RAD51 could be demonstrated for Pol θ (Ceccaldi et al., 2015). The plant homologue to Pol θ was originally called TEBICHI (TEB). TEB is involved in processes of DNA replication, recombination and gene expression, as well as in DNA repair (Inagaki et al., 2009; Klemm et al., 2017). In addition to the increased sensitivity of *Arabidopsis teb* mutants to genotoxins, only the uptake of T-DNA into the cell could be observed, but not its integration into the genome (van Kregten et al., 2016). As a result, Pol θ is assumed to play an essential role in T-DNA integration into the plant genome.

2.2 Genomic changes induced by NHEJ

c-NHEJ leads either to perfect restoration or small indels at the break site. Indels arise because not all DSBs repaired by c-NHEJ are perfect blunt ends. Sometimes the broken ends have to be processed by adding or deleting a few nucleotides to make them available for the direct re-ligation. Moreover, it is known that the +1 bp insertion often observed in Cas9 mutagenesis experiments results from Cas9 creating a 1-nt 5′ overhang that is filled in and then ligated. Nevertheless, if occurring in an ORF a deletion or insertion of only one or two nucleotides can result in a frameshift and therefore cause a complete knockout of gene function. However, a-NHEJ is a strongly error-prone pathway as the sequence information between the MHs is usually lost. Additionally, the pattern at the junction sites often reveals insertions. In the natural context, c-NHEJ is the prevalent pathway for DSB repair in somatic plant cells (Puchta, 2005; Schmidt et al., 2019) leading to no or only very small genomic changes. It is possible to shift the repair pattern to more and larger deletions by knockout of key enzymes involved in c-NHEJ. When analyzing NHEJ repair mutants of ku70, lig4 or polq in *Arabidopsis* and also *Drosophila*, the influence by those knockouts on repair patterns after DSB inductions could be shown (Schmidt et al., 2019; Shen et al., 2017; Yu and McVey, 2010).

The specific MH-based repair patterns of the a-NHEJ with large deletions and templated insertions can be observed in plant cells as well as for mammals (Qi et al., 2013; Shen et al., 2017). The formation of large deletions can be easily understood by the loss of the area between the MH used for the repair. However, the occurrence of the insertions, which usually shows sequence homologies to surrounding areas, is harder to comprehend.

In early experiments of DSB induction in tobacco cells, Gorbunova and Levy obtained quite large insertions originating from either genomic DNA or internal regions of transfected plasmid (Gorbunova and Levy, 1997). Shortly after, Salomon and Puchta reported insertions formed during DSB repair associated with unique as well as repetitive genomic sequences (Salomon and Puchta, 1998). Both groups found indications for annealing and priming during the formation of those insertions. Therefore, to explain the mechanism they suggested a model analogous to the SDSA model of HR (Section 3.1). Another model explaining the occurrence of direct or indirect repeats at the break site after repair was developed by Yu and McVey (2010). It is called synthesis-dependent MMEJ (SD-MMEJ) and assumed to involve POL θ for templated insertions. The mechanism is divided into two processes, loop-out and snap-back, which differ in the orientation of the MHs used (Fig. 2). With the loop-out SD-MMEJ, repair products with direct sequence repetitions are created, while with the snap-back mechanism, inverted sequence repeats can be found at the break point after the repair. These repeating sequence motifs have their origin in flanking areas in the immediate vicinity of the DSB. In addition to the simple templated insertions described, complex insertions were also obtained, which comprise several overlapping copies of the adjacent sequences. This class of complex insertion is also based on a repair model of SD-MMEJ and might results from several rounds of synthesis and dissociations from one or both sides of the DSB. DSBs can also arise if two SSBs are occurring on opposite strands at a shorter distance resulting in 3′ or 5′ overhangs at the DSB ends. As the CRISPR/Cas9 nuclease can be transformed by a simple mutation into a nickase such a scenario can be addressed experimentally (Ran et al., 2013; Schiml et al., 2014). In a recent study, two genomic SSBs were induced at different distances and in different regions of the *Arabidopsis* genome and the repair outcomes were analyzed (Schiml et al., 2016). SSBs on opposite DNA strands in intervals of 50–100 bp producing 5′ overhangs were shown to induce mutations in up to three-quarters of the analyzed reads and to be efficient for induction of heritable mutations.

In addition to deletions, tandem sequence duplications close to the break sites were also detected. Some repair outcomes seemed to be related to c-NHEJ, others involved MHs at the break site, indicating repair mediated by MMEJ. The

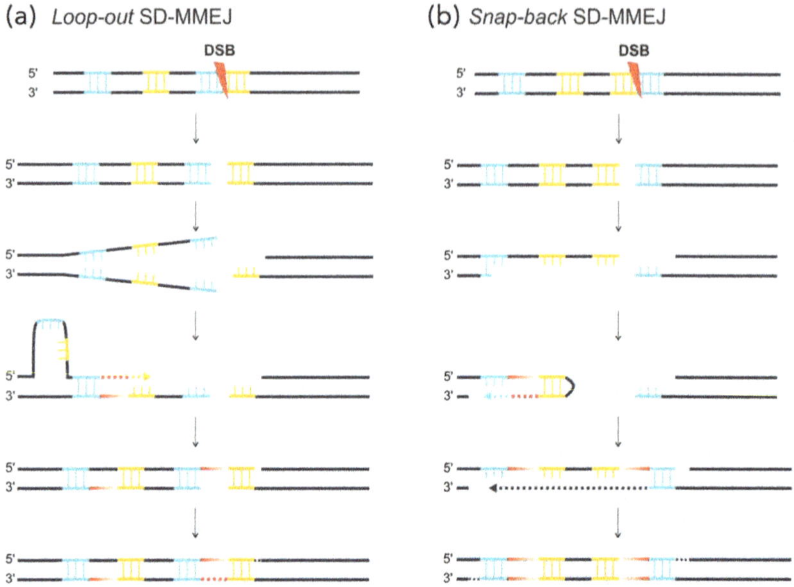

Figure 2 Mechanisms of synthesis-dependent MMEJ (SD-MMEJ). Two mechanisms are described here, which differ in the arrangement of the short homologies used: (a) in the loop-out SD-MMEJ, the sequence repeat pairs (yellow or light blue) are arranged alternately and one of the pairs spans the DSB. If the double strand of DNA is unwound, the blue sequence close to the break can pair with the homologous area, creating an outwardly curved loop (loop-out). After extending the DNA strand, the second homology (yellow) can link to the other end of the DSB, whereby the new synthesized area (red) leads to an insertion at the break point, (b) If the sequence repeats are inverted to one another, a hairpin structure can form after the ends of the break have been processed by folding over the single-stranded overhang. DNA synthesis is also started here based on the newly linked homology via a polymerase. The hairpin structure is then dissolved again and the elongated strand can pair with the opposite end of the break (snap-back). The sequence inserted in this way at the break (red) is in the opposite direction to the used template sequence.

most observed insertions originated from sequence context in close proximity. Therefore, Schiml et al. (2016) suggested two mechanisms responsible for the formation of tandem duplications after induction of paired SSBs on opposing strands (Fig. 3). In the absence of MHs at the break site, simultaneous degradation of the 5′ ends and synthesis from the 3′ ends followed by ligation of the blunt ends via c-NHEJ mechanisms can lead to tandem duplications (Fig. 3a). If MHs are present, a patch-mediated model for tandem duplications might explain repair (Fig. 3b) (Vaughn and Bennetzen, 2014). By hybridization of the single-stranded 5′ ends via MH a more or less stable intermediate structure is formed. Repair of the internal gaps by fill-in synthesis and removal of 5′ flaps at the same time give rise to staggered nicks that are finally ligated.

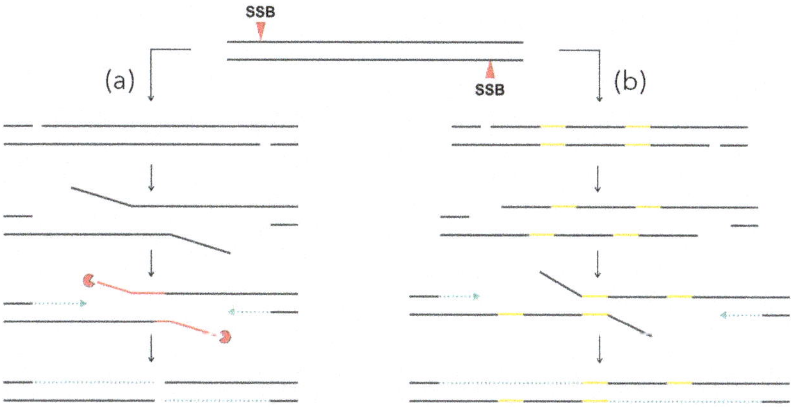

Figure 3 Repair of adjacent SSBs on different DNA strands can induce duplications. If two SSBs are induced on opposing DNA strands, a staggered DSB arises by unwinding of the complementary strands: (a) parallel resection from the 5′ end (red) and elongation synthesis starting at the 3′ end (green) leads to the formation of a blunt-ended break. Subsequent ligation via NHEJ results in the formation of MH-independent duplications, (b) if MHs are present, annealing of the distal MHs and subsequent DNA elongation leads to duplication of the sequence between the break sites and the homologies.

Interestingly, two SSBs on the same strand also induced similar alterations. However, the mutation frequencies for paired SSBs on the same strand were much lower than when on opposing strands with less than 1% compared to up to 77%. As high mutation frequencies correlated with the occurrence of MHs at the break site, the repair outcome and mutagenesis frequency seem to be highly dependent on the sequence context. Under natural conditions, adjacent SSBs might occur during base excision repair or through nicking errors in nucleotide excision repair. The presence of multiple short direct repeats in plant genomes (Vaughn and Bennetzen, 2014) indicates that the repair of adjacent SSBs seemed to have an important influence on the shaping of plant genomes during evolution.

SSBs occur during replication and are regularly induced during base and nucleotide excision repair (NER). As NER is a major repair pathway of UV-induced DNA damage, SSBs outnumber DSBs by orders of magnitudes in plant cells. Thus, although single SSBs themselves do not have the same potential as DSBs to induce repair-associated genomic changes these kinds of lesions have a significant influence on genome evolution, due to their sheer number.

In conjunction with the repair of adjacent SSBs, other molecular mechanisms are discussed as a source of genome variation in plants. Increased genome size is associated with polyploidization and transposable element proliferation (Adams and Wendel, 2005; Bennetzen and Wang, 2014). In contrast, transposon-mediated excision, dysploid reduction of chromosome number, loss of entire chromosomes as well as illegitimate recombination and replication slippage might lead to genome shrinkage (Devos et al., 2002; Schubert and Vu, 2016). Additionally, in smaller genomes of cotton, DNA removal during repair could counteract genome expansion through retrotransposons (Hawkins et al., 2009). In other studies, species-specific deletion sizes in combination with insertions were observed. Analyzing the formation of deletions in *Arabidopsis thaliana* and *Nicotiana tabacum*, the size of deletions was much smaller in the >20-fold larger tobacco genome (Kirik et al., 2000). Additionally, 40% of deletions in that species coincided with insertions but nearly no insertions accompanied the larger deletions in *Arabidopsis*. The same pattern could be shown in a comparison of DSB repair outcomes between *Arabidopsis* and barley (Vu et al., 2014, 2017). Here, large deletions were detected more frequently in *Arabidopsis*, whereas barley with its larger genome showed a significantly higher amount of insertions at imprecise repaired sequences. This correlation between genome size and DSB processing via NHEJ is thought to be related to species-specific differences in exonucleolytic degradation rates of free DSB ends before downstream processing (Cavalier-Smith, 2005; Orel et al., 2003). Concluding from this data, an underestimated contribution to genome size evolution could be the inaccurate repair of DSBs.

Usually cells can repair multiple breaks simultaneously through mechanisms that keep the correct broken ends in close proximity (Williams et al., 2010). Therefore, chromosomal aberrations are unlikely to arise. Yet, these mechanisms are not always accurate or can be impaired and rearrangements can result from the illegitimate joining of broken DNA ends. This misrepair can lead to many different changes like deletions, inversions and exchange of genomic sequences as well as recombinant chromosomes or chromosome fusions. The chromosomal rearrangements can result in changes in gene regulation, new gene functions or the suppression of meiotic recombination. They are linked to evolutionary processes such as the formation of new species and adaptive divergence. By comparative analysis of the genomes of different *Arabidopsis* ecotypes and of related species, many different rearrangements were found, which occurred naturally and shaped the karyotype of these *Brassicaceae* (Lysak et al., 2006; Zapata et al., 2016).

In genome engineering, the possibility of misjoining the broken ends is used to modify chromosomes in a targeted manner and create new combinations of the fragments. The induction of large chromosomal rearrangements (CRs) is based on the systematic introduction of more than one DSB at the same time. If two DSBs are induced on one chromosome, the region between can be deleted or inverted (Fig. 4a). Interchromosomal rearrangement results from the induction of DSBs on different chromosomes, which causes fragments to be exchanged and illegitimately mated (Fig. 4b). The induction of two

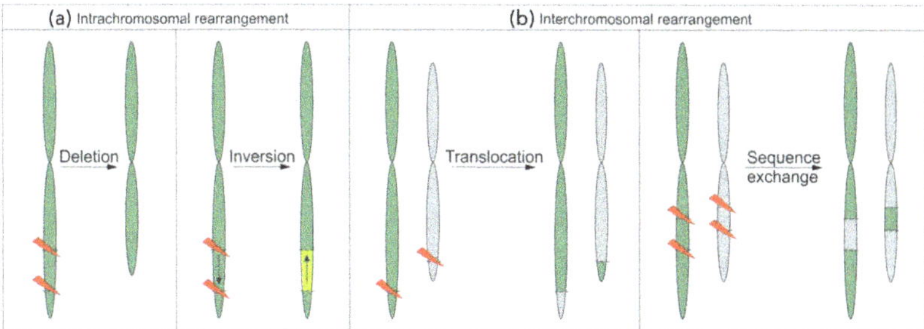

Figure 4 Induction of large chromosomal rearrangements (CRs): (a) if two DSBs are induced on one chromosome arm, intrachromosomal rearrangements can occur. The area between the fractions can either be deleted or inverted, (b) if breaks are induced on two heterologous chromosomes, the fragments can be exchanged. With one DSB on each chromosome arm, a translocation might be induced; if more breaks are located on the arms, it is possible to exchange the sequence information in between.

DSBs on heterologous chromosomes can lead to translocations. If four DSBs are induced, two on each chromosome, entire sequence regions can be exchanged.

The simplest form of a CR is the deletion. Here the sequence area between the breaks is removed and all information is lost. With this, entire genes, large gene clusters or non-coding regulatory sequences can be deleted and ultimately switched off; true loss-of-function mutant lines for functional studies can be generated without effort. Deletions of up to 1 kb are quite easy to induce both in *Arabidopsis* and in other plant species (Brooks et al., 2014; Gao et al., 2015; Kapusi et al., 2017; Upadhyay et al., 2013). In addition, deletions of up to 120 kb in *Arabidopsis* (Ordon et al., 2017; Wu et al., 2018) and up to a size of 245 kb in many other plant species such as rice, tobacco, tomato and *Medicago* can be achieved, but with a significantly lower frequency (Čermák et al., 2017; Zhou et al., 2014). Besides deletions, two breaks on one chromosome can result in the integration of the excised fragment in reverse orientations. In plants, targeted inversions up to 18 kb could be obtained using Cas9 (Gao et al., 2015; Schmidt et al., 2019; Zhang et al., 2017). In tobacco, the demonstration of the formation of DSB-induced reciprocal translocations has been achieved using the homing endonuclease I-*Sce*I whereby NHEJ or SSA joined the previously unlinked DSB ends (Pacher et al., 2007). Very recently, it was shown that by the use of CRISPR/Cas it is indeed possible to obtain heritable chromosomal translocations in *Arabidopsis* in a controlled way, opening the avenue of chromosome engineering (Beying et al., 2020).

To achieve controlled restructuring of plant genomes, it is necessary to consider the different repair mechanisms involved. In mammals it was shown, that both c-NHEJ and a-NHEJ can potentially contribute to the creation of rearrangements (Lupski and Stankiewicz, 2005). Interestingly, a-NHEJ was identified as the major mechanism for translocation formation in mouse embryonic stem cells and it is suppressed by c-NHEJ components such as Ku, Lig4 or Xrcc4 (Simsek and Jasin, 2010; Weinstock et al., 2007). In contrast, unlike mouse cells, translocations in human cells are generated by c-NHEJ (Ghezraoui et al., 2014). A drastic reduction in translocation frequency when both c-NHEJ and a-NHEJ components are lost could be detected in human cell lines (Brunet and Jasin, 2018), whereas murine cell lines showed an increase when cells are lacking Ku70 and Pol θ (Wyatt et al., 2016). Therefore, species-specific mechanisms could also apply to the formation of CRs in plants. In *Arabidopsis*, it was shown that under natural conditions large deletion and inversion formation was mediated by c-NHEJ. Yet, a significant increase for the frequency of inversion and translocation formation occurs when the key player of c-NHEJ, KU70, is depleted (Beying et al., 2020; Schmidt et al., 2019). Deep sequencing revealed repair patterns of the junctions with characteristics of a-NHEJ. This indicates that the a-NHEJ mechanism is significantly less selective to join the 'correct' ends during repair. Conversely, this also means that c-NHEJ is actively ensuring that the originally linked ends are re-ligated. Different factors are known in mammals that are involved in anchoring the two original broken ends during c-NHEJ and are thus able to stabilize the DSB. In this context, the KU heterodimer is essential. Two KU heterodimers initially bind to each end of the DSB. It has been possible to demonstrate biochemically, and with the aid of atomic force microscopy, the KU proteins already bound to the DNA can associate with each other (Cary et al., 1997; Ramsden and Gellert, 1998). The resulting model is based on the assumption that the two heterodimers combine to form a bridge complex, which then subsequently recruits the other repair factors. The two c-NHEJ factors XLF and PAXX, which both interact with the KU heterodimer, are certainly involved in connecting the two ends (Graham et al., 2018; Ochi et al., 2015). It was also shown that the heteromeric complex of XLF and XRCC4 is able to stably bridge two independent DNA molecules and that this bridge complex can move along the DNA. This observation suggests that in mammals the XRCC4-XLF complex forms a mobile, sleeve-like structure around the DNA that is able to hold broken ends together and that this complex is involved in the subsequent repair (Brouwer et al., 2016). So far, only homologs of the c-NHEJ factors KU70/KU80, LIG4 and XRCC4 have been identified in plants, but no homologous proteins to the two factors XLF and PAXX have been identified yet. Therefore, the XRCC4 protein alone might actually be able to support the cohesion of the two ends in plants.

In addition to the stabilization of a DSB, the location of the two broken ends in relation to one another also plays a decisive role, since the break ends should logically be in spatial proximity to be newly linked. In this context, two basic approaches as to how CRs can arise are differentiated, the 'contact-first' and the 'breakage first' theories (Sax, 1941; Serebrovski, 1929). In the 'contact-first' theory, the two chromosomal fragments, between which a structural rearrangement takes place, are already in close proximity, while the two chromosomal regions in the 'breakage-first' theory only come together after DSB induction. In the second model, the broken ends must be mobile and must be able to interact with one another. In yeast, it has been shown that chromosomal regions can move toward one another within minutes after DSB induction and that this leads to clustering of chromosome regions containing DSBs (Aten et al., 2004). A certain mobility of damaged DNA was also observed in mammalian cells, but DSBs are often described here as being more stable, which indicates that the predominant repair mechanism of c-NHEJ limits the mobility of DSBs (Lemaître et al., 2014; Robinett et al., 1996; Soutoglou et al., 2007). Increased mobility of chromosomal areas has so far been observed mainly in connection with the search for homologous areas in the context of HR (Aymard et al., 2017; Cho et al., 2014; Schrank et al., 2018). However, a recently published study also found a connection between PARP1-mediated ADP ribosylation, which leads to the formation of poly(ADP-ribose) (PAR) chains and the generation of dynamic areas in which the damaged DNA accumulates (Singatulina et al., 2019). It is assumed that movement happens at the beginning stages of DNA repair (Miné-Hattab and Rothstein, 2013).

In mammals, as well as in plants, PARP1 was shown to be involved in the repair of DSBs via a MH-based alternative end-joining repair pathway (Audebert et al., 2004; Jia et al., 2013; Wang et al., 2006). Such an accumulation of damaged DNA can have the advantage that repair factors are frequently present in these areas and thus a significantly faster and therefore more efficient repair can take place (Schrank et al., 2018; Singatulina et al., 2019). A disadvantage, however, could be that the likelihood of linking two break ends that were not originally connected to one another increases, and thus CR can also arise with a higher probability.

The same mechanisms might account for the increase of the likelihood of a chromosomal structural change if c-NHEJ is impaired in plants. On the one hand, in the absence of the KU heterodimer, no protein complex stabilizes the two break ends, which increases the chance of break ends being linked that were originally not connected. On the other hand, the clustering of damaged DNA regions that has been linked to the MMEJ in mammals could bring the two DSBs in close proximity to one another, which in turn increases the probability of a CR event occurring.

Thus, blocking c-NHEJ might be a general strategy to increase the efficiency of CRISPR/Cas-mediated plant chromosome engineering. If c-NHEJ is impaired, repair switches to the more error-prone pathway of a-NHEJ and mutations can be induced with higher frequencies. Unfortunately, this means that repair outcomes at the junctions of CRs are less predictable in those mutants as unforeseen mutations might form depending on the sequence context. As *ku* mutants also show other genomic instabilities such as telomere dysfunction and DNA repair defects (Bundock et al., 2002; Riha et al., 2002), the use of this mutant is not advisable for practical applications in plant breeding. Using an inducible knockdown of ku or different c-NHEJ mutants could be another possibility to increase the efficiency of chromosome engineering in plants by simultaneously minimizing side effects.

3 DSB repair using homologous sequences

In contrast to non-homologous end joining, HR occurs rarely in somatic cells and is mainly limited to the S- and G2-phases of the cell cycle (overview in Puchta, 2005). In meiotic tissue, on the other hand, HR is essential for pairing of chromosomes, recombination and exchange of genetic information. In principle, HR is based on the involvement of a homologous sequence region, of which the intact sequence information serves as a matrix for repairing the DSB. However, HR can be divided into conservative and non-conservative mechanisms, depending on whether the repaired DNA section matches the original sequence. HR comprises several different mechanisms, which differ in the repair process and the enzymes involved. These include the two repair mechanisms prevalent in somatic tissue, single-strand annealing (SSA) and synthesis-dependent strand annealing (SDSA), as well as the double-strand break repair (DSBR) and the dissolution pathway, both of which play an important role in meiosis (Fig. 5).

3.1 Basic mechanisms of HR

Common to all HR routes is the initial 5′-3′ resection, which is mediated by the MRN complex and generates single-stranded 3′ overhangs, which are protected from nucleolytic degradation by the attachment of the heterotrimeric RPA (replication protein A) complex (Eschbach and Kobbe, 2014).

If complementary sequences of more than about 20 bp are present in these overhangs, the DSB can be repaired using SSA (Fig. 5b). This repair route is mechanistically similar to a-NHEJ. After the homologies have been exposed, they can pair with one another and a chimeric double-stranded DNA molecule is formed. In yeast, RAD52 seems to be involved as it is recruited to the single-stranded DNA nucleoprotein complex after end resection (van Dyck et al., 2001). Also in humans, the interaction of RAD52 with RPA was demonstrated (Ma et al., 2017). For the *Arabidopsis* homologs of RAD52 there seems to be no involvement in this process (Samach et al., 2011), but the RAD51 paralogs XRCC2, RAD51B and RAD51D were also demonstrated to play a role in SSA (Serra et al., 2013). After priming, the overhanging ends are trimmed by a dimer

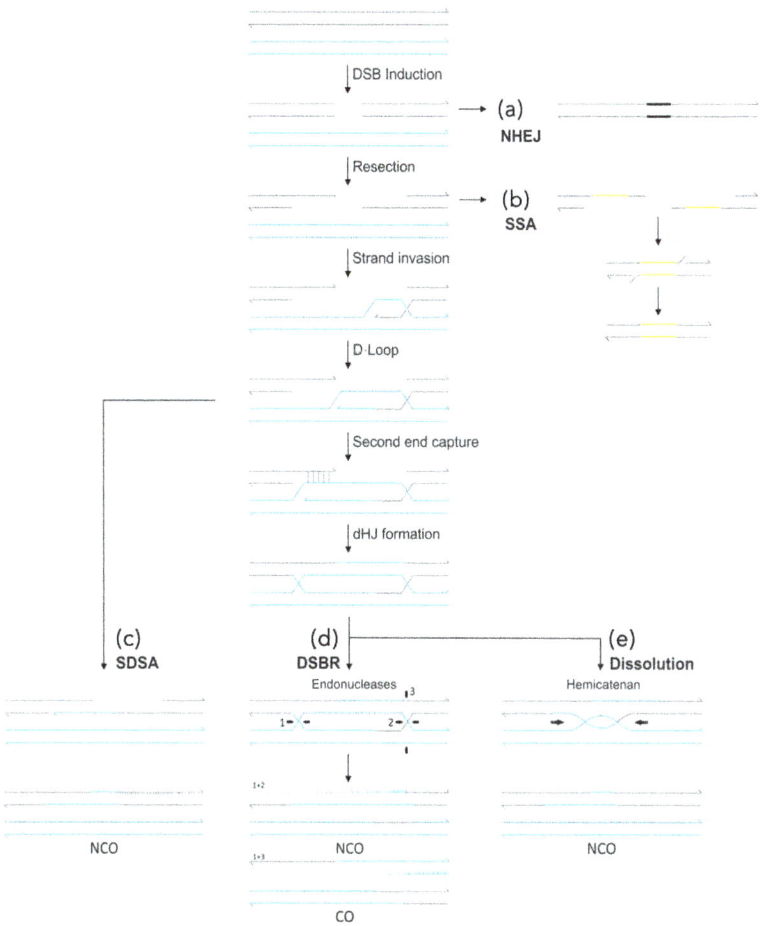

Figure 5 Overview of the repair mechanisms after a DSB: (a) If no homologous sequence is available, the break is repaired via NHEJ, (b) If homologous areas larger than 20 bp (yellow) are found on both sides of the DSB, the break can be repaired using single strand annealing (SSA). This leads to the loss of sequence information and results in mutagenic deletions. (c–e) Conservative homologous recombination (HR), which includes the three mechanisms of synthesis-dependent strand annealing (SDSA), double-strand break repair (DSBR) and dissolution, results in error-free repair of the DSB. After resection of the DSB ends and invasion of the 3′ overhang into the homologous area, a displacement loop (D-Loop) is formed. The three mechanisms differ in the resolution of this intermediate. In the SDSA pathway (c), the extended invading DNA strand is detached from the donor molecule and can pair with the complementary overhang of the second break end. The gaps are then closed by DNA synthesis, resulting in a non-crossover product (NCO). Due to further synthesis of the invaded strand, the D-Loop can be expanded leading to second end capture and a double Holliday junction (dHJ) is formed. This structure can be resolved in the dissolution pathway (e) by the formation of a hemicatenane via branch migration using the RTR complex, resulting in an NCO event. In the resolution pathway (d), structure-specific endonucleases can cleave the dHJ. In addition to NCO products, in DSBR, crossovers (COs) can arise depending on the cutting pattern.

of the proteins RAD1 and RAD10 (Dubest et al., 2002; Prado and Aguilera, 1995), and the gaps are filled by DNA synthesis. Finally, the DNA backbone is ligated. As a result, the sequence information between the homologies is lost, which is why this repair route is the only one assigned to non-conservative HR (overview in Knoll et al., 2014a).

If no repetitive homologies are available at the break site, an exogenous repair matrix is required for repair via the other mechanisms of HR. The search for such a template is mediated by RAD51. RAD51 replaces the RPA encasing the single strand and mediates the search for a repair template (Renkawitz et al., 2013). The free single strand can invade into a homologous sequence section and thereby displace one strand of the double-stranded donor molecule. The resulting structure is called a displacement loop (D-loop) (Fig. 5). The strand invasion is followed by an extension of the immigrated single strand, whereby the homologous strand is serving as a matrix. From here on, the invading strand can either be released or further extended. In SDSA, the D-loop is dissolved and due to the extension the invading strand can pair with the processed, opposite end of the DSB. As information is only copied to one strand, this repair always results in a non-crossover (NCO) product.

However, if the invading strand is greatly extended, hybridization can occur between the displaced strand of the D-loop and the now complementary end of the second single strand of the break, also referred to as second end capture. This process, which occurs mainly in meiosis, leads to the formation of a repair intermediate called the double Holliday junction (dHJ, Fig. 5) by ligating the break ends. The resolution of this dHJ structure can take place via the two HR repair routes of DSBR and dissolution. The predominant mechanism, the dissolution, maintains the integrity of the genome, since only NCO products are created and the two DNA molecules are thus separated without exchanging chromosome regions. The process of dissolution is catalyzed by the RTR complex, which is highly conserved in eukaryotes (Knoll et al., 2014b). It consists of a RECQ helicase, a type 1A topoisomerase as well as the structural proteins RMI1/RMI2. The helicase pushes the two junctions of the dHJ toward each other forming an intermediate structure called a hemicatenane. The topoisomerase relaxes the resulting super spiraling of the DNA and dissolves the resulting hemicatenane structure (Chen et al., 2014). In *Arabidopsis*, all proteins like RECQ4A, TOP3α and RMI1/RMI2 involved in the RTR complex were identified and were shown to interact in vivo (Bonnet et al., 2013; Dorn et al., 2018; Hartung et al., 2008; Röhrig et al., 2018; Schröpfer et al., 2014; Séguéla-Arnaud et al., 2015). The resolution of the dHJ is achieved in the DSBR model by structure-specific endonucleases, which are also called resolvases due to their function (Szostak et al., 1983). The induction of symmetrical cuts at the two crossing points of the dHJ leads to the separation of the two double strands. Depending on the cutting direction of the resolvases, this can result in NCO and crossover (CO) products (Fig. 5). dHJs mainly arise in meiosis and the CO products resulting from chromosome arm exchange are required for mixing parental genomes.

By using transgenic recombination substrates with different setups, it is possible to characterize the role of different factors in these pathways *in planta*. In those pathway-specific recombination traps, a DSB can be induced by the expression of a site-specific nuclease, which leads to the restoration of a marker by either SSA or SDSA (Orel et al., 2003). In early experiments, I-*Sce*I was used for break induction to restore a disrupted *GUS* gene. With this, the repair of one out of three DSBs induced by I-*Sce*I was shown to be mediated by SSA in tobacco, if the break site was flanked by homology stretches (Siebert and Puchta, 1998). In contrast, under similar conditions the SDSA pathway showed a five times reduction of efficiency compared to SSA (Orel et al., 2003). Additionally, the involvement of the aforementioned factors XRCC2, RAD51B and RAD51D as well as RAD1/RAD10 in SSA could be demonstrated (Dubest et al., 2002; Serra et al., 2013). Further proteins that might have a function in this pathway are the helicase FANCM and the nuclease MUS81 (Mannuss et al., 2010). As only minor involvement could be detected, other nucleases and helicases might be able to substitute each other, obstructing the detection of key factors in SSA. MUS81 seems to be also involved in SDSA, whereas mutants of proteins that are accounted for strand exchange only show an effect in repair of the SDSA recombination trap. As a result, in *Arabidopsis* an essential role for XRCC3 and RAD54 in addition to RAD51 in SDSA was postulated (Roth et al., 2012). Other ATPases like RAD5A, which plays a role in post replicative repair (Chen et al., 2008), and FANCM, which is also involved in control of meiotic rocombination (Knoll et al., 2012), are also involved in SDSA-mediated repair (Mannuss et al., 2010; Roth et al., 2012).

Next to SSA, it was demonstrated that only SDSA operates efficiently in somatic plant cells (Puchta, 1998). With an efficient repair via DSBR, somatic cells would endanger their genome stability, as this repair is prerequisite for crossover induction. The formation of crossovers between repeated ectopic sequences that are found all over plant genomes could lead to the formation of dicentric or acentric chromosomes. This is why DSBR occurs mainly during meiosis, where the exact pairing of homologous chromosomes is assured by the formation of the synaptonemal complex.

HR is dependent on the availability and accessibility of a homologous sequence. In SSA, the homologous sequence serving as a template has to be located close to the break site in direct orientation. Genomic regions harboring tandem repeats or other repetitive sequences are therefore preferentially repaired via SSA. The absolute frequency of SSA-mediated repair events was shown to be dependent on the distance between the reads. With increasing distance, longer end resection has to take place and results in a negative correlation between deletion size and frequency (Vu et al., 2017). As mentioned before, in small genomes like *Arabidopsis*, larger deletions are more common compared to the situation observed for tobacco (Kirik et al., 2000), barley (Vu et al., 2017) or human cells (van Overbeek et al., 2016). This leads to a higher abundance of SSA between repeats over increasing distances in *Arabidopsis*. In the other cases of HR, several different repair templates might be used (Fig. 6). In S- and G2-phases of the cell cycle, the sister chromatid is available for repair. In addition, homologous sequences on the same chromosome like transposable elements or gene duplications, an allelic sequence on the homologous chromosome or ectopic sequences located on a different chromosome can serve as a template. To test the efficiencies of the different repair templates, DSB induction by I-*Sce*I or transposons was generally used. Based on the restoration or loss of marker genes, frequencies could be quantified.

If homologies are available in close proximity to the break site, repair by HR pathways is quite efficient. Repair matrixes can be found either on the same chromosome or on the sister chromatid. If a DSB is repaired via the sister chromatid as a homologous template, no sequence change can be identified. Therefore, it is very difficult to assess frequencies of these repair events. Yet, Vu *et al.* were able to study repair involving the sister chromatid in barley by ethynyl-deoxyuridine-based staining showing frequencies of >81% (Vu et al., 2014). This unprecedented result leads to the conclusion that efficiencies for repair with this kind of template in S- and G2-phases are underrated. Intrachromosomal HR

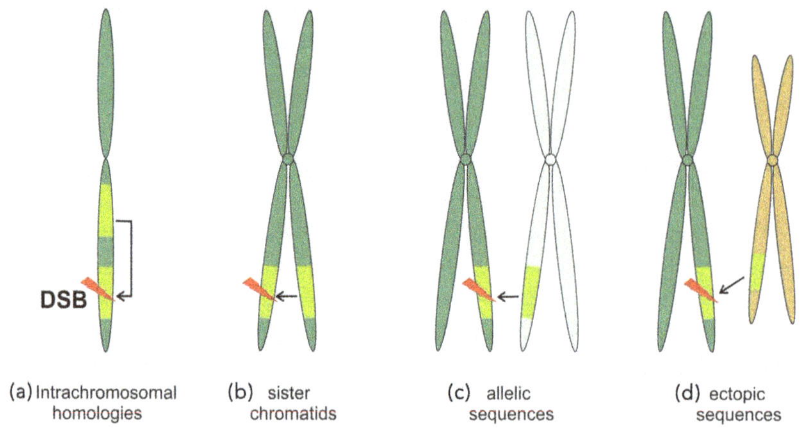

Figure 6 Possible templates for HR-mediated repair of a DSB. Depending on the cell-cycle phase, different repair templates are available for HR via the SDSA mechanism: (a) homologous sequences on the same chromosome are most frequently used for repair, as intrachromosomal regions are most accessible, (b) in G2- and S-phase, the sister chromatid is available and can be used efficiently for repair, (c) allelic sequences are primarily used during meiotic recombination, and only in very low frequencies in somatic cells, (d) ectopic sequences on another chromosome or on extrachromosomal material can also serve as templates for repair via HR.

using homology on the same chromosome as a template is the most abundant pathway of HR. In addition to direct repeats used by SSA as discussed above, other intrachromosomal homologies can serve as a template for DSB repair via SDSA. This was proven in many studies regarding intrachromosomal HR in plants (Chiurazzi et al., 1996; Mannuss et al., 2010; Orel et al., 2003; Roth et al., 2012; Siebert and Puchta, 2002).

In meiosis, mostly allelic regions function as the template for DSB repair. In somatic plant cells, allelic repair is detectable but was reported to not be significant for repair outcomes after DSB induction via I-*Sce*I, with an estimated low frequency of 10^{-4} in an initial study (Gisler et al., 2002). The same holds true for homologous sequences within an ectopic region. Using an ectopic repair template after DSB induction, similar efficiencies for HR in somatic tissue were detected (Puchta, 1999; Shalev and Levy, 1997). Therefore, the conclusion was drawn that homologies on a different chromosome are hard to access and so in somatic cells repair using them is negligible. Yet, the experimental setup from Gisler et al. (2002) to analyze allelic repair did not provide homologies directly at the cutting site of I-*Sce*I. In a recent study, another system was set up to test allele-dependent repair mediated by HR in tomato (Filler Hayut et al., 2017). In stark contrast to the earlier study, somatic HR was raised up to 14% per allele by DSB induction via Cas9. The higher efficiency might be because the homology was present directly at the break site in this later study. In the absence of DSB induction, somatic HR between homologous chromosomes is very low, as shown in earlier studies in tobacco (Carlson, 1974; Dulieu, 1975). However, in G2 somatic DSB can also be repaired by the sister chromatid. Indirect evidence indicates that this repair pathway is much more important than ectopic or allelic repair (Watanabe et al., 2009).

3.2 Application in gene targeting

In gene targeting (GT), HR is utilized to introduce desired changes in a precise and targeted manner. Unfortunately, low efficiencies of this method in higher plants made its application in basic research, plant biotechnology and breeding sparsely feasible. Therefore, many studies were conducted to improve frequencies of HR-based GT (for review: Huang and Puchta, 2019; Puchta and Fauser, 2013; Steinert et al., 2016; Sun et al., 2016; Wolter et al., 2018).

Following the first experiments applying GT showing these low frequencies (Beetham et al., 1999; Offringa et al., 1990; Paszkowski et al., 1988), initial improvements were made by inducing site-specific breaks at the target sequence. Using I-*Sce*I, GT frequencies could be enhanced by up to two orders of magnitude in tobacco (Puchta et al., 1996). In many other important crops such as maize, rice and wheat, DSB induction by site-specific nucleases, for example, the CRISPR/Cas system, was used to stimulate GT (Endo et al., 2016;

Gil-Humanes et al., 2017; Li et al., 2018a; Sun et al., 2016; Svitashev et al., 2015, 2016; Wang et al., 2017). Therefore, one strategy to improve GT is to adjust the efficiency and nature of the nuclease used for break induction. The most common used endonuclease is SpCas9 (Cas9 from *Streptococcus pyogenes*), but the application of *Staphylococcus aureus* Cas9 with its higher frequency of DSB induction (Steinert et al., 2015) was demonstrated to enhance GT frequencies over SpCas9 at least in *Arabidopsis* (Wolter et al., 2018). Both of these Cas9 orthologs induce blunt ends close to the PAM sequence. (Bothmer et al., 2017; Čermák et al., 2017). Furthermore, Cas12a which produces 5′ overhangs has been reported to enhance GT (Begemann et al., 2017; Li et al., 2020; Merker et al., 2020; Wolter and Puchta, 2019). This enzyme cleaves the target DNA on the PAM distal side. In contrast to the Cas9 enzymes, the break is induced far away from the seed sequence, which serves as nucleation site for base pairing between the crRNA and the target DNA (Fonfara et al., 2016). Thus, mismatches between crRNA and target DNA arising from NHEJ repair might not hinder further cleavage. As after each cut HR is again competing with NHEJ for DSB repair, the overall chances for HR rise overtime as long as the re-cutting of the NHEJ repaired site is still possible.

In GT, a repair template for HR-induced changes has to be provided (Fig. 7). The design of this template, its activation and availability, has to be considered for efficient GT. For efficient recombination the length of the homology for donor templates has to be considered. Although no systematic study of template length was performed for plants, a homology length of at least 500 bp was

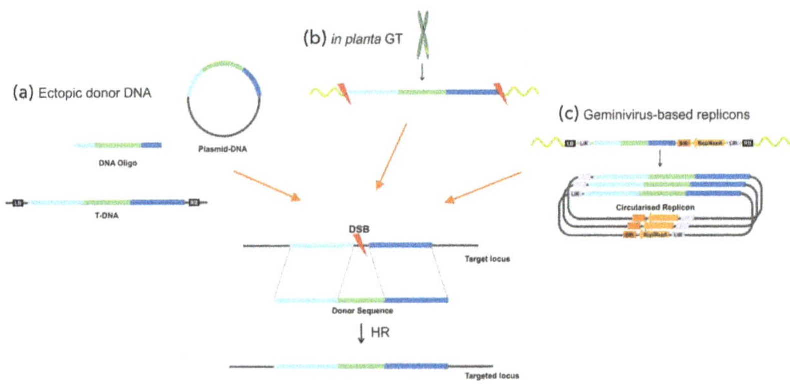

Figure 7 Gene targeting in plants and the applicable donor molecules. After DSB induction, gene targeting (GT) is mediated by HR. Therefore, the donor sequence needs to harbor homologies (blue) on both sides of the changes to be integrated (green): (a) possible templates for HR-based GT are ectopic sequences such as T-DNAs, plasmid DNA or DNA oligos that are transferred to the cell, (b) in the *in planta* GT approach, the template is already integrated into the genome and is activated through excision by nucleases, (c) the combination of *in planta* GT with geminivirus-based replicons enhances GT frequencies.

used in most studies. With sequence-specific DSBs as a very effective method to improve GT, the cutting of the repair template and the modified locus has to be prevented. In general, modification of the target sequence or for Cas9 the PAM region is the best way to abolish DNA cleavage by the nuclease. As SDSA is the main pathway responsible for GT in somatic plant cells, all changes that are supposed to be introduced in the genome should be located on one side of the DSB (Huang and Puchta, 2019). If the template is only provided transiently as an extrachromosomal repair template (Fig. 7a), degradation by the plants' intrinsic mechanisms reduces GT efficiencies. By stably integrating the donor DNA into the plant genome, the so-called '*in planta* GT' method is independent on transfected donor DNA and on transformation efficiencies as only one initial transformation event is sufficient (Fig. 7b) (Fauser et al., 2012; Schiml et al., 2014). Thus, this technique makes the application of GT in crop species that are difficult to transform more feasible. However, the template has to be activated via two cuts to release the linear sequence from the genome. These additional cleavage sites might be another reason as to why GT efficiencies are improved in this strategy. As discussed for the formation of chromosomal rearrangements, damaged DNA might accumulate in repair loci at the nuclear periphery (Caridi et al., 2018). This would bring the cut target site in close proximity to the excised repair template. Furthermore, by raising the copy number of the donor molecule, GT can be enhanced. This increase of the availability of the template was achieved using geminivirus replicons in combination with the *in planta* GT method (Fig. 7c) (Dahan-Meir et al., 2018). The successful application of geminivirus replicons for enhancing gene targeting was demonstrated before using ZFNs, TALENs and Cas9 (Baltes et al., 2014; Čermák et al., 2015).

In early experiments in plants, besides perfect events by HR, aberrant GT was also found (Puchta et al., 1996, reviewed in Puchta and Fauser, 2013). Those events showed combined NHEJ and HR repair outcomes, where one junction was repaired via HR and the other via NHEJ. The prevailing homologous repair mechanism behind GT in somatic plant cells is SDSA. Even though a homologous 3′ single strand of one end of the DSB invades into the donor, the possibility persists that the repair at the other end is conducted by NHEJ. Thus, it is necessary to check GT events for correct HR-induced repair at both junction sites via sequencing of the complete modified locus. Reduction of these one-sided integration events can be achieved by selecting against random NHEJ events with negative selectable markers. This was proved to be a useful way to obtain GT in rice (Terada et al., 2002), although this technique is quite laborious. Besides SDSA, SSA can be applied to induce genomic changes as well. Instead of introducing a foreign sequence into the genome, the removal of sequences between flanking homologies can be facilitated. In a recent study, SSA was used for the removal of inserted selection cassettes that contain flanking homologies, as a scarless alternative to the Cre-inducible recombination systems (Li et al., 2018b).

Another way of bypassing NHEJ-mediated repair and thus enhancing GT is the cell cycle phase or cell type-specific expression of the nuclease for break induction. While DSBs are repaired in G1 phase most frequently by NHEJ, in G2/M- and S-phase HR is more active compared to its near absence in G1. In yeast and other fungi, as well as in human pluripotent stem cells, higher GT efficiencies were detected in those cell-cycle phases (Tsakraklides et al., 2015; Yang et al., 2016). In mouse embryonic stem cells, GT frequencies of up to 100% could be demonstrated (Miura et al., 2018; Quadros et al., 2017). By egg-cell specific expression of the nuclease in *Arabidopsis*, promising frequencies of heritable GT events could recently be obtained (Miki et al., 2018; Wolter et al., 2018).

Additionally, suppressing NHEJ pathways or promoting the HR pathway was exploited to increase HR-mediated GT. In *Arabidopsis*, mutants of the two main factors in NHEJ, KU70 and LIG4 showed 16-times higher GT efficiencies (Qi et al., 2013). Similarly, it was reported that knockout of LIG4 via CRISPR/Cas9 in rice calli enhanced GT several folds (Endo et al., 2016). As an alternative strategy, it could be demonstrated that overexpression of some proteins involved in HR repair enhanced GT effectively, such as overexpressing the *S. cerevisiae* Rad54 protein in plants or expressing RAD54 specifically in egg cells (Even-Faitelson et al., 2011; Shaked et al., 2005). In *smc6b* mutants of *Arabidopsis*, both NHEJ-mediated mutagenesis and GT were found to be elevated at three loci (Qi et al., 2013). As in these mutants sister chromatid exchange is impaired, repair was either conducted via NHEJ or via HR-based GT with an extrachromosomal template present. Yet, DNA repair mutants and HR protein overexpression lines display genomic instabilities. Therefore, alternative ways of blocking NHEJ pathways or enhancing HR pathways utilizing the expanding toolkit of the CRISPR/Cas system, for example, are sought and might be applicable in the near future.

4 Perspectives

Harnessing the knowledge of DNA repair pathways and combining it with the increasing toolkit of different site-specific nucleases especially the CRISPR/Cas system, will open great possibilities in plant genome engineering. Editing of multiple loci simultaneously, restructuring of plant genomes, improving GT frequencies above percentage rates and many more applications are already being addressed. With this significant progress in plant breeding within sight, a new green revolution might become reality within the near future.

5 Where to look for further information

The following articles provide a good overview of the subject:

- Puchta, H. (2005). The repair of double-strand breaks in plants: mechanisms and consequences for genome evolution. *Journal of Experimental Botany* 56(409): 1–14.
- Bleuyard, J.-Y., Gallego, M. E. and White, C. I. (2006). Recent advances in understanding of the DNA double-strand break repair machinery of plants. *DNA Repair* 5(1): 1–12.
- Waterworth, W. M., Drury, G. E., Bray., C. M. et al. (2011). Repairing breaks in the plant genome: the importance of keeping it together. *The New Phytologist* 192(4): 805–822.
- Knoll, A., Fauser, F. and Puchta, H. (2014). DNA recombination in somatic plant cells: mechanisms and evolutionary consequences. *Chromosome Research: an International Journal on the Molecular, Supramolecular and Evolutionary Aspects of Chromosome Biology* 22(2): 191–201.
- Huang, T.-K. and Puchta, H. (2019). CRISPR/Cas-mediated gene targeting in plants: finally a turn for the better for homologous recombination. *Plant Cell Reports* 38(4): 443–453.
- Yeh, C. D., Richardson, C. D. and Corn, J. E. (2019). Advances in genome editing through control of DNA repair pathways. *Nature Cell Biology* 21(12): 1468–1478.

6 References

Adams, K. L. and Wendel, J. F. (2005). Polyploidy and genome evolution in plants. *Current Opinion in Plant Biology* 8(2): 135–141.

Ahmad, A., Robinson, A. R., Duensing, A., van Drunen, E., Beverloo, H. B., Weisberg, D. B., Hasty, P., Hoeijmakers, J. H. and Niedernhofer, L. J. (2008). ERCC1-XPF endonuclease facilitates DNA double-strand break repair. *Molecular and Cellular Biology* 28(16): 5082–5092.

Arnould, S., Perez, C., Cabaniols, J. P., Smith, J., Gouble, A., Grizot, S., Epinat, J. C., Duclert, A., Duchateau, P. and Pâques, F. (2007). Engineered I-CreI derivatives cleaving sequences from the human XPC gene can induce highly efficient gene correction in mammalian cells. *Journal of Molecular Biology* 371(1): 49–65.

Aten, J. A., Stap, J., Krawczyk, P. M., van Oven, C. H., Hoebe, R. A., Essers, J. and Kanaar, R. (2004). Dynamics of DNA double-strand breaks revealed by clustering of damaged chromosome domains. *Science (New York, N.Y.)* 303(5654): 92–95.

Audebert, M., Salles, B. and Calsou, P. (2004). Involvement of poly(ADP-ribose) polymerase-1 and XRCC1/DNA ligase III in an alternative route for DNA double-strand breaks rejoining. *The Journal of Biological Chemistry* 279(53): 55117–55126.

Aymard, F., Aguirrebengoa, M., Guillou, E., Javierre, B. M., Bugler, B., Arnould, C., Rocher, V., Iacovoni, J. S., Biernacka, A., Skrzypczak, M., Ginalski, K., Rowicka, M., Fraser, P. and Legube, G. (2017). Genome-wide mapping of long-range contacts unveils clustering of DNA double-strand breaks at damaged active genes. *Nature Structural and Molecular Biology* 24(4): 353–361.

Baltes, N. J., Gil-Humanes, J., Cermak, T., Atkins, P. A. and Voytas, D. F. (2014). DNA replicons for plant genome engineering. *The Plant Cell* 26(1): 151–163.

Beetham, P. R., Kipp, P. B., Sawycky, X. L., Arntzen, C. J. and May, G. D. (1999). A tool for functional plant genomics: chimeric RNA/DNA oligonucleotides cause in vivo gene-specific mutations. *Proceedings of the National Academy of Sciences of the United States of America* 96(15): 8774–8778.

Begemann, M. B., Gray, B. N., January, E., Gordon, G. C., He, Y., Liu, H., Wu, X., Brutnell, T. P., Mockler, T. C. and Oufattole, M. (2017). Precise insertion and guided editing of higher plant genomes using Cpf1 CRISPR nucleases. *Scientific Reports* 7(1): 11606.

Bennardo, N., Cheng, A., Huang, N. and Stark, J. M. (2008). Alternative-NHEJ is a mechanistically distinct pathway of mammalian chromosome break repair. *PLOS Genetics* 4(6): e1000110.

Bennetzen, J. L. and Wang, H. (2014). The contributions of transposable elements to the structure, function, and evolution of plant genomes. *Annual Review of Plant Biology* 65: 505–530.

Beying, N., Schmidt, C., Pacher, M., Houben, A. and Puchta, H. (2020). CRISPR-Cas9-mediated induction of heritable chromosomal translocations in Arabidopsis. *Nature Plants* 6(6): 638–645.

Black, S. J., Kashkina, E., Kent, T. and Pomerantz, R. T. (2016). DNA polymerase θ: a unique multifunctional end-joining machine. *Genes* 7(9): 67.

Bleuyard, J. Y., Gallego, M. E. and White, C. I. (2006). Recent advances in understanding of the DNA double-strand break repair machinery of plants. *DNA Repair* 5(1): 1–12.

Boch, J., Scholze, H., Schornack, S., Landgraf, A., Hahn, S., Kay, S., Lahaye, T., Nickstadt, A. and Bonas, U. (2009). Breaking the code of DNA binding specificity of TAL-type III effectors. *Science (New York, N.Y.)* 326(5959): 1509–1512.

Bonnet, S., Knoll, A., Hartung, F. and Puchta, H. (2013). Different functions for the domains of the Arabidopsis thaliana RMI1 protein in DNA cross-link repair, somatic and meiotic recombination. *Nucleic Acids Research* 41(20): 9349–9360.

Bothmer, A., Phadke, T., Barrera, L. A., Margulies, C. M., Lee, C. S., Buquicchio, F., Moss, S., Abdulkerim, H. S., Selleck, W., Jayaram, H., Myer, V. E. and Cotta-Ramusino, C. (2017). Characterization of the interplay between DNA repair and CRISPR/Cas9-induced DNA lesions at an endogenous locus. *Nature Communications* 8: 13905.

Britt, A. B. (1996). DNA damage and repair in plants. *Annual Review of Plant Physiology and Plant Molecular Biology* 47: 75–100.

Brooks, C., Nekrasov, V., Lippman, Z. B. and Van Eck, J. (2014). Efficient gene editing in tomato in the first generation using the clustered regularly interspaced short palindromic repeats/CRISPR-associated9 system. *Plant Physiology* 166(3): 1292–1297.

Brouwer, I., Sitters, G., Candelli, A., Heerema, S. J., Heller, I., de Melo, A. J., Zhang, H., Normanno, D., Modesti, M., Peterman, E. J. and Wuite, G. J. (2016). Sliding sleeves of XRCC4-XLF bridge DNA and connect fragments of broken DNA. *Nature* 535(7613): 566–569.

Brunet, E. and Jasin, M. (2018). Induction of chromosomal translocations with CRISPR-Cas9 and other nucleases: understanding the repair mechanisms that give rise to translocations. *Advances in Experimental Medicine and Biology* 1044: 15–25.

Bundock, P., van Attikum, H. and Hooykaas, P. (2002). Increased telomere length and hypersensitivity to DNA damaging agents in an Arabidopsis KU70 mutant. *Nucleic Acids Research* 30(15): 3395–3400.

Caridi, C. P., D'Agostino, C., Ryu, T., Zapotoczny, G., Delabaere, L., Li, X., Khodaverdian, V. Y., Amaral, N., Lin, E., Rau, A. R. and Chiolo, I. (2018). Nuclear F-actin and myosins drive relocalization of heterochromatic breaks. *Nature* 559(7712): 54–60.

Carlson, P. S. (1974). Mitotic crossing-over in a higher plant. *Genetical Research* 24(1): 109–112.
Cary, R. B., Peterson, S. R., Wang, J., Bear, D. G., Bradbury, E. M. and Chen, D. J. (1997). DNA looping by Ku and the DNA-dependent protein kinase. *Proceedings of the National Academy of Sciences of the United States of America* 94(9): 4267 -4272.
Cavalier-Smith, T. (2005). Economy, speed and size matter: evolutionary forces driving nuclear genome miniaturization and expansion. *Annals of Botany* 95(1): 147–175.
Ceccaldi, R., Liu, J. C., Amunugama, R., Hajdu, I., Primack, B., Petalcorin, M. I., O'Connor, K. W., Konstantinopoulos, P. A., Elledge, S. J., Boulton, S. J., Yusufzai, T. and D'Andrea, A. D. (2015). Homologous-recombination-deficient tumours are dependent on Polθ-mediated repair. *Nature* 518(7538): 258–262.
Ceccaldi, R., Rondinelli, B. and D'Andrea, A. D. (2016). Repair pathway choices and consequences at the double-strand break. *Trends in Cell Biology* 26(1): 52–64.
Čermák, T., Baltes, N. J., Čegan, R., Zhang, Y. and Voytas, D. F. (2015). High-frequency, precise modification of the tomato genome. *Genome Biology* 16: 232.
Čermák, T., Curtin, S. J., Gil-Humanes, J., Čegan, R., Kono, T. J. Y., Konečná, E., Belanto, J. J., Starker, C. G., Mathre, J. W., Greenstein, R. L. and Voytas, D. F. (2017). A multipurpose toolkit to enable advanced genome engineering in plants. *The Plant Cell* 29(6): 1196–1217.
Chan, S. H., Yu, A. M. and McVey, M. (2010). Dual roles for DNA polymerase theta in alternative end-joining repair of double-strand breaks in Drosophila. *PLoS Genetics* 6(7): e1001005.
Chang, H. H. Y., Pannunzio, N. R., Adachi, N. and Lieber, M. R. (2017). Non-homologous DNA end joining and alternative pathways to double-strand break repair. *Nature Reviews. Molecular Cell Biology* 18(8): 495–506.
Charbonnel, C., Gallego, M. E. and White, C. I. (2010). Xrcc1-dependent and Ku-dependent DNA double-strand break repair kinetics in Arabidopsis plants. *The Plant Journal: For Cell and Molecular Biology* 64(2): 280–290.
Chen, I. P., Mannuss, A., Orel, N., Heitzeberg, F. and Puchta, H. (2008). A homolog of ScRAD5 is involved in DNA repair and homologous recombination in Arabidopsis. *Plant Physiology* 146(4): 1786–1796.
Chen, K., Wang, Y., Zhang, R., Zhang, H. and Gao, C. (2019). CRISPR/Cas genome editing and precision plant breeding in agriculture. *Annual Review of Plant Biology* 70: 667–697.
Chen, S. H., Plank, J. L., Willcox, S., Griffith, J. D. and Hsieh, T. S. (2014). Top3α is required during the convergent migration step of double Holliday junction dissolution. *PLoS ONE* 9(1): e83582.
Chiurazzi, M., Ray, A., Viret, J. F., Perera, R., Wang, X. H., Lloyd, A. M. and Signer, E. R. (1996). Enhancement of somatic intrachromosomal homologous recombination in Arabidopsis by the HO endonuclease. *The Plant Cell* 8(11): 2057–2066.
Cho, N. W., Dilley, R. L., Lampson, M. A. and Greenberg, R. A. (2014). Interchromosomal homology searches drive directional ALT telomere movement and synapsis. *Cell* 159(1): 108–121.
Dahan-Meir, T., Filler-Hayut, S., Melamed-Bessudo, C., Bocobza, S., Czosnek, H., Aharoni, A. and Levy, A. A. (2018). Efficient in planta gene targeting in tomato using geminiviral replicons and the CRISPR/Cas9 system. *The Plant Journal: For Cell and Molecular Biology* 95(1): 5–16.

Deng, S. K., Gibb, B., Almeida, M. J. de, Greene, E. C. and Symington, L. S. (2014). RPA antagonizes microhomology-mediated repair of DNA double-strand breaks. *Nature Structural and Molecular Biology* 21(4): 405–412.

Devos, K. M., Brown, J. K. M. and Bennetzen, J. L. (2002). Genome size reduction through illegitimate recombination counteracts genome expansion in Arabidopsis. *Genome Research* 12(7): 1075–1079.

Dorn, A., Röhrig, S., Papp, K., Schröpfer, S., Hartung, F., Knoll, A. and Puchta, H. (2018). The topoisomerase 3α zinc-finger domain T1 of Arabidopsis thaliana is required for targeting the enzyme activity to Holliday junction-like DNA repair intermediates. *PLoS Genetics* 14(9): e1007674.

Dubest, S., Gallego, M. E. and White, C. I. (2002). Role of the AtRad1p endonuclease in homologous recombination in plants. *EMBO Reports* 3(11): 1049–1054.

Dulieu, H. L. (1975). Somatic variations on a yellow mutant in Nicotianatabacum L. (a1+/a1a2+/a2) II. Reciprocal genetic events occurring in leaf cells. *Mutation Research/Fundamental and Molecular Mechanisms of Mutagenesis* 28(1): 69–77.

Endo, M., Mikami, M. and Toki, S. (2016). Biallelic gene targeting in rice. *Plant Physiology* 170(2): 667–677.

Eschbach, V. and Kobbe, D. (2014). Different replication protein A complexes of Arabidopsis thaliana have different DNA-binding properties as a function of heterotrimer composition. *Plant and Cell Physiology* 55(8): 1460–1472.

Even-Faitelson, L., Samach, A., Melamed-Bessudo, C., Avivi-Ragolsky, N. and Levy, A. A. (2011). Localized egg-cell expression of effector proteins for targeted modification of the Arabidopsis genome. *The Plant Journal: For Cell and Molecular Biology* 68(5): 929–937.

Fauser, F., Roth, N., Pacher, M., Ilg, G., Sánchez-Fernández, R., Biesgen, C. and Puchta, H. (2012). In planta gene targeting. *Proceedings of the National Academy of Sciences of the United States of America* 109(19): 7535–7540.

Filler Hayut, S., Melamed Bessudo, C. and Levy, A. A. (2017). Targeted recombination between homologous chromosomes for precise breeding in tomato. *Nature Communications* 8: 15605.

Fonfara, I., Richter, H., Bratovič, M., Le Rhun, A. and Charpentier, E. (2016). The CRISPR-associated DNA-cleaving enzyme Cpf1 also processes precursor CRISPR RNA. *Nature* 532(7600): 517–521.

Friesner, J. and Britt, A. B. (2003). Ku80- and DNA ligase IV-deficient plants are sensitive to ionizing radiation and defective in T-DNA integration. *The Plant Journal: For Cell and Molecular Biology* 34(4): 427–440.

Gao, J., Wang, G., Ma, S., Xie, X., Wu, X., Zhang, X., Wu, Y., Zhao, P. and Xia, Q. (2015). CRISPR/Cas9-mediated targeted mutagenesis in Nicotiana tabacum. *Plant Molecular Biology* 87(1–2): 99–110.

Geisinger, J. M., Turan, S., Hernandez, S., Spector, L. P. and Calos, M. P. (2016). In vivo blunt-end cloning through CRISPR/Cas9-facilitated non-homologous end-joining. *Nucleic Acids Research* 44(8): e76.

Ghezraoui, H., Piganeau, M., Renouf, B., Renaud, J. B., Sallmyr, A., Ruis, B., Oh, S., Tomkinson, A. E., Hendrickson, E. A., Giovannangeli, C., Jasin, M. and Brunet, E. (2014). Chromosomal translocations in human cells are generated by canonical nonhomologous end-joining. *Molecular Cell* 55(6): 829–842.

Gil-Humanes, J., Wang, Y., Liang, Z., Shan, Q., Ozuna, C. V., Sánchez-León, S., Baltes, N. J., Starker, C., Barro, F., Gao, C. and Voytas, D. F. (2017). High-efficiency gene targeting in hexaploid wheat using DNA replicons and CRISPR/Cas9. *The Plant Journal: For Cell and Molecular Biology* 89(6): 1251–1262.

Gisler, B., Salomon, S. and Puchta, H. (2002). The role of double-strand break-induced allelic homologous recombination in somatic plant cells. *The Plant Journal: For Cell and Molecular Biology* 32(3): 277–284.

Gorbunova, V. and Levy, A. A. (1997). Non-homologous DNA end joining in plant cells is associated with deletions and filler DNA insertions. *Nucleic Acids Research* 25(22): 4650–4657.

Grabarz, A., Barascu, A., Guirouilh-Barbat, J. and Lopez, B. S. (2012). Initiation of DNA double strand break repair: signaling and single-stranded resection dictate the choice between homologous recombination, non-homologous end-joining and alternative end-joining. *American Journal of Cancer Research* 2(3): 249–268.

Graham, T. G. W., Carney, S. M., Walter, J. C. and Loparo, J. J. (2018). A single XLF dimer bridges DNA ends during nonhomologous end joining. *Nature Structural and Molecular Biology* 25(9): 877–884.

Guo, T., Feng, Y. L., Xiao, J. J., Liu, Q., Sun, X. N., Xiang, J. F., Kong, N., Liu, S. C., Chen, G. Q., Wang, Y., Dong, M. M., Cai, Z., Lin, H., Cai, X. J. and Xie, A. Y. (2018). Harnessing accurate non-homologous end joining for efficient precise deletion in CRISPR/Cas9-mediated genome editing. *Genome Biology* 19(1): 170.

Hartung, F., Suer, S., Knoll, A., Wurz-Wildersinn, R. and Puchta, H. (2008). Topoisomerase 3alpha and RMI1 suppress somatic crossovers and are essential for resolution of meiotic recombination intermediates in Arabidopsis thaliana. *PLoS Genetics* 4(12): e1000285.

Hawkins, J. S., Proulx, S. R., Rapp, R. A. and Wendel, J. F. (2009). Rapid DNA loss as a counterbalance to genome expansion through retrotransposon proliferation in plants. *Proceedings of the National Academy of Sciences of the United States of America* 106(42): 17811–17816.

Hogg, M., Sauer-Eriksson, A. E. and Johansson, E. (2012). Promiscuous DNA synthesis by human DNA polymerase θ. *Nucleic Acids Research* 40(6): 2611–2622.

Hogg, M., Seki, M., Wood, R. D., Doublié, S. and Wallace, S. S. (2011). Lesion bypass activity of DNA polymerase θ (POLQ) is an intrinsic property of the pol domain and depends on unique sequence inserts. *Journal of Molecular Biology* 405(3): 642–652.

Huang, T. K. and Puchta, H. (2019). CRISPR/Cas-mediated gene targeting in plants: finally a turn for the better for homologous recombination. *Plant Cell Reports* 38(4): 443–453.

Inagaki, S., Nakamura, K. and Morikami, A. (2009). A link among DNA replication, recombination, and gene expression revealed by genetic and genomic analysis of TEBICHI gene of Arabidopsis thaliana. *PLoS Genetics* 5(8): e1000613.

Jacquier, A. and Dujon, B. (1985). An intron-encoded protein is active in a gene conversion process that spreads an intron into a mitochondrial gene. *Cell* 41(2): 383–394.

Jasin, M. and Haber, J. E. (2016). The democratization of gene editing: insights from site-specific cleavage and double-strand break repair. *DNA Repair* 44: 6–16.

Jia, Q., den Dulk-Ras, A., Shen, H., Hooykaas, P. J. and de Pater, S. (2013). Poly(ADP-ribose) polymerases are involved in microhomology mediated back-up non-homologous end joining in Arabidopsis thaliana. *Plant Molecular Biology* 82(4–5): 339–351.

Jinek, M., Chylinski, K., Fonfara, I., Hauer, M., Doudna, J. A. and Charpentier, E. (2012). A programmable dual-RNA-guided DNA endonuclease in adaptive bacterial immunity. *Science (New York, N.Y.)* 337(6096): 816–821.

Kakarougkas, A. and Jeggo, P. A. (2014). DNA DSB repair pathway choice: an orchestrated handover mechanism. *The British Journal of Radiology* 87(1035): 20130685.

Kapusi, E., Corcuera-Gómez, M., Melnik, S. and Stoger, E. (2017). Heritable genomic fragment deletions and small indels in the putative ENGase gene induced by CRISPR/Cas9 in barley. *Frontiers in Plant Science* 8: 540.

Kent, T., Chandramouly, G., McDevitt, S. M., Ozdemir, A. Y. and Pomerantz, R. T. (2015). Mechanism of microhomology-mediated end-joining promoted by human DNA polymerase θ. *Nature Structural and Molecular Biology* 22(3): 230–237.

Kent, T., Mateos-Gomez, P. A., Sfeir, A. and Pomerantz, R. T. (2016). Polymerase θ is a robust terminal transferase that oscillates between three different mechanisms during end-joining. *eLife* 5: e13740.

Kim, Y. G., Cha, J. and Chandrasegaran, S. (1996). Hybrid restriction enzymes: zinc finger fusions to Fok I cleavage domain. *Proceedings of the National Academy of Sciences of the United States of America* 93(3): 1156–1160.

Kirik, A., Salomon, S. and Puchta, H. (2000). Species-specific double-strand break repair and genome evolution in plants. *The EMBO Journal* 19(20): 5562–5566.

Klemm, T., Mannuß, A., Kobbe, D., Knoll, A., Trapp, O., Dorn, A. and Puchta, H. (2017). The DNA translocase RAD5A acts independently of the other main DNA repair pathways, and requires both its ATPase and RING domain for activity in Arabidopsis thaliana. *The Plant Journal: For Cell and Molecular Biology* 91(4): 725–740.

Knoll, A., Fauser, F. and Puchta, H. (2014a). DNA recombination in somatic plant cells: mechanisms and evolutionary consequences. *Chromosome Research: An International Journal on the Molecular, Supramolecular and Evolutionary Aspects of Chromosome Biology* 22(2): 191–201.

Knoll, A., Higgins, J. D., Seeliger, K., Reha, S. J., Dangel, N. J., Bauknecht, M., Schröpfer, S., Franklin, F. C. and Puchta, H. (2012). The fanconi anemia ortholog FANCM ensures ordered homologous recombination in both somatic and meiotic cells in Arabidopsis. *The Plant Cell* 24(4): 1448–1464.

Knoll, A., Schröpfer, S. and Puchta, H. (2014b). The RTR complex as caretaker of genome stability and its unique meiotic function in plants. *Frontiers in Plant Science* 5: 33.

Koole, W., van Schendel, R., Karambelas, A. E., van Heteren, J. T., Okihara, K. L. and Tijsterman, M. (2014). A polymerase Theta-dependent repair pathway suppresses extensive genomic instability at endogenous G4 DNA sites. *Nature Communications* 5: 3216.

Lemaître, C., Grabarz, A., Tsouroula, K., Andronov, L., Furst, A., Pankotai, T., Heyer, V., Rogier, M., Attwood, K. M., Kessler, P., Dellaire, G., Klaholz, B., Reina-San-Martin, B. and Soutoglou, E. (2014). Nuclear position dictates DNA repair pathway choice. *Genes and Development* 28(22): 2450–2463.

Li, C., Zong, Y., Wang, Y., Jin, S., Zhang, D., Song, Q., Zhang, R. and Gao, C. (2018a). Expanded base editing in rice and wheat using a Cas9-adenosine deaminase fusion. *Genome Biology* 19(1): 59.

Li, S., Zhang, Y., Xia, L. and Qi, Y. (2020). CRISPR-Cas12a enables efficient biallelic gene targeting in rice. *Plant Biotechnology Journal* 18(6): 1351–1353.

Li, X., Bai, Y., Cheng, X., Kalds, P. G. T., Sun, B., Wu, Y., Lv, H., Xu, K. and Zhang, Z. (2018b). Efficient SSA-mediated precise genome editing using CRISPR/Cas9. *The FEBS Journal* 285(18): 3362–3375.

Liang, L., Deng, L., Nguyen, S. C., Zhao, X., Maulion, C. D., Shao, C. and Tischfield, J. A. (2008). Human DNA ligases I and III, but not ligase IV, are required for microhomology-mediated end joining of DNA double-strand breaks. *Nucleic Acids Research* 36(10): 3297–3310.

Lupski, J. R. and Stankiewicz, P. (2005). Genomic disorders: molecular mechanisms for rearrangements and conveyed phenotypes. *PLoS Genetics* 1(6): e49.

Lysak, M. A., Berr, A., Pecinka, A., Schmidt, R., McBreen, K. and Schubert, I. (2006). Mechanisms of chromosome number reduction in Arabidopsis thaliana and related Brassicaceae species. *Proceedings of the National Academy of Sciences of the United States of America* 103(13): 5224–5229.

Ma, C. J., Kwon, Y., Sung, P. and Greene, E. C. (2017). Human RAD52 interactions with replication protein A and the RAD51 presynaptic complex. *The Journal of Biological Chemistry* 292(28): 11702–11713.

Ma, Y. and Lieber, M. R. (2002). Binding of inositol hexakisphosphate (IP6) to Ku but not to DNA-PKcs. *The Journal of Biological Chemistry* 277(13): 10756–10759.

Mannuss, A., Dukowic-Schulze, S., Suer, S., Hartung, F., Pacher, M. and Puchta, H. (2010). RAD5A, RECQ4A, and MUS81 have specific functions in homologous recombination and define different pathways of DNA repair in Arabidopsis thaliana. *The Plant Cell* 22(10): 3318–3330.

Martin, G. M., Smith, A. C., Ketterer, D. J., Ogburn, C. E. and Disteche, C. M. (1985). Increased chromosomal aberrations in first metaphases of cells isolated from the kidneys of aged mice. *Israel Journal of Medical Sciences* 21(3): 296–301.

Masani, S., Han, L., Meek, K. and Yu, K. (2016). Redundant function of DNA ligase 1 and 3 in alternative end-joining during immunoglobulin class switch recombination. *Proceedings of the National Academy of Sciences of the United States of America* 113(5): 1261–1266.

Mateos-Gomez, P. A., Gong, F., Nair, N., Miller, K. M., Lazzerini-Denchi, E. and Sfeir, A. (2015). Mammalian polymerase θ promotes alternative NHEJ and suppresses recombination. *Nature* 518(7538): 254–257.

Merker, L., Schindele, P., Huang, T. K., Wolter, F. and Puchta, H. (2020). Enhancing in planta gene targeting efficiencies in Arabidopsis using temperature-tolerant CRISPR/LbCas12a. *Plant Biotechnology Journal* 18(12): 2382–2384.

Miki, D., Zhang, W., Zeng, W., Feng, Z. and Zhu, J. K. (2018). CRISPR/Cas9-mediated gene targeting in Arabidopsis using sequential transformation. *Nature Communications* 9(1): 1967.

Miné-Hattab, J. and Rothstein, R. (2013). DNA in motion during double-strand break repair. *Trends in Cell Biology* 23(11): 529–536.

Miura, H., Quadros, R. M., Gurumurthy, C. B. and Ohtsuka, M. (2018). Easi-CRISPR for creating knock-in and conditional knockout mouse models using long ssDNA donors. *Nature Protocols* 13(1): 195–215.

Ochi, T., Blackford, A. N., Coates, J., Jhujh, S., Mehmood, S., Tamura, N., Travers, J., Wu, Q., Draviam, V. M., Robinson, C. V., Blundell, T. L. and Jackson, S. P. (2015). DNA repair. PAXX, a paralog of XRCC4 and XLF, interacts with Ku to promote DNA double-strand break repair. *Science (New York, N.Y.)* 347(6218): 185–188.

Offringa, R., Groot, M. J. de, Haagsman, H. J., Does, M. P., van den Elzen, P. J. and Hooykaas, P. J. (1990). Extrachromosomal homologous recombination and gene targeting in plant cells after Agrobacterium mediated transformation. *The EMBO Journal* 9(10): 3077–3084.

Ordon, J., Gantner, J., Kemna, J., Schwalgun, L., Reschke, M., Streubel, J., Boch, J. and Stuttmann, J. (2017). Generation of chromosomal deletions in dicotyledonous

plants employing a user-friendly genome editing toolkit. *The Plant Journal: For Cell and Molecular Biology* 89(1): 155–168.
Orel, N., Kyryk, A. and Puchta, H. (2003). Different pathways of homologous recombination are used for the repair of double-strand breaks within tandemly arranged sequences in the plant genome. *The Plant Journal: For Cell and Molecular Biology* 35(5): 604–612.
Pacher, M., Schmidt-Puchta, W. and Puchta, H. (2007). Two unlinked double-strand breaks can induce reciprocal exchanges in plant genomes via homologous recombination and nonhomologous end joining. *Genetics* 175(1): 21–29.
Paszkowski, J., Baur, M., Bogucki, A. and Potrykus, I. (1988). Gene targeting in plants. *The EMBO Journal* 7(13): 4021–4026.
Perrin, A., Buckle, M. and Dujon, B. (1993). Asymmetrical recognition and activity of the I-SceI endonuclease on its site and on intron-exon junctions. *The EMBO Journal* 12(7): 2939–2947.
Prado, F. and Aguilera, A. (1995). Role of reciprocal exchange, one-ended invasion crossover and single-strand annealing on inverted and direct repeat recombination in yeast: different requirements for the Rad1, Rad10, and Rad52 genes. *Genetics* 139(1): 109–123.
Puchta, H. (1998). Repair of genomic double-strand breaks in somatic plant cells by one-sided invasion of homologous sequences. *The Plant Journal* 13(3): 331–339.
Puchta, H. (1999). Double-strand break-induced recombination between ectopic homologous sequences in somatic plant cells. *Genetics* 152(3): 1173–1181.
Puchta, H. (2005). The repair of double-strand breaks in plants: mechanisms and consequences for genome evolution. *Journal of Experimental Botany* 56(409): 1–14.
Puchta, H., Dujon, B. and Hohn, B. (1993). Homologous recombination in plant cells is enhanced by in vivo induction of double strand breaks into DNA by a site-specific endonuclease. *Nucleic Acids Research* 21(22): 5034–5040.
Puchta, H., Dujon, B. and Hohn, B. (1996). Two different but related mechanisms are used in plants for the repair of genomic double-strand breaks by homologous recombination. *Proceedings of the National Academy of Sciences of the United States of America* 93(10): 5055–5060.
Puchta, H. and Fauser, F. (2013). Gene targeting in plants: 25 years later. *The International Journal of Developmental Biology* 57(6–8): 629–637.
Qi, Y., Zhang, Y., Zhang, F., Baller, J. A., Cleland, S. C., Ryu, Y., Starker, C. G. and Voytas, D. F. (2013). Increasing frequencies of site-specific mutagenesis and gene targeting in Arabidopsis by manipulating DNA repair pathways. *Genome Research* 23(3): 547–554.
Quadros, R. M., Miura, H., Harms, D. W., Akatsuka, H., Sato, T., Aida, T., Redder, R., Richardson, G. P., Inagaki, Y., Sakai, D., Buckley, S. M., Seshacharyulu, P., Batra, S. K., Behlke, M. A., Zeiner, S. A., Jacobi, A. M., Izu, Y., Thoreson, W. B., Urness, L. D., Mansour, S. L., Ohtsuka, M. and Gurumurthy, C. B. (2017). Easi-CRISPR: a robust method for one-step generation of mice carrying conditional and insertion alleles using long ssDNA donors and CRISPR ribonucleoproteins. *Genome Biology* 18(1): 92.
Ramsden, D. A. and Gellert, M. (1998). Ku protein stimulates DNA end joining by mammalian DNA ligases: a direct role for Ku in repair of DNA double-strand breaks. *The EMBO Journal* 17(2): 609–614.

Ran, F. A., Hsu, P. D., Lin, C. Y., Gootenberg, J. S., Konermann, S., Trevino, A. E., Scott, D. A., Inoue, A., Matoba, S., Zhang, Y. and Zhang, F. (2013). Double nicking by RNA-guided CRISPR Cas9 for enhanced genome editing specificity. *Cell* 154(6): 1380–1389.

Ranjha, L., Howard, S. M. and Cejka, P. (2018). Main steps in DNA double-strand break repair: an introduction to homologous recombination and related processes. *Chromosoma* 127(2): 187–214.

Renkawitz, J., Lademann, C. A., Kalocsay, M. and Jentsch, S. (2013). Monitoring homology search during DNA double-strand break repair in vivo. *Molecular Cell* 50(2): 261–272.

Riha, K., Watson, J. M., Parkey, J. and Shippen, D. E. (2002). Telomere length deregulation and enhanced sensitivity to genotoxic stress in Arabidopsis mutants deficient in Ku70. *The EMBO Journal* 21(11): 2819–2826.

Robert, I., Dantzer, F. and Reina-San-Martin, B. (2009). Parp1 facilitates alternative NHEJ, whereas Parp2 suppresses IgH/c-myc translocations during immunoglobulin class switch recombination. *The Journal of Experimental Medicine* 206(5): 1047–1056.

Roberts, S. A., Strande, N., Burkhalter, M. D., Strom, C., Havener, J. M., Hasty, P. and Ramsden, D. A. (2010). Ku is a 5′-dRP/AP lyase that excises nucleotide damage near broken ends. *Nature* 464(7292): 1214–1217.

Robinett, C. C., Straight, A., Li, G., Willhelm, C., Sudlow, G., Murray, A. and Belmont, A. S. (1996). In vivo localization of DNA sequences and visualization of large-scale chromatin organization using lac operator/repressor recognition. *The Journal of Cell Biology* 135(6 Pt 2): 1685–1700.

Röhrig, S., Dorn, A., Enderle, J., Schindele, A., Herrmann, N. J., Knoll, A. and Puchta, H. (2018). The RecQ-like helicase HRQ1 is involved in DNA crosslink repair in Arabidopsis in a common pathway with the fanconi anemia-associated nuclease FAN1 and the postreplicative repair ATPase RAD5A. *The New Phytologist* 218(4): 1478–1490.

Roth, N., Klimesch, J., Dukowic-Schulze, S., Pacher, M., Mannuss, A. and Puchta, H. (2012). The requirement for recombination factors differs considerably between different pathways of homologous double-strand break repair in somatic plant cells. *The Plant Journal: For Cell and Molecular Biology* 72(5): 781–790.

Salomon, S. and Puchta, H. (1998). Capture of genomic and T-DNA sequences during double-strand break repair in somatic plant cells. *The EMBO Journal* 17(20): 6086–6095.

Samach, A., Melamed-Bessudo, C., Avivi-Ragolski, N., Pietrokovski, S. and Levy, A. A. (2011). Identification of plant RAD52 homologs and characterization of the Arabidopsis thaliana RAD52-like genes. *The Plant Cell* 23(12): 4266–4279.

Sax, K. (1941). The behavior of X-ray induced chromosomal aberrations in allium root tip cells. *Genetics* 26(4): 418–425.

Scheben, A. and Edwards, D. (2018). Towards a more predictable plant breeding pipeline with CRISPR/Cas-induced allelic series to optimize quantitative and qualitative traits. *Current Opinion in Plant Biology* 45(B): 218–225.

Schiml, S., Fauser, F. and Puchta, H. (2014). The CRISPR/Cas system can be used as nuclease for in planta gene targeting and as paired nickases for directed mutagenesis in Arabidopsis resulting in heritable progeny. *The Plant Journal: For Cell and Molecular Biology* 80(6): 1139–1150.

Schiml, S., Fauser, F. and Puchta, H. (2016). Repair of adjacent single-strand breaks is often accompanied by the formation of tandem sequence duplications in plant genomes.

Proceedings of the National Academy of Sciences of the United States of America 113(26): 7266–7271.
Schindele, A., Dorn, A. and Puchta, H. (2020). CRISPR/Cas brings plant biology and breeding into the fast lane. *Current Opinion in Biotechnology* 61: 7–14.
Schmidt, C., Pacher, M. and Puchta, H. (2019). Efficient induction of heritable inversions in plant genomes using the CRISPR/Cas system. *The Plant Journal: For Cell and Molecular Biology* 98(4): 577–589.
Schrank, B. R., Aparicio, T., Li, Y., Chang, W., Chait, B. T., Gundersen, G. G., Gottesman, M. E. and Gautier, J. (2018). Nuclear ARP2/3 drives DNA break clustering for homology-directed repair. *Nature* 559(7712): 61–66.
Schröpfer, S., Kobbe, D., Hartung, F., Knoll, A. and Puchta, H. (2014). Defining the roles of the N-terminal region and the helicase activity of RECQ4A in DNA repair and homologous recombination in Arabidopsis. *Nucleic Acids Research* 42(3): 1684–1697.
Schubert, I. and Vu, G. T. H. (2016). Genome stability and evolution: attempting a holistic view. *Trends in Plant Science* 21(9): 749–757.
Séguéla-Arnaud, M., Crismani, W., Larchevêque, C., Mazel, J., Froger, N., Choinard, S., Lemhemdi, A., Macaisne, N., Van Leene, J., Gevaert, K., De Jaeger, G., Chelysheva, L. and Mercier, R. (2015). Multiple mechanisms limit meiotic crossovers: TOP3α and two BLM homologs antagonize crossovers in parallel to FANCM. *Proceedings of the National Academy of Sciences of the United States of America* 112(15): 4713–4718.
Seol, J. H., Shim, E. Y. and Lee, S. E. (2018). Microhomology-mediated end joining: good, bad and ugly. *Mutation Research* 809: 81–87.
Serebrovski, A. S. (1929). A general scheme for the origin of mutations. *The American Naturalist* 63(687): 374–378.
Serra, H., Da Ines, O., Degroote, F., Gallego, M. E. and White, C. I. (2013). Roles of XRCC2, RAD51B and RAD51D in RAD51-independent SSA recombination. *PLOS Genetics* 9(11): e1003971.
Shaked, H., Melamed-Bessudo, C. and Levy, A. A. (2005). High-frequency gene targeting in Arabidopsis plants expressing the yeast RAD54 gene. *Proceedings of the National Academy of Sciences of the United States of America* 102(34): 12265–12269.
Shalev, G. and Levy, A. A. (1997). The maize transposable element Ac induces recombination between the donor site and an homologous ectopic sequence. *Genetics* 146(3): 1143–1151.
Shen, H., Strunks, G. D., Klemann, B. J. P. M., Hooykaas, P. J. and de Pater, S. (2017). CRISPR/Cas9-induced double-strand break repair in Arabidopsis nonhomologous end-joining mutants. *G3 (Bethesda, Md.)* 7(1): 193–202.
Shi, J. and Lai, J. (2015). Patterns of genomic changes with crop domestication and breeding. *Current Opinion in Plant Biology* 24: 47–53.
Shou, J., Li, J., Liu, Y. and Wu, Q. (2018). Precise and predictable CRISPR chromosomal rearrangements reveal principles of Cas9-mediated nucleotide insertion. *Molecular Cell* 71(4): 498–509.e4.
Siebert, R. and Puchta, H. (2002). Efficient repair of genomic double-strand breaks by homologous recombination between directly repeated sequences in the plant genome. *The Plant Cell* 14(5): 1121–1131.
Simsek, D. and Jasin, M. (2010). Alternative end-joining is suppressed by the canonical NHEJ component Xrcc4-ligase IV during chromosomal translocation formation. *Nature Structural and Molecular Biology* 17(4): 410–416.

Singatulina, A. S., Hamon, L., Sukhanova, M. V., Desforges, B., Joshi, V., Bouhss, A., Lavrik, O. I. and Pastré, D. (2019). PARP-1 activation directs FUS to DNA damage sites to form PARG-reversible compartments enriched in damaged DNA. *Cell Reports* 27(6): 1809–1821.e5.

Smith, J., Grizot, S., Arnould, S., Duclert, A., Epinat, J. C., Chames, P., Prieto, J., Redondo, P., Blanco, F. J., Bravo, J., Montoya, G., Pâques, F. and Duchateau, P. (2006). A combinatorial approach to create artificial homing endonucleases cleaving chosen sequences. *Nucleic Acids Research* 34(22): e149.

Soutoglou, E., Dorn, J. F., Sengupta, K., Jasin, M., Nussenzweig, A., Ried, T., Danuser, G. and Misteli, T. (2007). Positional stability of single double-strand breaks in mammalian cells. *Nature Cell Biology* 9(6): 675–682.

Steinert, J., Schiml, S., Fauser, F. and Puchta, H. (2015). Highly efficient heritable plant genome engineering using Cas9 orthologues from Streptococcus thermophilus and Staphylococcus aureus. *The Plant Journal: For Cell and Molecular Biology* 84(6): 1295–1305.

Steinert, J., Schiml, S. and Puchta, H. (2016). Homology-based double-strand break-induced genome engineering in plants. *Plant Cell Reports* 35(7): 1429–1438.

Strande, N., Roberts, S. A., Oh, S., Hendrickson, E. A. and Ramsden, D. A. (2012). Specificity of the dRP/AP lyase of Ku promotes nonhomologous end joining (NHEJ) fidelity at damaged ends. *The Journal of Biological Chemistry* 287(17): 13686–13693.

Sun, Y., Li, J. and Xia, L. (2016). Precise genome modification via sequence-specific nucleases-mediated gene targeting for crop improvement. *Frontiers in Plant Science* 7: 1928.

Svitashev, S., Schwartz, C., Lenderts, B., Young, J. K. and Mark Cigan, A. (2016). Genome editing in maize directed by CRISPR-Cas9 ribonucleoprotein complexes. *Nature Communications* 7: 13274.

Svitashev, S., Young, J. K., Schwartz, C., Gao, H., Falco, S. C. and Cigan, A. M. (2015). Targeted mutagenesis, precise gene editing, and site-specific gene insertion in maize using Cas9 and guide RNA. *Plant Physiology* 169(2): 931–945.

Szostak, J. W., Orr-Weaver, T. L., Rothstein, R. J. and Stahl, F. W. (1983). The double-strand-break repair model for recombination. *Cell* 33(1): 25–35.

Tamura, K., Adachi, Y., Chiba, K., Oguchi, K. and Takahashi, H. (2002). Identification of Ku70 and Ku80 homologues in Arabidopsis thaliana: evidence for a role in the repair of DNA double-strand breaks. *The Plant Journal: For Cell and Molecular Biology* 29(6): 771–781.

Terada, R., Urawa, H., Inagaki, Y., Tsugane, K. and Iida, S. (2002). Efficient gene targeting by homologous recombination in rice. *Nature Biotechnology* 20(10): 1030–1034.

Tomimatsu, N., Tahimic, C. G. T., Otsuki, A., Burma, S., Fukuhara, A., Sato, K., Shiota, G., Oshimura, M., Chen, D. J. and Kurimasa, A. (2007). Ku70/80 modulates ATM and ATR signaling pathways in response to DNA double strand breaks. *The Journal of Biological Chemistry* 282(14): 10138–10145.

Truong, L. N., Li, Y., Shi, L. Z., Hwang, P. Y., He, J., Wang, H., Razavian, N., Berns, M. W. and Wu, X. (2013). Microhomology-mediated End Joining and Homologous Recombination share the initial end resection step to repair DNA double-strand breaks in mammalian cells. *Proceedings of the National Academy of Sciences of the United States of America* 110(19): 7720–7725.

Tsakraklides, V., Brevnova, E., Stephanopoulos, G. and Shaw, A. J. (2015). Improved gene targeting through cell cycle synchronization. *PLoS ONE* 10(7): e0133434.

Upadhyay, S. K., Kumar, J., Alok, A. and Tuli, R. (2013). RNA-guided genome editing for target gene mutations in wheat. *G3 (Bethesda, Md.)* 3(12): 2233–2238.

van Dyck, E., Stasiak, A. Z., Stasiak, A. and West, S. C. (2001). Visualization of recombination intermediates produced by RAD52-mediated single-strand annealing. *EMBO Reports* 2(10): 905–909.

van Kregten, M., Pater, S. de, Romeijn, R., van Schendel, R., Hooykaas, P. J. and Tijsterman, M. (2016). T-DNA integration in plants results from polymerase-θ-mediated DNA repair. *Nature Plants* 2(11): 16164.

van Overbeek, M., Capurso, D., Carter, M. M., Thompson, M. S., Frias, E., Russ, C., Reece-Hoyes, J. S., Nye, C., Gradia, S., Vidal, B., Zheng, J., Hoffman, G. R., Fuller, C. K. and May, A. P. (2016). DNA repair profiling reveals nonrandom outcomes at Cas9-mediated breaks. *Molecular Cell* 63(4): 633–646.

Vaughn, J. N. and Bennetzen, J. L. (2014). Natural insertions in rice commonly form tandem duplications indicative of patch-mediated double-strand break induction and repair. *Proceedings of the National Academy of Sciences of the United States of America* 111(18): 6684–6689.

Voytas, D. F. (2013). Plant genome engineering with sequence-specific nucleases. *Annual Review of Plant Biology* 64: 327–350.

Vu, G. T. H., Cao, H. X., Fauser, F., Reiss, B., Puchta, H. and Schubert, I. (2017). Endogenous sequence patterns predispose the repair modes of CRISPR/Cas9-induced DNA double strand breaks in Arabidopsis thaliana. *The Plant Journal: For Cell and Molecular Biology* 92(1): 57–67.

Vu, G. T. H., Cao, H. X., Watanabe, K., Hensel, G., Blattner, F. R., Kumlehn, J. and Schubert, I. (2014). Repair of site-specific DNA double-strand breaks in barley occurs via diverse pathways primarily involving the sister chromatid. *The Plant Cell* 26(5): 2156–2167.

Wang, H. and Xu, X. (2017). Microhomology-mediated end joining: new players join the team. *Cell and Bioscience* 7: 6.

Wang, M., Lu, Y., Botella, J. R., Mao, Y., Hua, K. and Zhu, J. K. (2017). Gene targeting by homology-directed repair in rice using a geminivirus-based CRISPR/Cas9 system. *Molecular Plant* 10(7): 1007–1010.

Wang, M., Wu, W., Wu, W., Rosidi, B., Zhang, L., Wang, H. and Iliakis, G. (2006). PARP-1 and Ku compete for repair of DNA double strand breaks by distinct NHEJ pathways. *Nucleic Acids Research* 34(21): 6170–6182.

Watanabe, K., Pacher, M., Dukowic, S., Schubert, V., Puchta, H. and Schubert, I. (2009). The STRUCTURAL MAINTENANCE OF CHROMOSOMES 5/6 complex promotes sister chromatid alignment and homologous recombination after DNA damage in Arabidopsis thaliana. *The Plant Cell* 21(9): 2688–2699.

Watson, A., Ghosh, S., Williams, M. J., Cuddy, W. S., Simmonds, J., Rey, M. D., Asyraf Md Hatta, M., Hinchliffe, A., Steed, A., Reynolds, D., Adamski, N. M., Breakspear, A., Korolev, A., Rayner, T., Dixon, L. E., Riaz, A., Martin, W., Ryan, M., Edwards, D., Batley, J., Raman, H., Carter, J., Rogers, C., Domoney, C., Moore, G., Harwood, W., Nicholson, P., Dieters, M. J., DeLacy, I. H., Zhou, J., Uauy, C., Boden, S. A., Park, R. F., Wulff, B. B. H. and Hickey, L. T. (2018). Speed breeding is a powerful tool to accelerate crop research and breeding. *Nature Plants* 4(1): 23–29.

Weinstock, D. M., Brunet, E. and Jasin, M. (2007). Formation of NHEJ-derived reciprocal chromosomal translocations does not require Ku70. *Nature Cell Biology* 9(8): 978–981.

West, C. E., Waterworth, W. M., Jiang, Q. and Bray, C. M. (2000). Arabidopsis DNA ligase IV is induced by gamma-irradiation and interacts with an Arabidopsis homologue of the double strand break repair protein XRCC4. *The Plant Journal: For Cell and Molecular Biology* 24(1): 67–78.

West, C. E., Waterworth, W. M., Story, G. W., Sunderland, P. A., Jiang, Q. and Bray, C. M. (2002). Disruption of the Arabidopsis AtKu80 gene demonstrates an essential role for AtKu80 protein in efficient repair of DNA double-strand breaks in vivo. *The Plant Journal: For Cell and Molecular Biology* 31(4): 517–528.

Williams, G. J., Lees-Miller, S. P. and Tainer, J. A. (2010). Mre11-Rad50-Nbs1 conformations and the control of sensing, signaling, and effector responses at DNA double-strand breaks. *DNA Repair* 9(12): 1299–1306.

Wolter, F., Klemm, J. and Puchta, H. (2018). Efficient in planta gene targeting in Arabidopsis using egg cell-specific expression of the Cas9 nuclease of Staphylococcus aureus. *The Plant Journal: For Cell and Molecular Biology* 94(4): 735–746.

Wolter, F. and Puchta, H. (2019). In planta gene targeting can be enhanced by the use of CRISPR/Cas12a. *The Plant Journal: For Cell and Molecular Biology* 100(5): 1083–1094.

Wood, R. D. and Doublié, S. (2016). DNA polymerase θ (POLQ), double-strand break repair, and cancer. *DNA Repair* 44: 22–32.

Wu, R., Lucke, M., Jang, Y. T., Zhu, W., Symeonidi, E., Wang, C., Fitz, J., Xi, W., Schwab, R. and Weigel, D. (2018). An efficient CRISPR vector toolbox for engineering large deletions in Arabidopsis thaliana. *Plant Methods* 14: 65.

Wyatt, D. W., Feng, W., Conlin, M. P., Yousefzadeh, M. J., Roberts, S. A., Mieczkowski, P., Wood, R. D., Gupta, G. P. and Ramsden, D. A. (2016). Essential Roles for polymerase θ-Mediated End Joining in the Repair of Chromosome Breaks. *Molecular Cell* 63(4): 662–673.

Yang, D., Scavuzzo, M. A., Chmielowiec, J., Sharp, R., Bajic, A. and Borowiak, M. (2016). Enrichment of G2/M cell cycle phase in human pluripotent stem cells enhances HDR-mediated gene repair with customizable endonucleases. *Scientific Reports* 6: 21264.

Yu, A. M. and McVey, M. (2010). Synthesis-dependent microhomology-mediated end joining accounts for multiple types of repair junctions. *Nucleic Acids Research* 38(17): 5706–5717.

Zahn, K. E., Averill, A. M., Aller, P., Wood, R. D. and Doublié, S. (2015). Human DNA polymerase θ grasps the primer terminus to mediate DNA repair. *Nature Structural and Molecular Biology* 22(4): 304–311.

Zapata, L., Ding, J., Willing, E. M., Hartwig, B., Bezdan, D., Jiao, W. B., Patel, V., Velikkakam James, G., Koornneef, M., Ossowski, S. and Schneeberger, K. (2016). Chromosome-level assembly of Arabidopsis thaliana Ler reveals the extent of translocation and inversion polymorphisms. *Proceedings of the National Academy of Sciences of the United States of America* 113(28): E4052–E4060.

Zhang, C., Liu, C., Weng, J., Cheng, B., Liu, F., Li, X. and Xie, C. (2017). Creation of targeted inversion mutations in plants using an RNA-guided endonuclease. *The Crop Journal* 5(1): 83–88.

Zhou, H., Liu, B., Weeks, D. P., Spalding, M. H. and Yang, B. (2014). Large chromosomal deletions and heritable small genetic changes induced by CRISPR/Cas9 in rice. *Nucleic Acids Research* 42(17): 10903–10914.

Chapter 3

Advances in the generation of insertion-based genome edits in plants

Baike Wang and Juan Wang, Institute of Horticulture Crops – Xinjiang Academy of Agricultural Sciences, China; Shaoyong Huang, Institute of Horticulture Crops – Xinjiang Academy of Agricultural Sciences and College of Forestry and Horticulture – Xinjiang Agricultural University, China; and Yaping Tang, Ning Li, Shengbao Yang, Tao Yang and Qinghui Yu, Institute of Horticulture Crops – Xinjiang Academy of Agricultural Sciences, China

1 Introduction

Plant scientists and breeders strive to reduce poverty and undernourishment by increasing agricultural productivity and providing sufficient and nutritious food. The global human population is predicted to reach 10 billion by 2050, and crop production will need to be double that of 2005 (Tilman et al., 2011). However, agricultural productivity is faced with numerous constraints such as climate change, reduced arable land, and shortage of water resources. Therefore, there are increasing demands to generate new crop varieties with improved traits, such as disease resistance, stress tolerance, and efficient resource usage, in order to develop environmental-friendly sustainable agriculture, while maintaining high quality and nutrient-rich food for a better quality of life. To overcome these challenges, there is an urgent need to develop innovative and efficient crop breeding techniques to generate better varieties for increasing agricultural productivity.

The essence of plant breeding is to introduce or pyramid favorable traits (genes) and delete or knock-out unfavorable ones through various techniques. Over the past century, conventional plant breeding through self-, cross-, and

http://dx.doi.org/10.19103/AS.2020.0082.05

backcross-pollination strategies has been highly successful in generating new varieties and increasing agricultural productivity. For instance, the tremendous increases in wheat and rice yields have been promoted by the Green Revolution from the 1960s to the 2000s. The key factor in these yield improvements was the introduction of semi-dwarfing genes: the *Reduced height* (*Rht*) genes of wheat and the *semidwarf1* (*sd1*) gene of rice. The gene-carrying varieties were not only shorter with strong stalks that did not lodge but also had a much higher harvest index and nitrogen-use efficiency (NUE) (Hedden, 2003; Voss-Fels et al., 2019). Note: abbreviations are summarized in Table 2.

A new study has highlighted that genetic improvements in corn hybrids have resulted in an 89% increase in grain yields and a 73% increase in NUE in the United States over the past 70 years (Mueller et al., 2019). However, conventional breeding techniques have several inherent disadvantages, even when adopting marker-assisted selection approaches for enhancing selection efficiency. These disadvantages include (1) long generation times: it takes many years, even decades, to introduce expected alleles by cross breeding and increase variability by genetic recombination with many steps (Scheben et al., 2017); (2) linkage drags: it is difficult to eliminate undesirable alleles by genetic recombination if they are closely linked with the desirable alleles (it is difficult to break the linkage drag); (3) genetic bottleneck: owing to unintended selection during domestication and improvement sweeps over thousands of years, most parts of the genomes of main crops have been fixed and the genetic background has become greatly narrowed, limiting the potential to improve many desirable traits; and (4) natural reproductive isolation: reproductive isolation between cultivars and their wild relatives or between species leads to difficulties in breeding via conventional methods.

In the meantime, genetic modifications can solve these problems to some extent. However, there are concerns that the random and crude transformation of exogenous DNA into the genomes of recipient elite varieties may cause harmful effects. Therefore, many countries, such as the European Union (EU) countries, ban the cultivation of genetically modified (GM) crops. In contrast, some countries have embraced this technology, but the commercialization of GM crops can be a daunting task due to the long and costly regulatory evaluation processes (Chen et al., 2019).

Advances in molecular biology have allowed the exploitation and characterization of the genetic diversity present in both cultivated and wild crops. Thus, gene and allele discovery has been accelerated and a myriad of major genes and quantitative trait loci (QTLs) conferring various traits have been cloned. Together with the development of genome sequencing technologies, such as next generation sequencing (NGS) and third generation sequencing (TGS), it has been possible to fully sequence and decipher various plant genomes (Sun et al., 2018; Schmidt et al., 2017; Zhao et al., 2018).

This has opened up a new era of using genomes for crop improvement. The emergence of gene-editing technologies, such as zinc finger nucleases (ZFNs), transcription activator-like effector nucleases (TALENs), and clustered regularly interspaced short palindromic repeats/CRISPR-associated protein 9 (CRISPR/Cas9), has been crucial. These technologies enable the creation of genetic variation, which was previously done randomly using physical radiation and chemical mutagens. Moreover, these technologies enable the introduction of genomic modifications such as deletions, insertions, inversions, and base edits, at specific DNA positions in plant genomes with high precision.

Gene editing is a type of genetic engineering that mainly involves the use of sequence-specific nucleases (SSNs) to produce DNA double-stand breaks (DSBs) at specific gene sites. Random indels or repair templates are then inserted or replaced via one of the two main types of DNA repair system: non-homologous end joining (NHEJ) or homology-directed repair (HDR). There are several ways to cause DSBs and help the gene or repair template to insert into the specific locus. Meganucleases are sequence-specific endonucleases with large cleavage sites (>14 bp) that can be used for efficient homologous gene targeting in plants (Puchta et al., 1996). However, a limited number of meganucleases have been identified, and the probability of a homing site of one of these meganucleases being present in the target sequence is extremely low (Smith et al., 2006). Thus, researchers turned to ZFNs, which were designed to cut at specific DNA sequences (Durai et al., 2005). ZFNs are formed by combining the non-specific cleavage domain (N) of FokI endonuclease with zinc finger proteins (ZFPs). The ZFNs cleave the DNA in strict regions, offering a way to gain a site-specific DSB (Davis and Stokoe, 2010). Based on HDR, a gene insertion study was carried out using ZFNs to cleave at a specific site containing a 3′ partial *phosphinothricin acetyl transferase (PAT)* gene fragment and an Arabidopsis 4-*coumaryl synthase* (4-CoAS) intron; the 5′ PAT gene was then inserted into the 4-CoAS intron to generate a pre-integrated construct (Cai et al., 2009). When ZFN cleavage sites are present in both the genome and donor templates, NHEJ can be exploited to integrate the donor template into a specific genome site (Weinthal et al., 2013).

Another gene-editing tool has been derived from the TAL effectors (TALEs) of plant-pathogenic *Xanthomonas* spp., which have been identified as transcription activators in the plant cell (Boch et al., 2009; Moscou and Bogdanove, 2009). In TALEs, the central repeat domain (CRD) confers DNA binding and each repeat contains 34 amino acids that bind to one target nucleotide sequence. TALENs comprise a FokI endonuclease domain fused to a customizable TALE DNA binding domain and can thus be used to mediate DNA cleavage at a specific DNA sequence. Two TALENs-mediated site-specific *S* gene modifications have been made in rice to confer bacterial resistance (Li et al., 2012). However, DNA editing using ZFNs and TALENs has a high rate of failure and is laborious.

Thus, more recently, CRISPR has been adopted. CRISPR is a tool that has been discovered from bacteria, which use it against invading systems. CRISPR can be targeted to a specific genomic sequence by an easily engineered 20 bp-guide RNA (gRNA) sequence that binds to DNA target. The more recent development of the PITCh (precise integration into target chromosome) system, which is assisted by microhomology-mediated end joining (MMEJ), has provided a more efficient way of performing gene knock-ins (Sakuma et al., 2015).

The CRISPR system is particularly attractive to plant breeders because CRISPR cassettes can be removed and transgene-free plants can be obtained via genetic segregation during sexual reproduction. Moreover, CRISPR/Cas9 systems can be transiently expressed to deliver the mRNA and ribonucleoprotein components into plant cells, without the introduction of foreign DNA into the genome of the target plants (Mao et al., 2019). Since the first five almost simultaneous publications on CRISPR/Cas9 gene editing in plants (Feng et al., 2013; Li et al., 2013; Nekrasov et al., 2013; Shan et al., 2013b; Xie and Yang, 2013), the results of this type of research have shown explosive growth. The CRISPR/Cas9 system provides scientists and plant breeders with the potential ability to revolutionize crop improvement by developing new plant breeding techniques (NPBTs) (Zaidi et al., 2019; Shao et al., 2018), and ushers in a new era of highly efficient and precise crop improvement.

There are many excellent reviews that have summarized CRISPR/Cas9 gene editing in plants (Hua et al., 2019b; Zhang et al., 2018, 2019). In this chapter, we will focus on and present the technical advances aiming at the generation of insertion-based genome edits in plants using CRISPR/Cas9 systems. The topics covered include the insertion principle, insertion forms, insert template construction methods, and insertion efficiency, providing as much useful information to the readers as possible.

2 Outcomes of knock-ins

To date, a line of evidence has displayed that eukaryotes have three types of DSB repair pathways for producing knock-ins. These pathways include HDR and NHEJ (c-NHEJ), which are the most popular, and MMEJ (alt-NHEJ; Fig. 1) (Que et al., 2019). Commonly, the HDR pathway can be used for precisely targeted DNA sequence insertion or substitution that increases natural variability by promoting the exchange of homologous DNA fragments between parental chromosomes (Mao et al., 2019). To utilize this process for precise gene modification, a donor DNA or RNA template flanked by homologous sequences is usually introduced to the target region in reproductive cells (Ferenczi et al., 2017; Li et al., 2016, 2019b).

Due to the development and advantages of CRISPR-mediated gene editing, many plant species have been subjected to gene knock-ins. These

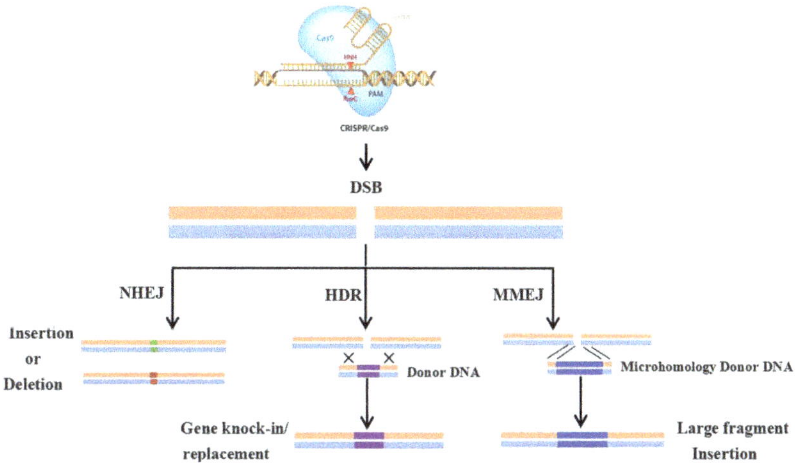

Figure 1 The DSB repair as effected by the non-homologous end joining (NHEJ) or the homologous recombination (HR) repair pathway. NHEJ involves the repair of DNA if homologous DNA is not available near a DSB resulting in insertions and deletions. For homology-directed repair (HDR) to occur, a homologous donor sequence is required, and this phenomenon leads to insertions and replacements. For MMEJ, a microhomologous donor sequence is required, and this phenomenon leads to insertions of large fragments.

knock-ins include the replacement of some alleles with others and the insertion of genes into particular genomic regions. There are four main classes of genome modifications based on the CRISPR technology, and some examples of these modifications and their outcomes are discussed below with relevant traits.

2.1 Gene insertion and substitution

Targeted DNA sequence insertion or replacement, also known as gene targeting, refers to the substitution of a DNA fragment or the insertion of a new sequence at a specific target site using precise gene-editing technology (Mao et al., 2019; Zhang et al., 2017). In the last few decades, ZFN and TALEN technologies have been used for gene insertion and substitution in model plant species such as Arabidopsis (Lloyd et al., 2005; Zhang et al., 2010), rice (Zhang et al., 2016), maize (Shukla et al., 2009), and tomato (Lor et al., 2014), and in other plant species, such as barely (Budhagatapalli et al., 2015) and potato (Forsyth et al., 2016). Despite the broad application of ZFNs and TALENs, CRISPR technology has been adopted as the preferred method for successful precise gene editing in multiple plant species. For example, in tomatoes, a strong promoter was inserted ahead of the *SlANT1* gene, which increased the accumulation of anthocyanin and produced purple tomatoes (Čermák et al., 2015). Furthermore, our group performed a precisely targeted gene replacement (T317A) using the HDR pathway to create a tomato line with a long shelf life (Yu et al., 2017). Moreover, another group used CRISPR/Cas9 to develop a smart tomato landing pad for targeted insertion in the promoter region of the *DFR* gene via HDR DNA repair. The successful insertion was visible as a purple-colored phenotype as soon as the plants were regenerated in vitro (Danilo et al., 2018).

Meanwhile, CRISPR-mediated gene knock-ins have been widely employed for crop improvement, such as for drought-tolerant maize (Shi et al., 2017), disease-resistant citrus (Jia et al., 2017), and herbicide-resistant soybean (Li et al., 2015), maize (Svitashev et al., 2015, 2016), rice (Endo et al., 2016; Sun et al., 2016), and flax (Sauer et al., 2016). These improvements were generated by introducing precision point substitutions, inserting new genes, or replacing elite promoters. It has even been demonstrated that CRISPR/Cas9 can be used for T-DNA integration in rice (Lee et al., 2019). Selectable marker insertion is one of the best examples of targeted gene insertion or replacement; the *hygromycin phosphotransferase* (*HPT*) gene was integrated into chromosome 4 of soybean via the HDR pathway using a donor fragment flanked by homologous sequences to the target site (Li et al., 2015).

2.2 Genomic rearrangements

The definition of a genomic rearrangement (GR) is the rearrangement of the linear sequence of chromosomes via transposition, duplication, deletion, inversion, or translocation of genomic DNA segments (Gray, 2000). Generally, the formation of GRs can be repaired by homologous recombination (HR) and NHEJ pathways (Puchta, 2005). The distribution of the rearrangement types are random and the same kind of events rarely occur at the same location in HR-mediated GRs (Lupski and Stankiewicz, 2005; Park et al., 2016). When both DSBs are far apart on the same chromosome, the outcome can be deletion, inversion, or duplication. In contrast, when the DSBs are on different chromosomes, a translocation or chromosome fusions event can be generated (Jansing et al., 2019).

ZFNs and TALENs have been successfully applied in plants to introduce a number of different GR events. In Arabidopsis, a large chromosomal deletion of approximately 9 Mb, and gene cluster inversions and duplications were performed using the ZFN method (Qi et al., 2013). Additionally, the inversion of a 1.3 kb DNA fragment was generated with a frequency of 1% using TALENs in rice (Shan et al., 2013a). More recently, because the CRISPR/Cas9 system is simpler and more efficient than the ZFN and TALEN systems, it has allowed the induction of multiple types of chromosomal changes in the genome (Cheong et al., 2018; Gratz et al., 2013).

In rice, large chromosomal excisions from 115 kb to 245 kb have been performed in protoplasts using CRISPR/Cas9 and the deletion of two clusters has been verified in regenerated T0 generation plants (Zhou et al., 2014). Additionally, an inversion of a fragment of approximately 300 bp has been generated between two DSBs using CRISPR/Cas9 in rice and Arabidopsis (Liang et al., 2016). In Arabidopsis, the duplication and meiotic transmission of a 2.3 kb DNA fragment has been induced (Lynagh et al., 2018). Recently, in an extensive study, efficient inversions were formed in Arabidopsis, of which the inversions of up to 18 kb were successfully transmitted to the next generation driven by an egg cell-specific promoter for Cas9 expression (Schmidt et al., 2019). GRs such as these break traditional breeding patterns and offer tremendous opportunities to remove entire gene clusters, create new linkages, and transfer traits from wild relatives to elite cultivars (Puchta, 2017).

2.3 Targeted nucleotide exchange

Base editors are a promising novel tool that enables precise nucleotide substitutions in a highly efficient and programmable manner, without the need for a foreign donor DNA or HDR (Chen et al., 2019; Jin et al., 2019; Komor et al., 2016). In recent years, the cytosine base editor (CBE) and adenine base editor (ABE) have been commonly used to efficiently generate precise point mutations (cytosine [C] to thymine [T] or adenine [A] to guanine [G]). The base editors have been used in plant genomes including cotton, rice, wheat, maize, and tomato (Qin et al., 2019; Shimatani et al., 2017; Zong et al., 2017). The CBE system comprises a cytidine deaminase, fused with a nickase Cas9 (nCas9) and a uracil glycosylase inhibitor (UGI), which can convert targeted cytosine in genomic DNA to uracil (Komor et al., 2016). BE3, BE4, and Target-AID are the most commonly used base editors and have been widely applied in various plant genomes. Modifications have been used to optimize BE3 to expand its target scope and to attain high fidelity (HF) (Bao et al., 2019). In rice, the efficiencies of the C to T (G to A) transition in BE3 and HF1-BE3 groups were higher than in the control and ABE groups. Similarly, Target-AID has been combined with dCas9 (endonuclease dead Cas9) or nCas9 recruited to PmCDA1 (cytidine deaminase from *Petromyzon marinus*) and single-guide RNAs (sgRNAs) to achieve multiple herbicide-resistance in rice and tomato genomes (Shimatani et al., 2017). Another CBE system, human APOBEC3A (A3A-PBE) fused with nCas9, has been widely used in wheat, rice, and potato and can efficiently convert C to T with a 17-nucleotide editing window (Zong et al., 2018).

Recently, an elite base editor, ABE7.10, has been developed, which is based on the fusion of a tRNA adenosine deaminase to a *Streptococcus pyogenes* Cas9 (SpCas9) nickase (D10A). One of the seven constructed ABE fusion proteins, PABE-7, has been used together with the enhanced sgRNA (esgRNA) to achieve efficient A•T to G•C substitutions at frequencies of up to

7.5% in both rice and wheat protoplasts, and of 59.1% in T0 regenerated rice (Li et al., 2018a). A newly engineered ABE system, ABE7.10-nSpCas9-NGv1, has been constructed to investigate the mutation patterns and efficiency at each target site by sequencing. Compared to the ABE7.10 system, the new system did not result in any undesired substitutions in rice calli, which indicated that ABE7.10-nSpCas9-NGv1 could expand the scope of A-G base editing and improves the conversion efficiency (Negishi et al., 2019).

2.4 Gene stacking

One of the outcomes of knock-ins is to achieve gene stacking, which refers to the sequential addition of multiple genes into the target plant genome, potentially enabling them at the same locus (Chen et al., 2019). Several studies have made advances in targeted gene integration, and different approaches have been used for gene stacking in different plant species. For example, in a previous study, ZFN was exploited to integrate a cassette carrying two exogenous target reporter genes with a flanking donor template into a tobacco (*Nicotiana tabacum*) genomic locus via HDR (Cai et al., 2009). In another study, a pSliceN'Excise vector was used for the constitutive expression of the blasticidin deaminase (*BSD*) marker, *GFP* reporter, and *SpCas9* gene to generate a marker-reporter-free Cre/Cas9 system in *Nannochloropsis gaditana* (Verruto et al., 2018). Stacking genes can also be used to alter multiple elite traits in a single crop variety to enable crop improvement. A re-engineered meganuclease was designed for the precise insertion of two additional herbicide tolerance genes (*hppd* and *epsps*) via HR into cotton (D'Halluin et al., 2013). In summary, targeted stacking could improve trait features and enhance the tolerance of plants to stresses or pathogeny via the sophisticated insertion of additional trait genes into plant genomes at a single locus.

3 The manner of targeting the integration of DNA fragments

3.1 Double-stranded donor repair templates

3.1.1 Donor repair templates with two long homology arms

In yeast, it is relatively easy to insert DNA strands that are several kilobases long via HDR. The efficiency of insertion is low in higher eukaryotes, but it has been demonstrated in mouse cells. The efficiency can be increased to one in ten thousand by inducing DSBs or by increasing homology. HDR usually occurs in the S and G2 phases of the cell cycle. Orthwein et al. (2015) have reported that the cell cycle controls the interaction of the DNA repair protein BRCA1 with PALB2-BRCA2 to constrain the function of BRCA2 to the S and G2 phases in human cells. In late S and G2, a sister chromatid with homologous DNA sequences that are hundreds of bases long is used for DSB repair (Branzei and Foiani, 2008; Mladenov et al., 2016).

To increase the efficiency of genome editing, the artificially introduced donor DNA, which contains approximately 0.5–2 kb of homology arms, is used instead of the sister chromatid. Previous experiments showed that a ZFN-induced DSB can provide an entry point for HDR, which led to the possibility of genome editing (Bibikova et al., 2001). In a later experiment, a site-specific cleavage that was created using ZFNs was repaired via NHEJ in the absence of donor DNA, while the site-specific cleavage was repaired via HDR in the presence of donor DNA (Bibikova et al., 2002). After that, ZFN was used to repair endogenous genes in human cells (Urnov et al., 2005).

The use of long stretches of homology in the form of homology arms of approximately 500–800 bp, and even from 200 bp to several kb sometimes, can be beneficial. In cultured cells, programmable nucleases and the HDR pathway have been used for the accurate integration of transgenes from 8 kb to 15 kb by using long stretches of homology (Jiang et al., 2013). Unfortunately, DNA insertion has a comparatively low efficiency, and it is difficult to insert a gene into the genome of a whole organism, especially on a large scale. Because of this, just a few targeted DNA bases have been inserted via HDR in some whole organisms. In most systems, NHEJ is the preferred pathway for repairing DSBs (Carroll et al., 2010).

To address whether the presence of a donor DNA would lead to gene replacement by HDR, Li et al. (2013) co-expressed plant codon-optimized SpCas9 (pcoCas9) and the sgRNA in *Nicotiana benthamiana* protoplasts, and simultaneously supplied a double-stranded DNA (dsDNA) donor. As the dsDNA donor contained a unique *AvrII* site, which was flanked by a 533 bp left homology arm and a 114 bp right homology arm to the *phytoene desaturase* (*NbPDS*) locus, it could precisely integrate *AvrII* into the target site. The Sanger sequencing results showed that the anticipated creation of the *AvrII* site in the target sequence without additional modifications was successful. In addition, the frequency of HDR-mediated gene replacement was 9.0% (Li et al., 2013).

In another experiment that used long homology arms and relied on HDR, the aim was to precisely replace the *ALC* gene in the target site with the *alc* gene. Yu et al. (2017) constructed a dsDNA donor template, which was inserted into the vector downstream of the sgRNA. A 23 bp sequence, containing 20 bp sgRNA and 3 bp protospacer adjacent motif (PAM), was chosen as the middle core sequence of the donor fragment as only one base substitution (T to A) was desired. This middle core sequence was flanked with a 235 bp left arm and a 450 bp right arm, which were consistent with the extension of the target sequence. The results showed that HDR-mediated repair events occurred in two plants; one plant possessed the heterozygous alleles while another plant possessed chimeric alleles, which were detected from 26 positive transgenic plants (Yu et al., 2017).

Another manner of targeting the integration of DNA fragments using two long homology arms is homology-mediated end joining (HMEJ). The HMEJ-mediated method is similar to MMEJ, but HMEJ involves homology arms of 800 bp, whereas MMEJ involves the use of 20 bp microhomology arms (Yao et al., 2017, 2018). In a detailed set of experiments, HMEJ was compared to MMEJ, NHEJ, and classical HDR in various cell types of mouse and monkey embryos. Even though the same efficiency was observed for HMEJ and MMEJ in a few cases, HMEJ was superior to other methods in almost all cases. The efficiency of DNA insertion by HMEJ is dependent on the cell type, ranging from 5% to 50%; in mouse embryos, the efficiency ranged from 15% to 30%. It is still unknown how well HMEJ performs when the insertions are larger than the few hundred base pairs of the mCherry reporter. The sequencing results showed that precise integrations occurred in almost all cases. It seems that HMEJ acted via HR, NHEJ, or perhaps a novel HMEJ pathway depending on the recipient cells. Further research is required to clarify the mechanism.

3.1.2 Donor repair templates with two short homology arms

Previous studies demonstrated that using linear donor molecules with just 50 bp of chromosomal homology was sufficient for efficient and targeted transgene insertion. However, classical HDR requires the use of a long homology arm of 0.5–2 kb (Orlando et al., 2010). Specifically, donor homology length is relatively unimportant when the integrated DNA is small, but as the length of the donor DNA increases, the donor homology length becomes more important. Using linear DNA sequences as repair templates will enhance HDR efficiency. When the genomic target site is cleaved by CRISPR/Cas9, homology arms of 30–100 bases in length have been shown to be able to facilitate the introduction of the donor DNA (Paix et al., 2014). Usually, the homology arms should be at the ends of the donor DNA, whether the donor DNA is a dsPCR fragment or a long single-stranded oligodeoxynucleotide (ssODN) (Paix et al., 2016). Richardson et al. (2016) have reported that the efficiency of inserting ssODNs via HDR into cultured human cells can be increased to nearly 60% if the homology arms are asymmetric. Double-stranded DNA is used for classical HR. However, using ssODNs instead of dsDNA, can increase the efficiency of insertion by as much as 500 fold. Thus, it was still previously intractable by making genome editing possible in organisms (Ferenczi et al., 2017). Please note that the procedure is simplified because DNA cloning does not require either ssODN or dsPCR fragments.

Many systems are more efficient and precise when the HDR donor has relatively short homology arms. Paix et al. (2016) have reported that the efficiency of edits for 35 base homology was 75%, for 60 base homology was 47%, and for 100 base homology was only 33% in *Caenorhabditis elegans*. It is therefore possible that the efficiency of site-specific integration decreases as the homology length increases (Paix et al., 2016). Liang et al. (2017) have reported that for ssODN insertion into mammalian cells, the optimal homology arm length is 30–40 bases (Liang et al., 2017). A case-study approach was chosen to ensure that short homology arms can also be applied to introduce dsPCR fragments. Paix et al. (2017a) designed a PCR fragment to insert mCherry near the *adenylyl cyclase 3* (*Adcy3*) locus in fertilized mouse eggs. The mCherry open reading frame (ORF) was 739 bp in length and was flanked by 36 bp homology arms. It was found that the editing efficiency was 31% and that the precision was 100% (no indels) (Paix et al., 2017a). This was similar to the efficiency associated with using long ssODNs (Quadros et al., 2017). In contrast, the insertional efficiency decreased when the donor DNA was more than 1 kb long with homology arms of 33 bases. It is noteworthy that irrelevant of whether the homology arms were composed of 33 or 518 bases, the insertion efficiency of a 714 bp donor in human cells was the same. A detailed protocol containing useful tips has recently been published for genome editing insertions and deletions using short homology arms (Paix et al., 2017b).

Compared with the direct integration of linear donor DNA, the introduction of DNA with short homology arms seems more likely to work through synthesis-dependent strand annealing (SDSA). Experiments have shown that the characteristics of DNA repair are consistent with those of gene conversion by SDSA. One junction can be precise, but the other junction can be imprecise because the accuracy of repair is asymmetric. These results suggest that the different homology requirements to initiate and resolve repair are consistent with those of SDSA for gene conversion (Paix et al., 2017a).

The experiments referred to above showed that short homology arms target DNA to a single genomic site for Cas9 cleavage through a sgRNA. However, a similar approach has been reported involving two gRNAs. This strategy has been used to introduce two *LoxP* sites into mice for conditional knock-outs by co-injecting Cas9 mRNA. Two directed RNAs (sgRNAs) and two ssDNAs (each with a *LoxP* site), flanked by 40–80 bases of homology provided insertion efficiencies of 2–5%, with only one early report of 16% efficiency (Bishop et al., 2016). This approach possibly has a low efficiency due to the deletions caused by NHEJ beside the desired two *LoxP* sites in *cis*. To solve this, Horii et al. (2017) investigated a method that involved the sequential injections of one sgRNA and one ssODN at different cell stages of mouse embryos (Horii et al., 2017). However, it is still bulky and not suitable for high throughput. Miura et al. (2015) used a two-step method to synthesize longer ssDNA molecules (514 bases) containing two *LoxP* sites flanked by homology arms. Using this method, the efficiency was significantly increased to 44% and the precision was 86% (Miura et al., 2015). Similarly, Lanza et al. (2017) reported a large-scale pilot study that showed that using a single longer ssODN is more efficient than inserting two ssODNs (Lanza et al., 2017). These results indicate that longer ssODNs are likely to be the best choice for scaled conditional null allele production.

In mammalian systems, Paix et al. (2017a) designed a PCR fragment to insert mCherry near the C terminus of the mouse *adenylyl cyclase 3* (*Adcy3*) locus to test whether PCR fragments with short homology arms can also function as donors. The editing efficiency was 31%; meanwhile the efficiency of a parallel editing experiment that used an mCherry supercoiled plasmid with 500 bp homology arms was only 20%. These results suggest that the PCR fragments have similar editing efficiencies in mouse embryos as ssDNAs and plasmids, and it is a convenience for the ease of synthesis, especially for long inserts. In addition, Paix et al. (2017a) decided to knock-in GFP at three loci in HEK293T cells to determine whether PCR fragments can also function in the genome editing of human cells. The complexes of PCR fragments and in vitro-assembled Cas9-gRNAs were introduced into HEK293T cells without selection. Furthermore, the editing efficiency was reduced by approximately 50% by reducing the molarity of the PCR fragments, but it was not increased by increasing the length of the homology arms. However, reducing the length of the homology arms to 15 bp led to decreases in efficiency. It could be deduced that PCR fragments function as more efficient donors in HEK293T cells than the plasmids with much longer homology arms. In addition, it is convenient to introduce the 35 bp homology arms by PCR amplification. Liang et al. (2017) have also reported that 30 bp homology arms are optimal for ssODNs (Liang et al., 2017). Furthermore, using 5′-modified dsDNA donors with 50 bp homology arms increased the knock-in rate in HEK293T cells by a factor of five, with 65% efficiency for a 700 bp insertion and 40% efficiency for a 2 500 bp insertion (Yu et al., 2020).

3.1.3 Donor repair templates with only the left homology arm

It has been shown that a donor repair template (DRT) carrying only the left homology arm can be used to successfully perform a precise targeted replacement of the *acetolactate synthase* (*ALS*) gene, in the same way that a DRT with two homology arms can (Li et al., 2018b). An LbCpf1 endonuclease and CRISPR RNA (crRNA) plant expression vector was designed, which contained a DRT with only a 97 bp left homology arm (left armed-DRT). The crRNAs could guide the LbCpf1 endonuclease to target and cut out a fragment from the wild-type *ALS* gene. The two expected mutations, which contained the 549 bp DRT, made rice plants resistant to multiple ALS-inhibiting herbicides. For plasmids with only the left homology arm, in 71 transformed plants, PCR-restriction enzyme (RE) and sequencing analyses were performed. It was shown that there were one heterozygous line and one chimeric line, with one allele containing the precise replacement as designed, and two lines with a partial HDR locus (Li et al., 2018b).

In the same way, an LbCpf1 and crRNA plant expression vector was designed that contained two homology arms (armed-DRT). This 670 bp armed-DRT was similar to the 549 bp left armed-DRT described above, but possessed a 121 bp right homology arm. The 94 plants transformed with the vector with two homology arms were subjected to PCR-RE and sequencing analyses. There were two chimeric genotypes with one allele that had the anticipated precise gene replacement, and five genotypes with partial replacements. These results showed that the left homology arm alone is sufficient for HDR. This finding will help us to further understand the exact HDR mechanism in plants and greatly simplifies the procedure of using DRTs for targeted allele replacement in crop plants (Li et al., 2018b).

3.1.4 Precise integration into the target chromosome (PITCh)

At present, the HR- or NHEJ-mediated methods are prevalent for CRISPR knock-ins. By contrast, PITCh, another method for inserting exogenous fragments into the genome works via MMEJ and utilizes toolkits including TALENs and CRISPR/Cas9 (Sakuma et al., 2015; Nakade et al., 2014). DSBs are generated in the PITCh vector and the genomic DNA using corresponding nucleases, and then the donor template is inserted into the genome via MMEJ, which is stimulated by the microhomologous DNA ends on the PITCh vector generated by the corresponding nucleases (Sakuma et al., 2015). Using PCR or the insertion of annealed oligonucleotides, the microhomologous DNA can be easily added to the PITCh vector. Because the targeting vectors of HR-mediated genome engineering harbor long (500–1000 bp) homology arms, microhomologous sequences are more convenient to construct (Cristea et al., 2013). Moreover, unlike for NHEJ-mediated methods, using MMEJ-mediated gene knock-ins obviates the need for the sticky ends produced by dimeric FokI-fused nucleases such as ZFNs and TALENs (Maresca et al., 2013).

3.1.5 Homology-independent targeted integration (HITI)

Recently, homology-independent targeted integration (HITI), which is an excellent system based on CRISPR-Cas9 and NHEJ, has been used for DNA insertion. In HITI, the sgRNA of CRISPR cleaves the same target sequence in the genome and flanking the gene of interest in the donor vector. Henceforth, the target gene is inserted into the genomic target site via NHEJ. Theoretically, both orientations are suitable for target gene insertion, but if the sgRNA target site is to be re-created, the donor DNA is presumably excised and reinserted until it is in the final stable orientation. When used for site-specific integration into proliferating HEK293 cells, the efficiency of HITI was 30–40%. Better efficiency will be achieved after optimization of the nuclear localization sequence that was used for Cas9. Although most of the cells had precise insertions, up to 25% exhibited imprecise insertions (Suzuki et al., 2016).

In cultured post-mitotic neurons, the HITI method also displayed high practical value for DNA site-specific integration with an efficiency of 0.58% of all cells and of 56% of the transformed cells. Moreover, a high precision of integration of approximately 90% was observed (Suzuki et al., 2016). HITI with adeno-associated viral (AAV) plasmids were injected into the cortex of adult mouse brains, and the efficiency of target-site editing of all cells was 3.5% and of infected cells was 10.6%. As a result of intravenous injection of HITI-AAV plasmids into mice, in the muscle, liver, heart, and other tissues, 3–10% of cells expressed nuclear GFP. This experiment had good precision compared to the HEK293 experiments (Suzuki et al., 2016). This suggests that the HITI-method of target-site editing using whole animals is a somatic, but not a germ line event.

3.2 Single-stranded donor repair templates

A 72 bp single-stranded DRT was designed to test HDR-mediated genome modification in rice protoplasts. The anticipated integration of *Kpn*I + *Eco*RI into *OsPDS* occurred in 2 of the 29 single colonies. Thus, the possibility of using ssDNA oligos, sgRNA, and Cas9 for HDR-mediated gene editing in plant cells was demonstrated (Shan et al., 2013b).

3.3 RNA donor repair templates

An RNA DRT was adopted for gene integration, and the DNase I, RNase H, and RNase A digestions of the DRT sequence were reacted in vitro. It was demonstrated that the ssRNA could be used as a DRT for HDR. The average efficiencies with ssDNA DRT at the two targets were 1.16% and 1.84%, yet the average efficiencies at these targets with ssRNA DRT were 0.13% and 0.07%, respectively. RNA transcripts alone could accomplish HDR, and the average efficiencies at the two targets were 0.09% and 0.06% (Li et al., 2019b). The vectors were introduced into rice calli by particle bombardment for in vivo transcript-templated (TT)-HDR. It was demonstrated that in rice, the RNA transcripts could be used as repair donors for HDR. This technology holds great promise for improving the agronomic traits of crops using CRISPR-mediated HDR (Li et al., 2019b).

4 Methods to increase the efficacy and accuracy of knock-ins

4.1 The structure of the gRNA

Cleavage efficiency is an important restraining factor in CRISPR-mediated gene editing and knock-in, and is mainly determined by the gRNA. Because the gRNA has a small size, multiple sgRNAs can stimulate the simultaneous editing of more than one target by Cas9. Minor changes in the gRNA sequence can be made in either the spacer or scaffold region of the gRNA. Adding one or two guanidine residues 50-upstream of the 20 nucleotides spacer for efficient in vitro transcription of T7 has been shown to improve the specificity (Cho et al., 2013). G-rich and A-rich sgRNAs are more stable than other sgRNAs. This may be because G-quadruplexes can form on the target DNA strand, stabilizing CRISPR-DNA interaction complexes (Moreno-Mateos et al., 2015). Optimizing the gRNAs to have high efficiencies and to locate in close proximity to the insertion site enhances the gene knock-in efficiency (Liang et al., 2017). Additionally, extending the sgRNA by 10 bp and mutating the continuous sequence of Ts at position 4 to C or G significantly enhances sgRNA efficiency (Dang et al., 2015). In another study, chimeric guide RNA (cgRNA) molecules were used to carry both the target-specific sequence and the repair template (Butt et al., 2017; Sun et al., 2016). Furthermore, crRNA modifications have shown a synergistic effect when combined with an extension of the crRNA-transactivating RNA (tracrRNA) duplex length of approximately 5 bp in the tetraloop region (Chen et al., 2013).

4.2 Virus-based vectors

By improving the delivery of CRISPR/Cas9, gRNAs, and repair templates into cells, virus-based vectors can help to increase the efficiency of gene knock-ins. Virus-mediated delivery methods can be separated into two categories: using RNA viruses to escape tissue culture and using geminiviruses through tissue culture genome engineering. Tobacco rattle virus (TRV) has been developed as a vehicle to deliver gRNA into cells of a Cas9 overexpression line to enable efficient gene editing in *N. benthamiana* and *Arabidopsis thaliana* (Ali et al., 2015). Recent research on delivering multiple gRNAs using pea early browning virus (PEBV) has shown that using PEBV results in a higher insertion efficiency than using TRV in tobacco and *Arabidopsis thaliana* (Ali et al., 2018). Similarly, beet necrotic yellow vein virus (BNYVV)-based vectors have been used to drive multiple genes or gRNAs into Cas9 overexpression lines and were able to permit the expression of large proteins (Jiang et al., 2019). However, the carrying capacity of RNA viruses has limitations with respect to delivering large DNA templates.

Geminiviruses are able to overcome those limitations. Bean yellow dwarf virus (BeYDV) as a DNA virus, can not only hold the gene-editing tools and repair templates but is also capable of plant transformation (Baltes et al., 2014). Promoting the abundance of the donor template using BeYDV led to target gene integration efficiencies of 6% and 8.5% in tomato and rice, respectively (Čermák et al., 2015; Wang et al., 2017). *Arabidopsis thaliana* (Col-0) and

N. benthamiana were employed as hosts of the beet severe curly top virus (BSCTV), and the overexpression of Cas9 and sgRNA was used to inhibit virus accumulation (Ji et al., 2015). In order to produce herbicide-resistant potatoes, geminivirus replicons were used to increase the copy number of CRISPR/Cas9 and a repair template to enhance the insertion efficiency (Butler et al., 2016). However, there is a limitation to using geminiviruses; these viruses cannot transfer genes from cell to cell, and so the transformation has to be performed through tissue culture.

4.3 Suppression of the NHEJ-mediated pathway

The major pathways for repairing DNA damage are NHEJ and HDR. HDR repairs an extensive range of DNA damage types using existing homologous sequences. Meanwhile, NHEJ is the main mechanism for repairing DSBs in mammalian cells in a homology-independent manner (He et al., 2018). Many researchers have tried to utilize the NHEJ mechanism to improve the knock-in efficiency of exogenous DNA. Numerous methods, such as synchronized expression, chemical modulation, and overlapping homology arms, arose by inhibiting NHEJ or enhancing HDR. There is a competitive relationship between NHEJ and HDR repair pathways. If the repairability of the NHEJ pathway is weakened, it will inevitably lead to an appropriate improvement of the HR repairability.

In order to modify a gene precisely, researchers focused on the suppression of the NHEJ-mediated pathway in order to accomplish enhanced HDR efficiency. Chu et al. demonstrated that the knockdown of KU70, KU80, and DNA ligase IV led to a shift from NHEJ to HDR, because they are core proteins involved in NHEJ. This gene silencing of KU70, KU80, or DNA ligase IV increased the HDR efficiency by a factor of 4–5. In addition, the expression of a degradation-promoting protein (AdE1B55K or AdE4orf6) increased the HDR efficiency by up to eight times (Chu et al., 2015). To promote HDR at the expense of NHEJ, Maruyama et al. reported that SCR7 treatment increased the efficiency of HDR-mediated genome editing by up to 19-fold. The inhibitor, SCR7, targets DNA ligase IV (Maruyama et al., 2015). Pinder et al. (2015) identified the small molecule, RS-1, as a potent enhancer of CRISPR-based genome editing. RS-1 enhanced HDR by 3–6-fold depending on the increasing expression of Rad51, a protein that plays a major role in the HDR pathway (Pinder et al., 2015). Meanwhile, Song et al. (2016) showed that the in vitro application of RS-1 increased knock-in efficiency by 2–5-fold at different loci, whereas the NHEJ inhibitor SCR7 had minimal effects (Song et al., 2016).

A large number of studies have revealed the important roles of key links in the regulation of NHEJ and HR pathways and their related genes; Lig4, DNA-PK, and XRCC6 are all central molecules in the NHEJ pathway. Lig4 is a

DNA ligase, which can directly reconnect the ends of DSBs. Lig4 and XRCC4 form complexes that form further complexes with DNA-PK, which are necessary for the NHEJ pathway. DNA-PK is a phosphatidylinositol-3-dependent silk/threonine protein kinase (Hollick et al., 2003). The auto-phosphorylation of DNA-PK changes its conformation and allows the end processing enzyme to approach the damaged end of DNA. XRCC6, also known as Ku70, forms a heterodimer with Ku80 that binds to the end of DNA double-strand damage. Studies have shown that inhibition of Lig4, DNA-PK, and XRCC6 can effectively promote HR repair. Therefore, inhibiting multiple key molecules of the NHEJ pathway using short hairpin RNA (shRNA) at the same time as gene editing may be an effective way to improve HR efficiency.

4.4 Cas9 variants

Thanks to the discovery of the CRISPR/Cas system, robust and cost-effective genome editing is possible. However, although SpCas9 generally shows high levels of activity, there are some limitations associated with using SpCas9 depending on the application. SpCas9 can cause mutagenesis at the target sequence and only recognizes the simple NGG PAM sequence (Cebrian-Serrano and Davies, 2017; Hua et al., 2019a). To overcome these shortcomings, multiple high-accuracy Cas9 variants have been developed (Casini et al., 2018; Chen et al., 2017; Kleinstiver et al., 2016; Slaymaker et al., 2016). Two recently engineered SpCas9 variants, namely xCas9 and Cas9-NG, may potentially improve the targeting specificity and range (Hua et al., 2019a; Wang et al., 2019).

Two reports have shown that in rice, xCas9 can recognize a broad range of PAM sequences, such as NG, GAA, and GAT, which has increased editing efficiencies (Li et al., 2019a; Wang et al., 2019). These two studies explored the PAM compatibility of codon-optimized xCas9 and further confirmed the feasibility of using xCas9 in the gene or base editing by generating stable lines with either targeted indels or expected single G to A nucleotide substitution. However, it is worth mentioning that the editing efficiencies in both experiments were lower using xCas9 (70.8% (46/65) or 2.08–12.5%) with PAM sites than when using SpCas9 (88.9% (56/63) or 0–4.17%).

Additionally, Cas9-NG was recently reported to be superior to xCas9 for targeting non-canonical NG PAMs (Nishimasu et al., 2018). Although Cas9-NGv1 and Cas9-NG have been successfully applied in plants (Endo et al., 2019; Nishimasu et al., 2018), a CBE based on Cas9-NGv1, ABE-NG, and CBE-NG were developed with low activity (Endo et al., 2019; Hua et al., 2019a). To improve the activity, an optimized Cas9-NG (D10A)-PmCDA1 was developed for C-T base editing (Zhong et al., 2019). The C to T conversion rates of xCas9 (D10A)-PmCDA1 (2%) were slightly higher than those of Cas9-WT (D10A)-PmCDA1 in rice protoplasts, whereas the conversion rate of Cas9-NG (D10A)-PmCDA1 was 20% (10-fold higher). Overall, the two Cas9 variants, along with their derived base editors provide a new efficacy and useful tools for plant genome-editing applications. Additionally, the CRISPR/Cas12a system, which is a Class 2 Type V-A system, recognizes PAMs rich in thymine 'TTTV'. The improved Cas12a and two crRNAs (one driven by the OsU6 and one by the OsU3 promoter) were used for efficient biallelic gene targeting in rice within one generation (Li et al., 2019c).

In other studies, two AsCpf1 variants carrying the mutations S542R/K607R and S542R/K548V/N552R, which recognize TYCV and TATV PAMs, respectively, displayed enhanced activities in vitro and were successfully applied in humans (Gao et al., 2017). Genome-wide assessment of off-target activity indicated that these variants retain high DNA-targeting specificity, and thus can be used as a new way to enhance the efficacy and accuracy of plant genome editing (Gao et al., 2017).

5 Conclusion and future trends

Genetic editing technology has become a very valuable technique and has brought unprecedented changes to the development of plant breeding technology that no other technology has been able to provide on such a scale. In particular, the CRISPR/Cas9 system has contributed greatly to gene-editing advances due to its simplicity, convenience, accuracy, and efficiency. Numerous events and research projects on plant genome editing have been carried out over the last 7 years (from 2013 to 2019) (Nogué et al., 2019). However, most of them have focused on gene knock-out and only a few on gene knock-in (gene replacements or insertions) events. A primary cause for this is that gene knock-ins are less efficient than gene knock-outs. Methods are now being developed to improve the efficiency of gene knock-ins, but such studies are more likely to involve mammalian cells than plant cells. Considering that the genetic repair systems associated with gene knock-ins are mostly conserved across eukaryotes, these methods developed for animals, in theory, should also apply to plants. Therefore, these methods can provide a basis for further research on increasing the efficiency of gene knock-ins in plants.

A recent study covalently tethered an ssODN to the Cas9-gRNA ribonucleoprotein (RNP) complex via a fused HUH endonuclease, thus spatially and temporally co-localizing the DSB machinery and donor DNA. Using this new method, the HDR efficiency was enhanced by up to 30-fold (Aird et al., 2018). A later study confirmed that Cas9 fusions, with HDR effectors such as CtIP, Rad52, and Mre11, promote HDR by up to two-fold in human cells and significantly reduce NHEJ events. Further studies confirmed that the Cas9-CtIP fusion and the MS2-CtIP system increased the ratio of HDR/NHEJ from 4.5-fold to 6-fold. In addition, optimal results were obtained by the combined use of Cas9-CtIP and MS2-CtIP, which increased the HDR/NHEJ ratio to 14.9-fold (Tran et al., 2019).

Despite the remarkable results of this technology, there is no doubt that more and more new ways to boost the generation of insertion-based genome edits in plants are forthcoming (Table 1). To avoid deleterious genome damage associated with the DSBs induced by the Cas9 nuclease, such as accidental deletions or rearrangement of chromosomes, some ongoing research is focusing on developing a new method for nucleotide substitution edits that do not generate DSBs. For example, Nakajima et al. (2018) confirmed that a combination of single nicks in the target gene and donor plasmid (SNGD) using Cas9-D10A nickase promotes efficient nucleotide substitution in human cells. The SNGD-mediated gene editing led to a markedly lower indel frequency than when using the DSB-mediated approach. More recently, two studies have reported on new CRISPR gene editing based on transposons (jumping genes). The CAST (CRISPR-associated transposase) and INTEGRATE (insertion of transposable elements by gRNA-assisted targeting) methods allow the insertion of DNA sequences into the genome without cutting the target DNA (Klompe et al., 2019; Strecker et al., 2019).

According to a new report issued by the World Bank and the United Nations, as the global population is predicted to increase by nearly 3 billion people by 2050, the global food needs will expand by 50% and the demand for animal-based products will swell by nearly 70%. Under these circumstances, if we do not find a way to grow far more food on less land, we'll need to covert nearly 600 million ha (an area nearly twice the size of India) of forests, savannas, peatlands, and other ecosystems to agricultural land area. The researchers also highlighted a list of 10 specific innovative technologies that could help boost food production. As expected, the first technology listed is genome editing tools like CRISPR, which should be used to unlock traits that boost yields (www.technologyreview.com/s/613979/gene-editing-will-help-far-more-than-organic-food-to-slow-global-warming/). Consequently, considering the high superiority of genome editing techniques in the application of plant science, there is no doubt that approaches based on CRISPR/Cas technology are increasingly imminent. Looking forward, such editing technologies will boost the development of new plant breeding techniques to generate sustainable varieties in order to increase agricultural productivity, while simultaneously reducing the costs to the environment tremendously .

6 Where to look for further information

In recent years, a variety of bioinformatics tools based on CRISPR system are emerging. The article 'Design and analysis of CRISPR-Cas experiments' by Ruth

Table 1 Application of knock-in in plants

Plant species	Targeted sequence(s)	Delivery method/main strategy
Tomato (*Solanum lycopersicum*)	Anthocyanin mutant 1 (ANT1)	Geminivirus replicons
Tomato (*Solanum lycopersicum*)	The carotenoid isomerase (CRTISO) and phytoene synthase 1 (PSY1) genes	CRISPR-Cas system combined with the bean yellow dwarf virus rolling circle replicon
Bread wheat (*Triticum aestivum* L.)	The TaGW2 gene	CRISPR/Cas9 ribonucleoproteins (RNPs)
Arabidopsis (Arabidopsis thaliana)	Defective *glabrous1* gene	In planta gene targeting (IPGT) approach
Arabidopsis (Arabidopsis thaliana)	Early embryo-specific DD45 gene promoter	A sequential transformation method
Banana cv. Rasthali	Two PDS genes (RAS-PDS1 and RAS-PDS2)	CRISPR/Cas9
Rice	5-enolpyruvylshikimate-3-phosphate synthase (EPSPS)	The non-homologous end joining (NHEJ) pathway/ the clustered regularly interspaced short palindromic repeats (CRISPR) – CRISPR-associated protein 9 (Cas9) system
Rice (*Oryza sativa* L)	Acetolactate synthase 1 (ALS1)	Harbors cassettes expressing Cas9 and two gRNAs also contained a donor fragment designed as a template for HDR
Maize *(Zea mays)*	Upstream of the liguleless1 (LIG1) gene, male fertility genes (Ms26 and Ms45), and acetolactate synthase (ALS) genes (ALS1 and ALS2)	(CRISPR)-associated (Cas)-guide RNA technology
Maize *(Zea mays)*	ARGOS8	CRISPR-Cas-enabled advanced breeding technology

Table 2 Abbreviations in this chapter

Abbreviation	Description	Reference
4-CoAS	4-coumaryl synthase	(Cai et al., 2009)
ABE	adenine base editor	(Negishi et al., 2019)
Adcy3	adenylyl cyclase 3	(Paix et al., 2017a)
ALS	acetolactate synthase	(Li et al., 2018b)
BeYDV	bean yellow dwarf virus	(Baltes et al., 2014)
BNYVV	beet necrotic yellow vein virus	(Jiang et al., 2019)
BSCTV	beet severe curly top virus	(Ji et al., 2015)
Cas	CRISPR-associated	(Strecker et al., 2019)
CAST	CRISPR-associated transposase	(Strecker et al., 2019)
CBE	cytosine base editors	(Komor et al., 2016)
cgRNA	chimeric guide RNA	(Butt et al., 2017)
CRD	central repeat domain	(Boch et al., 2009)
CRISPR/Cas9	clustered regularly interspaced short palindromic repeats/CRISPR-associated protein 9	(Ali et al., 2015)
crRNA	CRISPR RNA	(Li et al., 2018b)
dCas9	dead CRISPR-associated 9	(Shimatani et al., 2017)
DRT	donor repair template	(Li et al., 2018b)
DSBs	DNA double-stand breaks	(Carroll et al., 2010)
dsDNA	double-stranded DNA	(Li et al., 2013)
dsPCR	double-stranded PCR	(Paix et al., 2016)
GFP	green fluorescent protein	(Verruto et al., 2018)
GM	genetically modified	(Chen et al., 2019)
GRs	genomic rearrangement	(Puchta, 2005)
HDR	homology-directed repair	(Que et al., 2019)
HITI	homology-independent targeted integration	(Suzuki et al., 2016)
HITI-AAV	HITI-adeno associated viral	(Suzuki et al., 2016)
HPT	hygromycin phosphotransferase	(Li et al., 2015)
HR	homologous recombination	(Puchta, 2005)
INTEGRATE	insertion of transposable elements by guide RNA-assisted targeting	(Strecker et al., 2019)

(Continued)

Table 2 (*Continued*)

Abbreviation	Description	Reference
MMEJ	microhomology-mediated end joining	(Sakuma et al., 2015)
nCas9	CRISPR-associated 9 nickase	(Komor et al., 2016)
NGS	next generation sequencing	(Sun et al., 2018)
NHEJ	non-homologous end joining	(He et al., 2018)
NPBTs	new plant breeding techniques	(Zaidi et al., 2019)
NUE	nitrogen-use efficiency	(Voss-Fels et al., 2019)
ORF	open reading frames	(Paix et al., 2017a)
PDS	phytoene desaturase; preharvest sprouting	(Li et al., 2013)
PAM	protospacer adjacent motif	(Yu et al., 2017)
PAT	phosphinothricin acetyl transferase	(Cai et al., 2009)
pcoCas9	plant codon-optimized Cas9	(Li et al., 2013)
PEBV	pea early browning virus	(Ali et al., 2018)
PITCh	precise integration into target chromosome	(Sakuma et al., 2015)
Rht	reduced height	(Hedden, 2003)
RNP	RNA ribonucleoprotein	(Aird et al., 2018)
sd1	semidwarf1	(Hedden, 2003)
SDSA	synthesis-dependent strand annealing	(Paix et al., 2017a)
sgRNA	single-guide RNA	(Shimatani et al., 2017)
SNGD	single nicks in the target gene and donor plasmid	(Nakajima et al., 2018)
spCas9	Cas9 from *Streptococcus pyogenes*	(Li et al., 2018a)
SSNs	sequence specific nucleases	(Puchta et al., 1996)
ssODN	single-stranded oligodeoxynucleotide	(Paix et al., 2016)
TALEN	TAL effector nucleases	(Li et al., 2012)
TGS	third generation sequencing	(Zhao et al., 2018)
tracrRNA	trans-activating crRNA	(Chen et al., 2013)
TRV	tobacco rattle mosaic virus	(Ali et al., 2015)
UGI	uracil glycosylase inhibitor	(Komor et al., 2016)
xCas9	expanded PAM spCas9 variant	(Hua et al., 2019a)
ZFNs	zinc finger nucleases	(Durai et al., 2005)
ZFPs	zinc finger proteins	(Davis and Stokoe, 2010)

E. Hanna and John G. Doench (Hanna and Doench, 2020) aim to provide a practical guide to these tools that will allow both new and experienced users of CRISPR-Cas technologies to select and critically evaluate available and emerging software.

https://www.nature.com/articles/s41587-020-0490-7#article-info.

Convenient and useful websites for gene editing tools include:

- CRISPR-P 2.0: an improved CRISPR/Cas9 tool for genome editing in plants (http://cbi.hzau.edu.cn/cgi-bin/CRISPR).
- CRISPR-GE: A convenient software toolkit for CRISPR-based genome editing (http://skl.scau.edu.cn/).
- Cas-Designer: A bulge-allowed quick guide-RNA designer for CRISPR/Cas-derived RGENs (http://www.rgenome.net/cas-designer/).
- sgRNA-PSM: Predict sgRNAs on-target activity based on position-specific mismatch (http://bliulab.net/sgRNA-PSM/).
- Cas-OFFinder: A fast and versatile algorithm that searches for potential off-target sites of Cas9 RNA-guided endonucleases (http://www.rgenome.net/cas-offinder/).
- DSDecodeM: An updated web tool for rapid decoding of multiple superimposed sequencing chromatograms (http://skl.scau.edu.cn/dsdecode/#).
- EditR: A novel base editing quantification software using Sanger sequencing (https://moriaritylab.shinyapps.io/editr_v10/).
- CRISPRCasFinder (online): The program enables the easy detection of CRISPRs and cas genes (https://crisprcas.i2bc.paris-saclay.fr/CrisprCasFinder/Index)

7 References

Aird, E. J., Lovendahl, K. N., St. Martin, A., Harris, R. S. and Gordon, W. R. (2018). Increasing Cas9-mediated homology-directed repair efficiency through covalent tethering of DNA repair template, *Communications Biology* 1, 54.

Ali, Z., Abul-faraj, A., Li, L., Ghosh, N., Piatek, M., Mahjoub, A., Aouida, M., Piatek, A., Baltes, N. J., Voytas, D. F., Dinesh-Kumar, S. and Mahfouz, M. M. (2015). Efficient virus-mediated genome editing in plants using the CRISPR/Cas9 system, *Molecular Plant* 8(8), 1288–91.

Ali, Z., Eid, A., Ali, S. and Mahfouz, M. M. (2018). Pea early-browning virus-mediated genome editing via the CRISPR/Cas9 system in *Nicotiana benthamiana* and *Arabidopsis*, *Virus Research* 244, 333–7.

Baltes, N. J., Gil-Humanes, J., Cermak, T., Atkins, P. A. and Voytas, D. F. (2014). DNA replicons for plant genome engineering, *Plant Cell* 26(1), 151–63.

Bao, A., Burritt, D. J., Chen, H., Zhou, X., Cao, D. and Tran, L. P. (2019). The CRISPR/Cas9 system and its applications in crop genome editing, *Critical Reviews in Biotechnology* 39(3), 321–36.

Bibikova, M., Carroll, D., Segal, D. J., Trautman, J. K., Smith, J., Kim, Y. G. and Chandrasegaran, S. (2001). Stimulation of homologous recombination through targeted cleavage by Chimeric nucleases, *Molecular and Cellular Biology* 21(1), 289-97.

Bibikova, M., Golic, M., Golic, K. G. and Carroll, D. (2002). Targeted chromosomal cleavage and mutagenesis in drosophila using zinc-finger nucleases, *Genetics* 161(3), 1169-75.

Bishop, K. A., Harrington, A., Kouranova, E., Weinstein, E. J., Rosen, C. J., Cui, X. and Liaw, L. (2016). CRISPR/Cas9-mediated insertion of loxP sites in the mouse Dock7 gene provides an effective alternative to use of targeted embryonic stem cells, *G3: Genes, Genomes, Genetics* 6(7), 2051-61.

Boch, J., Scholze, H., Schornack, S., Landgraf, A., Hahn, S., Kay, S., Lahaye, T., Nickstadt, A. and Bonas, U. (2009). Breaking the code of DNA binding specificity of TAL-type III effectors, *Science* 326(5959), 1509-12.

Branzei, D. and Foiani, M. (2008). Regulation of DNA repair throughout the cell cycle, *Nature Reviews. Molecular Cell Biology* 9(4), 297-308.

Budhagatapalli, N., Rutten, T., Gurushidze, M., Kumlehn, J. and Hensel, G. (2015). Targeted modification of gene function exploiting homology-directed repair of TALEN-mediated double-strand breaks in Barley, *G3: Genes, Genomes, Genetics* 5(9), 1857-63.

Butler, N. M., Baltes, N. J., Voytas, D. F. and Douches, D. S. (2016). Geminivirus-mediated genome editing in potato (*Solanum tuberosum l.*) using sequence-specific nucleases, *Frontiers in Plant Science* 7, 1045.

Butt, H., Eid, A., Ali, Z., Atia, M. A. M., Mokhtar, M. M., Hassan, N., Lee, C. M., Bao, G. and Mahfouz, M. M. (2017). Efficient CRISPR/Cas9-mediated genome editing using a chimeric single-guide RNA molecule, *Frontiers in Plant Science* 8, 1441.

Cai, C. Q., Doyon, Y., Ainley, W. M., Miller, J. C., DeKelver, R. C., Moehle, E. A., Rock, J. M., Lee, Y. L., Garrison, R., Schulenberg, L., Blue, R., Worden, A., Baker, L., Faraji, F., Zhang, L., Holmes, M. C., Rebar, E. J., Collingwood, T. N., Rubin-Wilson, B., Gregory, P. D., Urnov, F. D. and Petolino, J. F. (2009). Targeted transgene integration in plant cells using designed zinc finger nucleases, *Plant Molecular Biology* 69(6), 699-709.

Carroll, D., Beumer, K. J. and Trautman, J. K. (2010). High-efficiency gene targeting in drosophila with zinc finger nucleases, *Methods in Molecular Biology* 649, 271-80.

Casini, A., Olivieri, M., Petris, G., Montagna, C., Reginato, G., Maule, G., Lorenzin, F., Prandi, D., Romanel, A., Demichelis, F., Inga, A. and Cereseto, A. (2018). A highly specific SpCas9 variant is identified by in vivo screening in yeast, *Nature Biotechnology* 36(3), 265-71.

Cebrian-Serrano, A. and Davies, B. (2017). CRISPR-Cas orthologues and variants: optimizing the repertoire, specificity and delivery of genome engineering tools, *Mammalian Genome* 28(7-8), 247-61.

Čermák, T., Baltes, N. J., Čegan, R., Zhang, Y. and Voytas, D. F. (2015). High-frequency, precise modification of the tomato genome, *Genome Biology* 16, 232.

Chen, B., Gilbert, L. A., Cimini, B. A., Schnitzbauer, J., Zhang, W., Li, G. W., Park, J., Blackburn, E. H., Weissman, J. S., Qi, L. S. and Huang, B. (2013). Dynamic imaging of genomic loci in living human cells by an optimized CRISPR/Cas system, *Cell* 155(7), 1479-91.

Chen, J. S., Dagdas, Y. S., Kleinstiver, B. P., Welch, M. M., Sousa, A. A., Harrington, L. B., Sternberg, S. H., Joung, J. K., Yildiz, A. and Doudna, J. A. (2017). Enhanced proofreading governs CRISPR-Cas9 targeting accuracy, *Nature* 550(7676), 407-10.

Chen, K., Wang, Y., Zhang, R., Zhang, H. and Gao, C. (2019). CRISPR/Cas Genome editing and precision plant breeding in agriculture, *Annual Review of Plant Biology* 70, 667–97.

Cheong, T. C., Blasco, R. B. and Chiarle, R. (2018). The CRISPR/Cas9 system as a tool to engineer chromosomal translocation in vivo. In Zhang, Y. (Ed.), *Chromosome Translocation, Advances in Experimental Medicine and Biology*, vol 1044. Springer, Singapore, 39–48.

Cho, S. W., Kim, S., Kim, Y., Kweon, J., Kim, H. S., Bae, S. and Kim, J. S. (2013). Analysis of off-target effects of CRISPR/Cas-derived RNA-guided endonucleases and nickases, *Genome Research* 24(1), 132–41.

Chu, V. T., Weber, T., Wefers, B., Wurst, W., Sander, S., Rajewsky, K. and Kühn, R. (2015). Increasing the efficiency of homology-directed repair for CRISPR-Cas9-induced precise gene editing in mammalian cells, *Nature Biotechnology* 33(5), 543–8.

Cristea, S., Freyvert, Y., Santiago, Y., Holmes, M. C., Urnov, F. D., Gregory, P. D. and Cost, G. J. (2013). In vivo cleavage of transgene donors promotes nuclease-mediated targeted integration, *Biotechnology and Bioengineering* 110(3), 871–80.

D'Halluin, K., Vanderstraeten, C., Van Hulle, J., Rosolowska, J., Van Den Brande, I., Pennewaert, A., D'Hont, K., Bossut, M., Jantz, D., Ruiter, R. and Broadhvest, J. (2013). Targeted molecular trait stacking in cotton through targeted double-strand break induction, *Plant Biotechnology Journal* 11(8), 933–41.

Dang, Y., Jia, G., Choi, J., Ma, H., Anaya, E., Ye, C., Shankar, P. and Wu, H. (2015). Optimizing sgRNA structure to improve CRISPR-Cas9 knockout efficiency, *Genome Biology* 16, 280.

Danilo, B., Perrot, L., Botton, E., Nogué, F. and Mazier, M. (2018). The DFR locus: a smart landing pad for targeted transgene insertion in tomato, *PLoS ONE* 13(12), e0208395.

Davis, D. and Stokoe, D. (2010). Zinc finger nucleases as tools to understand and treat human diseases, *BMC Medicine* 8, 42.

Durai, S., Mani, M., Kandavelou, K., Wu, J., Porteus, M. H. and Chandrasegaran, S. (2005). Zinc finger nucleases: custom-designed molecular scissors for genome engineering of plant and mammalian cells, *Nucleic Acids Research* 33(18), 5978–90.

Endo, M., Mikami, M., Endo, A., Kaya, H., Itoh, T., Nishimasu, H., Nureki, O. and Toki, S. (2019). Genome editing in plants by engineered CRISPR–Cas9 recognizing NG PAM, *Nature Plants* 5(1), 14–7.

Endo, M., Mikami, M. and Toki, S. (2016). Biallelic gene targeting in rice, *Plant Physiology* 170(2), 667–77.

Feng, Z., Zhang, B., Ding, W., Liu, X., Yang, D. L., Wei, P., Cao, F., Zhu, S., Zhang, F., Mao, Y. and Zhu, J. K. (2013). Efficient genome editing in plants using a CRISPR/Cas system, *Cell Research* 23(10), 1229–32.

Ferenczi, A., Pyott, D. E., Xipnitou, A. and Molnar, A. (2017). Efficient targeted DNA editing and replacement in Chlamydomonas reinhardtii using Cpf1 ribonucleoproteins and single-stranded DNA, *Proceedings of the National Academy of Sciences of the United States of America* 114(51), 13567–72.

Forsyth, A., Weeks, T., Richael, C. and Duan, H. (2016). Transcription activator-like effector nucleases (TALEN)-mediated targeted DNA insertion in Potato plants, *Frontiers in Plant Science* 7, 1572.

Gao, L., Cox, D. B. T., Yan, W. X., Manteiga, J. C., Schneider, M. W., Yamano, T., Nishimasu, H., Nureki, O., Crosetto, N. and Zhang, F. (2017). Engineered Cpf1 variants with altered PAM specificities, *Nature Biotechnology* 35(8), 789–92.

Gratz, S. J., Wildonger, J., Harrison, M. M. and O'Connor-Giles, K. M. (2013). CRISPR/Cas9-mediated genome engineering and the promise of designer flies on demand, *Fly (Austin)* 7(4), 249–55.

Gray, Y. H. M. (2000). It takes two transposons to tango: transposable-element-mediated chromosomal rearrangements, *Trends in Genetics* 16(10), 461–8.

Hanna, R. E., and Doench, J. G. (2020). Design and analysis of CRISPR-Cas experiments, *Nature Biotechnology*, 38, 813–23.

He, X., Li, Y. X. and Feng, B. (2018). New turns for high efficiency knock-in of large DNA in human pluripotent stem cells, *Stem Cells International* 2018, 9465028.

Hedden, P. (2003). The genes of the green revolution, *Trends in Genetics* 19(1), 5–9.

Hollick, J. J., Golding, B. T., Hardcastle, I. R., Martin, N., Richardson, C., Rigoreau, L. J., Smith, G. C. and Griffin, R. J. (2003). 2,6-Disubstituted pyran-4-one and thiopyran-4-one inhibitors of DNA-dependent protein kinase (DNA-PK), *Bioorganic and Medicinal Chemistry Letters* 13(18), 3083–6.

Horii, T., Morita, S., Kimura, M., Terawaki, N., Shibutani, M. and Hatada, I. (2017). Efficient generation of conditional knockout mice via sequential introduction of lox sites, *Scientific Reports* 7(1), 7891.

Hua, K., Tao, X., Han, P., Wang, R. and Zhu, J. K. (2019a). Genome engineering in rice using Cas9 variants that recognize NG PAM sequences, *Molecular Plant* 12(7), 1003–14.

Hua, K., Zhang, J., Botella, J. R., Ma, C., Kong, F., Liu, B. and Zhu, J. K. (2019b). Perspectives on the application of genome-editing technologies in crop breeding, *Molecular Plant* 12(8), 1047–59.

Jansing, J., Schiermeyer, A., Schillberg, S., Fischer, R. and Bortesi, L. (2019). Genome editing in agriculture: technical and practical considerations, *International Journal of Molecular Sciences* 20(12), 2888.

Ji, X., Zhang, H., Zhang, Y., Wang, Y. and Gao, C. (2015). Establishing a CRISPR-Cas-like immune system conferring DNA virus resistance in plants, *Nature Plants* 1, 15144.

Jia, H., Zhang, Y., Orbović, V., Xu, J., White, F. F., Jones, J. B. and Wang, N. (2017). Genome editing of the disease susceptibility gene CsLOB1 in citrus confers resistance to citrus canker, *Plant Biotechnology Journal* 15(7), 817–23.

Jiang, J., Jing, Y., Cost, G. J., Chiang, J. C., Kolpa, H. J., Cotton, A. M., Carone, D. M., Carone, B. R., Shivak, D. A., Guschin, D. Y., Pearl, J. R., Rebar, E. J., Byron, M., Gregory, P. D., Brown, C. J., Urnov, F. D., Hall, L. L. and Lawrence, J. B. (2013). Translating dosage compensation to trisomy 21, *Nature* 500(7462), 296–300.

Jiang, N., Zhang, C., Liu, J. Y., Guo, Z. H., Zhang, Z. Y., Han, C. G. and Wang, Y. (2019). Development of Beet necrotic yellow vein virus-based vectors for multiple-gene expression and guide RNA delivery in plant genome editing, *Plant Biotechnology Journal* 17(7), 1302–15.

Jin, S., Zong, Y., Gao, Q., Zhu, Z., Wang, Y., Qin, P., Liang, C., Wang, D., Qiu, J. L., Zhang, F. and Gao, C. (2019). Cytosine, but not adenine, base editors induce genome-wide off-target mutations in rice, *Science* 364(6437), 292–5.

Kleinstiver, B. P., Pattanayak, V., Prew, M. S., Tsai, S. Q., Nguyen, N. T., Zheng, Z. and Joung, J. K. (2016). High-fidelity CRISPR-Cas9 nucleases with no detectable genome-wide off-target effects, *Nature* 529(7587), 490–5.

Klompe, S. E., Vo, P. L. H., Halpin-Healy, T. S. and Sternberg, S. H. (2019). Transposon-encoded CRISPR-Cas systems direct RNA-guided DNA integration, *Nature* 571(7764), 219–25.

Komor, A. C., Kim, Y. B., Packer, M. S., Zuris, J. A. and Liu, D. R. (2016). Programmable editing of a target base in genomic DNA without double-stranded DNA cleavage, *Nature* 533(7603), 420–4.

Lanza, D. G., Gaspero, A., Lorenzo, I., Liao, L., Zheng, P., Wang, Y., Deng, Y., Cheng, C., Zhang, C., Rasband, M. N., Seavitt, J. R., DeMayo, F. J., Xu, J., Dickinson, M. E., Beaudet, A. L. and Heaney, J. D. (2017). Employing single-stranded DNA donors for the high-throughput production of conditional knockout alleles in mice, *bioRxiv*, 1-25.

Lee, K., Eggenberger, A. L., Banakar, R., Mccaw, M. E., Zhu, H., Main, M., Kang, M., Gelvin, S. B. and Wang, K. (2019). CRISPR / Cas9-mediated targeted T-DNA integration in rice, *Plant Molecular Biology* 99(4–5), 317–28.

Li, C., Zong, Y., Wang, Y., Jin, S., Zhang, D., Song, Q., Zhang, R. and Gao, C. (2018a). Expanded base editing in rice and wheat using a Cas9-adenosine deaminase fusion, *Genome Biology* 19(1), 59.

Li, J. F., Norville, J. E., Aach, J., McCormack, M., Zhang, D., Bush, J., Church, G. M. and Sheen, J. (2013). Multiplex and homologous recombination-mediated genome editing in Arabidopsis and *Nicotiana benthamiana* using guide RNA and Cas9, *Nature Biotechnology* 31(8), 688–91.

Li, J., Luo, J., Xu, M., Li, S., Zhang, J., Li, H., Yan, L., Zhao, Y. and Xia, L. (2019a). Plant genome editing using xCas9 with expanded PAM compatibility, *Journal of Genetics and Genomics* 46(5), 277–80.

Li, J., Meng, X., Zong, Y., Chen, K., Zhang, H., Liu, J., Li, J. and Gao, C. (2016). Gene replacements and insertions in rice by intron targeting using CRISPR-Cas9, *Nature Plants* 2, 16139.

Li, S., Li, J., He, Y., Xu, M., Zhang, J., Du, W., Zhao, Y. and Xia, L. (2019b). Precise gene replacement in rice by RNA transcript-templated homologous recombination, *Nature Biotechnology* 37(4), 445–50.

Li, S., Li, J., Zhang, J., Du, W., Fu, J., Sutar, S., Zhao, Y. and Xia, L. (2018b). Synthesis dependent repair of Cpf1-induced double strand DNA breaks enables targeted gene replacement in rice, *Journal of Experimental Botany* 69(20), 4715–21.

Li, S., Zhang, Y., Xia, L. and Qi, Y. (2019c). CRISPR-Cas12a enables efficient biallelic gene targeting in rice, *Plant Biotechnology Journal*, 18(6), 1351-3..

Li, T., Liu, B., Spalding, M. H., Weeks, D. P. and Yang, B. (2012). High-efficiency TALEN-based gene editing produces disease-resistant rice, *Nature Biotechnology* 30(5), 390–2.

Li, Z., Liu, Z. B., Xing, A., Moon, B. P., Koellhoffer, J. P., Huang, L., Ward, R. T., Clifton, E., Falco, S. C. and Cigan, A. M. (2015). Cas9-guide RNA directed genome editing in soybean, *Plant Physiology* 169(2), 960–70.

Liang, G., Zhang, H., Lou, D. and Yu, D. (2016). Selection of highly efficient sgRNAs for CRISPR/Cas9-based plant genome editing, *Scientific Reports* 6, 21451.

Liang, X., Potter, J., Kumar, S., Ravinder, N. and Chesnut, J. D. (2017). Enhanced CRISPR/Cas9-mediated precise genome editing by improved design and delivery of gRNA, Cas9 nuclease, and donor DNA, *Journal of Biotechnology* 241, 136–46.

Lloyd, A., Plaisier, C. L., Carroll, D. and Drews, G. N. (2005). Targeted mutagenesis using zinc-finger nucleases in Arabidopsis, *Proceedings of the National Academy of Sciences of the United States of America* 102(6), 2232–7.

Lor, V. S., Starker, C. G., Voytas, D. F., Weiss, D. and Olszewski, N. E. (2014). Targeted mutagenesis of the tomato PROCERA gene using transcription activator-like effector nucleases, *Plant Physiology* 166(3), 1288–91.

Lupski, J. R. and Stankiewicz, P. (2005). Genomic disorders: molecular mechanisms for rearrangements and conveyed phenotypes, *PLoS Genetics* 1(6), e49.

Lynagh, P. G., Inagaki, S., Amundson, K. R., Marimuthu, M. P. A., Pike, B. R., Henry, I. M., Tan, E. H. and Comai, L. (2018). Translocation and duplication from CRISPR-Cas9 editing in *Arabidopsis thaliana*, *bioRxiv* 400507.

Mao, Y., Botella, J. R., Liu, Y. and Zhu, J. K. (2019). Gene editing in plants: progress and challenges, *National Science Review* 6(3), 421-37.

Maresca, M., Lin, V. G., Guo, N. and Yang, Y. (2013). Obligate ligation-gated recombination (ObLiGaRe): custom-designed nuclease-mediated targeted integration through nonhomologous end joining, *Genome Research* 23(3), 539-46.

Maruyama, T., Dougan, S. K., Truttmann, M. C., Bilate, A. M., Ingram, J. R. and Ploegh, H. L. (2015). Increasing the efficiency of precise genome editing with CRISPR-Cas9 by inhibition of nonhomologous end joining, *Nature Biotechnology* 33(5), 538-42.

Miura, H., Gurumurthy, C. B., Sato, T., Sato, M. and Ohtsuka, M. (2015). CRISPR/Cas9-based generation of knockdown mice by intronic insertion of artificial microRNA using longer single-stranded DNA, *Scientific Reports* 5, 12799.

Mladenov, E., Magin, S., Soni, A. and Iliakis, G. (2016). DNA double-strand-break repair in higher eukaryotes and its role in genomic instability and cancer: cell cycle and proliferation-dependent regulation, *Seminars in Cancer Biology* 37-38, 51-64.

Moreno-Mateos, M. A., Vejnar, C. E., Beaudoin, J. D., Fernandez, J. P., Mis, E. K., Khokha, M. K. and Giraldez, A. J. (2015). CRISPRscan: designing highly efficient sgRNAs for CRISPR-Cas9 targeting in vivo, *Nature Methods* 12(10), 982-8.

Moscou, M. J. and Bogdanove, A. J. (2009). A simple cipher governs DNA recognition by TAL effectors, *Science* 326(5959), 1501.

Mueller, S. M., Messina, C. D. and Vyn, T. J. (2019). Simultaneous gains in grain yield and nitrogen efficiency over 70 years of maize genetic improvement, *Scientific Reports* 9(1), 9095.

Nakade, S., Tsubota, T., Sakane, Y., Kume, S., Sakamoto, N., Obara, M., Daimon, T., Sezutsu, H., Yamamoto, T., Sakuma, T. and Suzuki, K. T. (2014). Microhomology-mediated end-joining-dependent integration of donor DNA in cells and animals using TALENs and CRISPR/Cas9, *Nature Communications* 5, 5560.

Nakajima, K., Zhou, Y., Tomita, A., Hirade, Y., Gurumurthy, C. B. and Nakada, S. (2018). Precise and efficient nucleotide substitution near genomic nick via noncanonical homology-directed repair, *Genome Research* 28(2), 223-30.

Negishi, K., Kaya, H., Abe, K., Hara, N., Saika, H. and Toki, S. (2019). An adenine base editor with expanded targeting scope using SpCas9-NG v1 in rice, *Plant Biotechnology Journal* 17(8), 1476-8.

Nekrasov, V., Staskawicz, B., Weigel, D., Jones, J. D. G. and Kamoun, S. (2013). Targeted mutagenesis in the model plant *Nicotiana benthamiana* using Cas9 RNA-guided endonuclease, *Nature Biotechnology* 31(8), 691-3.

Nishimasu, H., Shi, X., Ishiguro, S., Gao, L., Hirano, S., Okazaki, S., Noda, T., Abudayyeh, O. O., Gootenberg, J. S., Mori, H., Oura, S., Holmes, B., Tanaka, M., Seki, M., Hirano, H., Aburatani, H., Ishitani, R., Ikawa, M., Yachie, N., Zhang, F. and Nureki, O. (2018). Engineered CRISPR-Cas9 nuclease with expanded targeting space, *Science* 361(6408), 1259-62.

Nogué, F., Vergne, P., Chèvre, A. M., Chauvin, J. E., Bouchabké-Coussa, O., Déjardin, A., Chevreau, E., Hibrand-Saint Oyant, L., Mazier, M., Barret, P., Guiderdoni, E., Sallaud, C., Foucrier, S., Devaux, P. and Rogowsky, P. M. (2019). Crop plants with improved

culture and quality traits for food, feed and other uses, *Transgenic Research* 28(Suppl 2), 65–73.
Orlando, S. J., Santiago, Y., DeKelver, R. C., Freyvert, Y., Boydston, E. A., Moehle, E. A., Choi, V. M., Gopalan, S. M., Lou, J. F., Li, J., Miller, J. C., Holmes, M. C., Gregory, P. D., Urnov, F. D. and Cost, G. J. (2010). Zinc-finger nuclease-driven targeted integration into mammalian genomes using donors with limited chromosomal homology, *Nucleic Acids Research* 38(15), e152.
Orthwein, A., Noordermeer, S. M., Wilson, M. D., Landry, S., Enchev, R. I., Sherker, A., Munro, M., Pinder, J., Salsman, J., Dellaire, G., Xia, B., Peter, M. and Durocher, D. (2015). A mechanism for the suppression of homologous recombination in G1 cells, *Nature* 528(7582), 422–6.
Paix, A., Folkmann, A., Goldman, D. H., Kulaga, H., Grzelak, M. J., Rasoloson, D., Paidemarry, S., Green, R., Reed, R. R., and Seydoux, G. (2017a). Precision genome editing using synthesis-dependent repair of Cas9-induced DNA breaks, *Proceedings of the National Academy of Sciences of the United States of America* 114(50), e10745.
Paix, A., Folkmann, A. and Seydoux, G. (2017b). Precision genome editing using CRISPR-Cas9 and linear repair templates in *C. elegans*, *Methods* 121–122, 86–93.
Paix, A., Schmidt, H. and Seydoux, G. (2016). Cas9-assisted recombineering in *C. elegans*: genome editing using in vivo assembly of linear DNAs, *Nucleic Acids Research* 44(15), e128.
Paix, A., Wang, Y., Smith, H. E., Lee, C. Y. S., Calidas, D., Lu, T., Smith, J., Schmidt, H., Krause, M. W. and Seydoux, G. (2014). Scalable and versatile genome editing using linear DNAs with microhomology to Cas9 sites in *Caenorhabditis elegans*, *Genetics* 198(4), 1347–56.
Park, C. Y., Sung, J. J. and Kim, D. W. (2016). Genome editing of structural variations: modeling and gene correction, *Trends in Biotechnology* 34(7), 548–61.
Pinder, J., Salsman, J. and Dellaire, G. (2015). Nuclear domain "knock-in" screen for the evaluation and identification of small molecule enhancers of CRISPR-based genome editing, *Nucleic Acids Research* 43(19), 9379–92.
Puchta, H. (2005). The repair of double-strand breaks in plants: mechanisms and consequences for genome evolution, *Journal of Experimental Botany* 56(409), 1–14.
Puchta, H. (2017). Applying CRISPR/Cas for genome engineering in plants: the best is yet to come, *Current Opinion in Plant Biology* 36, 1–8.
Puchta, H., Dujon, B. and Hohn, B. (1996). Two different but related mechanisms are used in plants for the repair of genomic double-strand breaks by homologous recombination, *Proceedings of the National Academy of Sciences of the United States of America* 93(10), 5055–60.
Qi, Y., Li, X., Zhang, Y., Starker, C. G., Baltes, N. J., Zhang, F., Sander, J. D., Reyon, D., Joung, J. K. and Voytas, D. F. (2013). Targeted deletion and inversion of tandemly arrayed genes in *Arabidopsis thaliana* using zinc finger nucleases, *G3* 3(10), 1707–15.
Qin, L., Li, J., Wang, Q., Xu, Z., Sun, L., Alariqi, M., Manghwar, H., Wang, G., Li, B., Ding, X., Rui, H., Huang, H., Lu, T., Lindsey, K., Daniell, H., Zhang, X. and Jin, S. (2019). High-efficient and precise base editing of C*G to T*A in the allotetraploid cotton (*Gossypium hirsutum*) genome using a modified CRISPR/Cas9 system, *Plant Biotechnology Journal* 18, 45–56.
Quadros, R. M., Miura, H., Harms, D. W., Akatsuka, H., Sato, T., Aida, T., Redder, R., Richardson, G. P., Inagaki, Y., Sakai, D., Buckley, S. M., Seshacharyulu, P., Batra, S. K., Behlke, M. A., Zeiner, S. A., Jacobi, A. M., Izu, Y., Thoreson, W. B., Urness, L. D.,

Mansour, S. L., Ohtsuka, M. and Gurumurthy, C. B. (2017). Easi-CRISPR: a robust method for one-step generation of mice carrying conditional and insertion alleles using long ssDNA donors and CRISPR ribonucleoproteins, *Genome Biology* 18(1), 92.

Que, Q., Chen, Z., Kelliher, T., Skibbe, D., Dong, S. and Chilton, M. D. (2019). Plant DNA repair pathways and their applications in genome engineering, *Methods in Molecular Biology* 1917, 3-24.

Richardson, C. D., Ray, G. J., DeWitt, M. A., Curie, G. L. and Corn, J. E. (2016). Enhancing homology-directed genome editing by catalytically active and inactive CRISPR-Cas9 using asymmetric donor DNA, *Nature Biotechnology* 34(3), 339-44.

Sakuma, T., Nakade, S., Sakane, Y., Suzuki, K. T. and Yamamoto, T. (2015). MMEJ-assisted gene knock-in using TALENs and CRISPR-Cas9 with the PITCh systems, *Nature Protocols* 11(1), 118-33.

Sauer, N. J., Narváez-Vásquez, J., Mozoruk, J., Miller, R. B., Warburg, Z. J., Woodward, M. J., Mihiret, Y. A., Lincoln, T. A., Segami, R. E., Sanders, S. L., Walker, K. A., Beetham, P. R., Schöpke, C. R. and Gocal, G. F. (2016). Oligonucleotide-mediated genome editing provides precision and function to engineered nucleases and antibiotics in plants, *Plant Physiology* 170(4), 1917-28.

Scheben, A., Wolter, F., Batley, J., Puchta, H. and Edwards, D. (2017). Towards CRISPR/CAS crops - bringing together genomics and genome editing, *New Phytologist* 216(3), 682-98.

Schmidt, C., Pacher, M. and Puchta, H. (2019). Efficient induction of heritable inversions in plant genomes using the CRISPR/Cas system, *Plant Journal: for Cell and Molecular Biology* 98(4), 577-89.

Schmidt, M. H., Vogel, A., Denton, A. K., Istace, B., Wormit, A., van de Geest, H., Bolger, M. E., Alseekh, S., Maß, J., Pfaff, C., Schurr, U., Chetelat, R., Maumus, F., Aury, J. M., Koren, S., Fernie, A. R., Zamir, D., Bolger, A. M. and Usadel, B. (2017). De novo assembly of a new *Solanum pennellii* accession using nanopore sequencing, *The Plant Cell* 29(10), 2336-48.

Shan, Q., Wang, Y., Chen, K., Liang, Z., Li, J., Zhang, Y., Zhang, K., Liu, J., Voytas, D. F., Zheng, X., Zhang, Y. and Gao, C. (2013a). Rapid and efficient gene modification in rice and Brachypodium using TALENs, *Molecular Plant* 6(4), 1365-8.

Shan, Q., Wang, Y., Li, J., Zhang, Y., Chen, K., Liang, Z., Zhang, K., Liu, J., Xi, J. J., Qiu, J. L. and Gao, C. (2013b). Targeted genome modification of crop plants using a CRISPR-Cas system, *Nature Biotechnology* 31(8), 686-8.

Shao, Q., Punt, M. and Wesseler, J. (2018). New plant breeding techniques under food security pressure and lobbying, *Frontiers in Plant Science* 9, 1324.

Shi, J., Gao, H., Wang, H., Lafitte, H. R., Archibald, R. L., Yang, M., Hakimi, S. M., Mo, H. and Habben, J. E. (2017). ARGOS8 variants generated by CRISPR-Cas9 improve maize grain yield under field drought stress conditions, *Plant Biotechnology Journal* 15(2), 207-16.

Shimatani, Z., Kashojiya, S., Takayama, M., Terada, R., Arazoe, T., Ishii, H., Teramura, H., Yamamoto, T., Komatsu, H., Miura, K., Ezura, H., Nishida, K., Ariizumi, T. and Kondo, A. (2017). Targeted base editing in rice and tomato using a CRISPR-Cas9 cytidine deaminase fusion, *Nature Biotechnology* 35(5), 441-3.

Shukla, V. K., Doyon, Y., Miller, J. C., Dekelver, R. C., Moehle, E. A., Worden, S. E., Mitchell, J. C., Arnold, N. L., Gopalan, S., Meng, X., Choi, V. M., Rock, J. M., Wu, Y. Y., Katibah, G. E., Zhifang, G., McCaskill, D., Simpson, M. A., Blakeslee, B., Greenwalt, S. A., Butler, H.

J., Hinkley, S. J., Zhang, L., Rebar, E. J., Gregory, P. D. and Urnov, F. D. (2009). Precise genome modification in the crop species zea mays using zinc-finger nucleases, *Nature* 459(7245), 437–41.

Slaymaker, I. M., Gao, L., Zetsche, B., Scott, D. A., Yan, W. X. and Zhang, F. (2016). Rationally engineered Cas9 nucleases with improved specificity, *Science* 351(6268), 84–8.

Smith, J., Grizot, S., Arnould, S., Duclert, A., Epinat, J. C., Chames, P., Prieto, J., Redondo, P., Blanco, F. J., Bravo, J., Montoya, G., Pâques, F. and Duchateau, P. (2006). A combinatorial approach to create artificial homing endonucleases cleaving chosen sequences, *Nucleic Acids Research* 34(22), e149.

Song, J., Yang, D., Xu, J., Zhu, T., Chen, Y. E. and Zhang, J. (2016). RS-1 enhances CRISPR/Cas9- and TALEN-mediated knock-in efficiency, *Nature Communications* 7, 10548.

Strecker, J., Ladha, A., Gardner, Z., Schmid-Burgk, J. L., Makarova, K. S., Koonin, E. V. and Zhang, F. (2019). RNA-guided DNA insertion with CRISPR-associated transposases, *Science* 365(6448), 48–53.

Sun, S., Zhou, Y., Chen, J., Shi, J., Zhao, H., Zhao, H., Song, W., Zhang, M., Cui, Y., Dong, X., Liu, H., Ma, X., Jiao, Y., Wang, B., Wei, X., Stein, J. C., Glaubitz, J. C., Lu, F., Yu, G., Liang, C., Fengler, K., Li, B., Rafalski, A., Schnable, P. S., Ware, D. H., Buckler, E. S. and Lai, J. (2018). Extensive intraspecific gene order and gene structural variations between Mo17 and other maize genomes, *Nature Genetics* 50(9), 1289–95.

Sun, Y., Zhang, X., Wu, C., He, Y., Ma, Y., Hou, H., Guo, X., Du, W., Zhao, Y. and Xia, L. (2016). Engineering herbicide-resistant rice plants through CRISPR/Cas9-mediated homologous recombination of acetolactate synthase, *Molecular Plant* 9(4), 628–31.

Suzuki, K., Tsunekawa, Y., Hernandez-Benitez, R., Wu, J., Zhu, J., Kim, E. J., Hatanaka, F., Yamamoto, M., Araoka, T., Li, Z., Kurita, M., Hishida, T., Li, M., Aizawa, E., Guo, S., Chen, S., Goebl, A., Soligalla, R. D., Qu, J., Jiang, T., Fu, X., Jafari, M., Esteban, C. R., Berggren, W. T., Lajara, J., Nuñez-Delicado, E., Guillen, P., Campistol, J. M., Matsuzaki, F., Liu, G. H., Magistretti, P., Zhang, K., Callaway, E. M., Zhang, K. and Belmonte, J. C. (2016). In vivo genome editing via CRISPR/Cas9 mediated homology-independent targeted integration, *Nature* 540(7631), 144–9.

Svitashev, S., Schwartz, C., Lenderts, B., Young, J. K. and Mark Cigan, A. (2016). Genome editing in maize directed by CRISPR-Cas9 ribonucleoprotein complexes, *Nature Communications* 7, 13274.

Svitashev, S., Young, J. K., Schwartz, C., Gao, H., Falco, S. C. and Cigan, A. M. (2015). Targeted mutagenesis, precise gene editing, and site-specific gene insertion in maize using Cas9 and guide RNA, *Plant Physiology* 169(2), 931–45.

Tilman, D., Balzer, C., Hill, J. and Befort, B. L. (2011). Global food demand and the sustainable intensification of agriculture, *Proceedings of the National Academy of Sciences of the United States of America* 108(50), 20260–4.

Tran, N. T., Bashir, S., Li, X., Rossius, J., Chu, V. T., Rajewsky, K. and Kühn, R. (2019). Enhancement of precise gene editing by the association of Cas9 with homologous recombination factors, *Frontiers in Genetics* 10, 365.

Urnov, F. D., Miller, J. C., Lee, Y. L., Beausejour, C. M., Rock, J. M., Augustus, S., Jamieson, A. C., Porteus, M. H., Gregory, P. D. and Holmes, M. C. (2005). Highly efficient endogenous human gene correction using designed zinc-finger nucleases, *Nature* 435(7042), 646–51.

Verruto, J., Francis, K., Wang, Y., Low, M. C., Greiner, J., Tacke, S., Kuzminov, F., Lambert, W., McCarren, J., Ajjawi, I., Bauman, N., Kalb, R., Hannum, G. and Moellering, E. R. (2018). Unrestrained markerless trait stacking in *Nannochloropsis gaditana* through combined genome editing and marker recycling technologies, *Proceedings of the National Academy of Sciences of the United States of America* 115(30), E7015-22.

Voss-Fels, K. P., Stahl, A. and Hickey, L. T. (2019). Q&A: modern crop breeding for future food security, *BMC Biology* 17(1), 18.

Wang, J., Meng, X., Hu, X., Sun, T., Li, J., Wang, K. and Yu, H. (2019). xCas9 expands the scope of genome editing with reduced efficiency in rice, *Plant Biotechnology Journal* 17(4), 709-11.

Wang, M., Lu, Y., Botella, J. R., Mao, Y., Hua, K. and Zhu, J. K. (2017). Gene targeting by homology-directed repair in rice using a geminivirus-based CRISPR/Cas9 system, *Molecular Plant* 10(7), 1007-10.

Weinthal, D. M., Taylor, R. A. and Tzfira, T. (2013). Nonhomologous end joining-mediated gene replacement in plant cells, *Plant Physiology* 162(1), 390-400.

Xie, K. and Yang, Y. (2013). RNA-guided genome editing in plants using a CRISPR-Cas system, *Molecular Plant* 6(6), 1975-83.

Yao, X., Wang, X., Hu, X., Liu, Z., Liu, J., Zhou, H., Shen, X., Wei, Y., Huang, Z., Ying, W., Wang, Y., Nie, Y. H., Zhang, C. C., Li, S., Cheng, L., Wang, Q., Wu, Y., Huang, P., Sun, Q., Shi, L. and Yang, H. (2017). Homology-mediated end joining-based targeted integration using CRISPR/Cas9, *Cell Research* 27(6), 801-14.

Yao, X., Wang, X., Liu, J., Shi, L., Huang, P. and Yang, H. (2018). CRISPR/Cas9-mediated targeted integration in vivo using a homology-mediated end joining-based strategy, *Journal of Visualized Experiments* 133(133), e56844.

Yu, Q. H., Wang, B. K., Li, N., Tang, Y. P., Yang, S. B., Yang, T., Xu, J., Guo, C., Yan, P., Wang, Q. and Asmutola, P. (2017). CRISPR/Cas9-induced targeted mutagenesis and gene replacement to generate long-shelf life tomato lines, *Scientific Reports* 7(1), 11874.

Yu, Y., Guo, Y., Tian, Q., Lan, Y., Yeh, H., Zhang, M., Tasan, I., Jain, S. and Zhao, H. (2020). An efficient gene knock-in strategy using 5′-modified double-stranded DNA donors with short homology arms, *Nature Chemical Biology*16(10), 387-90. doi: 10.1038/s41589-020-0470-8.

Zaidi, S. S., Vanderschuren, H., Qaim, M., Mahfouz, M. M., Kohli, A., Mansoor, S. and Tester, M. (2019). New plant breeding technologies for food security, *Science* 363(6434), 1390-1.

Zhang, F., Maeder, M. L., Unger-Wallace, E., Hoshaw, J. P., Reyon, D., Christian, M., Li, X., Pierick, C. J., Dobbs, D., Peterson, T., Joung, J. K. and Voytas, D. F. (2010). High frequency targeted mutagenesis in *Arabidopsis thaliana* using zinc finger nucleases, *Proceedings of the National Academy of Sciences of the United States of America* 107(26), 12028-33.

Zhang, H., Gou, F., Zhang, J., Liu, W., Li, Q., Mao, Y., Botella, J. R. and Zhu, J. K. (2016). TALEN-mediated targeted mutagenesis produces a large variety of heritable mutations in rice, *Plant Biotechnology Journal* 14(1), 186-94.

Zhang, H., Zhang, J., Lang, Z., Botella, J. R. and Zhu, J. K. (2017). Genome editing–principles and applications for functional genomics research and crop improvement, *Critical Reviews in Plant Sciences* 37, 102-12.

Zhang, Y., Massel, K., Godwin, I. D. and Gao, C. (2018). Applications and potential of genome editing in crop improvement, *Genome Biology* 19(1), 210.

Zhang, Y., Malzahn, A. A., Sretenovic, S. and Qi, Y. (2019). The emerging and uncultivated potential of CRISPR technology in plant science, *Nature Plants* 5(8), 778-94.

Zhao, Q., Feng, Q., Lu, H., Li, Y., Wang, A., Tian, Q., Zhan, Q., Lu, Y., Zhang, L., Huang, T., Wang, Y., Fan, D., Zhao, Y., Wang, Z., Zhou, C., Chen, J., Zhu, C., Li, W., Weng, Q., Xu,

Q., Wang, Z. X., Wei, X., Han, B. and Huang, X. (2018). Pan-genome analysis highlights the extent of genomic variation in cultivated and wild rice, *Nature Genetics* 50(2), 278–84.

Zhong, Z., Sretenovic, S., Ren, Q., Yang, L., Bao, Y., Qi, C., Yuan, M., He, Y., Liu, S., Liu, X., Wang, J., Huang, L., Wang, Y., Baby, D., Wang, D., Zhang, T., Qi, Y. and Zhang, Y. (2019). Improving plant genome editing with high-fidelity xCas9 and non-canonical PAM-targeting Cas9-NG, *Molecular Plant* 12(7), 1027–36.

Zhou, H., Liu, B., Weeks, D. P., Spalding, M. H. and Yang, B. (2014). Large chromosomal deletions and heritable small genetic changes induced by CRISPR/Cas9 in rice, *Nucleic Acids Research* 42(17), 10903–14.

Zong, Y., Song, Q., Li, C., Jin, S., Zhang, D., Wang, Y., Qiu, J. L. and Gao, C. (2018). Efficient c-to-t base editing in plants using a fusion of ncas9 and human apobec3a, *Nature Biotechnology* 36, 950–3.

Zong, Y., Wang, Y., Li, C., Zhang, R., Chen, K., Ran, Y., Qiu, J. L., Wang, D. and Gao, C. (2017). Precise base editing in rice, wheat and maize with a Cas9-cytidine deaminase fusion, *Nature Biotechnology* 35(5), 438–40.

Chapter 4

Viruses as vectors for the delivery of gene-editing reagents

Evan E. Ellison, James C. Chamness and Daniel F. Voytas, University of Minnesota, USA

1 Introduction

Viruses are natural vectors for nucleic acids, and both DNA and RNA plant viruses have been engineered to extend or replace conventional vectors for delivery of gene-editing reagents. This chapter reviews aspects of viral biology essential for engineering vectors, highlight landmark studies using viruses to overcome traditional limitations in gene editing, and outline important considerations for the use of viral vectors in new systems or for new targets.

Motivated by fundamental differences in both their infection modes and utility as vectors, DNA and RNA viruses are treated separately. DNA viruses are assessed as replicon vectors for efficient gene editing through homology-directed repair (HDR). The chapter then reviews RNA viruses as mobile vectors for gene-editing reagent delivery. The chapter includes key case studies as well as future trends in research.

http://dx.doi.org/10.19103/AS.2020.0082.28

2 DNA viruses: replicon vectors for efficient gene editing through homology-directed repair (HDR)

Since the first reports over 25 years ago, various plant DNA viruses have been sequenced, cloned, and modified to introduce exogenous sequences into plant cells. More recently, delivery of sequences encoding components for the major classes of site-specific nucleases (SSNs) has demonstrated the utility of DNA viral vectors for gene editing in a broad range of host species. Among both foundational studies in viral engineering and contemporary studies in gene editing, the vast majority describe work with *Geminiviridae*: single-stranded DNA (ssDNA) viruses that constitute the largest known family of DNA viruses in plants[1].

Engineered geminiviral clones may be delivered to plant cells by biolistics or, more commonly, by agroinfection. While some infectious clones retain mobility within the delivered tissue (and potentially to natural insect vectors)[2], most reports describe clones deconstructed into non-infectious, geminiviral replicons (GVRs): self-replicating, cell-autonomous DNA vectors, which transiently drive their genetic payload to high-copy number and overexpression in host cells, regardless of whether the source vectors integrate into the nuclear genome. Investigation of GVRs as overexpression vectors predates the widespread availability of gene-editing tools, but their utility for editing was quickly realized. In particular, GVRs offer key advantages for precision-editing strategies requiring homology-directed repair (HDR) from a donor template molecule to incorporate specific modifications into the genome.

When used to deliver SSNs and donor templates, GVRs dramatically increase the frequency of HDR in host cells[3]. Intuitively, stronger expression of SSNs may lead to more efficient double-stranded break induction. GVRs also increase the local concentration of donor template molecules available for recombination, and pleiotropic effects of viral replication proteins render the cell environment more amenable to HDR. In the following sections, we describe the basic structure and function of GVRs, consider vector design choices influencing their activity within plant cells, and review case studies using GVRs to drive efficient HDR in several species of both dicots and monocots.

3 Geminiviral replicons (GVRs): deconstructed geminiviruses that serve as replicating vectors in plants

3.1 Geminiviruses

Geminiviridae are a family of circular, ssDNA viruses, vectored by insects and infecting diverse host plants. The name derives from the twin-icosahedral shape of the encapsulated virion. The viral ssDNA genome is converted by host cell polymerases into a double-stranded DNA (dsDNA) intermediate, which serves as a template for rolling-circle replication and for transcription of viral gene products[1]. With small genomes, encoding just a few open-reading frames (ORFs), geminiviruses require host cell machinery to complete their life cycle; this reliance underscores complex interactions between viral and host factors[4], which are important to consider for the function of geminiviral vectors in plant cells. Geminiviruses encompass at least nine distinct taxa. Historically demarcated by host range or by variations in genome structure[1], more complex phylogenetic relationships are suggested due to viruses identified through high-throughput metagenomics[5,6], or isolates that challenge traditional structure-based classifications[7].

Despite the increasing availability of novel sequence isolates, the best-studied taxa remain *Begomoviruses* and *Mastreviruses*. The attention paid to these genera can be attributed to their status as significant pathogens of agricultural crops, including cereals, beans, cassava, and tomato. Most studies of the molecular biology of geminiviruses, and of their use as replicon vectors, have focused on three species of mastreviruses: *Maize Streak Virus* (MSV) and *Wheat Dwarf Virus* (WDV), infecting monocots (as do most mastreviruses), and *Bean Yellow Dwarf Virus* (BeYDV), from an outgroup infecting dicots. Begomoviruses may have monopartite or bipartite genomes; mastreviruses, with rare exceptions,[8] are monopartite. As nearly all GVRs used for gene editing are derived from WDV and BeYDV, this chapter focuses on mastreviruses.

3.2 Mastrevirus genome structure, function, and replication

The canonical mastrevirus genome is 2.6–2.8kb[1] (Fig. 1a). Two structural sequences termed the short (SIR) and long (LIR) intergenic regions separate reading frames in both orientations of the dsDNA intermediate, on the virion-sense (v-sense; +) and complementary-sense (c-sense; -) strands. A c-sense DNA primer molecule within the SIR, unique to mastreviruses, is thought to function as the origin for the first stage of viral replication: conversion of ssDNA to dsDNA by synthesis of the complementary strand. The origin for rolling-circle replication, conserved across nearly all geminiviruses, is a nonanucleotide stem loop sequence within the LIR that is nicked by the replication protein, Rep[9]. The LIR and SIR function as a bidirectional promoter and terminator, respectively, for the transcripts from both strands.

On the c-sense strand, alternative splicing of a 3′ intron gives rise to two protein-coding transcripts with overlapping reading frames: the unspliced 5′ transcript (RepA) and spliced transcript (Rep). Despite sharing much of their N-terminal sequence, studies in both monocot- and dicot-infecting mastrevirus species implicate these proteins in distinct and conserved functional roles. C-terminal deletion and splice-site mutants indicate the spliced transcript and minor product, Rep, is both necessary and sufficient for replication in host

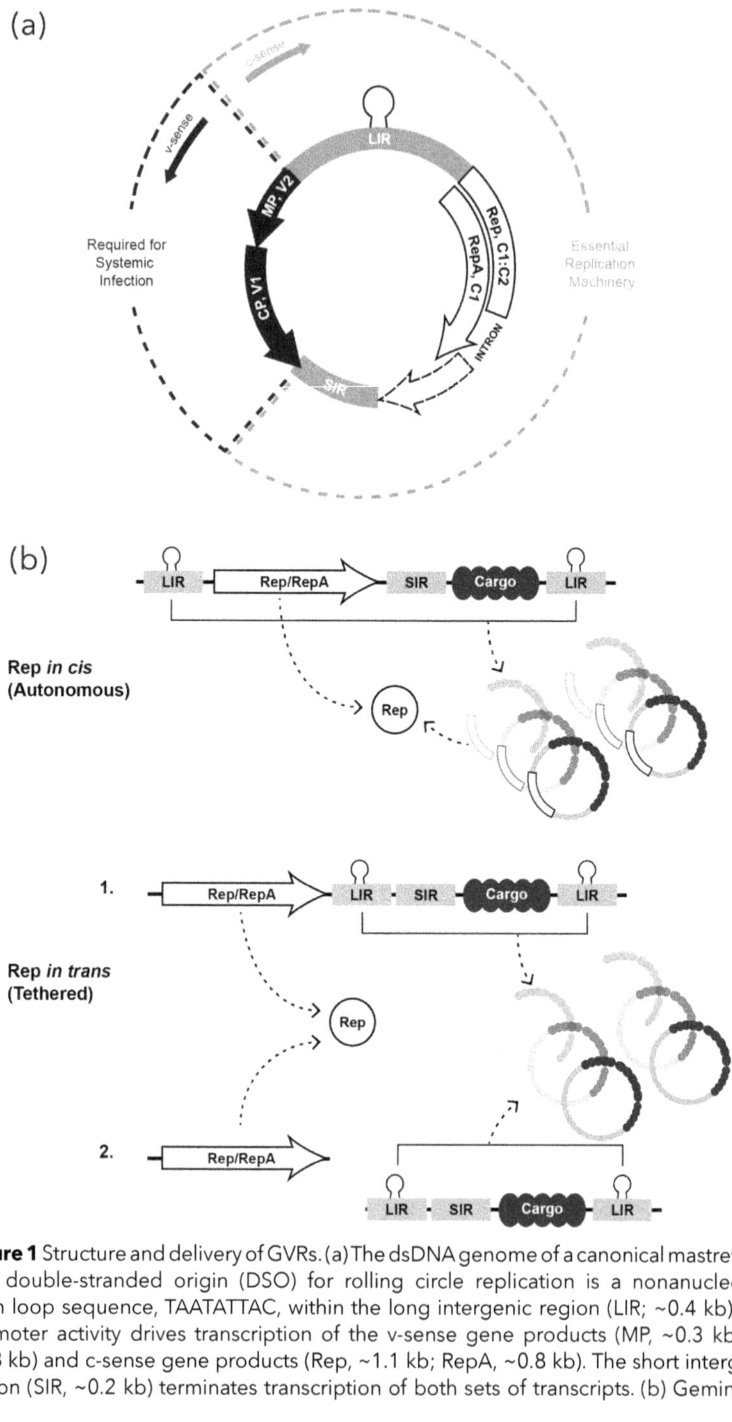

Figure 1 Structure and delivery of GVRs. (a) The dsDNA genome of a canonical mastrevirus. The double-stranded origin (DSO) for rolling circle replication is a nonanucleotide stem loop sequence, TAATATTAC, within the long intergenic region (LIR; ~0.4 kb). LIR-promoter activity drives transcription of the v-sense gene products (MP, ~0.3 kb; CP, ~0.8 kb) and c-sense gene products (Rep, ~1.1 kb; RepA, ~0.8 kb). The short intergenic region (SIR, ~0.2 kb) terminates transcription of both sets of transcripts. (b) Geminiviral

cells[10-12]. While not strictly required for replication, the unspliced transcript and major product, RepA, likely contributes to regulation of replication through interactions with both host and viral factors. First, RepA interacts with the host cell retinoblastoma-related protein, a highly conserved regulator of the cell cycle. By driving mitotically inactive somatic cells into a pseudo S-phase, mastreviruses upregulate the expression of host factors required for viral replication[4]. Second, RepA appears to be an obligate trans-activator for transcription of v-sense gene products. By supporting production of the capsid protein to encapsulate newly synthesized genomes (thus sequestering them from further replication), RepA may contribute to the downregulation of genome replication in live viruses[13]. Additional interactions between Rep and RepA, and their mechanistic roles in replication, have been the subject of considerable study, but are beyond the scope of this review. For an in-depth treatment of geminiviral replication, see Gutierrez[9,14,15].

On the v-sense strand, sequential reading frames encode the viral movement protein (MP) and capsid or coat protein (CP). These proteins are critical for infection. In addition to directing nuclear import of the virion to begin the replication cycle, the virion capsid is required for insect transmission and efficient systemic movement. The movement protein, which localizes to the cell periphery, mediates interactions between plasmodesmata and the virion capsid to facilitate movement among plant host cells[13]. In one classic example of maize agroinoculation with MSV clones, deletions of the CP and MP, or CP alone, abolished systemic movement, insect transmission, and typical disease symptoms versus a wild-type control. Cell-autonomous replication of these deletion vectors, however, was still detectable[2].

3.3 Deconstructing geminiviruses into replicon vectors

Experiments with mutants to elucidate wild-type gene function were soon followed by reports describing the addition of exogenous sequences, typically by replacing the v-sense gene products with reporter or marker gene cassettes. By maintaining components essential for viral replication (the LIR, SIR, and replication proteins) and eliminating components required for transmission

Figure 1 (*Continued*)
replicons (GVRs) can be delivered to cells in different architectures. When Rep is in its native position *in cis* to the replicon, driven by c-sense promoter activity of the LIR, it can be expressed from both the source vector and the high-copy replicon (hundreds to thousands of dsDNA intermediates per cell), creating a positive feedback loop for replication. When placed *in trans* to the linear LSL sector of the source vector, continued Rep expression depends upon persistent expression (i.e. from stable integration) of the source vector.

and infection (MP, CP), these clones function as cell-autonomous replicons. To date, replicons have been derived from mastreviruses including WDV[16,17], MSV[18,19], and BeYDV[20,21]. Although not a focus of this review, it should be noted that replicons have also been derived from begomoviruses including *Tomato Golden Mosaic Virus* (TGMV)[22,23], *Bean Dwarf Mosaic Virus* (BDMV)[24], *Tomato Leaf Curl Virus* (ToLCV)[25,26], and *Cabbage Leaf Curl Virus* (CaLCuV)[3]. In the context of plant viral expression vectors, replicons represent a transition from a 'full virus' strategy to a 'deconstructed virus' strategy[27], which has become the archetype for DNA viral expression vectors.

In moving away from a wild-type viral architecture, synthetic clones open up new design possibilities (Fig. 1b). First, natural mastrevirus genomes (~2.7 kb) are tightly size-constrained by the need to fit within the capsid, and for the virions to traverse plasmodesmata for systemic movement within the host. Removal of the MP and CP sequences alone (~1 kb) reduces the size of the replicon, but removal of requirements for encapsulation and cell-cell movement effectively shifts the upper bound for replicon size to the limit imposed by processive replication. This has the effect of greatly increasing vector 'payload' capacity versus infectious, full-virus clones. While no studies have systematically explored an upper limit to replicon size, evidence suggests that the efficiency of replication (and thus copy number) decreases with the replicon size[28]. In our own group, we have transformed functional BeYDV and WDV replicon vectors of approximately 15 kb. Finally, even independent of encapsulation, the size of plasmodesmata exerts a significant selective pressure on viral genome size, such that cell-cell mobility of larger plasmids is compromised[24].

Another design choice in the 'deconstructed' strategy is the placement of the replication proteins. In the native virus, both ORFs are driven by the c-sense promoter of the LIR, and so are expressed *in cis* to the replicon. However, these proteins can also be supplied *in trans*, by placing the ORFs under a heterologous promoter on the source vector or a second vector. This separation provides additional degrees of freedom for control over replicon activity, by 'tethering' expression of the replication protein to the source vector. For example, replication may be constrained by a chemically inducible promoter,[21] or by obligate co-delivery of two vectors.[3] Finally, as noted earlier, only Rep is strictly required for replication. While some synthetic clones purposed solely for overexpression eschew RepA,[21] the pleiotropic benefits for replication and HDR in somatic cells conferred by RepA have motivated its inclusion in all reports of gene-editing vectors to date.

3.4 Delivery and expression of replicons in plant cells

Replication-competent geminiviral clones can be delivered into plant cells as plasmids or as linear dsDNA fragments by traditional means of transformation:

PEG-treated protoplasts[29], biolistics[20], or agroinfection. In early reports describing agroinfection of begomovirus[30] and mastrevirus[31,32] clones, source vectors included full tandem repeats of the viral genome. Such repeats might enable escape and replication of a unit-length viral genome from the source vector by either of two, non-exclusive mechanisms: homologous recombination between repeats or by a replicative intermediate copied from the dsDNA source vector. It was later shown that duplication of just the v-sense ORI within the LIR is sufficient, implicating a rolling-circle replication mediated 'replicational release' mechanism[33] for viral escape. Subsequent reports largely follow a LIR-SIR-LIR (LSL) architecture for delivery of viral genomes from a dsDNA source vector. A benefit of this architecture and delivery mechanism is that the population of replicating viral genomes is both homogenous and predictable, consisting of the sequence between the origins within the duplicated LIRs. By contrast, circular viral replicons produced by homologous recombination, while observed, are far more likely to be heterogenous populations. Importantly, recombination of T-DNA source vectors may be stimulated by Rep-mediated viral replication, whether due to the availability of replicated sequences, or to potential crosstalk between replication and recombination machinery[34].

Upon delivery to plant cells, replicon copy number and cargo expression typically peak within a few days to a week after infection[3,35]. While this transient activity does not strictly require stable integration of the source vector (especially for autonomous replicons: see Fig. 1b), integration can lead to persistent replicon activity over time through mitotic and meiotic cell lineages. Studies have detected circular and high-copy replicons in the somatic tissues of T_0 rice and tomato plants regenerated through tissue culture[28,36], and there is evidence for persistence of a BeYDV replicon among the transgenic progeny of *Arabidopsis* lines as far as the T_3 generation[37].

4 Replicon vectors for efficient homology-directed repair

Precise gene targeting (GT), encompassing site-specific sequence alterations and targeted insertions, is typically achieved by HDR with a donor template molecule. Since induction of double-stranded breaks (DSBs) greatly increases the frequency of HDR in plant cells[38], SSNs including zinc-finger nucleases (ZFNs)[39], TAL-effector nucleases (TALENs)[40], and CRISPR/Cas nucleases have been used to stimulate HDR and recover stable GT events. However, the most common outcome from SSN-mediated DSB induction is mutagenesis, whereas GT events occur typically an order of magnitude less frequently. This is attributed to the predominance of alternative DNA repair pathways, namely microhomology-mediated end joining and non-homologous end joining, and to the additional requirement for HDR that DSB induction be coordinated with availability of the donor molecule.

GVRs contribute to higher frequencies of HDR in several ways. Perhaps the most significant is by greatly increasing availability of the donor molecule. With traditional T-DNA vectors, at most several copies of the donor are delivered to the cell. Biolistics can deliver a greater amount of donor, but the likelihood of multi-copy, random integration makes this approach less appealing for applications in breeding where segregation of integrated reagents is likely required. GVRs can be delivered to cells on T-DNA vectors, but increase the copy number of the donor molecule to hundreds or thousands within a single cell nucleus. Additionally, the activity of the replication proteins contributes to a more favorable cell environment for HDR. The prevalence of DNA repair pathways is highly influenced by the cell state, and HDR is particularly low in the mitotically inactive somatic cells which are primary targets for regeneration in many species. Even in the absence of DSB induction or a synthetic repair template, geminiviral infection was observed to enhance somatic HDR in a transgenic reporter line of *Arabidopsis*[41].

Besides the known influence of RepA on cell state, the replication proteins may also have evolved a role to promote recombination, due to an evolutionary selective pressure to generate new viral variants. For example, Rep and/or RepA may mediate recombination by their affinity for host factors. This could take the form of crosstalk between replication and recombination machinery[34], or even of more direct recruitment: one report from the begomovirus, *Mungbean yellow mosaic virus* (MYMIV), identified a protein-protein interaction between the MYMIV Rep and *Arabidopsis* RAD51, a highly conserved protein mediating homologous recombination in eukaryotes. While no such direct interactions have been described for mastreviruses, it is conceivable that analogous links between viral replication proteins and host recombination factors contribute to increasing the recombination potential of sequences carried on GVRs.

Regardless of mechanism, GVR vectors typically increase HDR frequencies by an order of magnitude. In the following case studies, we will highlight notable results and conclusions from reports establishing GVR-based editing strategies in dicots and monocots.

5 Case studies: GVR-based editing strategies in dicots and monocots

5.1 Baltes et al. (2014): DNA replicons for plant genome engineering[3]

Baltes et al. (2014) published the first report demonstrating the utility of GVRs for HDR. They used a previously established assay, leveraging a broken *nptII::gus* fusion gene present in a transgenic tobacco reporter line[42]. A ZFN pair targeting the middle of the broken fusion gene stimulates DSB induction, and templated repair completes the 5′ *nptII* and 3′ *gus* sequences, restoring the function of both markers. Positive GUS staining in agroinfiltrated leaves from the reporter line thus indicates HDR, and the density of positive cells within sectors of an infiltrated leaf provides a proxy for gene-targeting frequency. Using this assay, Baltes and colleagues found that the GVR system outperformed the standard T-DNA by as much as two orders of magnitude. They also tested what specific benefit might be attributed to enhanced SSN expression (putatively, via greater DSB induction), or to replication of the repair template, and found that gene-targeting frequencies were boosted only when the repair template was replicated. Removing the ZFNs entirely attenuated GUS staining, but placement of the ZFN pair inside or outside the LSL segment of the same T-DNA made no significant difference. Taken together, these results suggest DSB induction is essential but not rate-limiting to higher gene-targeting frequencies versus repair template copy number. Next, possible contributions from pleiotropic activity of the replication proteins were assessed. The gene-targeting frequency in a non-replicating T-DNA control was significantly increased by co-delivery of a plasmid expressing Rep and RepA, consistent with a role for these proteins in enhancing HDR independent from GVR replication. Finally, using kanamycin selection to identify cells in which the *nptII* marker was repaired through HDR, Baltes and colleagues were able to regenerate tobacco plants with sequence-confirmed HDR events, demonstrating the utility of the GVR-based approach to generate heritable events.

5.2 Cermak et al.[26]: High-frequency, precision modification of the tomato genome

Cermak et al.[26] used a similar strategy to recover edited tomato plants, targeting knock-in of a selectable marker. Unlike Baltes et al., however, they targeted an endogenous locus: *ANT1*, a transcription factor controlling anthocyanin accumulation. The purple pigmentation arising from *ANT1* overexpression provides a phenotype that can be scored non-destructively, unlike GUS, and is thus useful for selection in vivo. To create an HDR reporter, Cermak and colleagues targeted TALENs or CRISPR/Cas9 reagents to the *ANT1* promoter. Their repair template contained, in sequence, a *nptII* cassette, and an additional strong promoter to drive overexpression of the native *ANT1* gene. Vectors were delivered to tomato cotyledons by agroinfection. The fraction of cotyledons showing purple spots, normalized for transformation efficiency, provided an estimate for gene-targeting efficiency. The best treatment among their experiments was a single-vector GVR with Cas9 and sgRNA, which achieved a normalized gene-targeting frequency of 11.66%. This was an order of magnitude better than the conventional T-DNA treatment at 1.27%. Importantly, these experiments were carried out using kanamycin selection in the growth media; when cultured on non-selective media, no purple spots were observed, even with the best GVR treatment. While kanamycin was used in this initial phase of tissue culture to recover growing purple calli, visual selection alone proved sufficient to then regenerate plants with gene-targeting events; over two dozen plants were regenerated from several purple calli in the absence of kanamycin. Subsequent molecular analyses revealed events including perfect repair at both junctions, and one-sided targeting events that still gave rise to the phenotype. Most surprisingly, both PCR and Southern assays failed to detect the presence of the T-DNA or extrachromosomal replicon among any of the regenerated plants. This suggested that transient expression of the gene-editing reagents was sufficient, and nuclear integration was not required, to recover events. This is a notable result for some breeding systems where transgenesis may be undesirable, either for regulatory concerns, or due to difficulty in segregating integrated reagents from the desired edit (e.g. in clonally propagated species). Finally, Cermak and colleagues showed their gene-targeting events were heritable. T_1 progeny scored visually and genotyped by PCR approximated Mendelian expectation. A fitness penalty incurred from anthocyanin overproduction might explain the slight segregation bias against the edited allele, and might also have contributed to their failure to recover purple growths from transformed cotyledons in the absence of kanamycin selection.

5.3 *Dahan-Meier et al.[36]: Efficient* in planta *gene targeting in tomato using geminiviral replicons and the CRISPR/Cas9 system*

Both of the previous studies used HDR to activate chemical and phenotypic selectable markers, a requirement which limits utility in many breeding applications. Dahan-Meier et al.[36] developed a new strategy, demonstrating efficient and heritable gene targeting for a 'neutral' trait in tomato. Instead of selecting for their targeted knock-in, they created transgenic plants that carried a T-DNA with the editing reagents, combining an *in planta* approach to gene targeting[43] with a GVR vector. Their target was a mutant allele of the *CRTISO* locus, which encodes an enzyme in the carotenoid biosynthesis pathway. Whereas wild-type *CRTISO* plants have red fruits, their target line featured a 281 bp deletion in *CRTISO*. This recessive, loss-of-function mutation gives rise to orange fruits in homozygotes, which can be restored to red fruit through HDR-mediated correction of a single allele. While providing a scorable phenotype, the fruit color trait is only visible in mature plants; this contrasts to the anthocyanin marker used in Cermak et al., which permitted enrichment of events during the primary tissue culture step. The experimental T-DNA contained a *nptII* marker for transgenic selection, Cas9 and a sgRNA targeting the mutant allele, and a BeYDV GVR (with Rep *in trans*) to amplify the donor sequence. Among 36 regenerated T_0 plants, nine had entirely red fruit. Molecular evidence from junction PCR and Southern blots confirmed HDR in sampled leaf tissue. Of these nine T_0, eight fertile plants showed Mendelian segregation for the fruit color phenotype among their T_1 progeny, demonstrating heritability of these gene-targeting events. In contrast to the GVR treatment, only one, chimeric plant, with both red and orange fruits, was recovered among the 12 T_0 plants regenerated with a control construct that lacked the Rep cassette (thus disabling GVR replication). T_1 progeny from these red fruits segregated for the fruit color trait, indicative of a heterozygous knock-in, whereas no red fruits came from the orange T_0 fruits from the same chimeric plant.

The above results underscore a crucial point: in order to recover heritable gene-targeting events, the edit must occur in a germline cell, or germinal cell linage. The targeted *ANT1* tomato lines in the Cermak study were edited soon after transformation, as indicated by the purple phenotype already visible at the callus stage. In the Dahan-Meier study, the low rate of chimerism evidenced by uniform fruit color and Mendelian segregation of T_1 progeny in the replicon treatment similarly indicated that most of these events occurred early in development. By contrast, Shan et al.[37] failed to recover heritable gene-targeting events in *Arabidopsis* using a BeYDV GVR, despite a high frequency of somatic editing among transformed plants. In the previous studies, the transgenic tobacco or tomato plants were regenerated through tissue culture of somatic explants, involving organogenesis and establishment of a new germline. In *Arabidopsis*, the floral dip protocols used directly transformed existing germline cells. It is possible that established germline cells restrict viral activity more stringently than somatic cells. Alternatively, it may be that additional unknown barriers prevent efficient GVR-mediated HDR in *Arabidopsis*; another report failed to observe even efficient somatic gene targeting, again using a BeYDV replicon in Arabidopsis[44]. However, these results stand in contrast to the majority of published studies in other dicot species.

5.4 Gil-Humanes et al.[35] and Wang et al.[28]: Gene targeting in cereals with WDV

Soon after the first gene-editing reports with BeYDV, two studies described knock-ins in cereals using WDV. Gil-Humanes et al.[35] targeted fluorescent reporters to the wheat genome. Cas9 and a sgRNA targeted the endogenous ubiquitin locus, while a repair template included the coding sequence for GFP behind a T2A self-cleaving peptide linker. HDR of the break site placed the GFP gene in-frame with the native ubiquitin gene, leading to reporter expression driven by the native promoter. In protoplast transfections, green fluorescing cells were identified only in treatments placing all of the editing reagents on the WDV replicon; in the best case, with a strong promoter for Cas9, 3.8% of cells underwent the targeting event. Additional protoplast experiments demonstrated multiplexed gene targeting using a BFP tag at a second locus, and molecularly confirmed targeting at all three of the wheat ubiquitin homeoalleles. Despite these promising results, Gil-Humanes and colleagues were unable to recover plants with the gene-targeting event. They did observe reporter positives in biolistic transfection of embryonic scutella, a typical system for tissue culture of wheat. However, these cells could not be regenerated even under chemical selection for the biolistic vector or for an alternative knock-in of the *bar* selectable marker gene (Gil Humanes et al., unpublished).

Wang et al.[28] described gene targeting in rice using selectable and screenable markers. Similar to Gil-Humanes et al., they used Cas9 and a sgRNA to knock in markers behind the promoter of either of two highly expressed native loci. In this case, a *nptII::GFP* fusion gene provided means to select for HDR events in agroinfected-rice calli. Geneticin-resistant plantlets were recovered through tissue culture, and screened by PCR for correct repair at the repair template junctions. Two different rice lines were transformed with test constructs: a transgenic line constitutively expressing Cas9, and a wild-type line, for which Cas9 was included on the T-DNA vector but outside the replicon. Double-junction PCR positives were recovered among the T_0 plants of both lines transformed with WDV GVRs, at frequencies ranging from 4.7% (in the wild type line) to 19.4% (in the transgenic line). Positives were also recovered from the non-GVR control treatment in the Cas9 line at 6.8%. No non-GVR control treatment was reported for the wild-type line. T_1 progeny or other analyses demonstrating germinal transmission were also not reported.

Together, these results show that GVRs can enhance gene targeting in monocots, but do not yet present a convincing case that WDV vectors are simple to use for the recovery of targeted plants. Our own group and several collaborators have been unable to regenerate rice plants with knock-ins using WDV, even while targeting several different loci and applying selection for the T-DNA. We suspect Wang et al. succeeded in recovering transgenic plantlets largely by virtue of having imposed selection for the HDR event. The absence of additional reports reporting success suggests continued research is required to generalize the GVR strategy to cereals and other monocots.

6 GVR-based editing strategies: summary and future trends

In case studies described above and several additional reports, GVRs have been used to mediate HDR across species including tobacco, tomato, potato, cassava, wheat, and rice[3,26,28,35-37,45,46]. High-frequency, germline edits have been achieved in dicots with BeYDV, whereas WDV has shown promise for HDR in rice and in wheat cell culture. For these and other plant systems, additional research will be required to optimize GVR-based editing strategies. For example, the current BeYDV-and WDV-derived replicons may not function in all target crops: GVRs from other viruses have failed to replicate in some species, likely indicative of virus-specific host range requirements[25]. In monocots, more work is needed to expand the utility of GVRs within extant tissue culture regimes to enable heritable edits to be recovered. These challenges may be addressed by leveraging increasing viral sequence resources made available through metagenomics[5], or by more advanced engineering strategies for the delivery and control of GVRs.

7 RNA viruses: mobile vectors broadly applicable for gene-editing reagent delivery

RNA viral vectors are rapidly emerging as effective means for delivering genome engineering reagents. RNA viruses have small genomes with only a few essential genes required for replication, movement, and infection[47-50]. These viruses have been used as vectors for protein production, viral-induced gene silencing, and more recently, gene editing[51-53]. The earliest reports of their use for gene editing involved using them to express a ZFN to edit an integrated reporter[54]. Vectors then quickly adopted the CRISPR/Cas9 technology by expressing sgRNAs for editing in a transgenic plant that expresses Cas9[51]. This was first demonstrated in *Nicotiana benthamiana* and then expanded into several other species[55,56]. More recently, the technology was further improved to enable high frequencies of heritable editing in *N. benthamiana*[57]. Future work is required to expand heritable editing to other species and to develop vectors that no longer require transgenic plants for gene editing. Further advancement of the technology could allow transgene and tissue-culture-free gene editing of a variety of plant species. This would be a major advance for plant gene editing.

RNA viruses, or viruses that do not have a DNA intermediate in their replication cycle, have many advantageous properties for use as vectors, including compact genomes, systemic plant movement, non-host integrating replication cycle, and well-studied interactions with host defense systems. Regardless of whether RNA viral genomes are mono or multipartite, they are relatively simple, containing only a few coding regions essential to their replication, movement, and infection (Fig. 2). Replication of the viral genome and expression of sub-genomic RNAs, which are translated into proteins, is dependent on the RNA-dependent RNA polymerase (RdRP)[50,58]. Cell-to-cell and systemic movement of RNA viruses relies on the movement protein (MP). RNA-viral MPs guide the viral genome to the plasmodesmata[47] and often

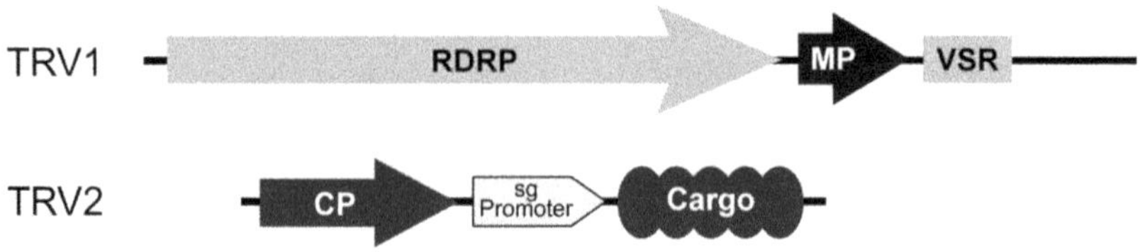

Figure 2 Tobacco Rattle Virus (TRV) vector. TRV is a bipartite positive-sense single-stranded RNA virus. TRV1 contains the RNA-dependent RNA polymerase (RdRP) which replicates both TRV1 and TRV2 and expresses subgenomic RNAs of other coding regions. TRV1 also encodes the movement protein (MP), responsible for enabling cell-to-cell movement, and the viral suppressor of RNA silencing (VSR), responsible for suppression of the host immune response. TRV2 contains the coat protein (CP) and can be modified for expression of heterologous sequences. The native TRV2 contains coding sequences to enable transmission between plants, which are removed in the vector. Heterologous sequences are expressed from a subgenomic promoter, either duplicated from the same virus or from a closely related virus.

increase the plasmodesmata size exclusion limit[48], permitting the full viral genome to pass from cell-to-cell. Systemic infection occurs by transport of the virus through the phloem, with unloading of the virus at distal sites, leading to systemic infection[49]. Some viruses require encapsidation, or coating of the viral genome with a viral encoded coat protein (CP), for systemic infection[59]. Encapsidation is also an important factor in vector transmission between plants by insects. Different plant RNA viruses may encode other factors involved in infection, transmission between plants, or other host interactions such as combating host defense.

The well-studied genomes of several RNA viruses make them ideal for engineering and repurposing for use as vectors. Indeed, shortly after some of the first infectious cDNA clones of RNA viruses were created, efforts were undertaken to use them as vectors for expression of heterologous proteins[53,60]. Early approaches involved expression of reporter genes[61], but vectors were quickly adapted for the expression of proteins of value for pharmaceutical[62] or agricultural applications[63].

7.1 Modifications of RNA viruses into RNA viral vectors

One of the first steps in creating a plant RNA viral vector is to generate a cDNA of the viral transcript and adapt it for compatibility with various molecular cloning techniques. The cDNA vectors are often placed behind standard plant promoters. When transferred to plant cells, a transcript is produced that serves as the RNA virus genome; the transcript is replicated and enables systemic infection throughout the plant. A second modification commonly used with viral vectors is to incorporate them into a backbone that enables replication in *Agrobacterium* and T-DNA transfer into plant cells. Finally, the viral genome itself is modified to express heterologous sequences (Fig. 2). This can involve viral gene replacement or gene insertion behind a subgenomic promoter[53]. For further reading on the construction of plant viral vectors we recommend Pasin et al.[64].

7.2 RNA viral vectors for virus-induced gene silencing

Perhaps one of the best-known applications of RNA viral vectors is for viral-induced gene silencing (VIGS). VIGS functions by taking advantage of the native plant post-transcriptional RNA silencing (PTGS) mechanism. By placing a small fragment of a gene of interest into an RNA viral vector, host PTGS, in an attempt to silence the virus, will also silence the native gene[52,65]. This has proven a valuable approach for reverse genetics to better understand gene function, and VIGS vectors have been developed for a large number of plant species[66]. A disadvantage of viral vectors for heterologous protein production or VIGS is that the effects of the assay are transient. Approaches to use viral vectors to create permanent genetic changes are outlined in the following sections.

8 Case studies: use viral vectors to create permanent genetic changes

8.1 Marton et al. 2010: Nontransgenic genome modification in plant cells[54]

With the advent of SSNs, RNA viral vectors were soon tested to determine if they could be used to express SSNs to create genetic modifications in infected plant tissues. One of the earliest examples of this is the use of Tobacco Rattle Virus (TRV) to express ZFNs for plant gene editing[54]. Traditionally, genome-editing reagents are delivered to plant cells through stable transformation, wherein the SSN is introduced into plant cells by *Agrobacterium* or biolistics. Transgenic cells are then regenerated into whole plants through tissue culture, and progeny are screened for the presence of the targeted genome modification. Marton and colleagues described a method for non-transgenic gene editing of plant cells using a TRV expression system. TRV is a bipartite virus with two RNAs: TRV1 and TRV2. TRV2 was modified to contain the Pea Early Browning Virus (PEBV) coat protein subgenomic promoter upstream of a multiple cloning site. This expression system was tested by first adding fluorescent reporter genes into the multiple cloning site. When plants were infected with these viruses, expression of the report (i.e., fluorescence) was observed in developing tissues throughout the plant. This indicated that TRV2 could be used to carry and express cargo in infected and growing portions of the plant. Next a ZFN monomer was added to TRV2 that targets a defective *gus* reporter integrated into the genomes of *N. benthamiana* and petunia. Mutagenesis by the ZFN should create a functional ***GUS*** gene. ***GUS*** expression was observed in infected tissue in both species, indicating ZFN-mediated editing of the reporter. Finally, *N. benthamiana* and petunia tissue infected with TRV2 expressing the ZFN was placed onto tissue-culture medium to regenerate plants from the infected tissue. ***GUS*** activity was observed in the regenerated tissue, and the edited target was stably inherited. The seedlings were also confirmed to be free of virus, verifying the method as a non-integrating delivery means of generating stably maintained, site-specific gene edits.

A longstanding goal for the use of RNA viral vectors for gene editing is to infect and edit endogenous genes in wild-type plants. This initial report of gene editing using RNA viruses was significant because it demonstrated editing in a plant by an RNA viral vector that expresses all the reagents required for gene editing. This was only made possible due to the relatively small size of a ZFN monomer. ZFN dimers are typically required to edit genes; however, the *gus* reporter had been engineered to carry two recognition sites for the ZFN monomer, thereby enabling gene editing by the monomer. Editing of endogenous genes is prohibited by the requirement of ZFN dimers for DNA cleavage and low-editing efficiencies[67]. Viral vectors for genome editing need improvements to effectively target endogenous loci.

Shortly after the previous case study was published, CRISPR/Cas9 technology was developed[68] which promised significantly improved ease of gene editing. One disadvantage of the CRISPR/Cas9 system is that while the targeting component of the site-specific nuclease is relatively small, namely the sgRNA, Cas9 is a large protein and well beyond the carrying capacity of RNA viral vectors developed to date. RNA viral vectors have a limited carrying capacity due to the error-prone RdRP, which often truncates non-essential viral components; shorter, wild-type viruses quickly outcompete modified vectors. Further, vectors modified to express large cargo are often unable to be encapsidated[53]. Ali and colleagues proposed a solution to this problem: they created a transgenic plant that expresses the Cas9 protein, and used the viral vector to express the sgRNA[51]. Their results are highlighted in the following case study.

8.2 Ali et al. 2015: Efficient virus-mediated genome editing in plants using the CRISPR/Cas9 system[51]

Recognizing delivery of genome engineering reagents to plant cells is a significant limitation to achieve gene editing, Ali and colleagues sought to provide a means to efficiently and quickly deliver CRISPR/Cas9 reagents using RNA viruses. They created transgenic *N. benthamiana* plants that express Cas9, and used TRV to deliver sgRNAs that target unique sequences in the genome. TRV2 was modified in a similar manner to Marton et al. 2010 by expressing the sgRNA from a PeBV-subgenomic promoter. To test the effectiveness of this vector in gene editing, a sgRNA-targeting *N. benthamiana* phytoene desaturase (PDS) was cloned into the TRV2 vector and subsequently agroinfected along with a TRV1 vector into the transgenic *N. benthamiana* plants that express Cas9. Gene editing was observed in the inoculated and systemically infected leaves. The method was further validated by targeting a second locus, PCNA, and editing was again observed in the inoculated and systemically infected leaves. Next, multiple TRV2 vectors, each expressing a unique sgRNA, were simultaneously tested to determine if they could achieve multiplexed editing. TRV2 vectors expressing sgRNAs targeting PDS or PCNA were co-infected into *N. benthamiana*, and editing was observed at both loci, although at a lower frequency than when inoculated separately.

The applicability of RNA viral vectors for gene editing would be greatly improved if the targeted mutation occurred in the plant germline. This would allow seeds containing the desired modification to be collected from the infected plant, thereby eliminating the need for tissue-culture in many gene-editing experiments and increasing ease and throughput. Ali and colleagues reported evidence of germinal transmission of a targeted mutation, although at a very low frequency. This indicated that future work was needed to increase germline edits to a frequency that is practical. Also, the approach needed to be expanded to species other than the *N. benthamiana* model.

8.3 Expanding RNA viral vector gene editing to other species

After Ali et al. demonstrated the ability of viral vectors to perform efficient gene editing in somatic cells of Cas9 expressing *N. benthamiana*, the same group and several others tested this approach for gene editing in other species and with other viral vectors. PEBV and Beet Necrotic Yellow Vein Virus (BNYVV) were demonstrated to perform gene editing in Cas9-expressing somatic cells of *N. benthamiana*[69], and Tobacco Mosaic Virus (TMV) was used for multiplexed gene editing[70]. TRV was also used to demonstrate gene editing in Cas9-expressing *Arabidopsis*[55], highlighting the wide host range and utility of this virus. In order to expand to monocot species, Foxtail Mosaic Virus (FoMV) was modified to contain a duplicated coat protein subgenomic promoter for expression of sgRNAs. This vector was able to perform gene editing in Cas9 expressing *N. benthamiana*, *Setaria viridis*, and maize[56]. Finally, a Barley Yellow Striate Mosaic Virus (BYSMV) vector was modified to express proteins in wheat, barley, foxtail millet, and maize. BYSMV is a negative-sense RNA virus, whereas all other virus vectors discussed are positive-sense RNA viruses. Negative-sense RNA viruses could provide a larger carrying capacity than the positive-sense RNA viruses, as the authors of this study demonstrated expression of both the sgRNA and

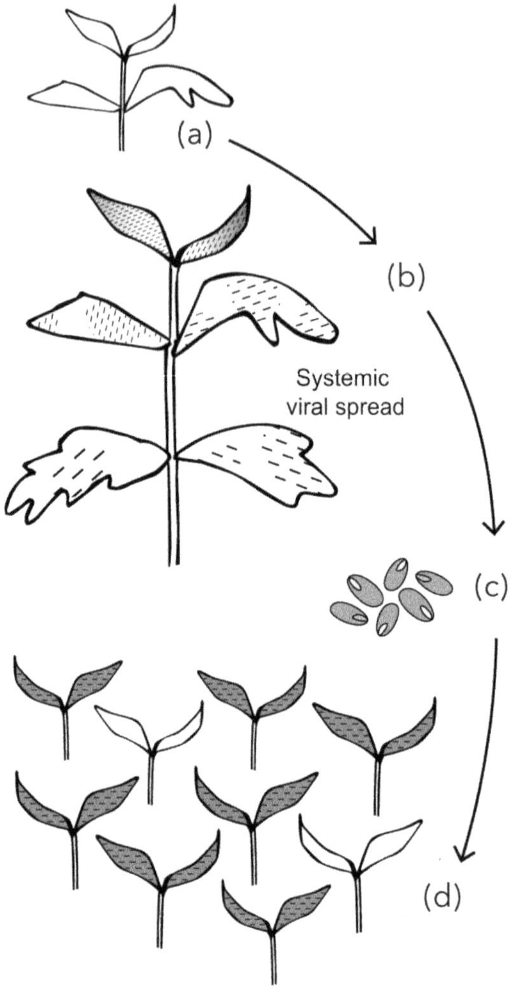

Figure 3 A method for infecting plants with RNA viral vectors and generating heritable genome edits. (a) Transgenic plants expressing Cas9 are infected with an RNA viral vector expressing a mobile sgRNA to increase genome-editing frequencies. (b) This vector systemically infects the plant, interacting with endogenously expressed Cas9 for high-frequency genome editing (future methods may express all the genome-editing reagents from the virus, allowing editing of wild-type plants). Dashed lines and shading indicate genome editing. (c) Seeds are collected from infected plants and progeny are screened for editing of the targeted site. (d) High percentages of the progeny contain the desired edit; shaded seedlings indicate the presence of targeted mutations.

Cas9 from the virus and editing at the infiltrated site in *N. benthamiana*; they did not report any systemic editing[71]. None of the previously mentioned studies demonstrated heritable editing without using tissue culture, the exception being Ali et al. 2015, in which, as mentioned above, heritable editing was observed at a very low frequency: only 1 in 438 seedlings tested. In order for viral vectors to be widely adopted as vectors for site-specific editing, the heritable editing frequency needs to be improved. Ellison and colleagues show improved heritable editing frequency in *N. benthamiana* by adding motifs that promote mobility of the sgRNAs expressed from TRV vectors (Fig. 3). Their results are described below.

8.4 Ellison et al. 2020: Multiplexed heritable gene editing using RNA viruses and mobile single guide RNAs[57]

Ellison and colleagues sought to improve the frequencies of heritable gene editing with RNA viral vectors, recognizing that editing through infection could eliminate or reduce the need for tissue-culture – a major bottleneck in applying biotechnology to plant science research. The authors hypothesized that by enabling viral or sgRNA sequences better access to the germline, gene edits would be transmitted to the next generation at higher frequencies. To enable mobility, a *Flowering Locus T* (FT) motif was added to the 3′ end of sgRNA sequences. The FT motif was previously shown to promote mobility of RNAs into the meristem[72]. sgRNA sequences targeting *N. benthamiana PDS* and augmented with this mobility sequence were cloned into the same TRV2 vector described in Ali et al. 2015. These vectors were agroinfected into Cas9 expressing *N. benthamiana* along with TRV1. Shortly after infection, phenotypes suggesting PDS mutagenesis emerged, and this was substantiated by determining that somatic editing frequencies were greater than 80%. Germinal transmission of these edits was observed by phenotyping (Fig. 4) and genotyping seedlings: up to 65% contained a mutation in at least one PDS allele. This high frequency of mutagenesis was demonstrated at a second locus, *AGAMOUS* (AG). In this case, up to 100% of seedlings contained a mutation in at least one AG allele. The mobility motif was not limited to FT: tRNA-like sequences have previously been shown to promote cell-to-cell mobility[73], and such tRNA-like sequences were also demonstrated to enhance frequencies of heritable mutagenesis when fused to sgRNAs. In addition, heritable multiplexed editing of native *N. benthamiana* genes was achieved. Multiple sgRNAs could be expressed from a single TRV2 vector to achieve heritable mutations in all targets. This report establishes the ability of RNA viral vectors to perform high efficiency, heritable gene editing. Further work needs to be performed to expand this method to other species and to develop RNA vectors that are not reliant on a transgenic plant expressing Cas9.

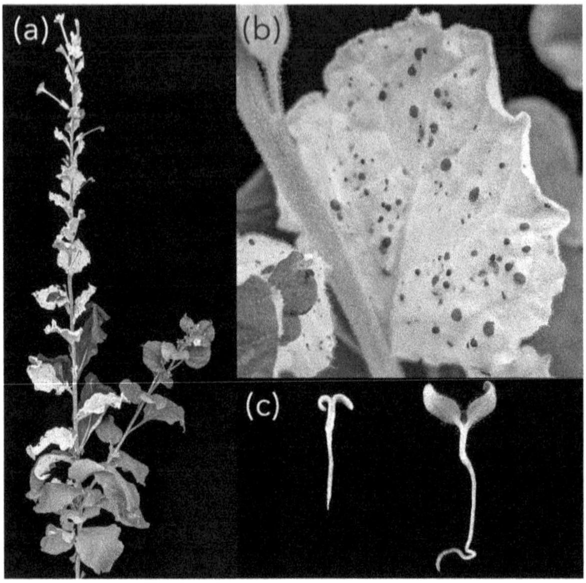

Figure 4 *Phytoene Desaturase* (PDS) phenotypes in Cas9-expressing *N. benthamiana* plants infected with Tobacco Rattle Virus (TRV) vectors. TRV expresses mobile sgRNAs targeting the PDS coding sequence, which results in a bleached phenotype due to chlorophyll photo-oxidation. (a) Infected plants show an increase in the PDS-knockout phenotype as the plant matures. In some cases, the entire upper portion of the plant is bleached, indicating a PDS mutation occurring in meristematic cells. (b) Systemic plant leaves contain high frequencies of editing from RNA viral vectors expressing mobile sgRNAs. The leaf shown here contains PDS mutations in nearly every cell, causing an almost completely bleached phenotype. (c) Seeds are collected from plants infected with the TRV vector expressing mobile *PDS* sgRNAs. Several of the resulting seedlings are entirely bleached, indicating germline mutagenesis of the *PDS* locus.

9 RNA viral vectors: conclusion and future trends

All three of the case studies described in this chapter illustrate the growing interest in the use of RNA viral vectors for plant gene editing. While there are considerable milestones yet to be achieved, these vectors provide considerable promise for tissue-culture-free gene editing. Future work to improve the technology should focus on expanding the species range of heritable gene editing. This will require optimization of vectors for each species along with the establishment of transgenic plant lines that express Cas9 at high levels and in the appropriate tissues. In order to eliminate the need for transgenesis, efforts should be made to test whether smaller Cas or other nucleases can be systemically transported by RNA viral vectors. Alternatively, it might be possible to increase the carrying capacity of virus vectors. Recently, Ma and colleagues developed vectors based on Sonchus Yellow Net Nucleorhabdovirus (SYNV), a negative-sense RNA virus.

Remarkably, these vectors could deliver both Cas9 and sgRNAs[74]. Gene-edited plants were recovered by regenerating infected plant tissue without selection, and over 90% of regenerated plants carried mutations. If SYNV could access the germline, perhaps through addition of RNA motifs that promote cell-to-cell mobility as described by Ellison et al.[57], it may be possible to recover heritable mutations solely through infection and without tissue culture and transgenesis.

RNA viral vector utility can also be expanded by using different Cas variants with different PAM requirements to increase the targeting range or to create unique editing outcomes. It should also be possible to move beyond non-homologous end joining or microhomology-mediated repair of the target locus to achieve more precise modifications through base editing[75-77] or prime editing[78]. Background work on RNA genome structure and function, followed by the development of vectors for protein expression, VIGS, and genome editing, will facilitate improvements and increase the scope, specificity, and broad applicability of RNA viral vectors.

10 Where to look for further information

Many viral vectors, including all of our group's published vectors, are available at Addgene, a non-profit plasmid repository. BeYDV and WDV replicons are among the available reagents, and they are compatible with the plant genome engineering toolkit described in Cermak et al.[79]. Lastly, viral sequences can sometimes be challenging to assemble and clone in bacterial hosts, due to structural features or toxicity of viral gene products; for a review of strategies for viral clone assembly, see Pasin et al.[64].

11 References

1. Zerbini, F. M., Briddon, R. W., Idris, A., Martin, D. P., Moriones, E., Navas-Castillo, J., Rivera-Bustamante, R., Roumagnac, P., Varsani, A. and Ictv Report Consortium. ICTV virus taxonomy profile: geminiviridae. *Journal of General Virology* 98(2), 131–133 (2017).
2. Lazarowitz, S. G., Pinder, A. J., Damsteegt, V. D. and Rogers, S. G. Maize streak virus genes essential for systemic spread and symptom development. *The EMBO Journal* 8(4), 1023-1032 (1989).
3. Baltes, N. J., Gil-Humanes, J., Cermak, T., Atkins, P. A. and Voytas, D. F. D. N. A. DNA replicons for plant genome engineering. *The Plant Cell* 26(1), 151–163 (2014).
4. Hanley-Bowdoin, L., Bejarano, E. R., Robertson, D. and Mansoor, S. Geminiviruses: masters at redirecting and reprogramming plant processes. *Nature Reviews. Microbiology* 11(11), 777–788 (2013).
5. Roossinck, M. J., Martin, D. P. and Roumagnac, P. Plant virus metagenomics: advances in virus discovery. *Phytopathology* 105(6), 716–727 (2015).
6. Fontenele, R. S., Alves-Freitas, D. M. T., Silva, P. I. T., Foresti, J., Silva, P. R., Godinho, M. T., Varsani, A. and Ribeiro, S. G. Discovery of the first maize-infecting Mastrevirus in

the Americas using a vector-enabled metagenomics approach. *Archives of Virology* 163(1), 263-267 (2018).
7. Varsani, A., Shepherd, D. N., Dent, K., Monjane, A. L., Rybicki, E. P. and Martin, D. P. A highly divergent South African geminivirus species illuminates the ancient evolutionary history of this family. *Virology Journal* 6, 36 (2009).
8. Kumar, J., Kumar, J., Singh, S. P. and Tuli, R. Association of satellites with a Mastrevirus in natural infection: complexity of wheat Dwarf India virus disease. *Journal of Virology* 88(12), 7093-7104 (2014).
9. Gutierrez, C. Geminivirus DNA replication. *Cellular and Molecular Life Sciences: CMLS* 56(3-4), 313-329 (1999).
10. Schalk, H. J., Matzeit, V., Schiller, B., Schell, J. and Gronenborn, B. Wheat dwarf virus, a geminivirus of graminaceous plants needs splicing for replication. *The EMBO Journal* 8(2), 359-364 (1989).
11. Wright, E. A., Heckel, T., Groenendijk, J., Davies, J. W. and Boulton, M. I. Splicing features in maize streak virus virion- and complementary-sense gene expression. *The Plant Journal: For Cell and Molecular Biology* 12(6), 1285-1297 (1997).
12. Liu, L., Davies, J. W. and Stanley, J. Mutational analysis of bean yellow dwarf virus, a geminivirus of the genus Mastrevirus that is adapted to dicotyledonous plants. *Journal of General Virology* 79(9), 2265-2274 (1998).
13. Boulton, M. I. Functions and interactions of Mastrevirus gene products. *Physiological and Molecular Plant Pathology* 60(5), 243-255 (2002).
14. Gutierrez, C. D. N. A.. DNA replication and cell cycle in plants: learning from geminiviruses. *The EMBO Journal* 19(5), 792-799 (2000).
15. Gutierrez, C. Strategies for geminivirus DNA replication and cell cycle interference. *Physiological and Molecular Plant Pathology* 60(5), 219-230 (2002).
16. Ugaki, M., Ueda, T., Timmermans, M. C., Vieira, J., Elliston, K. O. and Messing, J. Replication of a geminivirus derived shuttle vector in maize endosperm cells. *Nucleic Acids Research* 19(2), 371-377 (1991).
17. Matzeit, V., Schaefer, S., Kammann, M. and Gronenborna, B. Wheat dwarf virus vectors replicate and express foreign genes in cells of monocotyledonous plants *The Plant Cell* 3, 247-258 (1991).
18. Shen, W.-H. and Hohn, B. Amplification and expression of the B-glucuronidase gene in maize plants by vectors based on maize streak virus. *The Plant Journal* 5(2), 227-236 (1994).
19. Shen, W. H. and Hohn, B. Vectors based on maize streak virus can replicate to high copy numbers in maize plants. *Journal of General Virology* 76(4), 965-969 (1995).
20. Mor, T. S., Moon, Y. S., Palmer, K. E. and Mason, H. S. Geminivirus vectors for high-level expression of foreign proteins in plant cells. *Biotechnology and Bioengineering* 81(4), 430-437 (2003).
21. Zhang, X. and Mason, H. Bean Yellow Dwarf Virus replicons for high-level transgene expression in transgenic plants and cell cultures. *Biotechnology and Bioengineering* 93(2), 271-279 (2006).
22. Hayes, R. J., Petty, I. T. D., Coutts, R. H. A. and Buck, K. W. Gene amplification and expression in plants by a replicating geminivirus vector. *Nature* 334(6178), 179-182 (1988).
23. Hayes, R. J., Coutts, R. H. and Buck, K. W. Stability and expression of bacterial genes in replicating geminivirus vectors in plants. *Nucleic Acids Research* 17(7), 2391-2403 (1989).

24. Gilbertson, R. L., Sudarshana, M., Jiang, H., Rojas, M. R. and Lucas, W. J. Limitations on geminivirus genome size imposed by plasmodesmata and virus-encoded movement protein: insights into DNA trafficking. *The Plant Cell* 15(11), 2578–2591 (2003).
25. Pandey, P., Choudhury, N. R. and Mukherjee, S. K. A geminiviral amplicon (VA) derived from Tomato leaf curl virus (ToLCV) can replicate in a wide variety of plant species and also acts as a VIGS vector. *Virology Journal* 6, 152 (2009).
26. Čermák, T., Baltes, N. J., Čegan, R., Zhang, Y. and Voytas, D. F. High-frequency, precise modification of the tomato genome. *Genome Biology* 16, 232 (2015).
27. Gleba, Y., Marillonnet, S. and Klimyuk, V. Engineering viral expression vectors for plants: the 'full virus' and the 'deconstructed virus' strategies. *Current Opinion in Plant Biology* 7(2), 182–188 (2004).
28. Wang, M., Lu, Y., Botella, J. R., Mao, Y., Hua, K. and Zhu, J. K. Gene targeting by homology-directed repair in rice using a geminivirus-based CRISPR/Cas9 system. *Molecular Plant* 10(7), 1007–1010 (2017).
29. Woolston, C. J., Reynolds, H. V., Stacey, N. J. and Mullineaux, P. M. Replication of wheat dwarf virus DNA in protoplasts and analysis of coat protein mutants in protoplasts and plants. *Nucleic Acids Research* 17(15), 6029–6041 (1989).
30. Rogers, S. G., Bisaro, D. M., Horsch, R. B., Fraley, R. T., Hoffmann, N. L., Brand, L., Scott Elmer, J. and Lloyd, A. M. Tomato golden mosaic virus A component D replicates autonomously in transgenic plants. Cell 45(4), 593–600 (1986).
31. Grimsley, N., Hohn, B., Hohn, T. and Walden, R. "Agroinfection," an alternative route for viral infection of plants by using the Ti plasmid. *Proceedings of the National Academy of Sciences of the United States of America* 83(10), 3282–3286 (1986).
32. Woolston, C. J., Barker, R., Gunn, H., Boulton, M. I. and Mullineaux, P. M. Agroinfection and nucleotide sequence of cloned wheat dwarf virus DNA. *Plant Molecular Biology* 11(1), 35–43 (1988).
33. Stenger, D. C., Revington, G. N., Stevenson, M. C. and Bisaro, D. M. Replicational release of geminivirus genomes from tandemly repeated copies: evidence for rolling-circle replication of a plant viral DNA. *Proceedings of the National Academy of Sciences of the United States of America* 88(18), 8029–8033 (1991).
34. Zhao, X., Coats, I., Fu, P., Gordon-Kamm, B. and Lyznik, L. A. T-DNA recombination and replication in maize cells. *The Plant Journal: For Cell and Molecular Biology* 33(1), 149–159 (2003).
35. Gil-Humanes, J., Wang, Y., Liang, Z., Shan, Q., Ozuna, C. V., Sánchez-León, S., Baltes, N. J., Starker, C., Barro, F., Gao, C. and Voytas, D. F. High-efficiency gene targeting in hexaploid wheat using DNA replicons and CRISPR/Cas9. *The Plant Journal: For Cell and Molecular Biology* 89(6), 1251–1262 (2017).
36. Dahan-Meir, T., Filler-Hayut, S., Melamed-Bessudo, C., Bocobza, S., Czosnek, H., Aharoni, A. and Levy, A. A. Efficient *in planta* gene targeting in tomato using geminiviral replicons and the CRISPR/Cas9 system. *The Plant Journal: For Cell and Molecular Biology* 95(1), 5–16 (2018).
37. Shan, Q., Baltes, N. J., Atkins, P., Kirkland, E. R., Zhang, Y., Baller, J. A., Lowder, L. G., Malzahn, A. A., Haugner, J. C., Seelig, B., Voytas, D. F. and Qi, Y. ZFN, TALEN and CRISPR-Cas9 mediated homology directed gene insertion in Arabidopsis: a disconnect between somatic and germinal cells. *Journal of Genetics and Genomics = Yi Chuan Xue Bao* 45(12), 681–684 (2018).

38. Puchta, H., Dujon, B. and Hohn, B. Homologous recombination in plant cells is enhanced by *in vivo* induction of double strand breaks into DNA by a site-specific endonuclease. *Nucleic Acids Research* 21(22), 5034–5040 (1993).
39. Townsend, J. A., Wright, D. A., Winfrey, R. J., Fu, F., Maeder, M. L., Joung, J. K. and Voytas, D. F. High-frequency modification of plant genes using engineered zinc-finger nucleases. *Nature* 459(7245), 442–445 (2009).
40. Zhang, Y., Zhang, F., Li, X., Baller, J. A., Qi, Y., Starker, C. G., Bogdanove, A. J. and Voytas, D. F. Transcription activator-like effector nucleases enable efficient plant genome engineering. *Plant Physiology* 161(1), 20–27 (2013).
41. Richter, K. S., Kleinow, T. and Jeske, H. Somatic homologous recombination in plants is promoted by a geminivirus in a tissue-selective manner. *Virology* 452–453, 287–296 (2014).
42. Wright, D. A., Townsend, J. A., Winfrey, R. J., Irwin, P. A., Rajagopal, J., Lonosky, P. M., Hall, B. D., Jondle, M. D. and Voytas, D. F. High-frequency homologous recombination in plants mediated by zinc-finger nucleases. *The Plant Journal: For Cell and Molecular Biology* 44(4), 693–705 (2005).
43. Fauser, F., Roth, N., Pacher, M., Ilg, G., Sánchez-Fernández, R., Biesgen, C. and Puchta, H. In planta gene targeting. *Proceedings of the National Academy of Sciences of the United States of America* 109(19), 7535–7540 (2012).
44. Hahn, F., Eisenhut, M., Mantegazza, O. and Weber, A. P. M. Homology-directed repair of a defective glabrous gene in Arabidopsis With Cas9-based gene targeting. *Frontiers in Plant Science* 9, 424 (2018).
45. Butler, N. M., Baltes, N. J., Voytas, D. F. and Douches, D. S. Geminivirus-mediated genome editing in potato (Solanum tuberosum L.) using sequence-specific nucleases. *Frontiers in Plant Science* 7, 1045 (2016).
46. Hummel, A. W., Chauhan, R. D., Cermak, T., Mutka, A. M., Vijayaraghavan, A., Boyher, A., Starker, C. G., Bart, R., Voytas, D. F. and Taylor, N. J. Allele exchange at the EPSPS locus confers glyphosate tolerance in cassava. *Plant Biotechnology Journal* 16(7), 1275–1282 (2018).
47. Atabekov, J. G. and Taliansky, M. E. Expression of a plant virus-coded transport function by different viral genomes. *Advances in Virus Research* 38, 201–248 (1990).
48. Atabekov, J. G. and Dorokhov, Y. L. Plant virus-specific transport function and resistance of plants to viruses. *Advances in Virus Research* 29, 313–364 (1984).
49. Carrington, J. C., Kasschau, K. D., Mahajan, S. K. and Schaad, M. C. Cell-to-cell and long-distance transport of viruses in plants. *The Plant Cell* 8(10), 1669–1681 (1996).
50. Ortín, J. and Martín-Benito, J. The RNA synthesis machinery of negative-stranded RNA viruses. *Virology* 479–480, 532–544 (2015).
51. Ali, Z., Abul-faraj, A., Li, L., Ghosh, N., Piatek, M., Mahjoub, A., Aouida, M., Piatek, A., Baltes, N. J., Voytas, D. F., Dinesh-Kumar, S. and Mahfouz, M. M. Efficient virus-mediated genome editing in plants using the CRISPR/Cas9 system. *Molecular Plant* 8(8), 1288–1291 (2015).
52. Burch-Smith, T. M., Anderson, J. C., Martin, G. B. and Dinesh-Kumar, S. P. Applications and advantages of virus-induced gene silencing for gene function studies in plants. *The Plant Journal: For Cell and Molecular Biology* 39(5), 734–746 (2004).
53. Scholthof, H. B., Scholthof, K. B. and Jackson, A. O. Plant virus gene vectors for transient expression of foreign proteins in plants. *Annual Review of Phytopathology* 34, 299–323 (1996).

54. Marton, I., Zuker, A., Shklarman, E., Zeevi, V., Tovkach, A., Roffe, S., Ovadis, M., Tzfira, T. and Vainstein, A. Nontransgenic genome modification in plant cells. *Plant Physiology* 154(3), 1079–1087 (2010).
55. Ali, Z., Eid, A., Ali, S. and Mahfouz, M. M. Pea early-browning virus-mediated genome editing via the CRISPR/Cas9 system in Nicotiana benthamiana and Arabidopsis. *Virus Research* 244, 333–337 (2018).
56. Mei, Y., Beernink, B. M., Ellison, E. E., Konečná, E., Neelakandan, A. K., Voytas, D. F. and Whitham, S. A. Protein expression and gene editing in monocots using foxtail mosaic virus vectors. *Plant Direct* 3(11), e00181 (2019).
57. Ellison, E. E., Nagalakshmi, U., Gamo, M. E., Huang, P. J., Dinesh-Kumar, S. and Voytas, D. F. Multiplexed heritable gene editing using RNA viruses and mobile single guide RNAs. *Nature Plants* 6(6), 620–624 (2020).
58. Ahlquist, P., Noueiry, A. O., Lee, W. M., Kushner, D. B. and Dye, B. T. (Host). Host Factors in positive-strand RNA virus genome replication. *Journal of Virology* 77(15), 8181–8186 (2003).
59. Dall'Ara, M., Ratti, C., Bouzoubaa, S. E. and Gilmer, D. Ins and outs of multipartite positive-strand RNA plant viruses: packaging versus systemic spread. *Viruses* 8(8), 228 (2016).
60. Siegel, A. Plant-virus-based vectors for gene transfer may be of considerable use despite a presumed high error frequency during RNA synthesis. *Plant Molecular Biology* 4(5), 327–329 (1985).
61. Chapman, S., Kavanagh, T. and Baulcombe, D. Potato virus X as a vector for gene expression in plants. *The Plant Journal: For Cell and Molecular Biology* 2(4), 549–557 (1992).
62. Czapar, A. E. and Steinmetz, N. F. Plant viruses and bacteriophages for drug delivery in medicine and biotechnology. *Current Opinion in Chemical Biology* 38, 108–116 (2017).
63. McGarry, R. C., Klocko, A. L., Pang, M., Strauss, S. H. and Ayre, B. G. Virus-induced flowering: an application of reproductive biology to benefit plant research and breeding. *Plant Physiology* 173(1), 47–55 (2017).
64. Pasin, F., Menzel, W. and Daròs, J. A. Harnessed viruses in the age of metagenomics and synthetic biology: an update on infectious clone assembly and biotechnologies of plant viruses. *Plant Biotechnology Journal* 17(6), 1010–1026 (2019).
65. Ruiz, M. T., Voinnet, O. and Baulcombe, D. C. Initiation and maintenance of virus-induced gene silencing. *Plant Cell* 10(6), 937–946 (1998).
66. Lacomme, C. Strategies for altering plant traits using virus-induced gene silencing technologies. *Methods in Molecular Biology* 1287, 25–41. doi:10.1007/978-1-4939-2453-0_2 (2015).
67. Gaj, T., Gersbach, C. A., Barbas, C. F. ZFN, TALEN, and CRISPR/Cas-based methods for genome engineering. *Trends in Biotechnology* 31(7), 397–405 (2013).
68. Jinek, M., Chylinski, K., Fonfara, I., Hauer, M., Doudna, J. A. and Charpentier, E. A programmable dual-RNAGuided DNA endonuclease in adaptive bacterial immunity. *Science* 337(6096), 816–821 (2012).
69. Jiang, N., Zhang, C., Liu, J. Y., Guo, Z. H., Zhang, Z. Y., Han, C. G. and Wang, Y. Development of Beet necrotic yellow vein virus-based vectors for multiple-gene expression and guide RNA delivery in plant genome editing. *Plant Biotechnology Journal* 17(7), 1302–1315 (2019).

70. Cody, W. B., Scholthof, H. B. and Mirkov, T. E. Multiplexed gene editing and protein overexpression using a *tobacco mosaic virus* viral vector. *Plant Physiology* 175(1), 23-35 (2017).
71. Gao, Q., Xu, W. Y., Yan, T., Fang, X. D., Cao, Q., Zhang, Z. J., Ding, Z. H., Wang, Y. and Wang, X. B. Rescue of a plant Cytorhabdovirus as versatile expression platforms for planthopper and cereal genomic studies. *New Phytologist* 223(4), 2120-2133 (2019).
72. Li, C., Zhang, K., Zeng, X., Jackson, S., Zhou, Y. and Hong, Y. A cis element within flowering locus T mRNA determines its mobility and facilitates trafficking of heterologous viral RNA. *Journal of Virology* 83(8), 3540-3548 (2009).
73. Zhang, W., Thieme, C. J., Kollwig, G., Apelt, F., Yang, L., Winter, N., Andresen, N., Walther, D. and Kragler, F. tRNA-related sequences trigger systemic mRNA transport in plants. *The Plant Cell* 28(6), 1237-1249 (2016).
74. Ma, X., Zhang, X., Liu, H. and Li, Z. Highly efficient DNA-free plant genome editing using virally delivered CRISPRCas9. *Nature Plants* 6(7), 773-779 (2020).
75. Kang, B. C., Yun, J. Y., Kim, S. T., Shin, Y., Ryu, J., Choi, M., Woo, J. W. and Kim, J. S. Precision genome engineering through adenine base editing in plants. *Nature Plants* 4(7), 427-431 (2018).
76. Shimatani, Z., Kashojiya, S., Takayama, M., Terada, R., Arazoe, T., Ishii, H., Teramura, H., Yamamoto, T., Komatsu, H., Miura, K., Ezura, H., Nishida, K., Ariizumi, T. and Kondo, A. Targeted base editing in rice and tomato using a CRISPR-Cas9 cytidine deaminase fusion. *Nature Biotechnology* 35(5), 441-443 (2017).
77. Zong, Y., Wang, Y., Li, C., Zhang, R., Chen, K., Ran, Y., Qiu, J. L., Wang, D. and Gao, C. Precise base editing in rice, wheat and maize with a Cas9-cytidine deaminase fusion. *Nature Biotechnology* 35(5), 438-440 (2017).
78. Anzalone, A. V., Randolph, P. B., Davis, J. R., Sousa, A. A., Koblan, L. W., Levy, J. M., Chen, P. J., Wilson, C., Newby, G. A., Raguram, A. and Liu, D. R. Search-and-replace genome editing without double-strand breaks or donor DNA. *Nature* 576(7785), 149-157 (2019).
79. Čermák, T., Curtin, S. J., Gil-Humanes, J., Čegan, R., Kono, T. J. Y., Konečná, E., Belanto, J. J., Starker, C. G., Mathre, J. W., Greenstein, R. L. and Voytas, D. F. A multipurpose toolkit to enable advanced genome engineering in plants. *The Plant Cell* 29(6), 1196-1217 (2017).

Chapter 5

Progress in precise and predictable genome editing in plants with base editing

Sabine Fräbel, Shai J. Lawit, Jingyi Nie, David G. Schwark, Thomas J. Poorten and Nathaniel D. Graham, Pairwise Plants, USA

1 Introduction

The genome editing field has changed rapidly since the development of the first site-directed nucleases (SDNs) and zinc finger nucleases (ZFNs) in 1996 (Kim et al., 1996). Since this time, many tools have been developed that allow for the targeted change of genetic sequences, the most broadly used being CRISPR/Cas9 (Jinek et al., 2012). SDNs have allowed researchers to easily target sequences within the genome and introduce changes in a very specific manner across a broad range of organisms, including plants (Feng et al., 2013). The use of SDNs has led to a diverse array of new phenotypes in plants in the short time since their introduction.

The focus of early genome editing was mostly on gene knockouts, which are easily achieved by targeted nucleases. SDNs form double-strand breaks (DSBs) that are repaired by the host's native repair machinery. This typically results in a return to the original genomic sequence, or an insertion or deletion

http://dx.doi.org/10.19103/AS.2020.0082.06

(indel) at the genomic break site - although other repairs such as in-frame substitutions are possible. The mechanism of DSB repair, reviewed in Schmidt et al. (2018), is not fully understood. Patterns can be observed, often thought to be a result of microhomology at each end of the cut site, which causes the final repair outcome to be nonrandom (van Overbeek et al., 2016). Researchers often rely on the repair pattern that is commonly associated with a given nuclease, that is, single-base insertions with Cas9 (Allen et al., 2019), to inform a target location that will result in their desired outcome. For knockouts, the target region is generally in the 5′ end of a gene with the hope that the repair results in a frameshift that produces a premature stop codon, and thus loss of the gene product. Computational tools have been developed to attempt to predict these editing outcomes, reviewed in Molla and Yang (2020); however, plants obtained after editing with SDNs will still contain an array of editing outcomes.

While SDNs can be very precisely targeted to chosen genomic regions, making specific changes, such as polymorphism conversions, is difficult due to the still developing understanding of repair outcomes. Previously, such targeted conversions have been attempted using homology directed repair (HDR) strategies; however, the editing frequency with HDR has been extremely low and thus extremely resource intensive. The advent of base editors addressed the need for targeted conversion by combining a fully or partially deactivated nuclease with a deaminase domain. The deaminase domain causes deamination of bases exposed by the deactivated nuclease which are then converted to alternative bases by the cells' native repair machinery. Base editors convert nucleotide bases within an editing window, defined by the structure of the complex at the target site, at a specific genomic location defined by the nuclease targeting component. Depending on the deaminase used, the most common conversions are cytosine to thymine (C>T) and adenine to guanine (A>G), with multiple conversions being possible within an editing window that varies with each deaminase and nuclease used. While base editors are relatively new in the field of plant genome editing, they represent a rapidly growing toolset for predictable genome editing outcomes.

The RNA-guided CRISPR/Cas9 nuclease broadened the toolkit of targeted genome editing. The Cas9 protein contains two nuclease domains, the RuvC domain that cleaves the PAM-containing strand, and the HNH domain that cleaves the PAM-complement DNA. Either domain can be inactivated by a single amino acid mutation. Nuclease dead Cas9 (dCas9) contains a D10A mutation in the RuvC domain and a H840A mutation in the HNH domain retains sequence-specific DNA-binding activity (Jinek et al., 2012). The first base editor, (BE), fused a cytidine deaminase domain to nuclease dead Cas9 to enable delivery of the cytidine deaminase to sequence-specific sites in the genome. The next design (BE2) included a uracil DNA glycosylase inhibitor (UGI) domain, and dCas9 was replaced by Cas9 (D10A) nickase in BE3, to enhance the desired editing outcome. With these improvements, BE3 achieved an average of 37% editing efficiency across six loci in human cells. Further base editor architecture optimization resulted in BE4, which contained altered protein linkers and an additional UGI domain, and BE4-Gam, which additionally is fused to a bacteriophage MU protein, to improve editing efficiency and product purity (Komor et al., 2017). Following cytosine base editor (CBE) success, an adenosine base editor (ABE) was developed by replacing the cytidine deaminase with an engineered DNA adenine deaminase domain derived from multiple rounds of directed evolution and protein engineering (Gaudelli et al., 2017). Both CBE and ABE function with limited editing windows, and editing efficiency varies from locus to locus. In general, BE3 and ABE7.10 yield the highest editing efficiency, with an editing window from positions 4–8 and 4–7, respectively (counting the PAM as position 21–23).

The mechanisms of the CBE and ABE are similar, with both editors producing a deamination event that is ultimately resolved into an alternate base. To use BE3 as an example (Fig. 1), once the editing complex is delivered to the Cas9 binding site, the cytidine deaminase converts any cytidine base in the editing window, on the PAM-containing strand to uracil, creating a U:G mismatch. Which cytidine bases are targeted is mostly determined by the amount of ssDNA exposed and accessible to the deaminase domain by the deactivated nuclease. This mismatch can be repaired back to C:G or converted to U:T. In the design of BE3, the UGI domain inhibits repair of U back to C, and the nick on the PAM-complement strand stimulates replacement of G with A, using the PAM-containing strand that now contains a U as the template, hence a C:G to T:A conversion (Komor et al., 2016). In the case of the ABE, any adenine base in the editing window is first converted to inosine which will be converted to G following repair (Gaudelli et al., 2017).

2 Progress in mammalian systems

Base editing was initially designed specifically with therapeutic intent. As such, significant effort has been applied to optimize base editors for this purpose – resulting in a plethora of design variety which can be informative to base editing in plants. Base editing application in the human genome toward therapeutic benefits requires high specificity of the editing tool to edit the intended C or A at high efficiency while avoiding changes in bystander targets or off-targets. Introduction of four point-mutations (N497A, R661A, Q695A and Q926A) to Cas9 was shown to improve Cas9 binding specificity by eliminating nonspecific interactions between Cas9 and the target stand (Kleinstiver et al., 2016). The same mutations are also demonstrated to improve base editor specificity, leading to development of high-fidelity base editor (HF-BE3) (Rees et al., 2017).

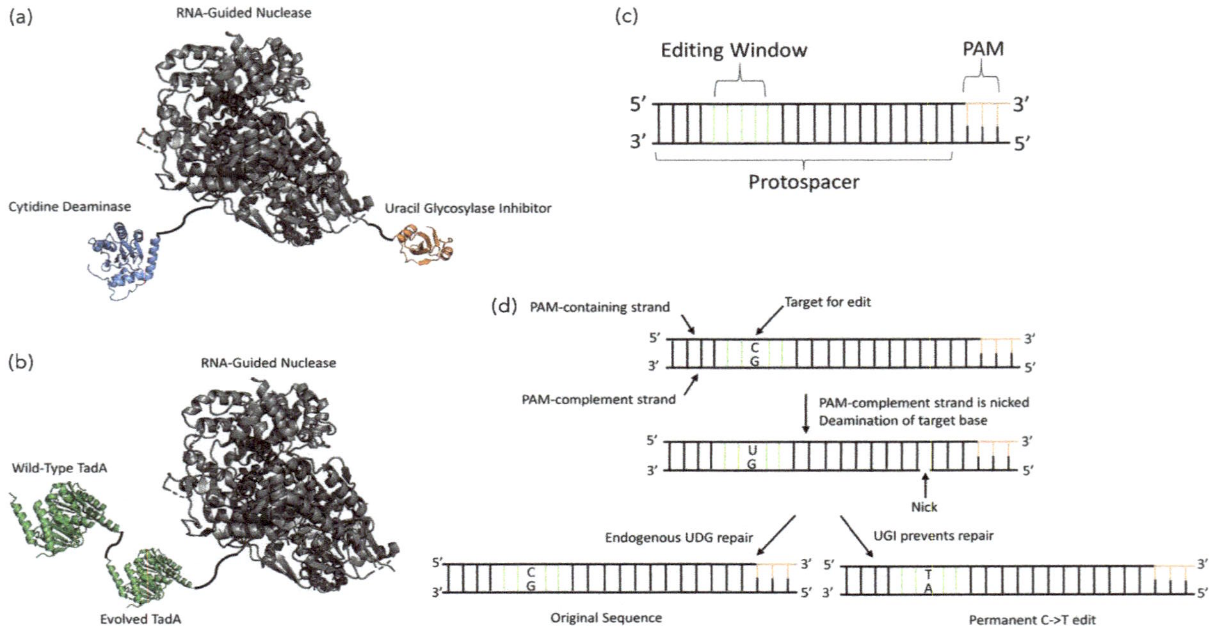

Figure 1 (a) An example of a C>T base editor composed of three parts. A cytidine deaminase (blue, PDB ID: 4XXO) and a uracil glycosylase inhibitor (UGI, orange, PDB ID: 1UGI) are fused to a deactivated RNA-guided nuclease such as SpCas9 (grey, PDB ID: 4CMP) (Bohn et al., 2015; Jinek et al., 2014). (b) An example of an A>G base editor composed of three parts. A wild-type TadA monomer and an engineered TadA monomer (green, PDB ID: 1Z3A) are fused to a deactivated RNA-guided nuclease such as SpCas9 (grey, PDB ID: 4CMP) (Jinek et al., 2014; Kim et al., 2006). (c) Example of a base editing window for a Cas9 cytosine base editor. The editing window is indicated in green, and the protospacer adjacent motif (PAM) is indicated in orange. (d) A simplified scheme of cytosine base editing. The PAM-complement strand of DNA is nicked by the nuclease, and the target cytosine is deaminated to uridine by the cytidine deaminase. Two options are then possible: (1) endogenous DNA repair machinery UDG can repair the DNA, resulting in the original sequence; (2) UGI prevents repair, resulting in a permanent C>T edit.

Cas9 variants that recognized different PAM sequences were also explored to expand target accessibility by base editors (Huang et al., 2019; Liu et al., 2019; Wang et al., 2019b; Zhang et al., 2017, 2020). In addition to the APOBEC1-based base editor, a variety of cytidine deaminases can be used to achieve flexible editing windows (Cheng et al., 2019). An engineered human APOBEC3A with higher sequence specificity is shown to effectively reduce bystander mutations when more than one C is present within the editing window (Gehrke et al., 2018). Continuous directed evolution has been applied to both the APOBEC domain and the Cas9 domain to achieve higher expression and activity (Wang et al., 2018a), as well as to broadened PAM compatibility and increased DNA specificity (Hu et al., 2018; Thuronyi et al., 2019). Editing window and editing efficiency details of various base editor versions is out of scope of this chapter; therefore, we direct interested readers to other papers (Kleinstiver et al., 2019; Rees and Liu, 2018).

In addition to engineering protein components of base editors, delivery methods and protein recruitment strategies are also effective ways to improve editing activity and specificity. For example, delivery of the base editor and sgRNA as ribonucleoprotein (RNP) complexes reduced off-target activity compared to plasmid delivery, while maintaining the same on-target activity level (Rees et al., 2017). Additionally, free UGI co-transfection with BE3 increased the ratio of desired base conversions over undesired indels (Wang et al., 2017). Further, instead of a direct fusion between Cas9 and the deaminase, GCN4 peptides fused to nCas9 allow recruitment of multiple copies of scFv-fused APOBEC-UGI-GB1 to achieve a broadened editing window (Jiang et al., 2018). Similar multi-copy recruitment of deaminases can also be achieved using RNA aptamers such as MS2 stem loops (Hess et al., 2016).

Base editing success in human cell laboratory lines, such as commonly used HEK293T cells, inspired application of these tools in other cell types and organisms. Nuclear localization signal and codon optimization (Koblan et al., 2018), linker optimization (Tan et al., 2019) and protein engineering (Liang et al., 2017) have proven to be effective ways of tuning editing efficiency and fidelity to the cell type of interest. The body of work in various human and animal cells provides useful guidance for adapting base editing tool to plant systems.

3 Cytosine base editing in plants

Different versions of CBEs have been applied to, and optimized for, targeted base modifications in plants. Table 1 provides an editing efficiency overview of base editor variants applied in plant gene editing to date. In contrast to the focus on specificity in therapeutic applications, plant base editing work has to date focused on increasing CBE activity.

Table 1 Examples of base editing in plant systems. Less accurate analysis methods, such as Sanger sequencing, are specifically noted

Species	Editor	Architecture	Frequency	Notes	Citation
Arabidopsis	Cytosine	BE3	2%	Analysis of T1 plants	(Chen et al., 2017)
Arabidopsis	Cytosine	BE3	2-8%	Sanger sequencing of T1 plants	(Xue et al., 2018)
Arabidopsis	Cytosine	Target-AID	50%	Measured in T1 plants	(Bastet et al., 2019)
Cotton	Cytosine	BE3	26–58%		(Qin et al., 2020a)
Maize	Cytosine	BE3	14%		(Li et al., 2019)
Maize	Cytosine	BE3	10%		(Zong et al., 2017)
Potato	Cytosine	PmCDA1	25%		(Veillet et al., 2019b)
Potato	Cytosine	A3A	7%		(Zong et al., 2018)
Rice	Cytosine	BE1	9%	Frequency estimated from Sanger sequencing	(Lu and Zhu, 2017)
Rice	Cytosine	BE3	0–80%	Tested with multiple Cas9 variants	(Hua et al., 2019)
Rice	Cytosine	BE3	25–89%	Includes efficiency increase due to sgRNA modifications	(Qin et al., 2020b)
Rice	Cytosine	BE3	5–72%	Tested with multiple Cas9 variants	(Qin et al., 2020a)
Rice	Cytosine	BE3	26%	Tested with Cas9-NG variant	(Endo et al., 2019)
Rice	Cytosine	BE3	40–65%		(Jin et al., 2019)
Rice	Cytosine	BE3	43%		(Zong et al., 2017)
Rice	Cytosine	BE3	2–44%	Measured with restriction enzyme test, not sequencing	(Li et al., 2017)
Rice	Cytosine	BE3	0–20%	Frequency estimated from Sanger sequencing	(Xu et al., 2019)
Rice	Cytosine	BE4	0–55%	Tested with multiple Cas9 variants	(Zeng et al., 2020)
Rice	Cytosine	BE4max	17–83%	Includes Anc689 as deaminase domain, tested with multiple Cas9 variants	(Wang et al., 2019a)
Rice	Cytosine	rBE5	8–73%	Uses hyperactive hAID mutant	(Ren et al., 2018)
Rice	Cytosine	PmCDA	2–9%	Tested with multiple Cas9 variants	(Qin et al., 2020b)
Rice	Cytosine	PmCDA1	38–50%		(Shimatani et al., 2017)
Rice	Cytosine	PmCDA1	0–90%	Tested with multiple Cas9 variants	(Wu et al., 2019)

Published by Burleigh Dodds Science Publishing Limited, 2021.

Rice	Cytosine	PmCDA1	5-100%	Frequency estimated from Sanger sequencing	(Xu et al., 2019)
Rice	Cytosine	PmCDA1	10–50%	Single transcript of CBE coding sequence and sgRNA	(Tang et al., 2019)
Rice	Cytosine	PmCDA1	30–45%	Tested with Cas9NG variant	(Zhong et al., 2019)
Rice	Cytosine	Target-AID	19–85%	Includes efficiency increase due to sgRNA modifications	(Qin et al., 2020b)
Rice	Cytosine	A3A	44%		(Zong et al., 2018)
Tomato	Cytosine	PmCDA1	12–81%		(Veillet et al., 2019b)
Watermelon	Cytosine	BE3	23%		(Tian et al., 2018)
Wheat	Cytosine	BE3	3%		(Zhang et al., 2019)
Wheat	Cytosine	BE3	1%		(Zong et al., 2017)
Wheat	Cytosine	A3A	16–82%		(Zong et al., 2018)
Arabidopsis	Adenine	ABE 7.10	85%	Measured in T1 plants	(Kang et al., 2018)
Rice	Adenine	ABE 7.10	0–66%	Tested with multiple Cas9 variants	(Hua et al., 2019)
Rice	Adenine	ABE 7.10	0–62 %		(Yan et al., 2018)
Rice	Adenine	ABE 7.10	5–61%	Tested with multiple Cas9 variants	(Hua et al., 2018)
Rice	Adenine	ABE 7.10	7–63%	Tested with multiple Cas9 variants	(Qin et al., 2020b)
Rice	Adenine	ABE 7.10	3–46%		(Li et al., 2018)
Rice	Adenine	ABE 7.10	21–60%		(Jin et al., 2019)
Rice	Adenine	ABE 7.10	0–6.5%	Tested with multiple Cas9 variants	(Zeng et al., 2020)
Rice	Adenine	ABE 7.10	7–16%		(Hao et al., 2019)
Rice	Adenine	ABEmax	8–48%		(Wang et al., 2019a)
Rice	Adenine	Simplified ABE 7.10	6–96%	Modified to contain only single TadA domain	(Hua et al., 2020b)
Wheat	Adenine	ABE 7.10	0.4–1%		(Li et al., 2018)
Rice	Cytosine/Adenine	STEME-1	(0.49–15.0%) C>T + A>G		(Li et al., 2020)
Rice	Cytosine/Adenine	hAID*Δ/TadA*	24–28%	Mixed CBE and ABE vectors with guide libraries	(Kuang et al., 2020)

Published by Burleigh Dodds Science Publishing Limited, 2021.

A major approach to increasing activity has been the substitution of deaminase variants, rAPOBEC1, hAPOBEC3A, PmCDA1 and hAID. The substitution of the rat ABOBEC1 subunit with human ABOBEC3A (A3A-PBE) (Zong et al., 2018), demonstrated in rice, wheat and potato, exemplifies successful CBE activity and editing window enhancement in plants. This optimized version comprises a codon-optimized N-terminal APOBEC3A fused to a SpCas9 D10A nickase and a C-terminal UGI flanked by two nuclear localization sequences (NLSs). Direct comparison of this optimized CBE to a plant-optimized BE3-version at multiple target sites in wheat, rice and potato showed a 11–13-fold higher C>T conversion efficiency. Moreover, the new A3A-PBE (plant base editor) had an extended 17-nt editing window compared to the limited 7-nt editing range of BE3. Through A3A-PBE application at different target sites in rice, wheat and potato, edited plants could be regenerated with editing frequencies ranging from 6.5% to 82.9%. Only very low indel frequency could be detected in protoplast assays and regenerated rice and potato mutants, demonstrating high A3A-PBE specificity. Another beneficial A3A-PBE feature is efficient C>T conversion independent of genomic context. In contrast to BE3, which showed no activity at three target sites in rice and wheat in a high CG context, editing efficiencies of up to 41.2% could be detected for A3A-PBE (Zong et al., 2018). Similarly, rAPOBEC1 substitution by PmCDA1 revealed higher editing efficiency at three out of four target sites in rice protoplast assays (Tang et al., 2019). While C>T conversions were below 10% using rAPOBEC1-nCas9-UGI, the nCas9-PmCDA1-UGI variant showed 10–50% editing efficiencies. However, the authors emphasized that the rAPOBEC1 sequence had not been codon optimized for this study; thus protein levels could contribute a substantial effect on editing efficacies (Tang et al., 2019). Similar results were obtained in transgenic rice calli with codon-optimized expression cassettes where rAPOBEC1 substitution by PmCDA1 led to 3–10-fold higher activity at five out of seven target sites (Xu et al., 2019).

To increase CBE efficiency on GC targets, a new variant was developed that carried a human AID triple mutant lacking a nuclear export signal (AID*Δ) (Ren et al., 2018). The deaminase domain was codon optimized for application in rice and fused to a nCas9 with no UGI domain (rBE5). This CBE variant functioned well on GC, AC, TC and CC target sites in rice protoplasts. Moreover, the AID*Δ-based CBE was able to efficiently catalyze C>T conversions at a genomic locus in transgenic rice lines that had been inaccessible for rAPOBEC1-CBE. It is worth emphasizing that 71.9% of transgenic lines also carried indel mutations in the target region. Further enzyme modification via UGI domain fusion reduced the indel ratio; however, frequencies were still higher than with rAPOBEC1-CBE (Ren et al., 2018).

Besides exchanging different CBE subunits, major improvements in C>T base conversion efficiencies were achieved by protein biosynthesis and nuclear transport optimization, both of which had been identified as key base editing bottlenecks in human cells (Koblan et al., 2018). In rice, CBE efficiency could be increased through codon optimization, integration of bipartite nuclear localization signals and substitution of the deaminase domain by Anc689, an ancestor of rAPOBEC1 previously identified through ancestral sequence reconstruction using APOBEC homologs (Koblan et al., 2018; Wang et al., 2019a).

The fidelity of rAPOBEC1 (BE3) and PmCDA1 (CDA) CBEs was further improved by increasing UGI concentration *in planta* (Qin et al., 2020a). A DNA sequence encoding three UGI-NLS copies each preceded by the self-cleaving peptide, 2A, was attached to the 3′ end of codon-optimized BE3 and CDA sequences (eBE3 and eCDA, respectively) leading to a polycistronic transcript and increased UGI amounts after translation. C>T conversion and indel formation was analyzed at five different target sites in transgenic rice plants. While editing efficiencies for both CBE variants did not differ significantly at most target sites (25–58.8%), indel frequency was drastically reduced from 12.5% to 25.0% for BE3 to 0% for the optimized variant. In addition, unwanted base changes were considerably lower with the optimized enzyme design (0–3.5%, eBE3) than for conventional BE3 (2.5–14.7%). This result could be reproduced in rice plants transformed with eCDA (Qin et al., 2020b).

Interestingly, a study using PmCDA1 in rice and tomato in a configuration that did not contain a UGI domain observed not only C>T conversions but also C>G changes (Shimatani et al., 2017), suggesting that a larger array of changes may be possible with CBEs when the UGI domain is removed. Indeed, similar results have been observed in subsequent studies in which a larger diversity of outcomes are observed. For instance, a study in *Arabidopsis* observed C>T, C>G and C>A changes when a PmCDA1 editor was utilized without a UGI domain (Bastet et al., 2019). While this configuration results in a larger array of editing outcomes, undesired changes can be removed through subsequent segregation or backcrossing, thus providing further utility to CBEs beyond C>T changes.

It is very challenging to define the best CBE design for every plant species. Due to the different experimental designs and analyses of each individual study, no direct conclusions can be drawn. However, researchers have identified some key factors that are generally important for successful application of CBEs in plants. As can be seen above, the choice of deaminase can have a dramatic effect on editing efficiency. While the A3A and PmCDA1 deaminases appear to be more efficient than the original rAPOBEC1, further studies directly comparing each design are required to determine the optimal deaminase in plants. One major advancement of CBE designs for plants includes the optimization of coding sequences to increase CBE concentrations in vivo (Koblan et al., 2018). Moreover, nuclear import can be enhanced by integration of two bipartite NLSs (Wang et al., 2019a). Finally, co-localization of additional UGI domains can increase CBE fidelity (Qin et al., 2020b).

4 Adenine base editing in plants

The recently developed adenine base editors have so far been applied in monocots (Hua et al., 2018, 2019, 2020b; Li et al., 2018; Wang et al., 2019a; Yan et al., 2018) as well as dicots (Kang et al., 2018). As with CBEs, various optimizations have been employed to achieve high gene expression and increase editing efficiencies in plants. Several ABE variants have been generated by fusion of evolved adenine deaminase heterodimer TadA-TadA*7.10 or TadA-TadA*7.8 to a SpCas9 nickase or a dead SpCas9 (Yan et al., 2018). However, only TadA-TadA*7.10 fused to nSpCas9 catalyzed desired A>G conversions in rice calli. Through substitution of nSpCas9 by its ortholog from *S. aureus* a new ABE version was created that efficiently generated desired base edits in rice and showed a broader base editing window (Hua et al., 2018; Qin et al., 2019). Moreover, replacement of SpCas9 by VQR-SpCas9 or Cas9-NG variants allowed efficient editing and multiplexing at NGA PAMs, which were not accessible with SpCas9-ABE (Hua et al., 2019; Zeng et al., 2020). Other optimizations of ABEs in plants focused on N- and C-terminal fusions of the ABE7.10 heterodimer to nSpCas9 and the number and position of nuclear localization signals (Li et al., 2018). Remarkably, integration of three NLSs into ABE led to significantly increased editing efficiencies in rice protoplasts compared to ABEs with only one or two NLSs. However, ABE variants in which TadA-TadA*7.10 was fused to the C-terminus of nCas9 were ineffective (Li et al., 2018). Further optimization of ABEs for application in plants was achieved through substitution of the TadA-TadA*7.10 heterodimer by the TadA*7.10 mutant, a strategy used for adenine base editing in bacteria (Gaudelli et al., 2017; Hua et al., 2020b). This simplified ABE version showed higher protein levels in rice calli and protoplasts, which was associated with an up to 1.9-fold increase of A>G edits at six out of seven target sites in two rice varieties (Hua et al., 2020b). Moreover, editing efficiencies of ABEs harboring different nCas9 variants (nSaCas9 and nSaKKHCas9) could be increased by adopting these enzymes to the simplified design (Hua et al., 2020b). Two major advantages of ABEs over CBEs are very low indel frequency of about 0.1% and precise base exchange in plants, respectively (Hua et al., 2018; Kang et al., 2018; Li et al., 2018).

5 Broadening protospacer adjacent motif (PAM) accessibility

One major limitation of targeted base modification with SpCas9-based tools is the prerequisite of an NGG PAM adjacent to the target site. To increase the number of accessible genomic sequences in plants, *S. pyogenes* Cas9 can be substituted by different enzyme variants (Qin et al., 2019). Extensive expansion of Cas9 PAM recognition has been achieved via rational design as well as directed evolution (Hu et al., 2018; Kleinstiver et al., 2015; Nishimasu et al., 2018). In addition to Cas9-based editors, Cas12a-based editors have been developed that recognize T-rich PAM sequences (Kleinstiver et al., 2016; Li et al., 2018; Sanson et al., 2019). While not all PAM variants of Cas9 and Cas12a have been adapted and optimized for base editing, the diversity of available Cas DNA-binding platforms indicates that a base editor should be able to access nearly any base in a genome (Fig. 2).

Several Cas9 variants have been tested in plants and demonstrated to function in base editing configurations. For instance, VQR-SpCas9n using PmCDA1 as the cytosine deaminase broadened the accessibility of SpCas9 CBEs to reach some NGA PAMs, most notably NGAG in rice (Wu et al., 2019). The same study also found that SpCas9n using PmCDA1 as the cytosine

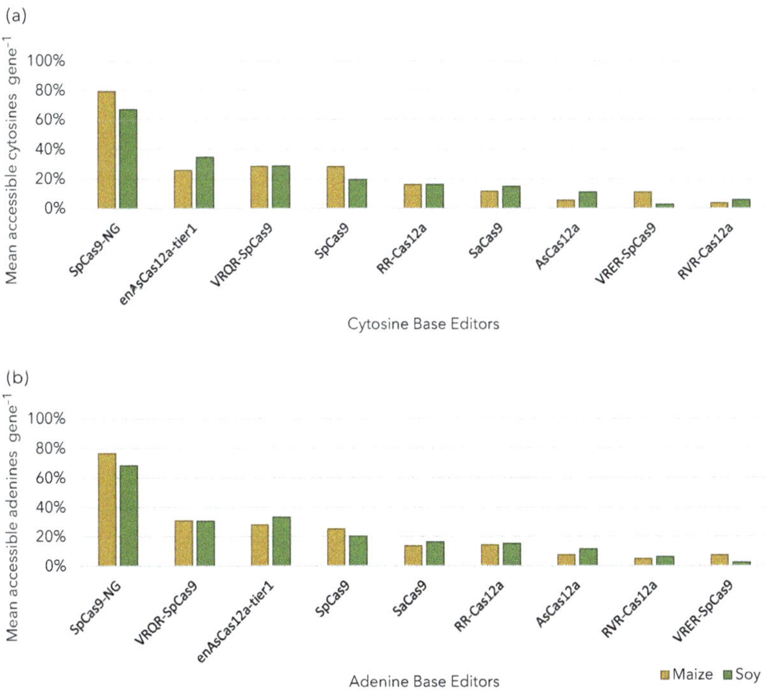

Figure 2 Base accessibility to CRISPR/Cas base editors in corn and soy genomes. (a) Mean cytosines accessible by various CRISPR/Cas cytosine base editors in maize and soy genes. (b) Mean adenines accessible by various CRISPR/Cas adenine base editors in maize and soy genes. Base editing window positions are as follows: SpCas9 and variants, 4 through 8; Cas12a and variants, 8 through 13; SaCas9, 2 through 12. The enAsCas12a-tier1 PAMs are defined in Kleinstiver et al. (2019). Base editing windows assume the use of rAPOBEC1 and ABE7.10 as the cytosine deaminase and adenine deaminase domains, respectively.

deaminase could effectively edit with an NAG PAM (Wu et al., 2019). Among the most promiscuous Cas9 variants is Cas9-NG, which recognizes noncanonical NG PAMs. In rice protoplasts, nCas9-NG-PmCDA-UGI showed high editing efficiency at most noncanonical PAMs, but low efficiency at NGG PAMs (Zhong et al., 2019). The strong deamination activity of PmCDA deaminase further complemented this broad PAM recognition with a wide editing window of 14 nt. The nickase xCas9 variant reported to have a NG PAM recognition fused to rAPOBEC1 demonstrated high editing efficiency at NGG PAMs comparable to nCas9-CBE, while a fusion to PmCDA1 increased C>T changes at C2 and C4 when compared in rice protoplasts (Zhong et al., 2019). However, other reports have shown xCas9 CBEs to be ineffective in rice (Zeng et al., 2020). In addition to Cas9-NG-CBE, an eCas9-NG variant CBE was developed to reduced off-target effects (Zeng et al., 2020). Both Cas9-NG-CBE and eCas9-NG-CBE enabled C-to-T changes at noncanonical PAM sites and increased the editing window width relative to Cas9-CBE.

Cas9 from other species also provide DNA-binding scaffolds to build base editors such as nSaCas9 and nSaKKHCas9 that recognize NNNRRT PAMs (Kim et al., 2017; Komor et al., 2017). Fusions were made with these Cas9 variants to rAPOBEC1 and either one or three copies of UGI. Three copies of UGI provided higher editing efficiency and increased the editing window at most target sites compared to the CBE with a single UGI moiety in rice (Qin et al., 2020b). However, replacing rAPOBEC1 with PmCDA1 in the nSaCas9 and nSaKKH CBEs resulted in lower editing efficiencies on average (Qin et al., 2020b).

6 Sequence diversification in crops

Crop improvements have long benefited from the introduction of genetic diversity and selective breeding. Chemical mutagens and radiation treatments have historically generated this diversity randomly in the genome. More targeted, semi-random indel diversity generation has been demonstrated in crops using CRISPR methods with dramatic phenotypic results from creating diversity in a known fruit size QTL and in promoters controlling inflorescence and plant architecture (Rodriguez-Leal et al., 2019). Similar spacer-dependent approaches in plant breeding with base editing would seem attractive. Base editing technologies also lend themselves to spacer-targeted sequence diversification within an editing window. Recently, simultaneous C and A base editing have been achieved in plants with effectors that incorporate both cytosine and adenine deamination activities in a single protein (Li et al., 2020). Pairing dual deaminases with NGG targeting Cas9 and with Cas9NG allowed access to 64% of all bases in a 1200-bp editing window and resulted in an overall 13% editing efficiency in rice. Moreover, C>G and C>A base changes were observed in addition to the expected C>T and A>G changes, generating additional sequence diversity (Li et al., 2020).

7 Off-target base editing

Despite a frequent desire to increase diversity in plants, there has been increased interest in whether additional variation beyond the designed edit (off-target editing) has occurred. The primary focus has been on guide-RNA dependent effects which occur when a base editor binds to a DNA site with similar, but not identical, sequence to the target site. Several strategies have been developed to reduce guide RNA-dependent off-target editing with nucleases, which are also effective with base editors. These strategies include (1) careful spacer design and off-target analysis to limit similarity to repeat sequences, (2) utilizing CRISPR effectors with mutations that increase DNA specificity, (3) reducing duration of editing activity by delivering the editor as an RNP complex and (4) modifying the guide RNA (Kim et al., 2019; Rees et al., 2017; Yeh et al., 2018).

Findings to date demonstrate that the off-target variation induced by base editors is several orders of magnitude less than the variation found in standard breeding programs. Some alternative deaminase domains have been shown (Doman et al., 2020) to reduce off-target editing, as well as reduce bystander base editing because of a smaller base editing window, likely both caused by lower catalytic efficiency. To assess deaminase effects on guide RNA-independent off-target effects, the Gao Lab (Jin et al., 2019) analyzed genome-wide indels and single-nucleotide variants (SNVs) induced by cytosine (rAPOBEC1 deaminase) and adenine (ABE 7.10) base editors in stably transformed rice. The finding was that the CBEs do increase C>T SNVs with an average <2-fold higher than the transformation control (means of ~500–630 SNVs vs. ~350 SNVs, respectively), but that no other genome wide variations increased with treatment by either editor class. However, even the highest detected whole genome C>T variation (~700 SNVs) is similar to previously observed somaclonal mutation frequency range (440–2600 mutations per plant in rice) (Li et al., 2016) and well below another similar study (~20 000) (Zhang et al., 2014). Thus, while increased background variation results from the presence of deaminase domains, it is less than variation typically introduced through the transformation process used to introduce editors to plants and vastly dwarfed by the millions of variants in breeding pools (Darracq et al., 2018; Li et al., 2014; Springer et al., 2009; Wang et al., 2018b; Zhao et al., 2018).

In their unaltered states, even when fused to a CRISPR effector, some deaminases can demonstrate affinity for RNA. The Joung Lab (Grünewald et al., 2019) performed an extensive study in human cell lines showing that both BE3 (rAPOBEC1) and ABEmax (TadA::TadA*) edit tens of thousands of individual

RNA base positions. We are not aware of any published studies in plants examining potential off-target editing of RNA by base editors. Nonetheless, as the same base editing machinery is used in plants as in human cell lines it is likely that the RNA-editing phenomenon is likely occurring in plants while CRISPR/Cas base editors are expressed. However, as the effect would be limited to short-lived RNA transcripts and the editing distributed amongst the whole transcriptome, there is little risk associated with this phenomenon in plants.

The ability to selectively breed many agricultural crops allows for base editor gene segregation after a heritable edit has been generated, which could additionally segregate away unintended off-target DNA edits and would stop the RNA editing process. For crops that are vegetatively propagated, transient delivery of the gene editing reagents or other alternatives to removal through breeding-cycle segregation will likely be preferred. Secondly, selecting for phenotypically normal plants, which is central to agricultural breeding programs, is expected to automatically select against any potential ill effects to plant health. Identifying and eliminating off-types has always been central to plant breeding programs, and regulatory agencies consider this a reliable process for ensuring food safety (Tieman et al., 2017).

8 Current applications of base editors in crops

While conventional breeding relies on the random mutagenesis of the plant genome through application of mutagenic agents or radiation, SDNs and base editor development has, for the first time, facilitated breeders to precisely modify plant genomes. The added precision of base editors allows breeders to fine-tune the genome like-never-before. As a result, in the short time since their introduction, there are several examples of successful trait improvement through base editing.

Herbicide tolerance is a powerful trait for crop improvement. One challenge of generating transgene-free edited plants is the selection of edited plants from the pool of non-edited regenerants. Herbicide-resistant gene base editing enables regeneration of edited but transgene-free plants on herbicide selection medium while simultaneously introducing additional traits through multiplexing (Veillet et al., 2019b). In this regard, one prominent target is the acetolactate synthase gene (ALS), which encodes the first enzyme in the biosynthetic pathway of branch chained amino acids (McCourt and Duggleby, 2006). Specific amino acid substitution can create resistance to sulfonylurea and imidazolinone herbicides and was introduced into crops including rice, wheat, tomato, potato and watermelon (Zhang et al., 2019). Competing with crops for essential resources in the field, weeds are a major risk in modern agriculture (Oerke and Dehne, 2004). In this regard, herbicide treatment has become an essential agricultural technique and herbicide resistance introduction into crops is an important commercial agronomic trait that has been demonstrated recently in rice through base editing for directed evolution of ALS (Kuang et al., 2020). In addition, herbicide-resistant rice plants could be generated through amino acid substitution via adenine base editing in the acetyl-coenzyme A carboxylase gene, a key enzyme in fatty acid biosynthesis (Li et al., 2018). Another example is the introduction of virus resistance by base editor application which has recently been shown in *A. thaliana* (Bastet et al., 2019). An amino acid substitution (N176K; caused by a C>G conversion), naturally occurring in pea (*Pisum sativum*), was introduced into *A. thaliana* translation initiation factor 4E by creating SNPs in the respective *eIF4E1* gene using a CBE lacking a UGI domain. Transgene-free T4 plants homozygous for desired edits were no longer susceptible to clover yellow vein virus (ClYVV). At the same time modified plants did not show any pleiotropic effects - in contrast to *eIF4E1* knockout mutants. The introduction of naturally occurring amino acid variations via base editing of susceptibility factors like *eIF4E* in relevant agricultural species offers great opportunities for crop improvement and control of economically devastating pests (Bastet et al., 2019). Manipulation and characterization of proteins involved in other traits, including quality, can be and have been performed (Veillet et al., 2019a).

Besides targeted amino acid residue modification of proteins, base editing tools can be deployed to modify mature mRNA architecture. In eukaryotes, precursor mRNA is processed to mature mRNA by spliceosomes that connect exons through removal of introns (Reddy et al., 2013). In plants, alternative splicing events occur in more than 60% of intron-containing genes and play an important role in increasing protein diversity (Reddy et al., 2013). Manipulation of mRNA splicing outcomes through introduction of SNPs via base editing enables functional investigation of specific protein variations while gene expression is still regulated by the endogenous promoter. Splicing site modification has been achieved in *Arabidopsis* and rice by application of cytosine and adenine base editors (Kang et al., 2018; Li et al., 2017; Xue et al., 2018). In *Arabidopsis*, specific splicing outcomes were enhanced while others could be inhibited through mutation of splicing sites by CBEs (Kang et al., 2018; Li et al., 2017; Xue et al., 2018). In this way, functions of HAB1 and RS31A protein variants contributing to ABA hypersensitive and genotoxic tolerant phenotypes, respectively, could be analyzed in planta (Kang et al., 2018; Li et al., 2017; Xue et al., 2018).

9 Conclusion

In a short period, base editors have been widely adopted as gene editing tools in plants. While early development of base editors occurred in non-plant systems, in planta editing has been shown to often be efficient (Table 1). Refinements

Published by Burleigh Dodds Science Publishing Limited, 2021.

for plant systems, such as codon optimization and plant-specific subcellular targeting sequences, have increased efficiency. To date, both adenine and CBEs have been demonstrated in plants with efficiencies over 80% and continue to be optimized. High-efficiency editing has been demonstrated in both monocot and dicot species showing that base editing should be possible in any transformable plant system. As can be seen in Fig. 2, accessibility varies depending on a given species genetic context, so base editor utility in plants is further increased by combining deaminase domains with CRISPR/Cas variants recognizing diverse PAM profiles.

A discussion on unintended, or off-target, editing often accompanies any conversation about genome editing; however, it is important to consider the use of editing tool in the context of its application. In plants, where increasing variation is central to the breeding process, background variation introduction is of little concern - especially in comparison to therapeutic uses in mammalian applications. Additionally, as base editors rely on the same RNA-guided machinery as traditional CRISPR/Cas nucleases, they maintain the same high-DNA specificity as SDNs, and off-target modification should be similarly low. The deaminase domain addition, which is specific to base editors, does add additional complexity as these domains have been shown to increase variation in both DNA and RNA sequences after delivery (Jin et al., 2019). While this would be a major concern in therapeutic applications, the variation extent observed is in line with traditional plant breeding practices and any off-types would be removed by standard agricultural breeding programs.

Recently, the development of prime editors (Anzalone et al., 2019) has shown promise as an additional tool that can result in predictable editing outcomes. While base editors are limited to changes of specific bases within a defined editing window, prime editors can replace entire portions of sequence in a base-independent manner, and with editing windows that are potentially dozens of bases long. While prime editors have the potential to be extremely powerful tools for gene editing, their efficiency in plants has thus far been low (Hua et al., 2020a; Lin et al., 2020). While some improvement in efficiency can be observed through the use of a surrogate selection system conferring herbicide resistance following a successful edit (Xu et al., 2020), it is clear that much more work needs to be done to increase the efficiency of this system to be widely applicable to plant applications.

The application of base editors to plant-specific applications has, to date, been limited. However, many aspects of the tools give base editors a great potential in plant gene editing. First, unlike traditional SDNs where the editing result is often a frame shift, base editors make nucleotide changes while maintaining the native reading frame. The result is a tool that is much better suited to sequence diversification for amino acid alterations. Additionally, the change that occurs as a base editing result is predictable, meaning that precisely targeted alterations can be recovered. By combining adenine and CBEs together, sequence diversification can be attempted - in a targeted and predictable manner. Looking forward, base editing has the potential to be a powerful tool in the plant breeding toolbox.

10 Where to look for further information

For an introduction to genome editing in plants Wada et al. (2020) give a comprehensive overview of the current state of the field. Rees and Liu (2018) provide a comprehensive review of base editing technology and is an excellent starting resource on this specific topic. The papers listed in Table 1 are the best starting points for base editing papers on a specific plant species.

Conferences on genome editing or DNA repair are often the best sources to find the latest developments in the field, some recurring meetings of note:

- International Conference on Base Editing (Deaminet).
- CRISPR and Plant Genome Editing Conference.
- Plant Genomics and Gene Editing Congress.

11 References

Allen, F., Crepaldi, L., Alsinet, C., Strong, A. J., Kleshchevnikov, V., De Angeli, P., Páleníková, P., Khodak, A., Kiselev, V., Kosicki, M., Bassett, A. R., Harding, H., Galanty, Y., Muñoz-Martínez, F., Metzakopian, E., Jackson, S. P. and Parts, L. (2019). Predicting the mutations generated by repair of Cas9-induced double-strand breaks. *Nature Biotechnology* 37(1): 64–72. DOI: 10.1038/nbt.4317.

Anzalone, A. V., Randolph, P. B., Davis, J. R., Sousa, A. A., Koblan, L. W., Levy, J. M., Chen, P. J., Wilson, C., Newby, G. A., Raguram, A. and Liu, D. R. (2019). Search-and-replace genome editing without double-strand breaks or donor DNA. *Nature* 576(7785): 149–157. DOI: 10.1038/s41586-019-1711-4.

Bastet, A., Zafirov, D., Giovinazzo, N., Guyon-Debast, A., Nogué, F., Robaglia, C. and Gallois, J. L. (2019). Mimicking natural polymorphism in eIF4E by CRISPR-Cas9 base editing is associated with resistance to potyviruses. *Plant Biotechnology Journal* 17(9): 1736–1750. DOI: 10.1111/pbi.13096.

Bohn, M. F., Shandilya, S. M. D., Silvas, T. V., Nalivaika, E. A., Kouno, T., Kelch, B. A., Ryder, S. P., Kurt-Yilmaz, N., Somasundaran, M. and Schiffer, C. A. (2015). The ssDNA mutator APOBEC3A is regulated by cooperative dimerization. *Structure* 23(5): 903-911. DOI: 10.1016/j.str.2015.03.016.

Chen, Y., Wang, Z., Ni, H., Xu, Y., Chen, Q. and Jiang, L. (2017). CRISPR/Cas9-mediated base-editing system efficiently generates gain-of-function mutations in Arabidopsis. *Science China. Life Sciences* 60(5): 520–523. DOI: 10.1007/s11427-017-9021-5.

Cheng, T. L., Li, S., Yuan, B., Wang, X., Zhou, W. and Qiu, Z. (2019). Expanding C–T base editing toolkit with diversified cytidine deaminases. *Nature Communications* 10(1): 3612. DOI: 10.1038/s41467-019-11562-6.

Darracq, A., Vitte, C., Nicolas, S., Duarte, J., Pichon, J. P., Mary-Huard, T., Chevalier, C., Bérard, A., Le Paslier, M. C., Rogowsky, P., Charcosset, A. and Joets, J. (2018). Sequence analysis of European maize inbred line F2 provides new insights into molecular and chromosomal characteristics of presence/absence variants. *BMC Genomics* 19(1): 119. DOI: 10.1186/s12864-018-4490-7.

Doman, J. L., Raguram, A., Newby, G. A. and Liu, D. R. (2020). Evaluation and minimization of Cas9-independent off-target DNA editing by cytosine base editors. *Nature Biotechnology* 38(5): 620–628. DOI: 10.1038/s41587-020-0414-6.

Endo, M., Mikami, M., Endo, A., Kaya, H., Itoh, T., Nishimasu, H., Nureki, O. and Toki, S. (2019). Genome editing in plants by engineered CRISPR–Cas9 recognizing NG PAM. *Nature Plants* 5(1): 14–17. DOI: 10.1038/s41477-018-0321-8.

Feng, Z., Zhang, B., Ding, W., Liu, X., Yang, D. L., Wei, P., Cao, F., Zhu, S., Zhang, F., Mao, Y. and Zhu, J. K. (2013). Efficient genome editing in plants using a CRISPR/Cas system. *Cell Research* 23(10): 1229–1232. DOI: 10.1038/cr.2013.114.

Gaudelli, N. M., Komor, A. C., Rees, H. A., Packer, M. S., Badran, A. H., Bryson, D. I. and Liu, D. R. (2017). Programmable base editing of A•T to G•C in genomic DNA without DNA cleavage. *Nature* 551(7681): 464–471. DOI: 10.1038/nature24644.

Gehrke, J. M., Cervantes, O., Clement, M. K., Wu, Y., Zeng, J., Bauer, D. E., Pinello, L. and Joung, J. K. (2018). An APOBEC3A-Cas9 base editor with minimized bystander and off-target activities. *Nature Biotechnology* 36(10): 977–982. DOI: 10.1038/nbt.4199.

Grünewald, J., Zhou, R., Iyer, S., Lareau, C. A., Garcia, S. P., Aryee, M. J. and Joung, J. K. (2019). CRISPR DNA base editors with reduced RNA off-target and self-editing activities. *Nature Biotechnology* 37(9): 1041–1048. DOI: 10.1038/s41587-019-0236-6.

Hao, L., Ruiying, Q., Xiaoshuang, L., Shengxiang, L., Rongfang, X., Jianbo, Y. and Pengcheng, W. (2019). CRISPR/Cas9-mediated adenine base editing in rice genome. *Rice Science* 26(2): 125–128. DOI: 10.1016/j.rsci.2018.07.002.

Hess, G. T., Frésard, L., Han, K., Lee, C. H., Li, A., Cimprich, K. A., Montgomery, S. B. and Bassik, M. C. (2016). Directed evolution using dCas9-targeted somatic hypermutation in mammalian cells. *Nature Methods* 13(12): 1036–1042. DOI: 10.1038/nmeth.4038.

Hu, J. H., Miller, S. M., Geurts, M. H., Tang, W., Chen, L., Sun, N., Zeina, C. M., Gao, X., Rees, H. A., Lin, Z. and Liu, D. R. (2018). Evolved Cas9 variants with broad PAM compatibility and high DNA specificity. *Nature* 556(7699): 57–63. DOI: 10.1038/nature26155.

Hua, K., Tao, X., Yuan, F., Wang, D. and Zhu, J. K. (2018). Precise A·T to G·C base editing in the rice genome. *Molecular Plant* 11(4): 627–630. DOI: 10.1016/j.molp.2018.02.007.

Hua, K., Tao, X. and Zhu, J. K. (2019). Expanding the base editing scope in rice by using Cas9 variants. *Plant Biotechnology Journal* 17(2): 499–504. DOI: 10.1111/pbi.12993.

Hua, K., Jiang, Y., Tao, X. and Zhu, J. K. (2020a). Precision genome engineering in rice using prime editing system. *Plant Biotechnology Journal* 18(11): 2167–2169. DOI: 10.1111/pbi.13395.

Hua, K., Tao, X., Liang, W., Zhang, Z., Gou, R. and Zhu, J. K. (2020b). Simplified adenine base editors improve adenine base editing efficiency in rice. *Plant Biotechnology Journal* 18(3): 770–778. DOI: 10.1111/pbi.13244.

Huang, T. P., Zhao, K. T., Miller, S. M., Gaudelli, N. M., Oakes, B. L., Fellmann, C., Savage, D. F. and Liu, D. R. (2019). Circularly permuted and PAM-modified Cas9 variants broaden the targeting scope of base editors. *Nature Biotechnology* 37(6): 626–631. DOI: 10.1038/s41587-019-0134-y.

Jiang, W., Feng, S., Huang, S., Yu, W., Li, G., Yang, G., Liu, Y., Zhang, Y., Zhang, L., Hou, Y., Chen, J., Chen, J. and Huang, X. (2018). BE-PLUS: a new base editing tool with broadened editing window and enhanced fidelity. *Cell Research* 28(8): 855–861. DOI: 10.1038/s41422-018-0052-4.

Jin, S., Zong, Y., Gao, Q., Zhu, Z., Wang, Y., Qin, P., Liang, C., Wang, D., Qiu, J. L., Zhang, F. and Gao, C. (2019). Cytosine, but not adenine, base editors induce genome-wide off-target mutations in rice. *Science* 364(6437): 292–295. DOI: 10.1126/science.aaw7166.

Jinek, M., Chylinski, K., Fonfara, I., Hauer, M., Doudna, J. A. and Charpentier, E. (2012). A programmable dual-RNA-guided DNA endonuclease in adaptive bacterial immunity. *Science* 337(6096): 816–821. DOI: 10.1126/science.1225829.

Jinek, M., Jiang, F., Taylor, D. W., Sternberg, S. H., Kaya, E., Ma, E., Anders, C., Hauer, M., Zhou, K., Lin, S., Kaplan, M., Iavarone, A. T., Charpentier, E., Nogales, E. and Doudna, J. A. (2014). Structures of Cas9 endonucleases reveal RNA-mediated conformational activation. *Science* 343(6176): 1247997. DOI: 10.1126/science.1247997.

Kang, B. C., Yun, J. Y., Kim, S. T., Shin, Y., Ryu, J., Choi, M., Woo, J. W. and Kim, J. S. (2018). Precision genome engineering through adenine base editing in plants. *Nature Plants* 4(7): 427–431. DOI: 10.1038/s41477-018-0178-x.

Kim, D., Kim, D. E., Lee, G., Cho, S. I. and Kim, J. S. (2019). Genome-wide target specificity of CRISPR RNA-guided adenine base editors. *Nature Biotechnology* 37(4): 430–435. DOI: 10.1038/s41587-019-0050-1.

Kim, J., Malashkevich, V., Roday, S., Lisbin, M., Schramm, V. L. and Almo, S. C. (2006). Structural and kinetic characterization of Escherichia coli TadA, the wobble-specific tRNA deaminase. *Biochemistry* 45(20): 6407–6416. DOI: 10.1021/bi0522394.

Kim, Y. B., Komor, A. C., Levy, J. M., Packer, M. S., Zhao, K. T. and Liu, D. R. (2017). Increasing the genome-targeting scope and precision of base editing with engineered Cas9-cytidine deaminase fusions. *Nature Biotechnology* 35(4): 371–376. DOI: 10.1038/nbt.3803.

Kim, Y. G., Cha, J. and Chandrasegaran, S. (1996). Hybrid restriction enzymes: zinc finger fusions to Fok I cleavage domain. *Proceedings of the National Academy of Sciences of the United States of America* 93(3): 1156–1160. DOI: 10.1073/pnas.93.3.1156.

Kleinstiver, B. P., Prew, M. S., Tsai, S. Q., Topkar, V. V., Nguyen, N. T., Zheng, Z., Gonzales, A. P., Li, Z., Peterson, R. T., Yeh, J. R., Aryee, M. J. and Joung, J. K. (2015). Engineered CRISPR-Cas9 nucleases with altered PAM specificities. *Nature* 523(7561): 481–485. DOI: 10.1038/nature14592.

Kleinstiver, B. P., Pattanayak, V., Prew, M. S., Tsai, S. Q., Nguyen, N. T., Zheng, Z. and Joung, J. K. (2016). High-fidelity CRISPR-Cas9 nucleases with no detectable genome-wide off-target effects. *Nature* 529(7587): 490–495. DOI: 10.1038/nature16526.

Kleinstiver, B. P., Sousa, A. A., Walton, R. T., Tak, Y. E., Hsu, J. Y., Clement, K., Welch, M. M., Horng, J. E., Malagon-Lopez, J., Scarfò, I., Maus, M. V., Pinello, L., Aryee, M. J. and Joung, J. K. (2019). Engineered CRISPR-Cas12a variants with increased activities and improved targeting ranges for gene, epigenetic and base editing. *Nature Biotechnology* 37(3): 276–282. DOI: 10.1038/s41587-018-0011-0.

Koblan, L. W., Doman, J. L., Wilson, C., Levy, J. M., Tay, T., Newby, G. A., Maianti, J. P., Raguram, A. and Liu, D. R. (2018). Improving cytidine and adenine base editors by expression optimization and ancestral reconstruction. *Nature Biotechnology* 36(9): 843–846. DOI: 10.1038/nbt.4172.

Komor, A. C., Kim, Y. B., Packer, M. S., Zuris, J. A. and Liu, D. R. (2016). Programmable editing of a target base in genomic DNA without double-stranded DNA cleavage. *Nature* 533(7603): 420–424. DOI: 10.1038/nature17946.

Komor, A. C., Zhao, K. T., Packer, M. S., Gaudelli, N. M., Waterbury, A. L., Koblan, L. W., Kim, Y. B., Badran, A. H. and Liu, D. R. (2017). Improved base excision repair inhibition and bacteriophage Mu Gam protein yields C:G-to-T:A base editors with higher efficiency and product purity. *Science Advances* 3(8): eaao4774. DOI: 10.1126/sciadv.aao4774.

Kuang, Y., Li, S., Ren, B., Yan, F., Spetz, C., Li, X., Zhou, X. and Zhou, H. (2020). Base-editing-mediated artificial evolution of OsALS1 in planta to develop novel herbicide-tolerant rice germplasms. *Molecular Plant* 13(4): 565–572. DOI: 10.1016/j.molp.2020.01.010.

Li, C., Zong, Y., Wang, Y., Jin, S., Zhang, D., Song, Q., Zhang, R. and Gao, C. (2018). Expanded base editing in rice and wheat using a Cas9-adenosine deaminase fusion. *Genome Biology* 19(1): 59. DOI: 10.1186/s13059-018-1443-z.

Li, C., Zhang, R., Meng, X., Chen, S., Zong, Y., Lu, C., Qiu, J. L., Chen, Y. H., Li, J. and Gao, C. (2020). Targeted, random mutagenesis of plant genes with dual cytosine and adenine base editors. *Nature Biotechnology* 38(7): 875–882. DOI: 10.1038/s41587-019-0393-7.

Li, J., Sun, Y., Du, J., Zhao, Y. and Xia, L. (2017). Generation of targeted point mutations in rice by a modified CRISPR/Cas9 system. *Molecular Plant* 10(3): 526–529. DOI: 10.1016/j.molp.2016.12.001.

Li, S., Zheng, Y. C., Cui, H. R., Fu, H. W., Shu, Q. Y. and Huang, J. Z. (2016). Frequency and type of inheritable mutations induced by γ rays in rice as revealed by whole genome sequencing. *Journal of Zhejiang University: Science B* 17(12): 905–915. DOI: 10.1631/jzus.B1600125.

Li, Y., Zhu, J., Wu, H., Liu, C., Huang, C., Lan, J., Zhao, Y. and Xie, C. (2019). Precise base editing of non-allelic acetolactate synthase genes confers sulfonylurea herbicide resistance in maize. *The Crop Journal* 8(3): 449–456. DOI: 10.1016/j.cj.2019.10.001.

Li, Y. H., Zhou, G., Ma, J., Jiang, W., Jin, L. G., Zhang, Z., Guo, Y., Zhang, J., Sui, Y., Zheng, L., Zhang, S. S., Zuo, Q., Shi, X. H., Li, Y. F., Zhang, W. K., Hu, Y., Kong, G., Hong, H. L., Tan, B., Song, J., Liu, Z. X., Wang, Y., Ruan, H., Yeung, C. K., Liu, J., Wang, H., Zhang, L. J., Guan, R. X., Wang, K. J., Li, W. B., Chen, S. Y., Chang, R. Z., Jiang, Z., Jackson, S. A., Li, R. and Qiu, L. J. (2014). De novo assembly of soybean wild relatives for pan-genome analysis of diversity and agronomic traits. *Nature Biotechnology* 32(10): 1045–1052. DOI: 10.1038/nbt.2979.

Liang, P., Sun, H., Sun, Y., Zhang, X., Xie, X., Zhang, J., Zhang, Z., Chen, Y., Ding, C., Xiong, Y., Ma, W., Liu, D., Huang, J. and Songyang, Z. (2017). Effective gene editing by high-fidelity base editor 2 in mouse zygotes. *Protein and Cell* 8(8): 601–611. DOI: 10.1007/s13238-017-0418-2.

Lin, Q., Zong, Y., Xue, C., Wang, S., Jin, S., Zhu, Z., Wang, Y., Anzalone, A. V., Raguram, A., Doman, J. L., Liu, D. R. and Gao, C. (2020). Prime genome editing in rice and wheat. *Nature Biotechnology* 38(5): 1–4. DOI: 10.1038/s41587-020-0455-x.

Liu, Z., Chen, M., Shan, H., Chen, S., Xu, Y., Song, Y., Zhang, Q., Yuan, H., Ouyang, H., Li, Z. and Lai, L. (2019). Expanded targeting scope and enhanced base editing efficiency in rabbit using optimized xCas9(3.7). *Cellular and Molecular Life Sciences : CMLS* 76(20): 4155–4164. DOI: 10.1007/s00018-019-03110-8.

Lu, Y. and Zhu, J. K. (2017). Precise editing of a target base in the rice genome using a modified CRISPR/Cas9 system. *Molecular Plant* 10(3): 523–525. DOI: 10.1016/j.molp.2016.11.013.

McCourt, J. A. and Duggleby, R. G. (2006). Acetohydroxyacid synthase and its role in the biosynthetic pathway for branched-chain amino acids. *Amino Acids* 31(2): 173–210. DOI: 10.1007/s00726-005-0297-3.

Molla, K. A. and Yang, Y. (2020). Predicting CRISPR/Cas9-induced mutations for precise genome editing. *Trends in Biotechnology* 38(2): 136–141. DOI: 10.1016/j.tibtech.2019.08.002.

Nishimasu, H., Shi, X., Ishiguro, S., Gao, L., Hirano, S., Okazaki, S., Noda, T., Abudayyeh, O. O., Gootenberg, J. S., Mori, H., Oura, S., Holmes, B., Tanaka, M., Seki, M., Hirano, H., Aburatani, H., Ishitani, R., Ikawa, M., Yachie, N., Zhang, F. and Nureki, O. (2018). Engineered CRISPR-Cas9 nuclease with expanded targeting space. *Science* 361(6408): 1259–1262. DOI: 10.1126/science.aas9129.

Oerke, E.-C. and Dehne, H.-W. (2004). Safeguarding production—losses in major crops and the role of crop protection. *Crop Protection* 23(4): 275–285. DOI: 10.1016/j.cropro.2003.10.001.

Qin, L., Li, J., Wang, Q., Xu, Z., Sun, L., Alariqi, M., Manghwar, H., Wang, G., Li, B., Ding, X., Rui, H., Huang, H., Lu, T., Lindsey, K., Daniell, H., Zhang, X. and Jin, S. (2020a). High-efficient and precise base editing of C•G to T•A in the allotetraploid cotton (Gossypium hirsutum) genome using a modified CRISPR/Cas9 system. *Plant Biotechnology Journal* 18(1): 45–56. DOI: 10.1111/pbi.13168.

Qin, R., Li, J., Li, H., Zhang, Y., Liu, X., Miao, Y., Zhang, X. and Wei, P. (2019). Developing a highly efficient and wildly adaptive CRISPR-SaCas9 toolset for plant genome editing. *Plant Biotechnology Journal* 17(4): 706–708. DOI: 10.1111/pbi.13047.

Qin, R., Liao, S., Li, J., Li, H., Liu, X., Yang, J. and Wei, P. (2020b). Increasing fidelity and efficiency by modifying cytidine base-editing systems in rice. *The Crop Journal* 8(3): 396–402. DOI: 10.1016/j.cj.2019.04.007.

Reddy, A. S. N., Marquez, Y., Kalyna, M. and Barta, A. (2013). Complexity of the alternative splicing landscape in plants. *The Plant Cell* 25(10): 3657–3683. DOI: 10.1105/tpc.113.117523.

Rees, H. A. and Liu, D. R. (2018). Base editing: precision chemistry on the genome and transcriptome of living cells. *Nature Reviews Genetics* 19(12): 770–788. DOI: 10.1038/s41576-018-0059-1.

Rees, H. A., Komor, A. C., Yeh, W. H., Caetano-Lopes, J., Warman, M., Edge, A. S. B. and Liu, D. R. (2017). Improving the DNA specificity and applicability of base editing through protein engineering and protein delivery. *Nature Communications* 8(1): 15790. DOI: 10.1038/ncomms15790.

Ren, B., Yan, F., Kuang, Y., Li, N., Zhang, D., Zhou, X., Lin, H. and Zhou, H. (2018). Improved base editor for efficiently inducing genetic variations in rice with CRISPR/Cas9-guided hyperactive hAID mutant. *Molecular Plant* 11(4): 623–626. DOI: 10.1016/j.molp.2018.01.005.

Rodriguez-Leal, D., Xu, C., Kwon, C. T., Soyars, C., Demesa-Arevalo, E., Man, J., Liu, L., Lemmon, Z. H., Jones, D. S., Van Eck, J., Jackson, D. P., Bartlett, M. E., Nimchuk, Z. L. and Lippman, Z. B. (2019). Evolution of buffering in a genetic circuit controlling plant stem cell proliferation. *Nature Genetics* 51(5): 786–792. DOI: 10.1038/s41588-019-0389-8.

Sanson, K. R., DeWeirdt, P. C., Sangree, A. K., Sanson, K. R., DeWeirdt, P. C., Sangree, A. K., Hanna, R. E., Hegde, M., Teng, T., Borys, S. M., Strand, C., Keith Joung, J., Kleinstiver, B. P., Pan, X., Huang, A. and Doench, J. G. (2019). Optimization of AsCas12a for combinatorial genetic screens in human cells. *bioRxiv*: 747170. DOI: 10.1101/747170.

Schmidt, C., Pacher, M. and Puchta, H. (2018). Transgenic plants, methods and protocols. *Methods in Molecular Biology* 1864: 237–266. DOI: 10.1007/978-1-4939-8778-8_17.

Shimatani, Z., Kashojiya, S., Takayama, M., Terada, R., Arazoe, T., Ishii, H., Teramura, H., Yamamoto, T., Komatsu, H., Miura, K., Ezura, H., Nishida, K., Ariizumi, T. and Kondo, A. (2017). Targeted base editing in rice and tomato using a CRISPR-Cas9 cytidine deaminase fusion. *Nature Biotechnology* 35(5): 441–443. DOI: 10.1038/nbt.3833.

Springer, N. M., Ying, K., Fu, Y., Ji, T., Yeh, C. T., Jia, Y., Wu, W., Richmond, T., Kitzman, J., Rosenbaum, H., Iniguez, A. L., Barbazuk, W. B., Jeddeloh, J. A., Nettleton, D. and Schnable, P. S. (2009). Maize inbreds exhibit high levels of copy number variation (CNV) and presence/absence variation (PAV) in genome content. *PLoS Genetics* 5(11): e1000734. DOI: 10.1371/journal.pgen.1000734.

Tan, J., Zhang, F., Karcher, D. and Bock, R. (2019). Engineering of high-precision base editors for site-specific single nucleotide replacement. *Nature Communications* 10(1): 439. DOI: 10.1038/s41467-018-08034-8.

Tang, X., Ren, Q., Yang, L., Bao, Y., Zhong, Z., He, Y., Liu, S., Qi, C., Liu, B., Wang, Y., Sretenovic, S., Zhang, Y., Zheng, X., Zhang, T., Qi, Y. and Zhang, Y. (2019). Single transcript unit CRISPR 2.0 systems for robust Cas9 and Cas12a mediated plant genome editing. *Plant Biotechnology Journal* 17(7): 1431–1445. DOI: 10.1111/pbi.13068.

Thuronyi, B. W., Koblan, L. W., Levy, J. M., Yeh, W. H., Zheng, C., Newby, G. A., Wilson, C., Bhaumik, M., Shubina-Oleinik, O., Holt, J. R. and Liu, D. R. (2019). Continuous evolution of base editors with expanded target compatibility and improved activity. *Nature Biotechnology* 37(9): 1070–1079. DOI: 10.1038/s41587-019-0193-0.

Tian, S., Jiang, L., Cui, X., Zhang, J., Guo, S., Li, M., Zhang, H., Ren, Y., Gong, G., Zong, M., Liu, F., Chen, Q. and Xu, Y. (2018). Engineering herbicide-resistant watermelon variety through CRISPR/Cas9-mediated base-editing. *Plant Cell Reports* 37(9): 1353–1356. DOI: 10.1007/s00299-018-2299-0.

Tieman, D., Zhu, G., Resende, M. F. R., Lin, T., Nguyen, C., Bies, D., Rambla, J. L., Beltran, K. S., Taylor, M., Zhang, B., Ikeda, H., Liu, Z., Fisher, J., Zemach, I., Monforte, A., Zamir, D., Granell, A., Kirst, M., Huang, S. and Klee, H. (2017). A chemical genetic roadmap to improved tomato flavor. *Science* 355(6323): 391–394. DOI: 10.1126/science.aal1556.

van Overbeek, M., Capurso, D., Carter, M. M., Thompson, M. S., Frias, E., Russ, C., Reece-Hoyes, J. S., Nye, C., Gradia, S., Vidal, B., Zheng, J., Hoffman, G. R., Fuller, C. K. and May, A. P. (2016). DNA repair profiling reveals nonrandom outcomes at Cas9-mediated breaks. *Molecular Cell* 63(4): 633–646. DOI: 10.1016/j.molcel.2016.06.037.

Veillet, F., Chauvin, L., Kermarrec, M. P., Sevestre, F., Merrer, M., Terret, Z., Szydlowski, N., Devaux, P., Gallois, J. L. and Chauvin, J. E. (2019a). The Solanum tuberosum GBSSI gene: a target for assessing gene and base editing in tetraploid potato. *Plant Cell Reports* 38(9): 1065–1080. DOI: 10.1007/s00299-019-02426-w.

Veillet, F., Perrot, L., Chauvin, L., Kermarrec, M. P., Guyon-Debast, A., Chauvin, J. E., Nogué, F. and Mazier, M. (2019b). Transgene-free genome editing in tomato and potato plants using Agrobacterium-mediated delivery of a CRISPR/Cas9 cytidine base editor. *International Journal of Molecular Sciences* 20(2): 402. DOI: 10.3390/ijms20020402.

Wada, N., Ueta, R., Osakabe, Y. and Osakabe, K. (2020). Precision genome editing in plants: state-of-the-art in CRISPR/Cas9-based genome engineering. *BMC Plant Biology* 20: 234.

Wang, L., Xue, W., Yan, L., Li, X., Wei, J., Chen, M., Wu, J., Yang, B., Yang, L. and Chen, J. (2017). Enhanced base editing by co-expression of free uracil DNA glycosylase inhibitor. *Cell Research* 27(10): 1289–1292. DOI: 10.1038/cr.2017.111.

Wang, M., Wang, Z., Mao, Y., Lu, Y., Yang, R., Tao, X. and Zhu, J. K. (2019a). Optimizing base editors for improved efficiency and expanded editing scope in rice. *Plant Biotechnology Journal* 17(9): 1697–1699. DOI: 10.1111/pbi.13124.

Wang, T., Badran, A. H., Huang, T. P. and Liu, D. R. (2018a). Continuous directed evolution of proteins with improved soluble expression. *Nature Chemical Biology* 14(10): 972–980. DOI: 10.1038/s41589-018-0121-5.

Wang, W., Mauleon, R., Hu, Z., Chebotarov, D., Tai, S., Wu, Z., Li, M., Zheng, T., Fuentes, R. R., Zhang, F., Mansueto, L., Copetti, D., Sanciangco, M., Palis, K. C., Xu, J., Sun, C., Fu, B., Zhang, H., Gao, Y., Zhao, X., Shen, F., Cui, X., Yu, H., Li, Z., Chen, M., Detras, J., Zhou, Y., Zhang, X., Zhao, Y., Kudrna, D., Wang, C., Li, R., Jia, B., Lu, J., He, X., Dong, Z., Xu, J., Li, Y., Wang, M., Shi, J., Li, J., Zhang, D., Lee, S., Hu, W., Poliakov, A., Dubchak, I., Ulat, V. J., Borja, F. N., Mendoza, J. R., Ali, J., Li, J., Gao, Q., Niu, Y., Yue, Z., Naredo, M. E. B., Talag, J., Wang, X., Li, J., Fang, X., Yin, Y., Glaszmann, J. C., Zhang, J., Li, J., Hamilton, R. S., Wing, R. A., Ruan, J., Zhang, G., Wei, C., Alexandrov, N., McNally, K. L., Li, Z. and Leung, H. (2018b). Genomic variation in 3,010 diverse accessions of Asian cultivated rice. *Nature* 557(7703): 43–49. DOI: 10.1038/s41586-018-0063-9.

Wang, Y., Liu, Y., Li, J., Yang, Y., Ni, X., Cheng, H., Huang, T., Guo, Y., Ma, H., Zheng, P., Wang, M., Sun, J. and Ma, Y. (2019b). Expanding targeting scope, editing window, and base transition capability of base editing in Corynebacterium glutamicum. *Biotechnology and Bioengineering* 116(11): 3016–3029. DOI: 10.1002/bit.27121.

Wu, Y., Xu, W., Wang, F., Zhao, S., Feng, F., Song, J., Zhang, C. and Yang, J. (2019). Increasing cytosine base editing scope and efficiency With engineered Cas9-PmCDA1 fusions and the modified sgRNA in rice. *Frontiers in Genetics* 10: 379. DOI: 10.3389/fgene.2019.00379.

Xu, R., Li, J., Liu, X., Shan, T., Qin, R. and Wei, P. (2020). Development of a plant prime editing system for precise editing in the rice genome. *Plant Communications* 1(3): 100043. DOI: 10.1016/j.xplc.2020.100043.

Xu, W., Song, W., Yang, Y., Wu, Y., Lv, X., Yuan, S., Liu, Y. and Yang, J. (2019). Multiplex nucleotide editing by high-fidelity Cas9 variants with improved efficiency in rice. *BMC Plant Biology* 19(1): 511. DOI: 10.1186/s12870-019-2131-1.

Xue, C., Zhang, H., Lin, Q., Fan, R. and Gao, C. (2018). Manipulating mRNA splicing by base editing in plants. *Science China. Life Sciences* 61(11): 1293–1300. DOI: 10.1007/s11427-018-9392-7.

Yan, F., Kuang, Y., Ren, B., Wang, J., Zhang, D., Lin, H., Yang, B., Zhou, X. and Zhou, H. (2018). Highly efficient A·T to G·C base editing by Cas9n-guided tRNA adenosine deaminase in rice. *Molecular Plant* 11(4): 631–634. DOI: 10.1016/j.molp.2018.02.008.

Yeh, W. H., Chiang, H., Rees, H. A., Edge, A. S. B. and Liu, D. R. (2018). In vivo base editing of post-mitotic sensory cells. *Nature Communications* 9(1): 2184. DOI: 10.1038/s41467-018-04580-3.

Zeng, D., Li, X., Huang, J., Li, Y., Cai, S., Yu, W., Li, Y., Huang, Y., Xie, X., Gong, Q., Tan, J., Zheng, Z., Guo, M., Liu, Y. G. and Zhu, Q. (2020). Engineered Cas9 variant tools expand targeting scope of genome and base editing in rice. *Plant Biotechnology Journal* 18(6): 1348–1350. DOI: 10.1111/pbi.13293.

Zhang, C., Xu, W., Wang, F., Kang, G., Yuan, S., Lv, X., Li, L., Liu, Y. and Yang, J. (2020). Expanding the base editing scope to GA and relaxed NG PAM sites by improved xCas9 system. *Plant Biotechnology Journal* 18(4): 884–886. DOI: 10.1111/pbi.13259.

Zhang, D., Wang, Z., Wang, N., Gao, Y., Liu, Y., Wu, Y., Bai, Y., Zhang, Z., Lin, X., Dong, Y., Ou, X., Xu, C. and Liu, B. (2014). Tissue culture-induced heritable genomic variation in rice, and their phenotypic implications. *PLoS ONE* 9(5): e96879. DOI: 10.1371/journal.pone.0096879.

Zhang, R., Liu, J., Chai, Z., Chen, S., Bai, Y., Zong, Y., Chen, K., Li, J., Jiang, L. and Gao, C. (2019). Generation of herbicide tolerance traits and a new selectable marker in wheat using base editing. *Nature Plants* 5(5): 480–485. DOI: 10.1038/s41477-019-0405-0.

Zhang, Y., Qin, W., Lu, X., Xu, J., Huang, H., Bai, H., Li, S. and Lin, S. (2017). Programmable base editing of zebrafish genome using a modified CRISPR-Cas9 system. *Nature Communications* 8(1): 118. DOI: 10.1038/s41467-017-00175-6.

Zhao, Q., Feng, Q., Lu, H., Li, Y., Wang, A., Tian, Q., Zhan, Q., Lu, Y., Zhang, L., Huang, T., Wang, Y., Fan, D., Zhao, Y., Wang, Z., Zhou, C., Chen, J., Zhu, C., Li, W., Weng, Q., Xu, Q., Wang, Z. X., Wei, X., Han, B. and Huang, X. (2018). Pan-genome analysis highlights the extent of genomic variation in cultivated and wild rice. *Nature Genetics* 50(2): 278–284. DOI: 10.1038/s41588-018-0041-z.

Zhong, Z., Sretenovic, S., Ren, Q., Yang, L., Bao, Y., Qi, C., Yuan, M., He, Y., Liu, S., Liu, X., Wang, J., Huang, L., Wang, Y., Baby, D., Wang, D., Zhang, T., Qi, Y. and Zhang, Y. (2019). Improving plant genome editing with high-fidelity xCas9 and non-canonical PAM-targeting Cas9-NG. *Molecular Plant* 12(7): 1027–1036. DOI: 10.1016/j.molp.2019.03.011.

Zong, Y., Wang, Y., Li, C., Zhang, R., Chen, K., Ran, Y., Qiu, J. L., Wang, D. and Gao, C. (2017). Precise base editing in rice, wheat and maize with a Cas9-cytidine deaminase fusion. *Nature Biotechnology* 35(5): 438–440. DOI: 10.1038/nbt.3811.

Zong, Y., Song, Q., Li, C., Jin, S., Zhang, D., Wang, Y., Qiu, J. L. and Gao, C. (2018). Efficient C-to-T base editing in plants using a fusion of nCas9 and human APOBEC3A. *Nature Biotechnology* 36(10): 950–953. DOI: 10.1038/nbt.4261.

Chapter 6

Advances in guide RNA design for editing plant genomes using CRISPR-Cas systems

Kaiyuan Chen, Hao Liu and Kabin Xie, Huazhong Agricultural University, China; Muhammad Tahir ul Qamar and Ling-Ling Chen, Huazhong Agricultural University, and Guangxi University, China

1 Introduction

Clustered regularly interspaced short palindromic repeats (CRISPR) and the CRISPR-associated (Cas) protein constitute the prokaryotic adaptive immune system that specifically degrades foreign nucleic acids. Since the characterization and use of engineered *Streptococcus pyogenes* Cas9 (SpCas9) for genome editing (Cong et al., 2013; Mali et al., 2013b), targeted gene modification in different organisms has become simpler than ever before. The CRISPR-Cas9 genome editing technology requires two components, Cas9 nuclease and a single guide RNA (sgRNA). The sgRNA for SpCas9 includes a conserved scaffold sequence at the 3′-end and a target-specific 20-nt guide sequence with perfect complementarity to the target DNA at the 5′-end (Fig. 1). The sgRNA directs SpCas9 to cognate DNA sequences adjacent to the protospacer-adjacent-motif (PAM; 5′-NGG-3′). SpCas9 also weakly recognizes 5′-NAG-3′ as the PAM (Meng et al., 2018). The CRISPR-Cas9 system has been engineered to develop many innovative tools for precise genetic manipulation, including targeted gene knockout, targeted gene replacement and knock-in, transcriptional activation and suppression, epigenetic editing, gene imaging, three-dimensional (3D) genome manipulation, and targeted base editing.

http://dx.doi.org/10.19103/AS.2020.0082.07

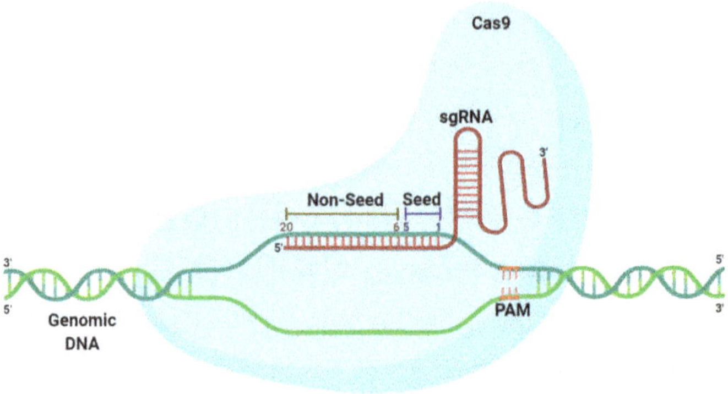

Figure 1 Schematic representation of the pairing between sgRNA and the target DNA sequence. The seed region, located proximal to the PAM, is critical for sgRNA binding.

The PAM requirement constrains the targeting space of SpCas9 in the genome. To address this technical limitation, many Cas9 nucleases from different sources, including *Staphylococcus aureus* (SaCas9), *Streptococcus thermophiles* (StCas9), and *Neisseria meningitides* (NmCas9), have been characterized and engineered for genome editing in the past 7 years (Cebrian-Serrano and Davies, 2017; Chen et al., 2019a). These Cas9 nucleases recognize various PAM sequences, thereby expanding the targeting sites for genome editing. Moreover, SpCas9 has been modified to generate different variants that recognize various PAMs, including VQR-SpCas9 (5′-NGA-3′), EQR-SpCas9 (5′-NGAG-3′), VRER-SpCas9 (5′-NGCG-3′), xCas9 (5′-NG-3′, 5′-GAA-3′, and 5′-GTA-3′), and SpCas9-NG (5′-NG-3′) (Chen et al., 2019a).

In addition to Cas9, variants of the class II type V-A Cas enzyme, Cpf1 (CRISPR from *Prevotella* and *Francisella*, also known as Cas12a), from *Francisella novicida* (FnCpf1), *Acidaminococcus* spp. (AsCpf1), and *Lachnospiraceae bacterium* (LbCpf1) have also been utilized as genome editing tools (Zetsche et al., 2015; Zhong et al., 2018). Cpf1 have several advantages over Cas9. Generally, Cpf1 recognizes a T-rich (5′-TTTV-3′) PAM sequence, thus expanding the target sites beyond those of SpCas9, which prefers a G-rich PAM. The Cpf1 cleavage site is located distal to and downstream of the PAM sequence. Native Cas9 uses a duplex of CRISPR RNA (crRNA) and transactivating crRNA (tracrRNA) as a guide, whereas the native Cpf1 requires only one crRNA molecule as the guide (Zetsche et al., 2015). More importantly, Cpf1 is able to process a crRNA array that simultaneously targets multiple sites (Zetsche et al., 2017).

Off-target effects are a major concern for CRISPR-Cas technologies. Because the native SpCas9 tolerates mismatches between sgRNA and target DNA, DNA sequences that partially match the sgRNA sequence might be cleaved by Cas9, resulting in undesirable off-target editing. For example, in mammalian cells, Cas9 recognizes an off-target site that pairs with a sgRNA containing five mismatches distal to the PAM (Wu et al., 2014). The targeting specificities of different Cas nucleases and their artificial engineered variants are variable. Extensive research on off-target effects of SpCas9 and other Cas nucleases (Hsu et al., 2013; Kleinstiver et al., 2016; Kuscu et al., 2014; Mali et al., 2013a; Pattanayak et al., 2013) suggests that the off-target sites of a given sgRNA can be predicted in organisms with a sequenced genome. Besides off-target editing, the on-target efficiency of SpCas9 (Doench et al., 2014) and some other Cas nucleases (Kim et al., 2016) has been studied extensively and estimated in human cells.

To date, many bioinformatics tools have been developed to facilitate the design of sgRNAs for CRISPR-based genome editing in eukaryotes. These programs compare the sgRNA sequences and predict off-target sites, sgRNA editing efficiencies (based on high-throughput editing experiments), and non-homologous end joining (NHEJ) outcomes (Molla and Yang, 2020). In this chapter, we discuss the general rules for selecting target sites for genome editing using the CRISPR-Cas technology and summarize the bioinformatics tools that can be used to design sgRNA sequences.

2 Designing sgRNAs

The first step in a CRISPR-Cas genome editing experiment is the design of an appropriate sgRNA. To design an sgRNA, some general rules must be followed to maximize on-target editing efficiency and minimize off-target mutation risk.

2.1 General considerations for sgRNA design

2.1.1 Off-target prediction

Because the CRISPR system can tolerate partial mismatches between the target DNA and sgRNA sequence, it is important to avoid or reduce off-target effects. Recognition of off-target sites by Cas nucleases depends on the number of mismatches and the positions of these mismatches relative to the PAM. Normally, Cas nucleases cannot tolerate mismatches in the region proximal to the PAM (referred to as the seed region, Fig. 1). For example, one mismatch adjacent to the PAM can abolish the sequence recognition ability of SpCas9 (Hsu et al., 2013). Off-target sites with insertions/deletions (InDels), known as bulge-type sites, are also related to spCas9 specificity and cleavage activity. In comparison with SpCas9, Cpf1 shows higher fidelity and is less tolerant to mismatches between the sgRNA sequence and the target DNA. Moreover, Cpf1 also recognizes a seed region in the target DNA sequence. Studies show that one mismatch at the seed region between AsCpf1 RNA guide and the off-targeting sequence strongly reduces mutation frequency (Kim et al., 2017a). In rice (*Oryza sativa*), Cpf1 tolerates two consecutive mismatches distal to the PAM but cannot recognize a single mismatch in the seed region (Zhong et al., 2018). These studies suggest that off-target effects of a given sgRNA can be minimized by comparing its sequence with the whole genome sequence.

2.1.2 On-target efficiency prediction

In addition to off-target mutations, on-target editing efficiency is also an important factor in sgRNA design. Many factors affect the editing efficiency of sgRNA, including the sequence context, GC content, and chromatin status of the target site (Xu et al., 2015; Doench et al., 2016). For example, in human and mouse cells, SpCas9 prefers G over T at nucleotide position -1 relative to the PAM (Doench et al., 2014). Graf and colleagues showed that two short sequence motifs (TT- and GCC) at the 3′-end of the target sequence are almost always present in inefficient sgRNAs, based on published sgRNA activity data sets (Graf et al., 2019). In the CRISPR-Cpf1 system, although FnCpf1 can recognize a TTN sequence as the PAM, it does not cleave sites with GTTA and GTTC sequences as the PAM. In fact, FnCpf1 is more likely to cleave TTTV PAM sites (Zhong et al., 2018). These findings indicate that the on-target editing efficiency of the CRISPR-Cas system is affected not only by the sgRNA sequence but also by the context of the target sequence. Although a few bioinformatics tools offer on-target efficiency prediction, generating a reliable prediction is challenging because the editing outcome is affected by many factors such as the relative concentration of Cas9/sgRNA complex in cells. As more and more CRISPR-Cas genome editing data become available, bioinformatics programs could be trained to provide an accurate prediction of on-target editing efficiency.

2.2 Specific considerations for sgRNA design

2.2.1 Targeted deletion of a DNA fragment via DNA repair pathways

The sgRNA guides the Cas9 protein to the target site and Cas9 produces a double-strand break (DSB) in the target DNA. A DSB is usually repaired through NHEJ, which often leads to InDels and, consequently, frameshift mutations that affect gene translation. The DSB is also repaired, albeit less commonly, by microhomology-mediated end joining (MMEJ). MMEJ usually employs short (5–25 bp) homologous DNA fragments on both sides of the cleaved site to expose microhomologous single chains through end processing. The two ends are then joined by homologous pairing. Usually, this results in sequence deletion between the two microhomologous regions (McVey and Lee, 2008).

The MMEJ pathway can be engineered to precisely delete a short fragment at a DSB (Shen et al., 2018). Recently, Tan et al. (2020) demonstrated that MMEJ-assisted Cas9 editing can delete large DNA fragments up to 20 667 bp. Therefore, analysis of microhomologous sequences at target sites is useful for Cas9-mediated DNA deletion.

2.2.2 Editing window for base editing

Targeted base editing using the CRISPR-Cas system has attracted much attention in recent years. In a base-editing study, the effective editing window is important for sgRNA design (Table 1). Different base editors have different editing windows, which are often located distal to the PAM. In the cytosine base editor (CBE) BE3 (Komor et al., 2016), which converts C to T, the editing window is located 4–8 nt of the guide sequence (here, the position of PAM is 21th–23th nt). In the adenine base editor (ABE) ABE7.10, which converts A to G (Gaudelli et al., 2017), the editing window is similar to BE3. Recently, many CBEs and ABEs (such as BE3, A3A-BE3, Target-AID, and ABE7.10) have been optimized for plants (Molla and Yang, 2019). These plant-optimized base editors have slightly different editing windows than the original base editors. However, BE3 and its derived based editor, but not ABE, induce substantial off-target mutations in rice and mouse embryos (Jin et al., 2019; Zuo et al., 2019). The recently reported base editor YEE-BE3 might be used to avoid such non-sequence specific off-target editing (Kim et al., 2017b). In summary, the targeted base should be included in the editing window when designing sgRNA sequences for base editing experiments.

2.2.3 Target activation and repression

Catalytically inactive Cas9 (dCas9) can be fused with a transcriptional activation domain to enhance expression of a target gene, or can be fused with a repression domain to suppress the expression of a target gene. In targeted transcriptional activation and suppression experiments, the dCas9-sgRNA complex normally targets the promoter of the target gene. Multiple dCas9-sgRNA complexes are required to synergistically target promoter regions to efficiently activate or suppress gene expression. Transcriptional activation of the target gene is maximized when the sgRNA targets a region located from −400 to −50 bp upstream of the transcription start site (TSS) (Gilbert et al., 2014; Konermann et al., 2015). On the contrary, strong transcriptional repression is observed when the sgRNA targets DNA from −50 to +300 bp relative to the TSS (Gilbert et al., 2014). In addition, activation and suppression efficiencies likely depend on the expression level of the target gene (Konermann et al., 2015; Chavez et al., 2015).

Table 1 Base editors and their editing windows

Base editor	Architecture[a]	Editing window[b]	Reference	Applications in plants
BE3	APOBEC1 – 16 aa – nCas9 (D10A) – 4 aa – UGI	5' 1 … 4–8 … 20 NGG 3'	Komor et al. (2016)	Rice (Li et al., 2017; Lu and Zhu, 2017; Zong et al., 2017); maize and wheat (Zong et al., 2017)
A3A-BE3	hAPOBEC3A – 16 aa – SpCas9 (D10A) – 4 aa – UGI	5' 1 … 2–13 … 20 NGG 3'	Wang et al. (2018)	Rice, wheat, and potato (Zong et al., 2018)
Target-AID	SpCas9 (D10A) – 100 aa – CDA1 – 9 aa – UGI	5' 1 … 2–6 … 20 NGG 3'	Nishida et al. (2016)	Rice and potato (Shimatani et al., 2017)
ABE7.10	TadA – 32 aa – TadA* – 32 aa – SpCas9 (D10A)	5' 1 … 4–8 … 20 NGG 3'	Gaudelli et al. (2017)	Rice (Hua et al., 2018; Li et al., 2018); wheat (Li et al., 2018)

[a] SpCas9 (D10A), SpCas9 variant with nickase activity generated by substituting aspartic acid (D) with alanine (A) at amino acid position 10; TadA, tRNA adenine deaminase; CDA1, cytidine deaminase 1; UGI, uracil DNA glycosylase inhibitor.
[b] Data are based on experiments in mammalian cells.

2.3 Guidelines for designing sgRNAs

Given the number of bioinformatics tools available currently, designing sgRNAs for SpCas9 is straightforward. These bioinformatics tools output a number of sgRNAs for each gene. Following is a list of some empirical parameters that should be considered when designing sgRNAs for SpCas9 and Cpf1-mediated genome editing:

1 **PAM**: SpCas9 recognizes 5′-NGG-3′ and 5′-NAG-3′ sequences as PAM, but recognizes 5′-NAG-3′ sequence with much less efficiency. Therefore, NAG PAM site is only considered in off-target prediction and is not used for gene editing experiment. Among the commonly used Cpf1 enzymes, FnCpf1 and LbCpf1 recognize 5′-TTN-3′ and 5′-TTTV-3′ as PAM sequences, respectively. To achieve efficient target site editing, 5′-NGG-3′ and 5′-TTTV-3′ are usually the preferred PAM sequences for SpCas9 and FnCpf1, respectively. However, the non-preferred PAMs should be taken into account when predicting off-target sites.

2 **Specificity**: Highly specific sgRNAs and crRNAs can be designed using bioinformatics tools (summarized below). Additionally, sgRNA specificity can be manually estimated using sequence alignment programs such as BLASTN.

3 **Position**: To completely knockout gene function, the target site should not be close to the stop codon. Ideally, the target site should be located at the 5′-end of the coding sequence or should encompass nucleotides that encode certain amino acids essential for gene function. If the objective is to delete a DNA fragment, a pair of target sites with microhomologous sequences is preferred to increase the deletion efficiency.

4 **Target number**: In most cases, one target per gene is sufficient; however, using two target sites per gene simultaneously will greatly increase gene knockout efficiency.

5 **GC content:** The GC content of sgRNAs should ideally range from 40 to 80%.

6 **Avoid potential transcriptional terminator:** The inclusion of a stretch of five thymine nucleotides (TTTTT) in the sgRNA should be avoided since this sequence causes RNA polymerase III (Pol III) to terminate transcription.

7 (Optional) Check the secondary structure of the sgRNA using programs such as RNA-fold (http://rna.tbi.univie.ac.at/cgi-bin/RNAWebSuite/RNAfold.cgi); note that sgRNAs with three proper stem-loop hairpins that mimics the structure of crRNA-tracrRNA duplex are preferred.

8 (Optional) To facilitate rapid screening of genome-edited lines, choose a sgRNA that causes mutation of a restriction endonuclease (RE) site;

Table 2 List of web-based bioinformatics tools used to design sgRNAs and predict off-target sites in plants

Tool	sgRNA design	Off-target prediction	SpCas9 system	Other Cas9 systems	Cpf1 systems	Application
CRISRP-P 2.0	✓	✓	✓	✓	✓	>60 plant species
CRISRP-GE	✓	✓	✓	✓	✓	39 plant species and other organisms
CRISRP-Plant	✓	✓	✓	✓	✓	Eight plant species (*Arabidopsis thaliana, Medicago truncatula, Solanum lycopersicum, Glycine max, Brachypodium distachyon, Oryza sativa, Sorghum bicolor, Zea mays)*
Breaking-Cas	✓	✓	✓	✓	✓	1435 Ensembl genomes, including 71 plant species
CRISPR MultiTargeter	✓	Use Cas-OFFinder	✓	✓	✓	12 organisms including three plant species (*A. thaliana, Oryza sativa* ssp. *japonica, Z. mays*)
Cas-Designer and Cas-OFFinder	✓	✓	✓	✓	✓	Vertebrates, insects, >80 plant species, and other organisms
Cas-Database and Cpf1-Database	✓	✓	✓	✓	✓	Five plant species (*A. thaliana, Musa acuminata, Vitis vinifera, S. lycopersicum, G. max*) and other organisms
CHOPCHOP V3	✓	✓	✓	✓	✓	30 plant species and other organisms
E-CRISPR	✓	✓	✓	✓	✗	Eight plant species (*A. thaliana, B. distachyon, Hordeum vulgare, Oryza sativa* ssp. *indica, Oryza nivara, O. sativa, V. vinifera, Z. mays*) and other organisms
CGAT	✓	✓	✓	✗	✗	Six plant species (*H. vulgare, Z. mays, Arachis duranensis, Arachis ipaensis, O. sativa, G. max*)
CRISPOR	✓	✓	✓	✓	✓	>30 plant species and other organisms
CRISPRdirect	✓	✓	✓	✗	✗	>50 plant species and other organisms

the genome-edited lines can then be easily screened using PCR and restriction digestion assays.

3 Bioinformatics programs for designing sgRNAs

Designing highly specific sgRNAs using bioinformatics programs can help to minimize off-target specificity and increases the likelihood of editing efficiency using CRISPR-Cas and CRISPR-Cpf1 systems. Many online and stand-alone tools have been developed for designing sgRNAs for target genes in plant genomes. Examples of online tools include CRISRP-P (Lei et al., 2014; Liu et al., 2017), CRISRP-GE (Xie et al., 2017), CRISRP-PLANT (Xie et al., 2014a), Breaking-Cas (Oliveros et al., 2016), CRISPR MultiTargeter (Prykhozhij et al., 2015), Cas-Designer (Park et al., 2015), Cas-OFFinder (Bae et al., 2014), Cas-Database (Park et al., 2016), Cpf1-Database (Park and Bae, 2018), CRISPOR (Haeussler et al., 2016), CHOPCHOP (Labun et al., 2019; Montague et al., 2014), E-CRISP (Heigwer et al., 2014), CGAT (Brazelton et al., 2015), and CRISPRdirect (Naito et al., 2015). Examples of stand-alone tools include CRISPRseek (Zhu et al., 2014) and CRISRP-Local (Sun et al., 2019). Details of each of these tools are described below; the features of online bioinformatics tools are listed in Table 2.

3.1 CRISRP-P and CRISRP-local

CRISRP-P (http://crispr.hzau.edu.cn/) is one of the most popular bioinformatics tools used for sgRNA design in plants (Lei et al., 2014). CRISRP-P predicts off-target loci for each potential sgRNA and provides RE sites for the selected sgRNA (Lei et al., 2014). The revised version of CRISRP-P, CRISRP-P 2.0, provides more advanced functions for more than 60 plant genomes: (1) it employs a modified scoring method to evaluate the on-target efficiency and off-target risk of sgRNAs for SpCas9; (2) it supports sgRNA design for various Cas9 endonucleases and Cpf1; (3) it provides the GC content, RE sites, microhomologous sequences flanking the target site, and secondary structure of the selected sgRNA; (4) it provides sgRNA for sequences uploaded by users (Liu et al., 2017).

CRISPR-Local is an extension of CRISRP-P, which was developed for genome-wide sgRNA design for non-reference plant genomes (Sun et al., 2019). CRISPR-Local enables high-throughput sgRNA design for 71 plant genomes for both CRISPR-Cas9 and CRISPR-Cpf1 systems (Sun et al., 2019). CRISPR-Local can be run on the local server with command-line or graphical interface modes, and sgRNAs capable of targeting multiple genes or a unique gene can be screened. Users can input multiple homologous or non-homologous genes. CRISPR-Local outputs common sgRNAs for multiple genes, and unique sgRNAs for a specific gene.

CRISPR-Local and CRISPR-P use the same on-target and off-target algorithms. However, CRISPR-P predicts off-targets in the whole genome

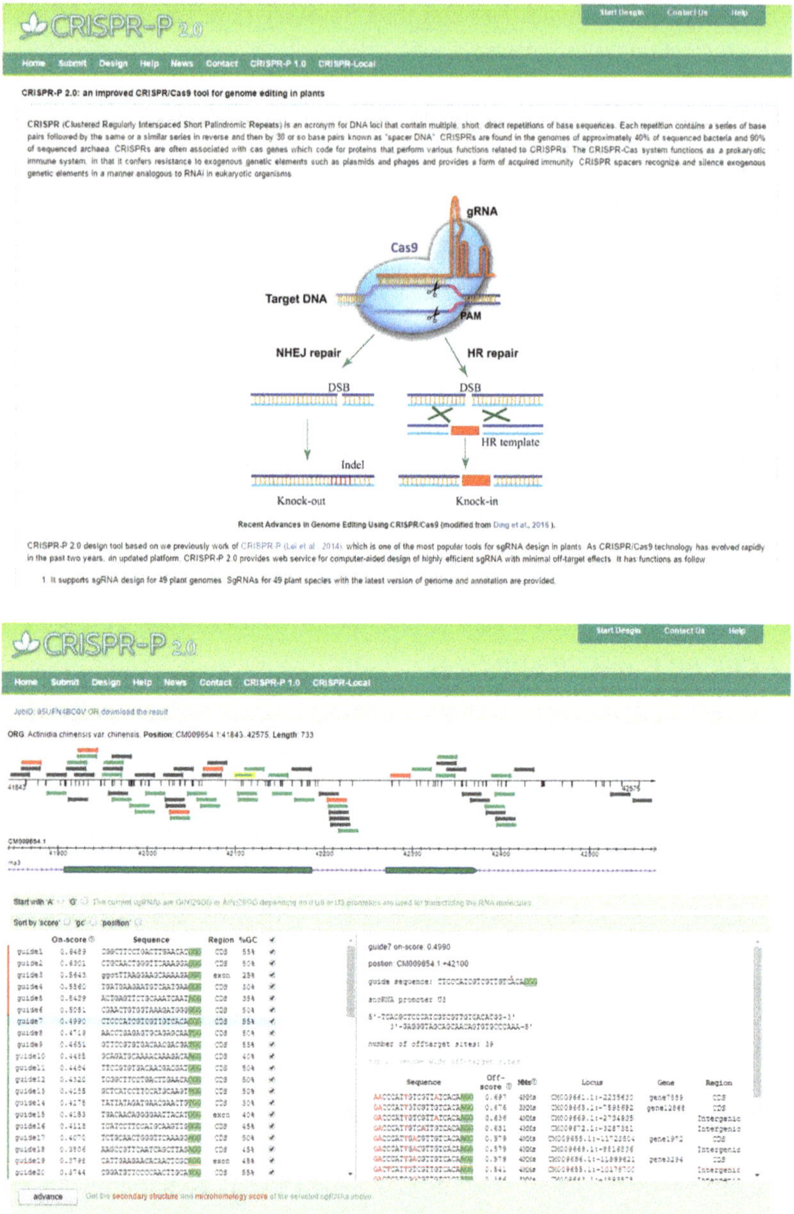

Figure 2 Web interface and outputs of CRISPR-P 2.0.

Figure 3 Web interface of CRISPR-local.

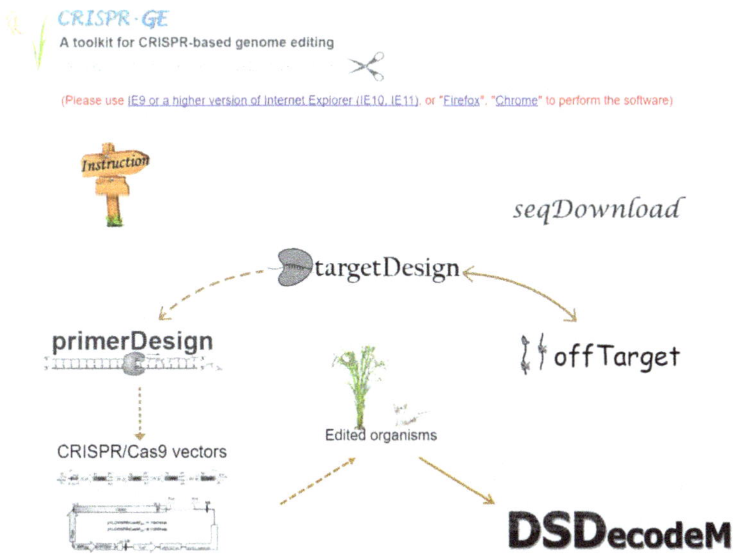

Figure 4 Web interface of CRISPR-GE.

including exons, introns, and intergenic regions, whereas CRISPR-Local only predicts off-targets in genic regions. In addition, CRISPR-P can output at most 20 off-targets, whereas CRISPR-Local only lists 1–2 off-target sites (Figs 2 and 3).

3.2 CRISRP-GE

CRISPR-GE (http://skl.scau.edu.cn/) is a comprehensive toolkit for CRISPR-based genome editing in plants and other organisms (Xie et al., 2017). CRISPR-GE contains five tools, each with unique functions: targetDesign for designing target sgRNAs; offTarget for predicting off-targets; primerDesign for designing primers to construct sgRNA expression cassettes and amplify target sequences; DSDecodeM for determining mutant sequences from the chromatograms of PCR products containing target sequences; and seqDownload for downloading target genomic sequences from reference genomes. To construct the CRISPR-Cas9 sgRNA vector system, U3 and U6, small nuclear RNA promoters (four from rice and four from *Arabidopsis thaliana*), are provided to express the target sgRNAs (Xie et al., 2017) (Fig. 4).

3.3 CRISRP-plant

CRISRP-PLANT (www.genome.arizona.edu/crispr) designs sgRNAs for eight plant species including *Arabidopsis*, *Brachypodium distachyon*, alfalfa (*Medicago truncatula*), maize (*Zea mays*), rice, sorghum (*Sorghum bicolor*), soybean (*Glycine max*), and tomato (*Solanum lycopersicum*). The identified sgRNA candidates are further classified into different groups depending on their off-target mutation risk, which is determined based on genome-wide sequence comparisons (Xie et al., 2014a). The revised version, CRISRP-PLANT v2 (www.genome.arizona.edu/crispr2) utilizes an improved off-target scoring method that can identify 1–4 mismatches and 1–2 InDels (bulge-type sites) in genome-wide prediction of off-target sites (Minkenberg et al., 2019) (Fig. 5).

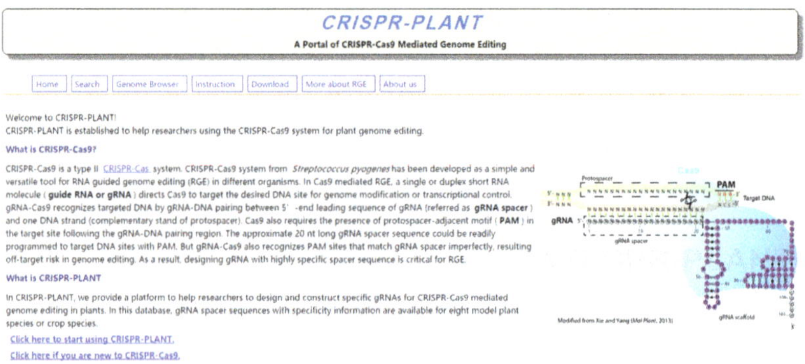

Figure 5 Web interface of CRISPR-PLANT.

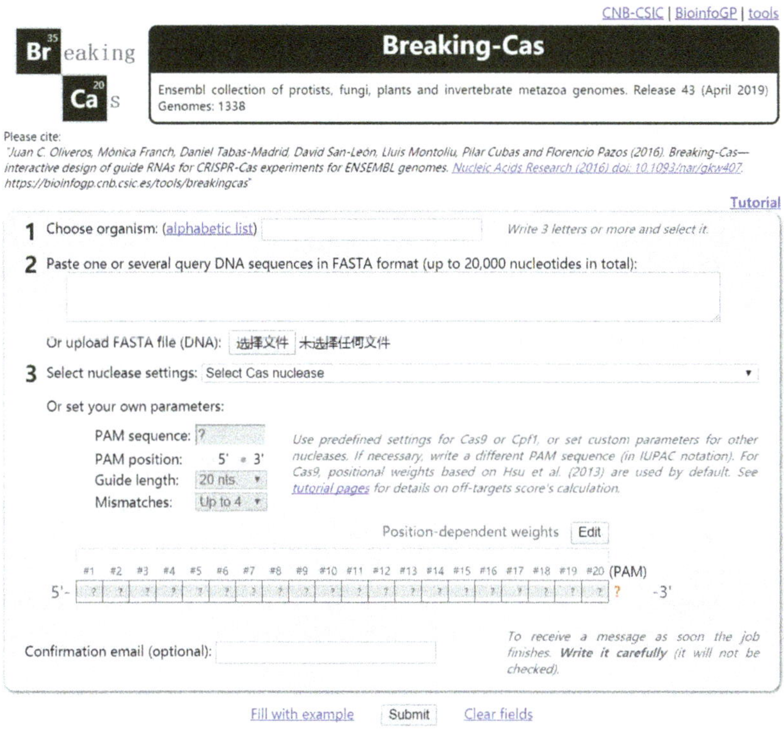

Figure 6 Web interface of Breaking-Cas.

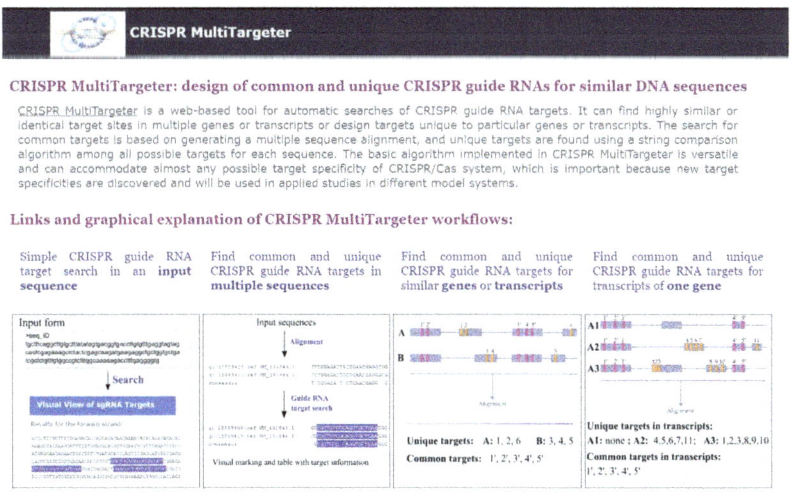

Figure 7 Web interface of CRISPR multitargeter.

3.4 Breaking-Cas

Breaking-Cas (http://bioinfogp.cnb.csic.es/tools/breakingcas) is a web server capable of designing sgRNAs for all available eukaryotic genomes, including plants, protists, fungi, and metazoa, in ENSEMBL. The recent version (Release 46) of Breaking-Cas contains 1435 genomes including 71 plant genomes and provides a user-friendly interface for detection of potential off-target sites. In addition, Breaking-Cas predefines parameters for popular Cas9 and Cpf1 proteins, while allowing the user to specify certain features and parameters (Oliveros et al., 2016). Users can also set their own parameters such as PAM position, PAM sequence, sgRNA length, and mismatch number (Fig. 6).

3.5 CRISPR multitargeter

The most important feature of CRISPR MultiTargeter (http://www.multicrispr.net/) is its ability to design common and unique sgRNAs for multiple sequences. The common sgRNA design is based on generating a multiple sequence alignment, while unique sgRNAs are obtained based on a string comparison algorithm for all possible targets. MultiTargeter (Prykhozhij et al., 2015) uses Cas-OFFinder (http://www.rgenome.net/cas-offinder/) and GT-Scan (http://gt-scan.braembl.org.au/gt-scan/) to predict off-targets, and can accommodate most CRISPR-Cas9 systems. Currently, 12 eukaryotic genomes including three plant genomes are included in CRISPR MultiTargeter. Users can also upload sequences less than 50 kb to design sgRNAs (Fig. 7).

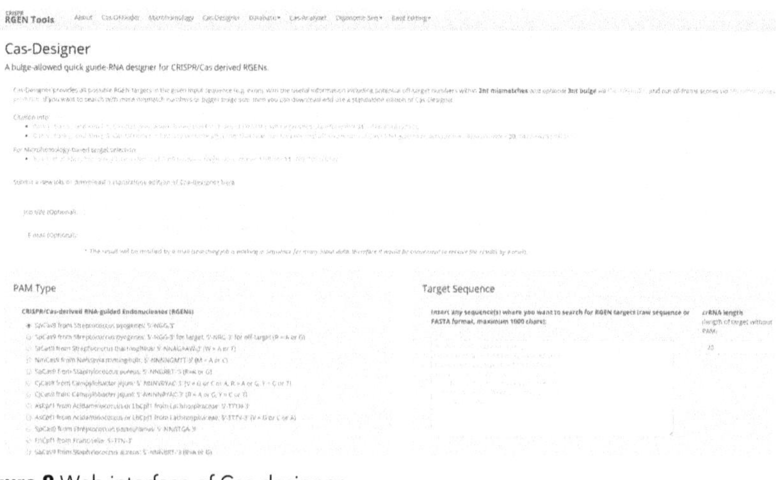

Figure 8 Web interface of Cas-designer.

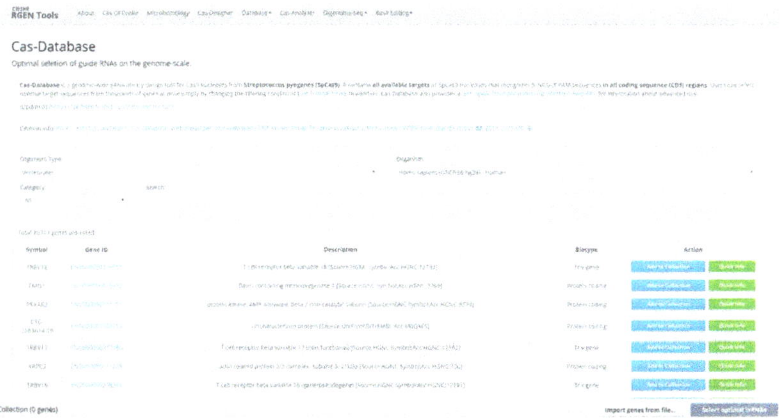

Figure 9 Web interface of Cas-database.

3.6 Cas-designer, Cas-OFFinder, Cas-database, and Cpf1-database

Cas-Designer (http://www.rgenome.net/cas-designer/) is a sgRNA design tool used for many type II Cas endonucleases (Park et al., 2015). Cas-Designer provides candidate sgRNAs based on an input DNA sequence, and predicts potential off-target sites, including bulge-type sites, in the selected genome (Park et al., 2015). Cas-OFFinder (http://www.rgenome.net/cas-offinder/) identifies potential off-target sites in a given genome or sequences submitted by the user (Bae et al., 2014).

Based on Cas-Designer and Cas-OFFinder, Park and colleagues further developed a genome-wide sgRNA library design tool for SpCas9, known as Cas-Database (http://www.rgenome.net/cas-database/), which allows users to select optimal sgRNAs via a convenient web interface. Cas-Database allows the selection of optimal target sequences from thousands of genes simultaneously for creating a genome-wide library and provides a web

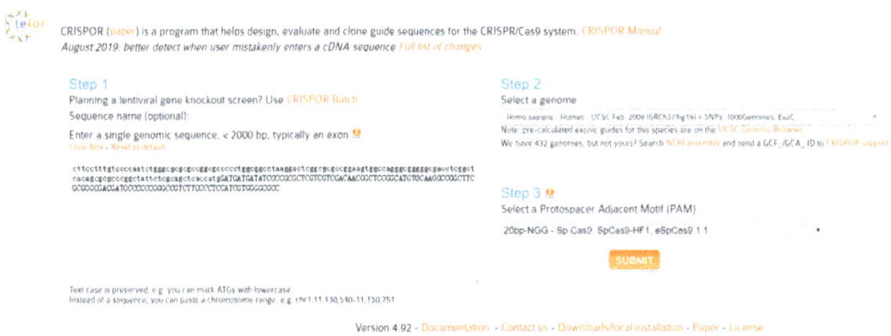

Figure 10 Web interface of CRISPOR.

application programming interface (API) (Park et al., 2016). Furthermore, a genome-wide sgRNA library design tool for LbCpf1 and AsCpf1 nucleases, known as Cpf1-Database (http://www.rgenome.net/cpf1-database/), has also been constructed. The Cpf1-Database contains DNA sequences with 5′-TTTN-3′ PAM at their 5′ ends in coding regions. Cpf1-Database provides a convenient way to design genome-wide sgRNAs for AsCpf1 and LbCpf1 nucleases. Users can easily access the data using the web API, and design sgRNAs for thousands of genes simultaneously (Park and Bae, 2018) (Figs 8 and 9).

3.7 CRISPOR

CRISPOR (http://crispor.tefor.net/) can help design efficient sgRNA sequences and predict off-targets for more than 120 genomes, including more than 30 plant genomes, using different Cas9 and Cpf1 proteins (Haeussler et al., 2016). CRISPOR utilizes several on-target efficiency prediction methods based on available data sets, and eight SpCas9 off-target methods. Then, CRISPOR generates a three-part output containing the following information: (1) annotation of the input sequence; (2) list of sgRNAs along with their specificity scores, efficiency prediction, out-of-frame scores, off-target score, and other parameters; (3) a list of primers (Fig. 10).

3.8 CHOPCHOP

Similar to other tools, CHOPCHOP (https://chopchop.cbu.uib.no/) accepts a wide range of inputs, including gene identifiers, genomic regions, or nucleotide sequences, and provides advanced options for target selection. CHOPCHOP uses efficient sequence alignments to reduce search times and predicts off-target sites of sgRNAs. For the query sequence, CHOPCHOP provides an

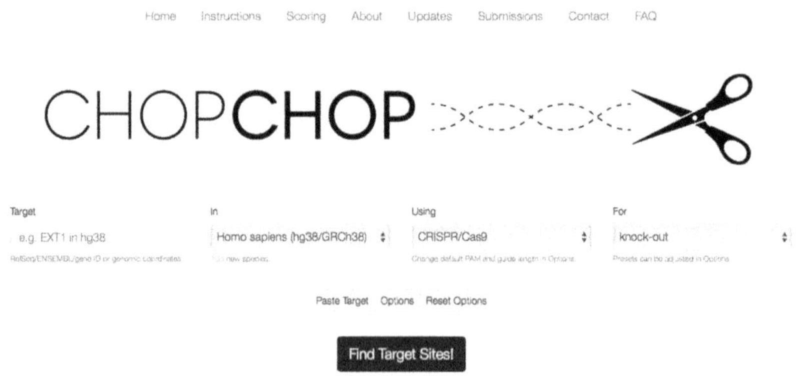

Figure 11 Web interface of CHOPCHOP.

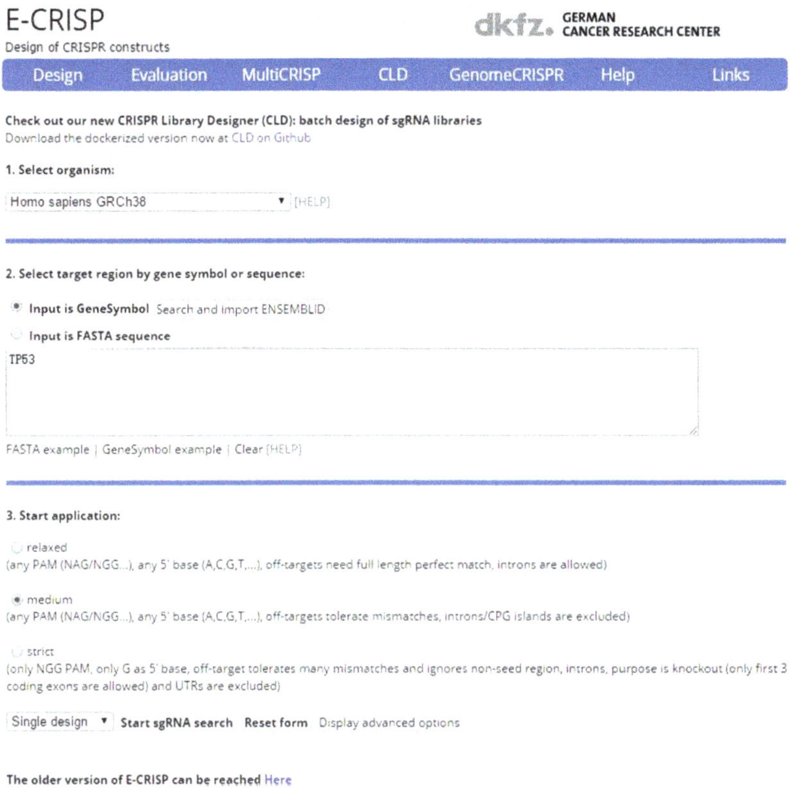

Figure 12 Web interface of E-CRISP.

interactive visualization of the gene region, with candidate target sites displayed according to their quality scores. In addition, CHOPCHOP provides a visual display of RE sites and suggests primer candidates for possible target sites, which facilitates subsequent generation and validation of mutations (Montague et al., 2014). While the original version of CHOPCHOP contained the widely used gene knockout function, the revised version, CHOPCHOP V3, expanded this tool to include the gene knock-in function. Moreover, CHOPCHOP V3 incorporates new DNA-targeting modes including CRISPRa, CRISPRi, targeted enrichment of loci for long-read sequencing, and Cas9 repair outcome prediction (Labun et al., 2019). CHOPCHOP V3 includes genome sequences of 30 plant species (Fig. 11).

3.9 E-CRISPR

E-CRISP (http://www.e-crisp.org/E-CRISP/) is a web-based tool used to design sgRNAs for the CRISPR-Cas9 system. E-CRISP utilizes a fast indexing

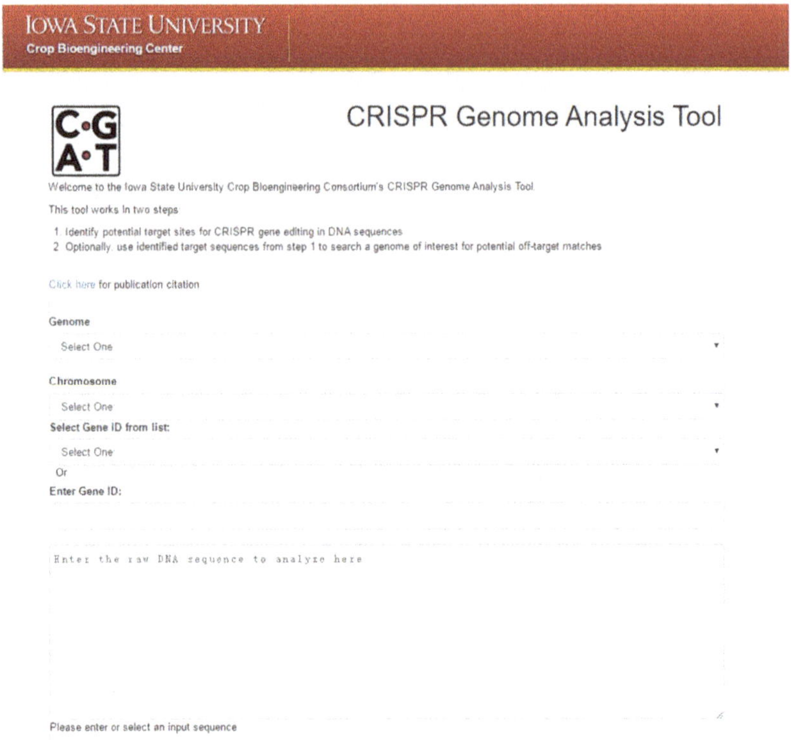

Figure 13 Web interface of CGAT.

method for identifying binding sites and a binary interval tree for annotating candidate sgRNA target sites rapidly. Using these methods, it is feasible to construct genome-wide sgRNA libraries for some species in a short time (Heigwer et al., 2014). E-CRISP provides a user-defined output and experiment-oriented design parameters, and can be used to design sgRNAs for eight plant species. Recently, Heigwer and colleagues further developed an integrated bioinformatics command-line application, CRISPR library designer (CLD, https://github.com/boutroslab/cld), which can be used to construct custom sgRNA libraries for any annotated organism (Heigwer et al., 2016) (Fig. 12).

3.10 CGAT

CGAT (http://cbc.gdcb.iastate.edu/cgat/) is another web-based sgRNA design tool, which uses gene name or sequences as the input, and generates an output containing sgRNAs for Cas9 and their predicted off-target sites (Brazelton et al., 2015). CGAT can be used to design sgRNAs for six plant genomes including

Arachis duranensis, *Arachis ipaensis*, barley (*Hordeum vulgare*), maize, rice, and soybean (Fig. 13).

3.11 CRISPRdirect

CRISPRdirect (http://crispr.dbcls.jp/) is also a web-based server used to select suitable CRISPR-Cas9 sgRNAs from the input sequence or gene accession number (Naito et al., 2015). Users can easily view the list of sgRNA candidates and select optimal target sequences with minimal off-target sites (Naito et al., 2015). CRISPRdirect can identify off-target candidates containing both mismatches and bulge-type sites (InDels). The CRISPRdirect web server currently contains whole genome sequences of more than 50 plant species (Fig. 14).

3.12 CRISPRseek

Unlike other online tools with web service, CRISPRseek (http://www.bioconductor.org) is an open source package that can be downloaded to the local server. CRISPRseek can identify sgRNAs that target an input sequence while minimizing off-target mutations. CRISPRseek identifies potential sgRNAs only for the CRISPR-Cas9 system and generates a cleavage score for potential off-targets using published or user-defined weight matrices. The identified sgRNAs can be filtered further to select those occurring in paired orientations or those overlapping RE sites. CRISPRseek ranks the identified sgRNAs for discriminating them between two related sequences (Zhu et al., 2014) (Fig. 15).

Among all the bioinformatics tools described above, CRISPR-P, CRISPR-Local, and CRISPR-PLANT are designed specifically for plants, while most of the other tools can be used for plants and other organisms. All sgRNA design tools perform the same basic functions, that is, identifying sgRNAs and off-target sites. E-CRISPR, CGAT, and CRISPRdirect identify sgRNAs for a few PAMs, while most of the other tools can find sgRNAs for many PAMs. Six programs (CRISRP-P 2.0, CRISRP-GE, Breaking-Cas, Cas-Designer, CRISPOR, and CHOPCHOP) design sgRNAs for the CRISPR-Cpf1 system.

When designing an effective sgRNA, it is important to evaluate its on-target efficiency and off-target effects. However, no plant-based algorithm has been developed to date. Nonetheless, most of the abovementioned tools, except CRISPR-PLANT, Cas-Designer, CGAT, and CRISPRdirect, provide evaluation algorithms for designing sgRNAs and predicting their off-target effects. Some tools provide additional useful functions, such as primer design, sgRNA RE sites, DNA and RNA bulge alignment, crRNA secondary structure, and off-target site location (e.g. exon, intron, or intergenic region); this information is highly convenient for users.

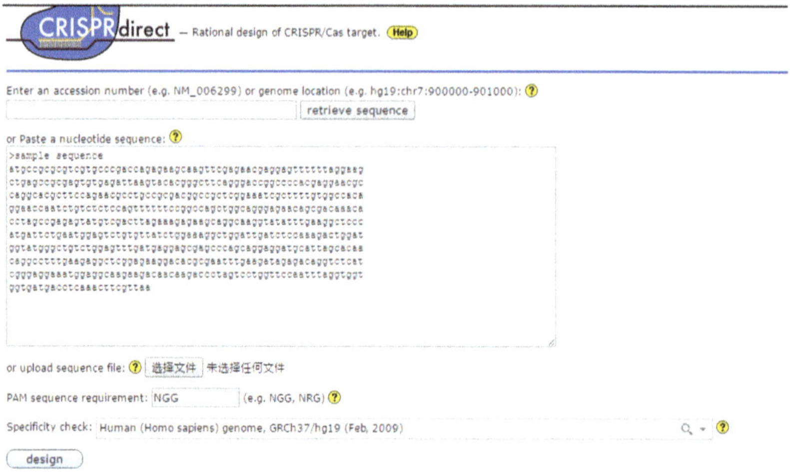

Figure 14 Web interface of CRISPRdirect.

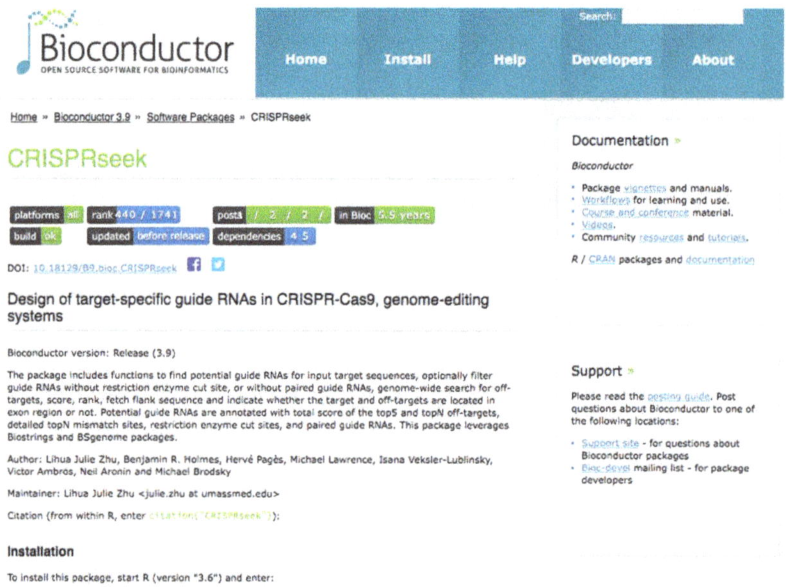

Figure 15 Description of CRISPRseek.

4 Predicting the outcome of CRISPR-mediated genome editing

Although the CRISPR-Cas9 system has been used widely to edit the genomes of various species, the editing outcomes seem variable and unpredictable. The process of genome editing by the Cas9 protein involves cutting the DNA sequence at a specific site, leading to DSBs. These DSBs are repaired by various endogenous DNA repair pathways such as NHEJ and MMEJ, which do not require a donor sequence, or homology-directed repair, which requires a donor template at the site of the DSB (Knott and Doudna, 2018). Both NHEJ and MMEJ are error-prone repair pathways but show clear differences in repair outcomes. Often, NHEJ induces InDels at the cleavage site when ligating the broken DNA ends. By contrast, MMEJ requires microhomologous sequences ranging from one to several nucleotides within the ends for annealing, resulting in the deletion of the intervening sequence (Chiruvella et al., 2013; Sfeir and Symington, 2015). Therefore, the outcomes of NHEJ are generally random and more unpredictable than those of MMEJ. However, some studies show that the outcomes of genome editing are strongly dependent on the targeted sequence, indicating that mutations caused by DNA repair occur in a nonrandom manner, and that gene editing results may be predictable (Lemos et al., 2018; van Overbeek et al., 2016).

To gain a comprehensive understanding of the mechanism underlying Cas9-induced mutation, a large number of Cas9-generated mutations were profiled recently and analyzed in human cell lines (Allen et al., 2019; Chakrabarti et al., 2019; Chen et al., 2019b; Leenay et al., 2019; Shen et al., 2018). However, similar research is lacking in plants. Allen and colleagues quantified more than 10^9 mutational outcomes generated by more than 40 000 sgRNAs using synthetic constructs, and showed that these mutations were of three main types: single-nucleotide insertions, short deletions, and long microhomology (MH)-mediated deletions (Allen et al., 2019). In another study, Shen and colleagues quantified a large data set of SpCas9-mediated editing outcomes at 1 872 target sites in the human genome and found the mutation types generated by NHEJ repairing were predictable (Shen et al., 2018). Three main classes of SpCas9-induced mutations included MH deletions, MH-less deletions, and single-base insertions, which accounted for 80–95% of all InDel patterns (Shen et al., 2018). Chen and coworkers profiled ~1.16 million mutational outcomes resulting from CRISPR-based genome editing, followed by NHEJ-mediated repair of 6 872 synthetic target sequences in human cells (Chen et al., 2019b). The authors found that repair patterns for a given target were reproducible but varied among different targets, and that the sequence context around the DSB site played an important role in shaping repair outcomes (Chen et al., 2019b). Additionally, insertion mutations were predominantly 1–2 bp long, which was highly consistent with the sequence upstream of the DSB but not consistent with the sequence downstream of the cleaved site (Chen et al., 2019b). These results suggest that cleavage of a DNA sequence by Cas9 does not always result in blunt ends, and 1–2-bp insertions may be templated by the nucleotide immediately upstream of the cleavage site. Shou et al. reported that Cas9 accurately cleaves the complementary strand at −3-bp upstream of

Table 3 Machine-learning models for predicting CRISPR-based genome editing outcomes

Machine-learning model	Cell type	Training data	Website
FORECasT	Human K562	>4000 synthetic targets	https://partslab.sanger.ac.uk/FORECasT
inDelphi	mESCs and human U2OS	1872 genomic sites	https://indelphi.giffordlab.mit.edu/
Lindel	Human (HEK) 293 T cells	6872 synthetic targets	https://lindel.gs.washington.edu/Lindel/
SPROUT	Human CD4+ T cells	1656 genomic sites	https://zou-group.github.io/SPROUT

mESCs, mouse embryonic stem cells.

the PAM but sometimes cleaves the non-complementary strand at multiple sites upstream of −3-bp position (Shou et al., 2018). Thus, Cas9 produces not only blunt ends but also staggered ends with 1–3-nt 5′ overhangs. Chen et al. proposed a model in which Cas9 produces staggered ends with 1–2-bp 5′ overhangs, which are repaired by gap-filling and ligation. Therefore, the results of DNA repair showed that the inserted nucleotides were added based on a template sequence located immediately upstream of the cleaved site (Chen et al., 2019b). Lemos and colleagues proposed a similar model based on NHEJ-mediated insertions in yeast (Lemos et al., 2018). Accumulating evidence shows that the repair outcomes largely depend on the sequence context around the cleavage site, and mutations generated by Cas9 can be predicted based on sequence alone.

Several recent studies suggest that the outcome of CRISPR-based genome editing can be predicted using machine-learning models. A large data set of Cas9-mediated gene editing outcomes is used to develop and train the machine-learning models. Several effective machine-learning algorithms have been developed for predicting the results of Cas9-induced DSBs, including FORECasT (Allen et al., 2019), inDelphi (Shen et al., 2018), Lindel (Chen et al., 2019b), and SPROUT (Leenay et al., 2019). The machine-learning model, inDelphi, was built on the basis of SpCas9-mediated editing outcomes at 1872 target sites in the human genome. The inDelphi model consists of three modules: a two neural network models that predict deletions induced by the MMEJ pathway or MH-less deletions, and a third module that uses *K*-nearest neighbors to predict single-base insertions. Three major editing outcomes, including MH deletions, MH-less deletions, and single-base insertions, can be comprehensively and accurately predicted by inDelphi (Shen et al., 2018). Shen and colleagues further used inDelphi to predict and identify suitable targets for template-free Cas9-mediated genome editing of pathogenic human alleles. Consistent with inDelphi predictions, 183 disease-related microduplication alleles and 508 pathogenic frameshift alleles were corrected with high efficiency (≥50%). Similar to inDelphi, three other machine-learning models can predict DNA repair outcomes with high accuracy. Table 3 lists a few machine-learning models used widely for predicting CRISPR-mediated genome editing outcomes. These tools can assist researchers in designing target sites for achieving specific editing outcomes.

5 Conclusions

Advances in the CRISPR-Cas genome editing technology have led to the development of new tools and methods for basic molecular research and genetic breeding of plants. Designing a sgRNA is a crucial step for CRISPR-based genome editing. Several tools and websites have been developed for designing sgRNAs. Users can choose different CRISPR systems, depending on their objective, and design appropriate sgRNAs. It is important to evaluate off-target and on-target efficiencies of each sgRNA. Improving the on-target efficiency of sgRNAs and reducing their off-target effects require continuous effort. To increase the precision of genome editing, more on-target and off-target data are needed to optimize sgRNA design programs.

6 Where to look for further information

Several review papers provide a comprehensive overview of the CRISPR-Cas9 and CRISPR-Cpf1 genome editing technologies and sgRNA design in plants (Aglawe et al., 2018; Ahmad et al., 2020; Chen et al., 2019a; Kumlehn et al., 2018; Satheesh et al., 2019). In addition, many CRISPR genome editing reagents are available in the Addgene CRISPR resources (http://www.addgene.org/crispr/). Moreover, the plant genome-editing database (PGED, http://plantcrispr.org/cgi-bin/crispr/index.cgi) provides information about gene-edited plant lines generated using the CRISPR-Cas9 technology.

7 References

Aglawe, S. B., Barbadikar, K. M., Mangrauthia, S. K. and Madhav, M. S. 2018. New breeding technique "genome editing" for crop improvement: applications, potentials and challenges. *3 Biotech* 8(8):336.

Ahmad, N., Rahman, M. U., Mukhtar, Z., Zafar, Y. and Zhang, B. 2020. A critical look on CRISPR-based genome editing in plants. *J. Cell. Physiol.* 235(2):666–82.

Allen, F., Crepaldi, L., Alsinet, C., Strong, A. J., Kleshchevnikov, V., De Angeli, P., Páleníková, P., Khodak, A., Kiselev, V., Kosicki, M., Bassett, A. R., Harding, H., Galanty, Y., Muñoz-Martínez, F., Metzakopian, E., Jackson, S. P. and Parts, L. 2019. Predicting the mutations generated by repair of Cas9-induced double-strand breaks. *Nat. Biotechnol.* 37:64–72.

Bae, S., Park, J. and Kim, J. S. 2014. Cas-OFFinder: a fast and versatile algorithm that searches for potential off-target sites of Cas9 RNA-guided endonucleases. *Bioinformatics* 30(10):1473–5.

Brazelton, V. A., Jr., Zarecor, S., Wright, D. A., Wang, Y., Liu, J., Chen, K., Yang, B. and Lawrence-Dill, C. J. 2015. A quick guide to CRISPR sgRNA design tools. *GM Crops Food* 6(4):266-76.

Cebrian-Serrano, A. and Davies, B. 2017. CRISPR-Cas orthologues and variants: optimizing the repertoire, specificity and delivery of genome engineering tools. *Mamm. Genome* 28(7-8):247-61.

Chakrabarti, A. M., Henser-Brownhill, T., Monserrat, J., Poetsch, A. R., Luscombe, N. M. and Scaffidi, P. 2019. Target-specific precision of CRISPR-mediated genome editing. *Mol. Cell* 73(4):699-713.e6.

Chavez, A., Scheiman, J., Vora, S., Pruitt, B. W., Tuttle, M., P R Iyer, E., Lin, S., Kiani, S., Guzman, C. D., Wiegand, D. J., Ter-Ovanesyan, D., Braff, J. L., Davidsohn, N., Housden, B. E., Perrimon, N., Weiss, R., Aach, J., Collins, J. J. and Church, G. M. 2015. Highly efficient Cas9-mediated transcriptional programming. *Nat. Methods* 12(4):326-8.

Chen, K., Wang, Y., Zhang, R., Zhang, H. and Gao, C. 2019a. CRISPR/Cas genome editing and precision plant breeding in agriculture. *Annu. Rev. Plant Biol.* 70:667-97.

Chen, W., McKenna, A., Schreiber, J., Haeussler, M., Yin, Y., Agarwal, V., Noble, W. S. and Shendure, J. 2019b. Massively parallel profiling and predictive modeling of the outcomes of CRISPR/Cas9-mediated double-strand break repair. *Nucleic Acids Res.* 47(15):7989-8003.

Chiruvella, K. K., Liang, Z. and Wilson, T. E. 2013. Repair of double-strand breaks by end joining. *Cold Spring Harb. Perspect. Biol.* 5(5):a012757.

Cong, L., Ran, F. A., Cox, D., Lin, S., Barretto, R., Habib, N., Hsu, P. D., Wu, X., Jiang, W., Marraffini, L. A. and Zhang, F. 2013. Multiplex genome engineering using CRISPR/Cas systems. *Science* 339(6121):819-23.

Doench, J. G., Fusi, N., Sullender, M., Hegde, M., Vaimberg, E. W., Donovan, K. F., Smith, I., Tothova, Z., Wilen, C., Orchard, R., Virgin, H. W., Listgarten, J. and Root, D. E. 2016. Optimized sgRNA design to maximize activity and minimize off-target effects of CRISPR-Cas9. *Nat. Biotechnol.* 34(2):184-91.

Doench, J. G., Hartenian, E., Graham, D. B., Tothova, Z., Hegde, M., Smith, I., Sullender, M., Ebert, B. L., Xavier, R. J. and Root, D. E. 2014. Rational design of highly active sgRNAs for CRISPR-Cas9-mediated gene inactivation. *Nat. Biotechnol.* 32(12):1262-7.

Gaudelli, N. M., Komor, A. C., Rees, H. A., Packer, M. S., Badran, A. H., Bryson, D. I. and Liu, D. R. 2017. Programmable base editing of A*T to G*C in genomic DNA without DNA cleavage. *Nature* 551(7681):464-71.

Gilbert, L. A., Horlbeck, M. A., Adamson, B., Villalta, J. E., Chen, Y., Whitehead, E. H., Guimaraes, C., Panning, B., Ploegh, H. L., Bassik, M. C., Qi, L. S., Kampmann, M. and Weissman, J. S. 2014. Genome-scale CRISPR-mediated control of gene repression and activation. *Cell* 159(3):647-61.

Graf, R., Li, X., Chu, V. T. and Rajewsky, K. 2019. sgRNA sequence motifs blocking efficient CRISPR/Cas9-mediated gene editing. *Cell Rep.* 26(5):1098-1103.e3.

Haeussler, M., Schönig, K., Eckert, H., Eschstruth, A., Mianné, J., Renaud, J. B., Schneider-Maunoury, S., Shkumatava, A., Teboul, L., Kent, J., Joly, J. S. and Concordet, J. P. 2016. Evaluation of off-target and on-target scoring algorithms and integration into the guide RNA selection tool CRISPOR. *Genome Biol.* 17(1):148.

Heigwer, F., Kerr, G. and Boutros, M. 2014. E-CRISP: fast CRISPR target site identification. *Nat. Methods* 11(2):122-3.

Heigwer, F., Zhan, T., Breinig, M., Winter, J., Brügemann, D., Leible, S. and Boutros, M. 2016. CRISPR library designer (CLD): software for multispecies design of single guide RNA libraries. *Genome Biol.* 17:55.

Hsu, P. D., Scott, D. A., Weinstein, J. A., Ran, F. A., Konermann, S., Agarwala, V., Li, Y., Fine, E. J., Wu, X., Shalem, O., Cradick, T. J., Marraffini, L. A., Bao, G. and Zhang, F. 2013. DNA targeting specificity of RNA-guided Cas9 nucleases. *Nat. Biotechnol.* 31(9):827–32.

Hua, K., Tao, X., Yuan, F., Wang, D. and Zhu, J. K. 2018. Precise A.T to G. C Base editing in the rice genome. *Mol. Plant* 11(4):627–30.

Jin, S., Zong, Y., Gao, Q., Zhu, Z., Wang, Y., Qin, P., Liang, C., Wang, D., Qiu, J. L., Zhang, F. and Gao, C. 2019. Cytosine, but not adenine, base editors induce genome-wide off-target mutations in rice. *Science* 364(6437):292–5.

Kleinstiver, B. P., Pattanayak, V., Prew, M. S., Tsai, S. Q., Nguyen, N. T., Zheng, Z. and Joung, J. K. 2016. High-fidelity CRISPR-Cas9 nucleases with no detectable genome-wide off-target effects. Nature 529(7587):490–5.

Kim, D., Kim, J., Hur, J. K., Been, K. W., Yoon, S. H. and Kim, J. S. 2016. Genome-wide analysis reveals specificities of Cpf1 endonucleases in human cells. *Nat. Biotechnol.* 34(8):863–8.

Kim, H. K., Song, M., Lee, J., Menon, A. V., Jung, S., Kang, Y. M., Choi, J. W., Woo, E., Koh, H. C., Nam, J. W. and Kim, H. 2017a. In vivo high-throughput profiling of CRISPR-Cpf1 activity. *Nat. Methods* 14(2):153–9.

Kim, Y. B., Komor, A. C., Levy, J. M., Packer, M. S., Zhao, K. T. and Liu, D. R. 2017b. Increasing the genome-targeting scope and precision of base editing with engineered Cas9-cytidine deaminase fusions. *Nat. Biotechnol.* 35(4):371–6.

Knott, G. J. and Doudna, J. A. 2018. CRISPR-Cas guides the future of genetic engineering. *Science* 361(6405):866–9.

Komor, A. C., Kim, Y. B., Packer, M. S., Zuris, J. A. and Liu, D. R. 2016. Programmable editing of a target base in genomic DNA without double-stranded DNA cleavage. *Nature* 533(7603):420–4.

Konermann, S., Brigham, M. D., Trevino, A. E., Joung, J., Abudayyeh, O. O., Barcena, C., Hsu, P. D., Habib, N., Gootenberg, J. S., Nishimasu, H., Nureki, O. and Zhang, F. 2015. Genome-scale transcriptional activation by an engineered CRISPR-Cas9 complex. *Nature* 517(7536):583–8.

Kumlehn, J., Pietralla, J., Hensel, G., Pacher, M. and Puchta, H. 2018. The CRISPR/Cas revolution continues: from efficient gene editing for crop breeding to plant synthetic biology. *J. Integr. Plant Biol.* 60(12):1127–53.

Kuscu, C., Arslan, S., Singh, R., Thorpe, J. and Adli, M. 2014. Genome-wide analysis reveals characteristics of off-target sites bound by the Cas9 endonuclease. *Nat. Biotechnol.* 32(7):677–83.

Labun, K., Montague, T. G., Krause, M., Torres Cleuren, Y. N., Tjeldnes, H. and Valen, E. 2019. CHOPCHOP v3: expanding the CRISPR web toolbox beyond genome editing. *Nucleic Acids Res.* 47(W1):W171–4.

Leenay, R. T., Aghazadeh, A., Hiatt, J., Tse, D., Roth, T. L., Apathy, R., Shifrut, E., Hultquist, J. F., Krogan, N., Wu, Z., Cirolia, G., Canaj, H., Leonetti, M. D., Marson, A., May, A. P. and Zou, J. 2019. Large dataset enables prediction of repair after CRISPR-Cas9 editing in primary T cells. *Nat. Biotechnol.* 37(9):1034–7.

Lei, Y., Lu, L., Liu, H. Y., Li, S., Xing, F. and Chen, L. L. 2014. CRISPR-P: a web tool for synthetic single-guide RNA design of CRISPR-system in plants. *Mol. Plant* 7(9):1494–6.

Lemos, B. R., Kaplan, A. C., Bae, J. E., Ferrazzoli, A. E., Kuo, J., Anand, R. P., Waterman, D. P. and Haber, J. E. 2018. CRISPR/Cas9 cleavages in budding yeast reveal templated insertions and strand-specific insertion/deletion profiles. *Proc. Natl. Acad. Sci. U. S. A.* 115(9):E2040–7.

Li, C., Zong, Y., Wang, Y., Jin, S., Zhang, D., Song, Q., Zhang, R. and Gao, C. 2018. Expanded base editing in rice and wheat using a Cas9-adenosine deaminase fusion. *Genome Biol.* 19(1):59.

Li, J., Sun, Y., Du, J., Zhao, Y. and Xia, L. 2017. Generation of targeted point mutations in rice by a modified CRISPR/Cas9 system. *Mol. Plant* 10(3):526–9.

Liu, H., Ding, Y., Zhou, Y., Jin, W., Xie, K. and Chen, L. L. 2017. CRISPR-P 2.0: an improved CRISPR-Cas9 Tool for genome editing in plants. *Mol. Plant* 10(3):530–2.

Lu, Y. and Zhu, J. K. 2017. Precise editing of a target base in the rice genome using a modified CRISPR/Cas9 system. *Mol. Plant* 10(3):523–5.

Mali, P., Aach, J., Stranges, P. B., Esvelt, K. M., Moosburner, M., Kosuri, S., Yang, L. and Church, G. M. 2013a. CAS9 transcriptional activators for target specificity screening and paired nickases for cooperative genome engineering. *Nat. Biotechnol.* 31(9):833–8.

Mali, P., Yang, L., Esvelt, K. M., Aach, J., Guell, M., DiCarlo, J. E., Norville, J. E. and Church, G. M. 2013b. RNA-guided human genome engineering via Cas9. *Science* 339(6121):823–6.

McVey, M. and Lee, S. E. 2008. MMEJ repair of double-strand breaks (director's cut): deleted sequences and alternative endings. *Trends Genet.* 24(11):529–38.

Meng, X., Hu, X., Liu, Q., Song, X., Gao, C., Li, J. and Wang, K. 2018. Robust genome editing of CRISPR-Cas9 at NAG PAMs in rice. *Sci. China Life Sci.* 61(1):122–5.

Minkenberg, B., Zhang, J., Xie, K. and Yang, Y. 2019. CRISPR-PLANT v2: an online resource for highly specific guide RNA spacers based on improved off-target analysis. *Plant Biotechnol. J.* 17(1):5–8.

Molla, K. A. and Yang, Y. 2019. CRISPR/cas-mediated base editing: technical considerations and practical applications. *Trends Biotechnol.* 37(10):1121–42.

Molla, K. A. and Yang, Y. 2020. Predicting CRISPR/Cas9induced mutations for precise genome editing. *Trends Biotechnol.* 38(2):136–41.

Montague, T. G., Cruz, J. M., Gagnon, J. A., Church, G. M. and Valen, E. 2014. CHOPCHOP: a CRISPR/Cas9 and TALEN web tool for genome editing. *Nucleic Acids Res.* 42(Web Server issue):W401–7.

Naito, Y., Hino, K., Bono, H. and Ui-Tei, K. 2015. CRISPRdirect: software for designing CRISPR/Cas guide RNA with reduced off-target sites. *Bioinformatics* 31(7):1120–3.

Nishida, K., Arazoe, T., Yachie, N., Banno, S., Kakimoto, M., Tabata, M., Mochizuki, M., Miyabe, A., Araki, M., Hara, K. Y., Shimatani, Z. and Kondo, A. 2016. Targeted nucleotide editing using hybrid prokaryotic and vertebrate adaptive immune systems. *Science* 353(6305):aaf8729.

Oliveros, J. C., Franch, M., Tabas-Madrid, D., San-León, D., Montoliu, L., Cubas, P. and Pazos, F. 2016. Breaking-Cas-interactive design of guide RNAs for CRISPR-Cas experiments for Ensembl genomes. *Nucleic Acids Res.* 44(W1):W267–71.

Park, J. and Bae, S. 2018. Cpf1-Database: web-based genome-wide guide RNA library design for gene knockout screens using CRISPR-Cpf1. *Bioinformatics* 34(6):1077–9.

Park, J., Bae, S. and Kim, J. S. 2015. Cas-Designer: a web-based tool for choice of CRISPR-Cas9 target sites. *Bioinformatics* 31(24):4014–6.

Park, J., Kim, J. S. and Bae, S. 2016. Cas-Database: web-based genome-wide guide RNA library design for gene knockout screens using CRISPR-Cas9. *Bioinformatics* 32(13):2017–23.

Pattanayak, V., Lin, S., Guilinger, J. P., Ma, E., Doudna, J. A. and Liu, D. R. 2013. High-throughput profiling of off-target DNA cleavage reveals RNA-programmed Cas9 nuclease specificity. *Nat. Biotechnol.* 31(9):839–43.

Prykhozhij, S. V., Rajan, V., Gaston, D. and Berman, J. N. 2015. CRISPR multitargeter: a web tool to find common and unique CRISPR single guide RNA targets in a set of similar sequences. *PLoS ONE* 10(3):e0119372.

Satheesh, V., Zhang, H., Wang, X. and Lei, M. 2019. Precise editing of plant genomes - Prospects and challenges. *Semin. Cell Dev. Biol.* 96:115–23.

Sfeir, A. and Symington, L. S. 2015. Microhomology-mediated end joining: a back-up survival mechanism or dedicated pathway? *Trends Biochem. Sci.* 40(11): 701–14.

Shen, M. W., Arbab, M., Hsu, J. Y., Worstell, D., Culbertson, S. J., Krabbe, O., Cassa, C. A., Liu, D. R., Gifford, D. K. and Sherwood, R. I. 2018. Predictable and precise template-free CRISPR editing of pathogenic variants. *Nature* 563(7733):646–51.

Shimatani, Z., Kashojiya, S., Takayama, M., Terada, R., Arazoe, T., Ishii, H., Teramura, H., Yamamoto, T., Komatsu, H., Miura, K., Ezura, H., Nishida, K., Ariizumi, T. and Kondo, A. 2017. Targeted base editing in rice and tomato using a CRISPR-Cas9 cytidine deaminase fusion. *Nat. Biotechnol.* 35(5):441–3.

Shou, J., Li, J., Liu, Y. and Wu, Q. 2018. Precise and predictable CRISPR chromosomal rearrangements reveal principles of Cas9-mediated nucleotide insertion. *Mol. Cell* 71(4):498–509.e4.

Sun, J., Liu, H., Liu, J., Cheng, S., Peng, Y., Zhang, Q., Yan, J., Liu, H. J. and Chen, L. L. 2019. CRISPR-Local: a local single-guide RNA (sgRNA) design tool for non-reference plant genomes. *Bioinformatics* 35(14):2501–3.

Tan, J., Zhao, Y., Wang, B., Hao, Y., Wang, Y., Li, Y., Luo, W., Zong, W., Li, G., Chen, S., Ma, K., Xie, X., Chen, L., Liu, Y. G. and Zhu, Q. 2020. Efficient CRISPR/Cas9-based plant genomic fragment deletions by microhomology-mediated end joining. *Plant Biotechnol. J.* doi: 10.1111/pbi.13390.

van Overbeek, M., Capurso, D., Carter, M. M., Thompson, M. S., Frias, E., Russ, C., Reece-Hoyes, J. S., Nye, C., Gradia, S., Vidal, B., Zheng, J., Hoffman, G. R., Fuller, C. K. and May, A. P. 2016. DNA repair profiling reveals nonrandom outcomes at Cas9-mediated breaks. *Mol. Cell* 63(4):633–46.

Wang, X., Li, J., Wang, Y., Yang, B., Wei, J., Wu, J., Wang, R., Huang, X., Chen, J. and Yang, L. 2018. Efficient base editing in methylated regions with a human APOBEC3A-Cas9 fusion. *Nat. Biotechnol.* 36(10):946–9.

Wu, X., Scott, D. A., Kriz, A. J., Chiu, A. C., Hsu, P. D., Dadon, D. B., Cheng, A. W., Trevino, A. E., Konermann, S., Chen, S., Jaenisch, R., Zhang, F. and Sharp, P. A. 2014. Genome-wide binding of the CRISPR endonuclease Cas9 in mammalian cells. *Nat. Biotechnol.* 32(7):670–6.

Xie, K., Zhang, J. and Yang, Y. 2014. Genome-wide prediction of highly specific guide RNA spacers for CRISPR-Cas9-mediated genome editing in model plants and major crops. *Mol. Plant* 7(5):923–6.

Xie, X., Ma, X., Zhu, Q., Zeng, D., Li, G. and Liu, Y. G. 2017. CRISPR-GE: a convenient software toolkit for CRISPR-based genome editing. *Mol. Plant* 10(9):1246–9.

Xu, H., Xiao, T., Chen, C. H., Li, W., Meyer, C. A., Wu, Q., Wu, D., Cong, L., Zhang, F., Liu, J. S., Brown, M. and Liu, X. S. 2015. Sequence determinants of improved CRISPR sgRNA design. *Genome Res.* 25(8):1147-57.

Zetsche, B., Gootenberg, J. S., Abudayyeh, O. O., Slaymaker, I. M., Makarova, K. S., Essletzbichler, P., Volz, S. E., Joung, J., van der Oost, J., Regev, A., Koonin, E. V. and Zhang, F. 2015. Cpf1 is a single RNA-guided endonuclease of a class 2 CRISPR-Cas system. *Cell* 163(3):759-71.

Zetsche, B., Heidenreich, M., Mohanraju, P., Fedorova, I., Kneppers, J., DeGennaro, E. M., Winblad, N., Choudhury, S. R., Abudayyeh, O. O., Gootenberg, J. S., Wu, W. Y., Scott, D. A., Severinov, K., van der Oost, J. and Zhang, F. 2017. Multiplex gene editing by CRISPR-Cpf1 using a single crRNA array. *Nat. Biotechnol.* 35(1):31-4.

Zhong, Z., Zhang, Y., You, Q., Tang, X., Ren, Q., Liu, S., Yang, L., Wang, Y., Liu, X., Liu, B., Zhang, T., Zheng, X., Le, Y., Zhang, Y. and Qi, Y. 2018. Plant genome editing using FnCpf1 and LbCpf1 nucleases at redefined and altered PAM sites. *Mol. Plant* 11(7):999-1002.

Zhu, L. J., Holmes, B. R., Aronin, N. and Brodsky, M. H. 2014. CRISPRseek: a bioconductor package to identify target-specific guide RNAs for CRISPR-Cas9 genome-editing systems. *PLoS ONE* 9(9):e108424.

Zong, Y., Song, Q., Li, C., Jin, S., Zhang, D., Wang, Y., Qiu, J. L. and Gao, C. 2018. Efficient C-to-T base editing in plants using a fusion of nCas9 and human APOBEC3A. *Nat. Biotechnol.* doi: 10.1038/nbt.4261.

Zong, Y., Wang, Y., Li, C., Zhang, R., Chen, K., Ran, Y., Qiu, J. L., Wang, D. and Gao, C. 2017. Precise base editing in rice, wheat and maize with a Cas9-cytidine deaminase fusion. *Nat. Biotechnol.* 35(5):438-40.

Zuo, E., Sun, Y., Wei, W., Yuan, T., Ying, W., Sun, H., Yuan, L., Steinmetz, L. M., Li, Y. and Yang, H. 2019. Cytosine base editor generates substantial off-target single-nucleotide variants in mouse embryos. *Science* 364(6437):289-92.

Chapter 7

Advances in assembling gRNA/Cas9 constructs in genome editing of plants

Marta Vázquez Vilar, Sara Selma, Asun Fernández del Carmen and Diego Orzáez, Instituto de Biología Molecular and Celular de Plantas de Valencia (CSIC-UPV), Spain

1 Introduction

The technical simplicity of CRISPR/Cas editing technology explains its rapid adoption in many laboratories, including small labs for which previous editing tools were inaccessible. Unlike preceding tools, CRISPR/Cas specificity resides in an easy-to-recode small RNA molecule, a feature that simplifies enormously the design of nucleases directed to specific genomic sites (Chen et al., 2019). However, the main technical obstacle for efficient editing in eukaryotes in general, and particularly in plants, now resides in the ability to deliver the CRISPR/Cas editing complex to cells. There are two main delivery options:

- Direct transfer; and
- Indirect encoding.

The first option is to transfer the CRISPR/Cas complex directly as a ribonucleoprotein to the cell (Banakar et al., 2019; Liang et al., 2017). The second option is indirectly encoding in RNA or DNA forms. In the latter

http://dx.doi.org/10.19103/AS.2020.0082.10

case, transgenic constructs encoding the nuclease and the guide RNAs can be transferred via viral vectors, biolistics, nanocarriers or employing agrobacterium-mediated transformation (Zaidi and Mansoor, 2017; Kumar et al., 2020; Ran et al., 2017).

Indirect strategies often involve a transgenic intermediate plant, which is produced upon stable integration of the CRISPR/Cas-containing transgene in the genome, which can be later segregated to create transgene-free edited plants. The production of transgenic intermediaries has certain regulatory drawbacks that could affect commercialization of edited plant products in certain countries. Despite this, agrobacterium-mediated transformation remains the most popular approach for genome editing in plants, including many crops. Several factors contribute to this popularity. In the first place, it is unclear that the final shape of regulatory frameworks in many countries will make distinctions among the different delivery procedures, provided that the final plant or plant product is devoid of exogenous DNA sequences. Secondly, agrobacterium transformation is a well-established methodology for many crop species, requiring relatively low-tech equipment. Finally, intermediary transgene constructs bring all the advantages of multigene cloning in binary vectors, which greatly expand the editing capabilities.

An important advantage derived from multigene cloning in the context of CRISPR/Cas editing is the possibility of creating arrays of gRNAs in a single T-DNA, thus enabling simultaneous targeting of several genes in a single transformation experiment. This strategy, known as multiplexing, can significantly accelerate the breeding process, particularly when trait pyramiding is required. Furthermore, multiplexing facilitates achieving loss-of-function phenotypes in cases where the gene function depends on families of homologous genes with high functional redundancy, as occurs often in polyploid genomes. Another advantage of the association of Cas9 with multigenic cloning is the ability to exploit its DNA-binding activity when devoid of its nuclease activity by mutagenesis (creating a so-called dead Cas9 or dCas9). By associating dCas9 with additional functional modules such as transcriptional regulators, enzymes, epigenetic modulators and so on, it is possible to expand the range of Cas9 activities while maintaining their DNA binding capacity, opening the door to a multitude of additional applications in plant breeding and biotechnology (Puchta, 2017).

In this chapter, we will review the principles of CRISPR cloning in binary vectors and the different methods and elements employed, including the nucleases alternative to Cas9. We pay special attention to modular cloning strategies and multiplexing tools as well as the engineering of expanded Cas activities. Finally, we include a case study of the cloning of a nine gRNA multiplex construct and the analysis of its transformants in tobacco plants.

2 Modular cloning tools for assembling CRISPR/Cas constructs

Numerous cloning strategies have been described to assemble constructs for editing, from the use of multicloning sites for type II rare cutters restriction enzymes (REs), type IIS REs or Gateway cloning (see Table 1). Overlapping PCR strategies such as Gibson assembly (Gibson et al., 2009), despite their power for facilitating seamless assembly of large constructs, are rarely used in CRISPR editing. This is probably due to the concurrence of many small elements to be combined, often repetitive and rich in secondary structure, all features for which overlapping-based techniques are not well adapted. As we will describe in detail below, even the most basic of the editing constructs contains three transcriptional units (TUs) comprising a minimum of nine functional elements. Although most of these elements can be considered constant, adaptation to different experimental scenarios requires that any cloning tool used for creating such constructs demonstrates maximum versatility when combining functional elements. These features make modular cloning probably the most suitable assembly strategy for building CRISPR constructs in plants.

Modular cloning is probably the most important legacy of plant biotechnologists to synthetic biology (Engler et al., 2009). It employs the Golden Gate cloning technology, based itself on the use of type IIS restriction enzymes, to create standard lego-like DNA parts that are used to build multigene constructs. The high efficiency of Golden Gate cloning enables the simultaneous assembly of several DNA parts in a one-tube multipartite reaction. Golden Gate also favors the adoption of common assembly rules known as assembly 'syntax', which are now widely adopted in the plant biotech community (Patron et al., 2015). The so-called 'phytobricks syntax' allows each of the basic elements in a transcriptional unit (TU) to be defined in standard terms, combined and readily exchanged among research groups without further adaptation, a feature that facilitates fair comparisons. Finally, modular cloning is neutral toward the use of repetitive elements, a key feature that favors its adoption for CRISPR constructs.

There is an increasing number of toolkits that offer standard DNA elements for plant CRISPR editing based on modular cloning. Many of them can be accessed via the Addgene repository. The recent OpenMTA initiative may also favor the exchange of elements among the scientific community, and this may favor the exchange of DNA parts and foster the development of the technique (Kahl et al., 2018). A non-exhaustive list of parts and collections can be consulted in Table 1. In our laboratory we have developed a collection of CRISPR elements based on the GoldenBraid (GB) modular cloning system (Sarrión et al., 2011; Sarrion-Perdigones et al., 2013). A distinctive feature of GB cloning is that it is iterative, allowing endless growth of multigene constructs and maximizing

Table 1 List of plasmid kits and collections for gene editing in plants

Resource	Laboratory	Cloning method	Details	Reference
A multipurpose editing toolkit	Voytas, DF	GoldenGate	Validated in tomato, tobacco. *Medicago truncatula*, wheat and barley	(Lowder et al., 2015)
pKAMA-ITACHI vectors	Higashiyama, T	GoldenGate	*Arabidopsis*	(Tsutsui and Higashiyama, 2017)
pBUN toolkit for multiplex genome editing	Qi-Jun Chen	GoldenGate	pGreen or pCAMBIA backbone	(Xing et al., 2014)
Dicot editing kit, chromosomal Deletions	Stuttmann J	GoldenGate	dicotyledonous	(Ordon et al., 2017)
Base editing kit	Gao, C	GoldenGate	Rice, wheat and maize	(Li et al., 2018)
Base editing kit	Kondo, A	GoldenGate	Rice and tomato	(Shimatani et al., 2017)
Ligation-independent cloning kit	Ali, GS	In fusion		(Khan et al., 2017)
Modular cloning toolkit for genome editing	Nekrasov, V	GoldenGate/ Moclo	Rice protoplasts	(Hahn et al., 2020)
Multiplexed plant genome editing and transcriptional regulation	Yiping Qi	GoldenGate/Gateway	Includes transcriptional regulation	(Lowder et al., 2015)
GoldenBraid genome editing toolkit	Orzaez, D	GoldenGate/ GoldenBraid	Tomato, tobacco, *N. benthamiana*, *Arabidopsis*, *Brassica napus*, eggplant	gbcloning.com

exchangeability of constructs between labs, or among different researchers within the same organization. Furthermore, GB constructs can be created online using our web-based assembly tools (https://gbcloning.upv.es/), with detailed wetlab protocols created for each in-silico made construct (Vazquez-Vilar et al., 2017). The GB system includes a section devoted to genome engineering with cloning webtools adapted for creating CRISPR constructs. All standard DNA parts required for building such a construct are listed in the GB database and deposited in the Addgene repository.

3 The basics of CRISPR/Cas9 cloning

The basic genome editing construct comprises at least three TUs, namely a Cas9-expressing TU, a TU containing the guide RNA and, typically, a third unit encoding a negative selection marker (depicted in Fig. 1). Additionally, other TUs can be included to add more elements as additional guide RNAs or positive selection markers.

The Cas9 TU is usually a constitutively expressed unit, driven by a strong promoter such as 35SCaMV or ubiquitin, and ending in a strong terminator (e.g., T35S, Tnos). Tissue-specific promoters are eventually employed, for example, in *Arabidopsis* to favor expression in relevant tissues for floral dip transformation (Zlobin et al., 2020). As far as Cas9 is concerned, there is a wide assortment of nuclease coding regions, many of which have been successfully tested in plants. *Streptococcus pyogenes* (SpCas9) is the most commonly used enzyme, although other engineered nucleases derived from other species are also employed (Ren et al., 2019; Wolter et al., 2018). The original bacterial Cas9 needs to be fused to a nuclear localization signal (NLS) and it is often codon optimized for plants, although some humanized Cas9 versions have excellent activity in plants and

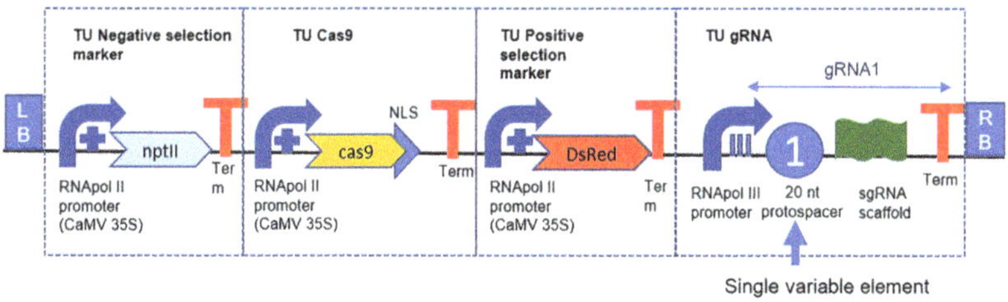

Figure 1 Basic structure of a CRISPR/Cas9 editing construct. CRISPR constructs typically contain four transcriptional units represented as boxes. Three are essential, namely the gRNA, the Cas9 and the negative selection marker. A fourth unit encoding positive selection marker is often included, here exemplified with a DsRed fluorescent marker. All DNA elements are depicted using visual SBOL-adapted symbols. Term, terminator; RB, right border; LB, left border; nptII, neomycin phosphotransferase II selection marker; NLS, nuclear localization signal.

are also widely used. Comparisons among different Cas9 versions have been reported elsewhere (Johnson et al., 2015; Raitskin et al., 2019).

Next to the Cas9 TU, the unit containing the gRNA transcript(s) has the most demanding design requirements, as it largely determines the efficiency and the extent of the editing. Like any TU, gRNA-TU consists of three main elements, the promoter, the gRNA transcript itself, and a transcription terminator. It should be noted that, in typical CRISPR editing constructs, the gRNA transcript constitutes the fusion in a single structure of two bacterial native RNAs acting in trans. For the expression of gRNAs, the use of RNA pol III promoters, which endogenously transcribe small nuclear RNAs, is highly recommended. The reason for this choice is twofold: on one hand, it ensures that the initiation of transcription takes place in a precise position, avoiding the subsequent processing (Cap, polyadenylation) associated with the transcripts of the RNA pol II. On the other hand, higher levels of accumulation in the nucleus are favored using pol III promoters. In return, it should be mentioned that the promoters of RNA pol III do not admit regulation.

The typical RNA pol III promoters used in CRISPR constructs are those from U3 and U6 snRNA genes. The *Arabidopsis* U6-26 promoter is standard for dicot plants, whereas the rice U3 snRNA promoter is also most commonly used for monocots. It is important to note that RNA pol III promoters impose strict requirements on the first nucleotide of the transcribed sequence. U6-26 imposes a G, whereas rice U3 imposes an A as initial nucleotide. Since the 20 nucleotide protospacer sequence starts immediately at the 5′ end of the gRNA transcript, these nucleotide requirements can pose a limitation in the design of protospacers, which need to be restricted to stretches of 20 nucleotides next to a PAM sequence and containing an initial G or A, respectively. Some results suggest that the presence of an additional nucleotide at the 5′ end of the gRNA has an influence on transcription efficiency, partially lifting the restriction on the starting G/A nucleotide (Gao et al., 2017). As will be explained below, the introduction of RNA processing signals in polycistronic constructs has also greatly reduced this limitation.

The next element in a basic gRNA TU is the gRNA transcript itself. This consists of a variable element, the 20 nucleotides protospacer, and a constant element, often referred to as the gRNA scaffold, which is necessary for anchoring the guide structure to the Cas9 protein. Finally, the construct typically concludes with a terminator element, which is added next to the scaffold to ensure proper finalization of the transcription. It is important to note that the gRNA scaffold and the terminator are two constant elements laying adjacently one to another and, although functionally different, the conjunct can be regarded as a single invariant structure for cloning purposes.

Editing constructs are often completed with additional TUs serving as selection markers. Negative selection markers such as antibiotic or herbicide-resistant genes are important in stable transformation for associating cell proliferation with the presence of an active Cas9 construct. The introduction of a positive selection marker (such as GFP or DsRed) is highly recommended to facilitate the segregation of T-DNA-free plant lines in the T1 generation. Figure 2A and B shows all the elements (phytobricks) typically employed in the assembly of a basic editing construct with the GoldenBraid system.

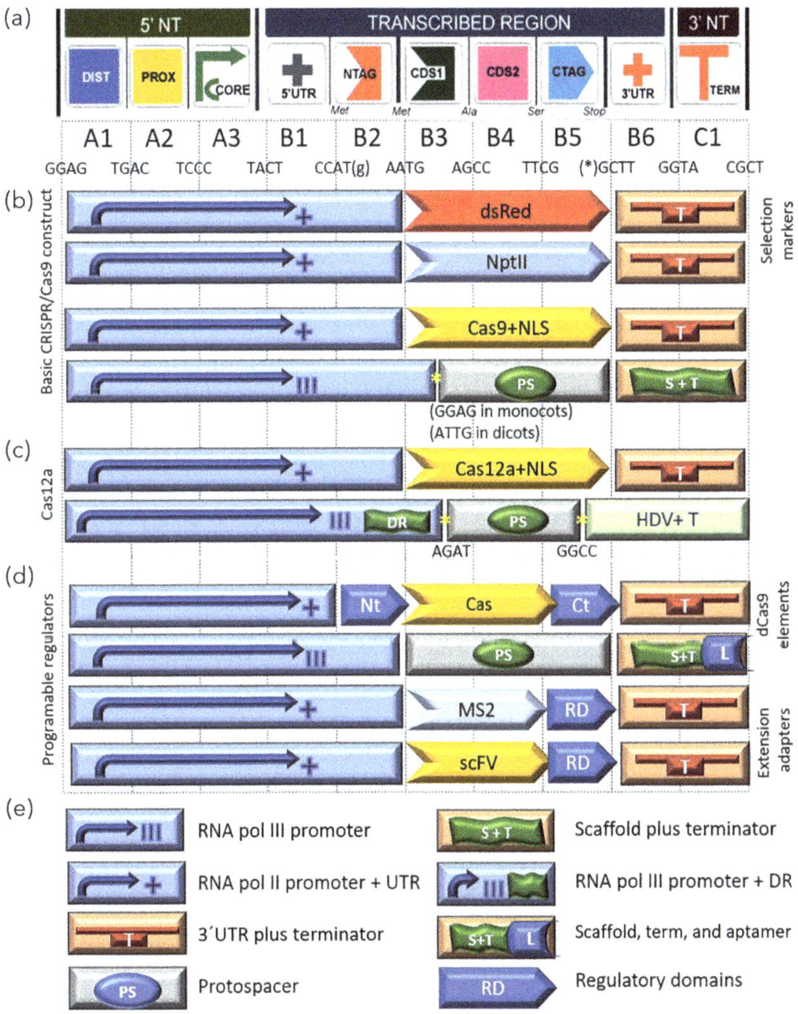

Figure 2 Standard elements in a basic modular cloning CRISPR construct. (a) Phytobricks syntax. (b) DNA standard elements employed in the assembly of all four TUs comprising a basic editing construct using spCas9. (c) TUs specific for Cas12 editing constructs. (d) DNA standard elements employed in the assembly of dCas9-based programmable regulators. (e) The standard elements employed in each unit are represented using SBOL-adapted symbols.

4 CRISPR-Cas12 specific cloning requirements

As mentioned above, a relatively large number of Cas9 variants have been reported, either derived from different bacterial species or engineered for different purposes such as different PAM recognition sites, increased precision and so on. Nevertheless, as far as cloning strategies are concerned, the basic structure described above is conserved for all Cas9 variants. However, this is not the case for Cas12a endonucleases discovered later (earlier known as Cpf1), such as AsCas12a, FnCas12a or LbCas12a (Zetsche et al., 2015). Distinctive features of Cas12a, such as its smaller size, its A/T reach PAM and its cleavage signature which produces staggered ends, makes it interesting for a number of applications in plant biotechnology. In contrast to Cas9, the native RNA constant scaffold element in the Cas12a group (comprising direct repeats or DRs with a compacted secondary structure) is located at the 5′ end of the transcript, whereas the protospacer occupies the last 20–23 nucleotides at the 3′ end. Consequently, the strictest design constraints are displaced to the 3′ end, where extra nucleotides (coming from, for example, a terminator sequence) could affect protospacer binding efficiency.

A preferred design option is the introduction of a self-cleavage ribozyme at the 3′ end of the transcript, followed by an oligo T terminator. Consequently, the final Cas12a gRNA setup typically comprises a pol III promoter, followed by a protospacer, a ribozyme and a terminator, the latter two often combined for modular cloning in a single constant element. It has been reported that the presence of the self-cleavage ribozyme increased the editing efficiency of the Cas12a constructs when assayed against the same target in transient expression experiments (Bernabé-Orts et al., 2019). Figure 2C also shows the standard elements used for cloning Cas12a constructs using GB cloning system.

5 Expanding CRISPR/Cas9 activities with modular cloning

Next to the gene editing applications derived from its site-directed nuclease activity, the gRNA-directed DNA-binding activities of Cas proteins can be used for a large number of additional applications. Once devoid of its nuclease activity, dCas variants can be translationally fused to an endless number of protein modules, creating new 'programmable' functions. These functions range from base modifiers (base editors, epimutators) to transcriptional regulators, fluorescent tags and so on. Modular cloning greatly facilitates the addition of new modules to dCas, while conserving the basic structure of the gRNA-TUs described above. It serves as a 'program' to direct the functionalized dCas to a specific genomic locus. For modular cloning of dCas protein fusions, a modified dCas syntax category is usually employed (B3–B4, instead to B3-B5, see Fig. 2D). This facilitates the entry of new elements as C-terminal fusions.

A powerful variation of the basic transcriptional fusion is the so-called SunTag strategy, where a multiepitope peptide is translationally attached to the Cas9 protein. A new TU is added to the construct encoding a minimal antibody fragment (scFv) fused to a functional domain (e.g., a transcriptional activator). In this way, the final structure comprises a multiepitope-extended Cas9 on which several functional domains are attached (see Fig. 3), which can reinforce their activity in a cooperative fashion (Tanenbaum et al., 2014).

The expansion of CRISPR/Cas activities can also be achieved by adding new functional modules to the gRNA. Two modifications of the basic cloning setup need to be incorporated to enable this to happen. First, the basic gRNA structure requires the addition of an aptamer, a small RNA structure with the ability to bind a protein domain. Aptamers can be added in two different permissive locations without affecting gRNA function. One location is the 3′ end of the gRNA, and the second site is in the second loop of the scaffold, which protrudes from the Cas9 when the ribonucleoprotein complex is formed. The second additional element is a TU encoding the aptamer-binding protein,

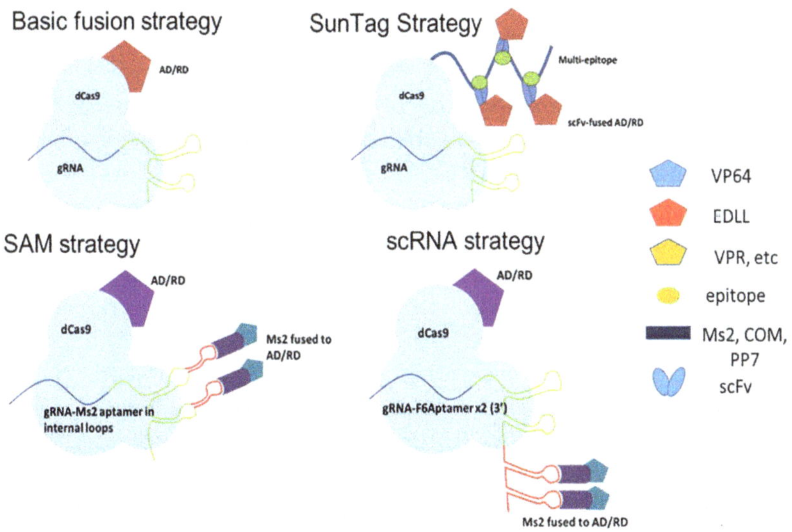

Figure 3 Structure of programmable regulators based on dCas9. Four strategies available to add functional domains to dCas9 are represented. Functional elements can be attached as protein fusions to dCas9 alone (basic fusion and multiepitope SunTag strategies), or combined with a second anchoring point consisting of an RNA aptamer engineered in the gRNA scaffold (SAM and scRNA strategies). In SAM, the aptamer is located at an RNA loop protruding from the Cas9 structure. In scRNA the aptamer is attached to the 3′ end of the RNA scaffold. Different aptamers are available, each one binding a different adapter protein (Ms2, COM, PP7).

which is then translationally fused to a functional protein domain as, for example, a transcriptional activator.

Finally, it is possible to combine translational fusion and RNA-aptamer strategies in a single construct, multiplying the functional capacities of Cas9 (Konermann et al., 2015, Mali et al., 2013). This has been demonstrated for the design of potent transcriptional activators (Lowder et al., 2018). Modular cloning facilitates the creation of multiple functional combinations. Several functional domains are listed in the GB database and can be obtained via the Addgene repository (see also Fig. 2). A dedicated set of webtools for the assembly of expanded Cas9 activities, focused on transcriptional regulators but adaptable to other activities, is available on the GB cloning page (https://gbcloning.upv.es/tools/grna_regulation/).

6 The polycistronic strategy for gRNA multiplexing

The most obvious advantage derived from the adoption of modular cloning strategies is the extended ability to create constructs for multiplexing. Different strategies have been put in place to achieve simultaneous editing of several loci, or several sites in a locus, in a single attempt. The most straightforward strategy is to add more gRNA TUs, each one comprising the same basic elements as described above. Whereas this is not feasible for some basic CRISPR-adapted binary plasmids, others, especially those based in modular cloning, are prepared to accept a prefixed number of gRNAs TUs. Iterative pipelines such as GoldenBraid also allow unlimited addition of new targets using this approach, at least as long as the cargo capacity of the plasmid backbone allows it (Fig. 4A).

In addition to this one-dimension multiplexing, a second dimension arises from the ability to produce polycistronic TUs carrying several gRNAs in a single transcript. This is possible by introducing precise processing signals so that individual gRNAs with precise boundaries can be released from the polycistronic RNA. Two main strategies have been described and successfully applied in plants. The first one makes use of Csy4 processing signals between individual gRNAs. In this approach, a Csy4 RNase needs to be added exogenously, usually incorporated in an additional TU. Upon transcription, the complex RNA is processed, releasing fully active individual gRNAs. This strategy, initially described in mammalian cells, was later adapted to plants and has subsequently been used in several demanding multiplexing approaches, successfully producing plants with several genes targeted simultaneously (Lowder et al., 2015; Qi et al., 2012).

A second strategy, initially described in plants by Xie and colleagues (Xie et al., 2015), relies on the introduction of tRNA sequences separating the gRNAs in a polycistronic transcript. The tRNAs are processed by endogenous plant tRNases, which are constitutively produced and targeted to the nucleus by the plant cell machinery. A precise design of tRNA-gRNA boundaries leads to the

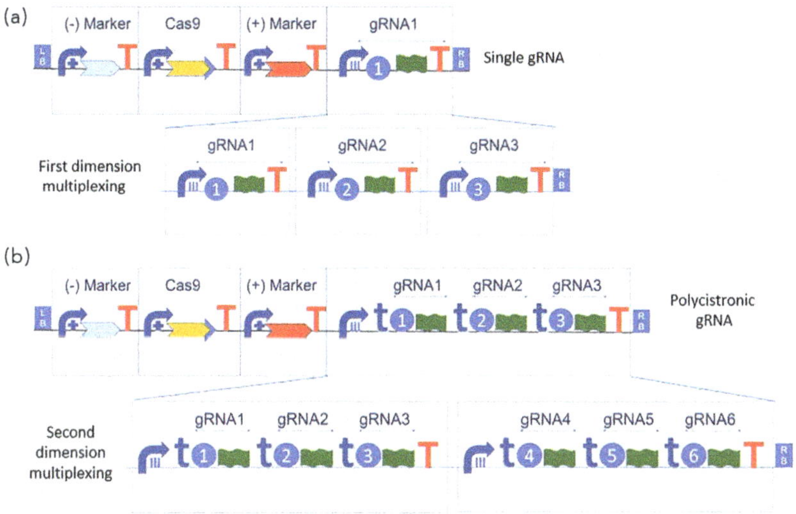

Figure 4 Multiplexing strategies using modular cloning. (a) Schema of a single gRNA construct and one-dimension multiplexing strategy that takes advantage of the capacity of modular cloning to build multigene assemblies incorporating several gRNAs TUs. (b) The two-dimension multiplexing strategy involves creating polycistronic TUs, each comprising several gRNAs, which are post-transcriptionally cleaved by RNases at specific internal sites to release individual gRNA units. The cleaving enzymes can be supplied externally (Csy4 strategy, explained in the text) or recruit endogenous enzymes (tRNA strategy, depicted here). A lower case t in the images represents tRNAs that will be processed by endogenous RNases. Symbols are based on SBOL standards with some adaptations.

generation of individual, fully active gRNA molecules (Fig. 4B). In this second approach, no transgenic RNases need to be introduced in the construct. This has the advantage of reducing cloning requirements. However, there is less control on the expression levels of the cleavage enzymes. A detailed description of this strategy is included in the case study section.

Finally, Cas12a type of nucleases represents a special case in the cloning design of polycistronic constructs for multiplexing. Cas12a has the native ability to process tandemly arrayed gRNA modules. Neither processing signals nor in trans RNase activities, therefore, need to be supplemented exogenously or recruited endogenously for efficient processing. The absence of repeated elements such as tRNAs or Csy4 RNase sites also facilitates the use of custom gene synthesis as a direct method for creating the tandem arrays. A suggested multiplex cloning strategy for Cas12a consists in the direct synthesis of the tandemly repeated variable elements, flanked with appropriated overhangs, which are later combined with a constant element comprising the pol III promoter and the DR sequence to create a complete multiplexing Cas 12 gRNA TU. The simplicity of this approach can be an additional advantage to favor Cas12a over Cas9 in multiplexing approaches in plants. A GB cloning tool supports design and cloning of multiplex Cas12a constructs: https://gbcloning.upv.es/tools/cas12multiplexing/.

7 Case study: multiple gene targeting in tobacco with GoldenBraid cloning

This section describes as a case study the cloning procedure followed to knock out a group of genes of the Squamosa-promoter binding protein-like (SPL) family in tobacco cv. K326 using Cas9-tRNA multiplexing strategy. Briefly, SPL genes are transcriptional regulators involved, among other aspects of plant biology, in the regulation of flowering time, branching and juvenility. In the context of improving tobacco plants as biofactories, the SPL-KO strategy aimed at generating new variability in growth habits, which could be later screened and selected for biofactory use. In this approach, a specific group of related SPLs, all targets of microRNA 156, were targeted for knock out using CRISPR/Cas9. A total of 10 SPL homologues were selected, and two protospacers were designed for each gene distributed mainly in exons 1 and 2. When possible, gRNAs were designed to target several genes simultaneously (100% match in the protospacer sequence). On-target scores below 51 (Doench et al., 2016) were discarded. Off-target analysis were also performed to discard the presence of high-score off-targets in SPL genes or somewhere else in the genome. As a result, a total of nine protospacer sequences were selected, which were distributed in three multiplexing constructs, containing three gRNAs each. All cloning procedures were performed using the GoldenBraid system and standards, aided by GB cloning software available at: https://gbcloning.upv.es/tools/cas9multiplexing_editing/.

The GB cloning procedure involves the combination of a number of pre-made constant elements, all available in the GB collection, with the custom-made elements (protospacers), which are specific for each editing experiment. The GB pipeline is designed to minimize the effort required to create the custom elements, which is reduced to the synthesis of a pair of 25 nt oligonucleotides per target site. The cloning process followed in this experiment used the following steps.

First, custom targets were 'domesticated' using GB software (step 1 in the overall cloning procedure). The domestication process generates pairs of partially overlapping oligonucleotides containing the 20 protospacers, but extended with 4 additional nucleotides employed in the cloning process. The additional nucleotides will serve as overhangs, matching the cloning sites of the remaining elements in the cloning strategy. The GB software also discards sequences that could include internal BsaI or BsmBI recognition sites. All protospacers are domesticated in the same form and can be incorporated in any position of the multiplexing TU. The domesticated elements need to be uploaded to the GB database to proceed with the cloning process.

Once the nine pairs of nucleotides were domesticated and uploaded to the GB database, they were distributed in three TUs, corresponding to the TUg1_3x + TUg2_3x + TUg3_3x two-dimension multiplexing strategy described above. A similar procedure was applied for each TU, as follows. First, the number of gRNAs to be included in each TU was selected. This selection imposes constraints on the pre-made cloning parts (aka level 1 and level 0 GB parts) to be subsequently employed. The GB collection of pre-made parts comprises tRNA elements, either maintained separately or combined with other constant elements such as the constant part of the gRNA or the transcriptional terminator. All pre-made parts are cloned in an entry plasmid (pVD) and contain appropriate overhangs that are released upon BsmBI digestion. The GB collection contains the pre-made elements required for creating multiplexing assemblies ranging from one to six gRNAS per TU.

As depicted in Fig. 5, creating a three gRNAs TU requires the combination of three custom elements (created in step 1, see above), with four pre-made

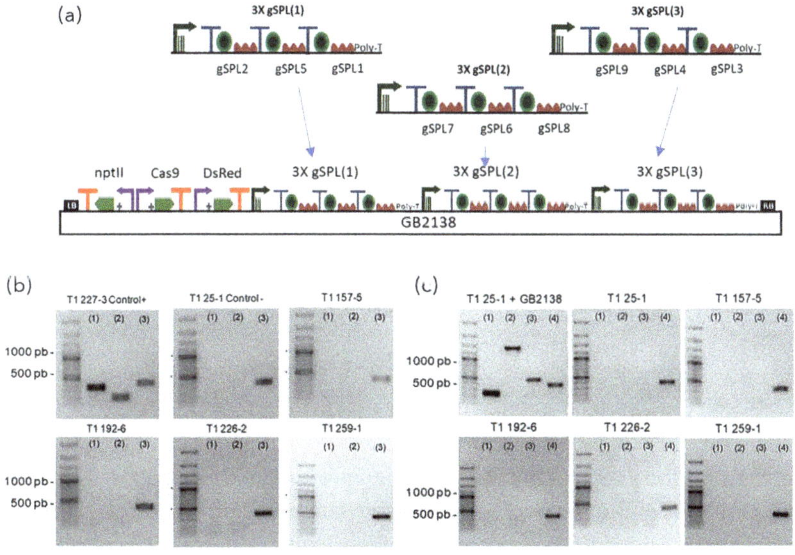

Figure 5 Case study of SPL editing in tobacco. (a) Final construct transformed in tobacco K326 cultivar comprising nine gRNAs. (b) Analysis of presence of T-DNA and in the genome-selected plants in T1 line. Genomic DNAs from each plant were amplified with primer pairs amplifying 35S (1), Tnos (2) or an endogenous gene (3). (c) Analysis of presence of vector backbone and in the genome-selected plants in T1 line. Genomic DNA from each plant was amplified with plasmid pairs amplifying different regions of the vector backbone (1–3) or an endogenous gene (4). T1-277 is a plant line that did not segregate the T-DNA (DsRED+). T25-1 is a non-transformed plant. The up-left gel in (c) shows amplification of a T1 25-1 plant gDNA preparation supplemented with GB2138 plasmid backbone in quantity equivalent to a genomic DNA copy.

elements in two consecutive Golden Gate cloning steps, which alternate BsaI and BsmBI enzymes. Briefly, the initial cloning reaction (step 2 in the overall cloning procedure) creates individual tRNA-protospacer-scaffold structures using BsmBI, which are later combined among them and with the promoter in a subsequent step using BsaI (step 3). The GB cloning tool https://gbcloning.upv.es/tools/cas9multiplexing_editing/ provides a stepwise protocol for each TU assembly made using GB software. Detailed protocols for each of the three TUs assembled in this case study are produced by the program.

The next step (step 4) consists in the combination of each gRNA TU between themselves and with the remaining TUs used in the experiment (Cas9, positive and negative markers). In this case study, we used a human codon-optimized Cas9 under the control of the 35S promoter (GB0639). As a negative selection marker, we employed a standard nptII TU providing KanR (GB0226) and, as a positive selection marker, we employed a P35S:DsRED:T35S TU (GB0360). TU assemblies in GB proceed following a binary scheme, with only two elements combined in each step and alternating BsaI and BsmBI reactions. The cloning pipeline in this study comprised the following binary reactions distributed in three rounds. Each round represents a group of assembly reactions that can be carried out in parallel.

Round 1:
(BR1): TUg1_3x + TUg2_3x
(BR2): TU_Cas9 + TU_DsRED

Round 2:
(BR3): (TUg1_3x-TUg2_3x) + TUg3_3x
(BR4): rTU_NptII + (TU_Cas9-TU_DsRED)

Round 3:
(BR5): (rTU_NptII-TU_Cas9-TU_DsRED) + (TUg1_3x-TUg2_3x-TUg3_3x)

Detailed protocols for each binary reaction are produced as output files when using the GoldenBraid binary assembly software tool (https://gbcloning.upv.es/do/bipartite/).

It is important to note that all GB TUs and multigene constructs are re-usable, meaning that, once assembled, they can be incorporated in virtually any construct. Here we have described all assembly steps, starting from individual TUs into a final 6xTUs multigene assembly (Fig. 5). However, pre-made 2xTUs and 3xTUs combinations comprising Cas9 and selection markers are available in our lab and many GB labs for reuse, significantly reducing the number of cloning steps required for obtaining the desired final combination. For instance, most GB labs are supplied with a pre-assembled rTU_NptII-TU_Cas9-TU_DsRED construct; therefore, reactions BR2 and BR4 can be omitted.

All individual GB parts and pre-made TUs used in this case study are available in the Addgene repository and have been tested thoroughly. Additional TUs with equivalent functions, as alternative selection markers, are also available. Users can also build their own TUs (e.g. carrying different Cas9 versions) by combining standard elements with the TU assembler tool (https://gbcloning.upv.es/do/multipartite/). Briefly, this tool guides users in the combination of multiple basic (aka level 0) elements, such as promoters, CDS, terminators and so on, to create a new TU (aka level 1 element) cloned into a binary destination plasmid. Destination plasmids in GB cloning are always binary plasmids. Currently, the GB collection includes three plasmid series derived from pGREEN, pCAMBIA and pLX_B backbones (Pasin et al., 2017). During the multipartite TU assembly process, users can decide the future orientation of the newly assembled TU by selecting the appropriate destination plasmid (forward or reverse) among those available in the GB collection. It is important to note, for example, that rTU_NptII displays a reverse orientation in the final multiplexing assembly (see Fig. 5). This orientation was opted to ensure divergent transcription from adjacent promoters, and was created during the assembly of this TU by selecting a reverse-orientation destination plasmid. The 5′ leftmost position of rTU_NptII was selected to ensure full integration of the T-DNA.

The final assembly in Fig. 5a was transformed into *Agrobacterium tumefaciens* LBA4404 strain and used in the stable transformation of tobacco plants. The resulting KanR, DsRED+ T0 generation plants were genotyped by amplicon sequencing followed by ICE analysis, and the results are depicted in Table 2. As shown, several loci showed frequent putative biallelic mutations (ICE efficiency >90%), whereas some genes, particularly the SPL9b subfamily, showed very low mutation frequencies. The best four edited lines (NFLv0.1_157, 192, 226, and 259) were selfed to the T1 generation. From each of them, four transgene-free (DsRed-) T1 plants were grown in the greenhouse. T1 plants were then genotyped for both gRNAs targeting each gene. The number of mutated targets in these plants ranged between six and seven, with two lines (T1 157-5 and T1 259-1) showing biallelic mutations for seven out of the ten targeted SPL genes. These two lines were further analyzed by PCR, confirming that no T-DNA or backbone vector sequences remained in the genome (Fig. 5B). The phenotypic analysis of the new population diversity created by multiplex SPL editing in tobacco is currently under review and will be described in detail elsewhere. Preliminary results indicate that SPL knockouts are associated with delayed flowering time and leaf juvenility.

This case study illustrates the power of multiplex editing in the generation of new genetic diversity in plants, by targeting specific loci as gene families. We

Table 2 Edition efficiencies in TO measured by ICE method. In green are efficiencies above 40%, efficiencies above 90% are considered homozygotic or biallelic

	Nitab4.5 _0001538g0080.1	Nitab4.5 _0000991g0020.1	Nitab4.5 _0003572g0010.1	Nitab4.5 _0000016g0300.1	Nitab4.5 _0003348g0050.1	Nitab4.5 _0003942g0050.1	Nitab4.5 _0000638g0040.1	Nitab4.5 _0001752g0040.1	Nitab4.5 _0007487g0020.1	Nitab4.5 _0002219g0060.1
Line	gSPL5	gSPL5	gSPL1	gSPL1	gSPL6	gSPL6	gSPL3	gSPL3	gSPL8	gSPL8
P157	5	0	59	78	53	7	96	97	87`	97
P178	1	2	44	75	38	3	90	93	91	98
P179	0	2	16	75	17	3	46	49	45	47
P180	1	2	1	62	4	1	2	2	1	3
P182	1	1	57	78	4	2	1	2	2	2
P185	1	75	74	68	0	1	2	1	2	36
P188	1	0	6	81	0	1	1	2	0	1
P192	1	0	58	63	69	43	98	93	7	96
P193	0	0	15	77	0	1	1	3	2	1
P195	1	72	60	76	40	3	98	40	95	91
P197	1	1	79	71	59	9	98	96	98	42
P225	1	0	6	74	2	8	79	86	40	97
P226	1	74	73	71	23	34	98	97	40	99
P258	50	0	14	79	8	5	70	95	91	94
P271	0	0	49	48	16	19	98	98	99	99

have shown that, in a single transformation experiment, we were able to recover up to seven biallelic mutants out of ten assessed in the T1 generation, with a number of intermediate mutants that can be used for phenotype association studies. The remaining three targets which were not mutated responded to gRNAs that showed low or null editing efficiency. Further improvements in multiplex editing will need to arise from a more accurate in silico prediction of gRNA efficiency.

8 References

Banakar, R., Eggenberger, A. L., Lee, K., Wright, D. A., Murugan, K., Zarecor, S., Lawrence-Dill, C. J., Sashital, D. G. and Wang, K. 2019. High-frequency random DNA insertions upon co-delivery of CRISPR-Cas9 ribonucleoprotein and selectable marker plasmid in rice. *Sci. Rep.* 9(1), 19902.

Bernabé-Orts, J. M., Casas-Rodrigo, I., Minguet, E. G., Landolfi, V., Garcia-Carpintero, V., Gianoglio, S., Vázquez-Vilar, M., Granell, A. and Orzaez, D. 2019. Assessment of Cas12a-mediated gene editing efficiency in plants. *Plant Biotechnol. J.* 17(10), 1971–1984.

Chen, K., Wang, Y., Zhang, R., Zhang, H. and Gao, C. 2019. CRISPR/Cas genome editing and precision plant breeding in agriculture. *Annu. Rev. Plant Biol.* 70, 667–697.

Doench, J. G., Fusi, N., Sullender, M., Hegde, M., Vaimberg, E. W., Donovan, K. F., Smith, I., Tothova, Z., Wilen, C., Orchard, R., Virgin, H. W., Listgarten, J. and Root, D. E. 2016. Optimized sgRNA design to maximize activity and minimize off-target effects of CRISPR-Cas9. *Nat. Biotechnol.* 34(2), 184–191.

Engler, C., Gruetzner, R., Kandzia, R. and Marillonnet, S. 2009. Golden Gate shuffling: a one-pot DNA shuffling method based on type IIs restriction enzymes. *PLoS ONE* 4(5), e5553.

Gao, Z., Harwig, A., Berkhout, B. and Herrera-Carrillo, E. 2017. Mutation of nucleotides around the +1 position of type 3 polymerase III promoters: the effect on transcriptional activity and start site usage. *Transcription* 8(5), 275–287.

Gibson, D. G., Young, L., Chuang, R. Y., Venter, J. C., Hutchison, C. A. and Smith, H. O. 2009. Enzymatic assembly of DNA molecules up to several hundred kilobases. *Nat. Meth.* 6(5), 343–345.

Hahn, F., Korolev, A., Sanjurjo Loures, L. and Nekrasov, V. 2020. A modular cloning toolkit for genome editing in plants. *BMC Plant Biol.* 20(1), 179.

Johnson, R. A., Gurevich, V., Filler, S., Samach, A. and Levy, A. A. 2015. Comparative assessments of CRISPR-Cas nucleases' cleavage efficiency in planta. *Plant Mol. Biol.* 87(1–2), 143–156.

Kahl, L., Molloy, J., Patron, N., Matthewman, C., Haseloff, J., Grewal, D., Johnson, R. and Endy, D. 2018. Opening options for material transfer. *Nat. Biotechnol.* 36(10), 923–927.

Khan, A. A., El-Sayed, A., Akbar, A., Mangravita-Novo, A., Bibi, S., Afzal, Z., Norman, D. J. and Ali, G. S. 2017. A highly efficient ligation-independent cloning system for CRISPR/Cas9 based genome editing in plants. *Plant Methods* 13, 86.

Konermann, S., Brigham, M. D., Trevino, A. E., Joung, J., Abudayyeh, O. O., Barcena, C., Hsu, P. D., Habib, N., Gootenberg, J. S., Nishimasu, H., Nureki, O. and Zhang, F. 2015.

Genome-scale transcriptional activation by an engineered CRISPR-Cas9 complex. *Nature* 517(7536), 583–588. doi: 10.1038/nature14136. Epub 2014 Dec 10. PMID: 25494202; PMCID: PMC4420636.

Kumar, S., Nehra, M., Dilbaghi, N., Marrazza, G., Tuteja, S. K. and Kim, K. H. 2020. Nanovehicles for plant modifications towards pest- and disease-resistance traits. *Trends Plant Sci.* 25(2), 198–212.

Li, C., Zong, Y., Wang, Y., Jin, S., Zhang, D., Song, Q., Zhang, R. and Gao, C. 2018. Expanded base editing in rice and wheat using a Cas9-adenosine deaminase fusion. *Genome Biol.* 19(1), 59.

Liang, Z., Chen, K., Li, T., Zhang, Y., Wang, Y., Zhao, Q., Liu, J., Zhang, H., Liu, C., Ran, Y. and Gao, C. 2017. Efficient DNA-free genome editing of bread wheat using CRISPR/Cas9 ribonucleoprotein complexes. *Nat. Commun.* 8, 14261.

Lowder, L. G., Zhang, D., Baltes, N. J., Paul, J. W., Tang, X., Zheng, X., Voytas, D. F., Hsieh, T. F., Zhang, Y. and Qi, Y. 2015. A CRISPR/Cas9 toolbox for multiplexed plant genome editing and transcriptional regulation. *Plant Physiol.* 169(2), 971–985.

Lowder, L. G., Zhou, J., Zhang, Y., Malzahn, A., Zhong, Z., Hsieh, T. F., Voytas, D. F., Zhang, Y. and Qi, Y. 2018. Robust transcriptional activation in plants using multiplexed CRISPR-Act2.0 and mTALE-act systems. *Mol. Plant* 11(2), 245–256.

Mali, P., Yang, L., Esvelt, K. M., Aach, J., Guell, M., DiCarlo, J. E., Norville, J. E. and Church, G. M. 2013. RNA-guided human genome engineering via Cas9. *Science* 339(6121), 823–826.

Ordon, J., Gantner, J., Kemna, J., Schwalgun, L., Reschke, M., Streubel, J., Boch, J. and Stuttmann, J. 2017. Generation of chromosomal deletions in dicotyledonous plants employing a user-friendly genome editing toolkit. *Plant J.* 89(1), 155–168.

Pasin, F., Bedoya, L. C., Bernabé-Orts, J. M., Gallo, A., Simón-Mateo, C., Orzaez, D. and García, J. A. 2017. Multiple T-DNA delivery to plants using novel mini binary vectors with compatible replication origins. *ACS Synth. Biol.* 6(10), 1962–1968.

Patron, N. J., Orzaez, D., Marillonnet, S., Warzecha, H., Matthewman, C., Youles, M., Raitskin, O., Leveau, A., Farré, G., Rogers, C., Smith, A., Hibberd, J., Webb, A. A. R., Locke, J., Schornack, S., Ajioka, J., Baulcombe, D. C., Zipfel, C., Kamoun, S., Jones, J. D. G., Kuhn, H., Robatzek, S., Van Esse, H. P., Sanders, D., Oldroyd, G., Martin, C., Field, R., O'Connor, S., Fox, S., Wulff, B., Miller, B., Breakspear, A., Radhakrishnan, G., Delaux, P. M., Loqué, D., Granell, A., Tissier, A., Shih, P., Brutnell, T. P., Quick, W. P., Rischer, H., Fraser, P. D., Aharoni, A., Raines, C., South, P. F., Ané, J. M., Hamberger, B. R., Langdale, J., Stougaard, J., Bouwmeester, H., Udvardi, M., Murray, J. A. H., Ntoukakis, V., Schäfer, P., Denby, K., Edwards, K. J., Osbourn, A. and Haseloff, J. 2015. Standards for plant synthetic biology: a common syntax for exchange of DNA parts. *New Phytol.* 208(1), 13–19.

Puchta, H. 2017. Applying CRISPR/Cas for genome engineering in plants: the best is yet to come. *Curr. Opin. Plant Biol.* 36, 1–8.

Qi, L., Haurwitz, R. E., Shao, W., Doudna, J. A. and Arkin, A. P. 2012. RNA processing enables predictable programming of gene expression. *Nat. Biotechnol.* 30(10), 1002–1006.

Raitskin, O., Schudoma, C., West, A. and Patron, N. J. 2019. Comparison of efficiency and specificity of CRISPR-associated (Cas) nucleases in plants: an expanded toolkit for precision genome engineering. *PLoS ONE* 14(2), e0211598.

Ran, Y., Liang, Z. and Gao, C. 2017. Current and future editing reagent delivery systems for plant genome editing. *Sci. China Life Sci.* 60(5), 490–505.

Ren, B., Liu, L., Li, S., Kuang, Y., Wang, J., Zhang, D., Zhou, X., Lin, H. and Zhou, H. 2019. Cas9-NG greatly expands the targeting scope of the genome-editing toolkit by recognizing NG and other atypical PAMs in rice. *Mol. Plant* 12(7), 1015-1026.

Sarrión, A., Juárez, P., Fernández del Carmen, A., Granell, A. and Orzaez, D. 2011. GoldenBraid: a modular, semi-dempotent system for the assembly of reciclable genetic pieces in plant synthetic biology. *PLoS ONE*.

Sarrion-Perdigones, A., Vazquez-Vilar, M., Palaci, J., Castelijns, B., Forment, J., Ziarsolo, P., Blanca, J., Granell, A. and Orzaez, D. 2013. GoldenBraid 2.0: a comprehensive DNA assembly framework for plant synthetic biology. *Plant Physiol.* 162(3), 1618-1631.

Shimatani, Z., Kashojiya, S., Takayama, M., Terada, R., Arazoe, T., Ishii, H., Teramura, H., Yamamoto, T., Komatsu, H., Miura, K., Ezura, H., Nishida, K., Ariizumi, T. and Kondo, A. 2017. Targeted base editing in rice and tomato using a CRISPR-Cas9 cytidine deaminase fusion. *Nat. Biotechnol.* 35(5), 441-443.

Tanenbaum, M. E., Gilbert, L. A., Qi, L. S., Weissman, J. S. and Vale, R. D. 2014. A protein-tagging system for signal amplification in gene expression and fluorescence imaging. *Cell* 159(3), 635-646.

Tsutsui, H. and Higashiyama, T. 2017. pKAMA-ITACHI vectors for highly efficient CRISPR/Cas9-mediated gene knockout in Arabidopsis thaliana. *Plant Cell Physiol.* 58(1), 46-56.

Vazquez-Vilar, M., Quijano-Rubio, A., Fernandez-Del-Carmen, A., Sarrion-Perdigones, A., Ochoa-Fernandez, R., Ziarsolo, P., Blanca, J., Granell, A. and Orzaez, D. 2017. GB3.0: a platform for plant bio-design that connects functional DNA elements with associated biological data. *Nucleic Acids Res.* 45(4), 2196-2209.

Wolter, F., Klemm, J. and Puchta, H. 2018. Efficient in planta gene targeting in Arabidopsis using egg cell-specific expression of the Cas9 nuclease of Staphylococcus aureus. *Plant J.* 94(4), 735-746.

Xie, K., Minkenberg, B. and Yang, Y. 2015. Boosting CRISPR/Cas9 multiplex editing capability with the endogenous tRNA-processing system. *PNAS* 112(11), 3570-3575.

Xing, H. L., Dong, L., Wang, Z. P., Zhang, H. Y., Han, C. Y., Liu, B., Wang, X. C. and Chen, Q. J. 2014. A CRISPR/Cas9 toolkit for multiplex genome editing in plants. *BMC Plant Biol.* 14, 327.

Zaidi, S. S.-A. and Mansoor, S. 2017. Viral vectors for plant genome engineering. *Front. Plant Sci.* 8, 539.

Zetsche, B., Gootenberg, J. S., Abudayyeh, O. O., Slaymaker, I. M., Makarova, K. S., Essletzbichler, P., Volz, S. E., Joung, J., van der Oost, J., Regev, A., Koonin, E. V. and Zhang, F. 2015. Cpf1 is a single RNA-guided endonuclease of a class 2 CRISPR-Cas system. *Cell* 163(3), 759-771.

Zlobin, N. E., Lebedeva, M. V. and Taranov, V. V. 2020. CRISPR/Cas9 genome editing through in planta transformation. *Crit. Rev. Biotechnol.* 40(2), 153-168.

Chapter 8

Strategies for CRISPR/Cas9-mediated genome editing: from delivery to production of modified plants

William Gordon-Kamm, Pierluigi Barone, Sergei Svitashev, Jeffry D. Sander, Sandeep Kumar and Todd Jones, Corteva Agriscience, USA

1 Introduction

Genome editing technology offers tremendous potential for agricultural improvements including consumer health, increased productivity, and alleviating the growing food security crisis. Progress in developing and implementing site-directed nucleases (SDNs), in particular CRISPR/Cas9, has been rapid and exciting. In addition, many new approaches have been concurrently developed for delivery of CRISPR/Cas9 components to plant cells, from *Agrobacterium* infection to nanoparticles (Chen et al., 2019). However, advances in genome editing and delivery technologies continue to outpace our ability to efficiently produce edited events due to the lack of robust methods for regeneration of fertile plants for many crops (Altpeter et al., 2016). Historically, recovering transgenic plants, even at low frequency, has been considered a major success for many crops. However, genome

http://dx.doi.org/10.19103/AS.2020.0082.12

editing applications comprising template-based modifications via homology-directed repair (HDR), often require production of hundreds of regenerated events to recover plants with the desired edit(s). This limitation in regeneration efficiency hinders full adoption of CRISPR/Cas9-mediated genome editing in agriculture.

The challenge of efficient regeneration influences many experimental options to develop a successful genome editing protocol and ultimately depends on how amenable a species is to *in vitro* manipulation. Given this prerequisite, easily cultured species such as *Arabidopsis thaliana* and *Nicotiana benthamiana* became the model for demonstrating multiple genome editing applications, using delivery to a variety of plant explants ranging from protoplasts (Li et al., 2013) to *in planta* (Maher et al., 2020). Conversely, many important row crop species (e.g. cereals, legumes, and woody perennials, among others) are difficult to manipulate *in vitro*, limiting the opportunities for successful application of genome editing technologies.

Recalcitrant species, which typically exhibit consistently low-regeneration frequencies, are usually suitable for simple mutagenesis-based knockouts, but not for more complex (and low frequency) template-based edits. While recalcitrance for a given species can sometimes be overcome by perseverance and patience through exploring tissue culture conditions, new approaches such as the use of morphogenic genes can be an important adjunct, aiding in both accelerating progress in certain species and expanding the range of accessible plant varieties.

To ensure success, choosing the optimal combination of CRISPR/Cas9 delivery method and tissue culture strategy requires thoughtful consideration for any given plant species. In this chapter we discuss delivery options, cell type or plant explant, and the potential benefits of using morphogenic genes.

2 Delivery of genome editing components into plant cells

Competence and regenerability are critical considerations in selecting an explant to be suitable for genome modification. Competence reflects the ability of a cell/tissue to be genetically modified by receiving exogenous biomolecules (e.g. DNA, RNA, protein etc.). Regenerability can be described as the plant cell's ability to switch toward a regenerative pathway resulting in the production of a full fertile plant capable of transmitting the intended genome modification to its progenies (Delporte et al., 2012).

Options for transformable explant material range from single cells, such as protoplasts or microspores, to embryonic tissues such as meristem, scutellum or cotyledons, to seedling-derived tissues such as apical/axillary meristems, hypocotyl or leaves, and finally *in planta* alternatives. In general, these tend to progress from methods involving extended lengths of *in vitro* culture (e.g.

protoplast, embryogenic/organogenic callus), toward those that capitalize on more rapid plant development, including direct somatic embryo or meristem/shoot formation, and finally remaining entirely *in planta*. Following this trend is a concomitant reduction in the time required from delivery of DNA, RNA, and/or protein to recovery of a fertile genome-edited plant, conditions that have long been recognized as desirable to minimize tissue culture and reduce somaclonal variation (Côte et al., 2001). The ideal scenario would be a simple *in planta* method that worked across all plants, but currently many species require long *in vitro* manipulation while others, including many agronomically important crops, remain recalcitrant to any of the generic strategies outlined in Table 1.

The cell types or tissues used for successful CRISPR/Cas9-mediated editing include examples such as lettuce protoplasts (Woo et al., 2015), leaf tissue in tobacco (Gao et al., 2015), immature embryos in cereals (Svitashev et al., 2015, 2016; Shi et al., 2017; Liang et al., 2017; Hamada et al., 2018), embryogenic callus in rice (Xu et al., 2015; Sun et al., 2016), and immature cotyledons in soybean (Li et al., 2015). Many alternative explants such as callus, cotyledons, true leaves, microspores, axillary meristems, and hypocotyl (or stem) have been used to produce transgenic plants (Table 1) and remain a viable starting material for future CRISPR/Cas9-mediated genome editing. For many crops, explant preference can change over time with concomitant reductions in tissue culture. Cowpea is one example of a recalcitrant crop in which the choice of explant and culture route has had a profound impact on the outcome. Accordingly, researchers have recognized the benefit of moving away from organogenic or embryogenic cultures as a transformation target, which results in prolonged periods of tissue culture. Instead, they have shifted to T-DNA delivery into such explants as cotyledons (Raveendar and Ignacimuthu, 2010) or cotyledonary nodes (Chaudhury et al., 2007), and the more direct route of *Agrobacterium*-mediated delivery into the embryo axis (Popelka et al., 2006; Bett et al., 2017), which reduced time in culture. Finally, by targeting the apical meristem (Bett et al., 2017) or cotyledonary/axillary meristems (Che et al., 2019), followed by direct shoot formation, the selection of transgenic events in cowpea has become simpler (Bett et al., 2017) or both simpler and more efficient (Che et al., 2019).

In planta methods that take advantage of normal growth and reproduction with a minimal perturbation to the system, are the quintessence of transformation approaches for all plant species. Floral-dip in Arabidopsis has been widely used for transformation over many years, and more recently has been extended to delivery of CRISPR/Cas9 components for mutagenesis and template-based genome editing (Table 1). In a more recent *in planta* report, the axillary meristem has been targeted for *Agrobacterium*-mediated delivery to demonstrate genome editing (Maher et al., 2020).

Table 1 Explants and growth strategies used to recover transgenic and/or edited plants

			Route to heritable transmission						
General strategy	Target explant	Delivery method	Callus	Regenerable structure	Regenerated plantlets	Fertile plant	Transgenic	**Genome editing**	Illustrative references
In vitro via somatic embryo	Protoplasts	TF	Yes	Somatic Embryo (SE)	Limited**	Limited**	Yes**	**Mut, Ins**	**Puchta 1996, Wright 2005, Li 1990, Park 2019, Lin 2018, Murovec 2018, Woo 2015**
	Callus	Agro	Yes	SE	Yes	Yes	Yes	-	Bevan 1994, Komari 1990, Hatanaka 1999, Sidorov 2006, Du 2019
	Zygotic embryo	PG, Agro	Yes	SE	Yes	T0	Yes	**Mut, Edit & Ins**	Wan 1995, Lowe 2016, **Svitashev 2015, others**
	Cotyledon	PG, Agro	Yes	SE	Yes	T0	Yes	-	Parrott 1998, Yan 2000
	True leaf	PG, Agro	Yes	SE	Yes	T0	Yes	-	Matsuda 2005, Lowe 2016
	Zygotic embryo	PG, Agro	⇨	SE	Yes	T0	Yes	**Deletion**	Lowe 2018, Hoerster 2020, **Gao 2020**
	Somatic embryo	PG, Agro	⇨	Secondary SE	Yes	T0	Yes	**Mut, Ins**	Li 2009, **Jacobs 2015, Li 2015**
	Microspore	PG, Agro	⇨	SE	Yes	Limited**	Yes**	-	Brew-Appiah 2013

In vitro via shoots	Callus	PG, Agro	⇨	Shoot	Yes	T0	Yes	-	Hong 2007
	Cotyledon node	PG, Agro	⇨	Shoot	Yes	T0	Yes	-	Meurer 1998, Olhoft 2007
	Embryo axis (meristem)	PG, Agro	⇨	Shoot	Yes	T0	Yes	**Mut**	Pigeaire 1997, Dutt 2007, Paz 2006, Sitther 2018, **Curtin 2011, Haun 2014, Demorest 2016**
	Isolated Axil MERISTEM	Agro	⇨	Shoot	Yes	T0	Yes	-	Matsuda 2005, Fujita 2009, Qi 2014, Mayavan 2015
	Hypocotyl/ stem	PG, Agro	⇨	Shoot	Yes	T0	Yes	-	Cardoza 2003, Zhao 2017
In planta	Floral dip	Agro	⇨	T1 Seed	⇨	T1	Yes	**Mut, Edit & Ins**	**Lloyd 2005, Zhang 2010, de Pater 2009 2013, Forner 2015**
	Axillary meristem	Agro	⇨	Shoot	⇨	T0	-	**Mut**	**Maher 2020**

Arrows indicate a step in the process that has been eliminated for a particular method.
References in bold font indicate nuclease-mediated modifications.
TF = polyethylene glycol- or electroporation-mediated transfection.
PG = Particle gun delivery.
Mut = mutagenesis.
Edit = Oligonucleotide-mediated edit.
Ins = targeted insertion.
NT* = Nanotubules have been successfully used in delivery of DNA.
** = Restricted genotype range.
HDR = Homology-Dependent Repair.

3 Delivery methods for genome editing reagents: delivery into single cells

The presence of a rigid glycan-rich cell wall surrounding the cell membrane makes the task of delivering exogenous material to plant cells extremely challenging. Current delivery options for genome editing reagents can be divided into three main categories: physical, chemical, and biological (Que et al., 2019). The method by which genome modification reagents are introduced is dictated by the explant type and its regenerative capacity for any given plant species. For most species, more than one delivery system can be used. In turn, these available alternatives allow delivery of different genome editing reagents (e.g. DNA, RNA, proteins, or mixtures). Among physical delivery options, particle bombardment or biolistic (Klein et al., 1987) and electroporation (Fromm et al., 1985; Shillito et al., 1985) are most common, predominantly used for delivery into either whole cells/tissue or into protoplasts, respectively. Polyethylene glycol (PEG) has been the most widely used chemical agent for protoplast transformation (Kofer et al., 1998). In terms of biological delivery, all mainstream methods rely on *Agrobacterium tumefaciens* (Gelvin, 2003; 2017), but other methods such as the use of viral particles have been attempted (Grimsley et al., 1987).

Protoplasts are plant cells from which the cell wall has been mechanically (Von Klercker, 1892) or enzymatically removed (Cocking, 1960; Takebe et al., 1968; Power and Cocking, 1970). They can be isolated directly from different parts of *in vitro* or greenhouse-grown plants, with the young leaf mesophyll being the preferred tissue. Alternatively, protoplasts can be isolated from *in vitro* cultured tissues, with cell suspension cultures derived from friable embryogenic callus being the most common source. Plant growing conditions and plant age exert a major influence on yield and viability of the isolated protoplasts (Shepard and Totten, 1977; Kao and Michayluk, 1980). In addition, *in vitro* cultured material provides a more reliable source for consistent high-quality protoplasts if the cells can be maintained at maximum growth rates and utilized at an early log phase (Ishii, 1988). The viability of freshly isolated protoplasts is also influenced by enzyme composition and concentration, pH, and osmotic pressure used for isolation (Uchimiya and Murashige, 1974). Both electroporation and PEG-mediated delivery are efficacious, with the success depending on the species and tissue source used to produce protoplasts as well as the physical parameters used during delivery.

Electroporation methods have been developed for protoplasts from a wide range of both monocotyledonous and dicotyledonous plants (Hauptmann et al., 1987). Biological parameters known to impact electroporation efficiency include cell diameter and cell density (Rouan et al., 1991; Rao et al., 1995). Other important factors include strength and duration of the electric pulse (Fisk and Dandekar, 2004), ion types, and their concentration in the buffer (Saunders et al., 1995; Niedz et al., 2003).

Chemical transformation of protoplasts commonly relies on the use of PEG (Krens et al., 1982; Paszkowski et al., 1984). PEG is an inert hydrophilic polymer of ethylene oxide and in combination with various salts facilitates the uptake of molecules into cells. Factors that influence the transfection efficiency of this process include the concentration and molecular weight of PEG, the pH of the transfection buffer, and the type and concentrations of the cations (Bittencourt et al., 1995; Maas and Werr, 1989).

Both electroporation and PEG-mediated delivery into protoplasts have been successfully used for transient and stable CRISPR/Cas9-mediated genome editing in multiple crop species (Feng et al., 2013; Xie and Yang, 2013; Gao et al., 2015; Bhowmik et al., 2018). The use of protoplasts for transient evaluation of editing reagents represents the latest application of this versatile single cell-based platform allowing the production of thousands of independent events and high-throughput screening (Dlugosz et al., 2016; Lin et al., 2018).

Isolated protoplasts are theoretically capable of dedifferentiation, re-entry into the cell cycle, and development into callus from which plants may be regenerated by somatic embryogenesis or organogenesis (Xu and Xue, 1999). The first report of plants regenerated from protoplasts was in *Nicotiana tabacum* (Takebe et al., 1971) followed by multiple other crops (Baset et al., 1991; Binding et al., 1978; Rhodes et al., 1988). The ability to regenerate plants has been demonstrated to be under genetic control in different species (Koornneef et al., 1987; Mórocz et al., 1990; Chupeau et al., 2013) and recalcitrance to regeneration has also been associated to physiological factors like the increase of intracellular level of polyamine (Papadakis et al., 2005) or oxidative stress induced during protoplast isolation and culture (Benson and Roubelakis-Angelakis, 1994; de Marco and Roubelakis-Angelakis, 1996; Papadakis et al., 2001). Improvement of regeneration efficiency has been achieved in some species by manipulation of the culture systems (Pati et al., 2005; Niedz, 2006; Kiełkowska and Adamus, 2012) and media manipulation (Grzebelus et al., 2012). However, regeneration can be efficiently obtained only in few genotypes (Eeckhaut et al., 2013; Hu et al., 1999) limiting the practical utility of protoplast-based systems for crop improvement (Table 1).

4 Delivery methods for genome editing reagents: delivery into intact tissues

Delivery of genome editing reagents to intact plant tissues has relied predominantly on particle bombardment and *Agrobacterium* with emphasis inexorably shifting toward the latter option. In addition, use of *Agrobacterium*-mediated delivery of viral-based replicons into leaf tissue of various dicot species is also increasing the efficiency of CRISPR/Cas9 gene modification (Baltes et al., 2014; Čermák et al., 2015; Butler et al., 2016). Direct delivery of Cas9 and guide RNA (gRNA) ribonucleoprotein (RNP) complexes into rice zygotes has also been reported (Toda et al., 2019). Most recently, the use of nanoparticles is bringing greater attention back to physical delivery methods. Specific examples from within these broad topics are presented in the next section.

4.1 Particle bombardment

The biolistic method relies on the ability of heavy-metal carriers (e.g. tungsten or gold) coated with cargo biomolecules to penetrate plant cells when accelerated using high-pressure helium in a partial vacuum. Once inside the cell the cargo can elute from the microparticles and either integrate into the genome (stable expression) or remain as extrachromosomal material (transient expression). This transformation system can transform not only the nucleus but also plastid and mitochondria (Sanford, 2000). Furthermore, biolistics can be used with diverse cell types and delivers a high number of biomolecules (Altpeter et al., 2005). One additional advantage of this delivery system is that multiple genome editing components can be co-delivered on separate plasmids, simplifying vector design and assembly. The success of biolistic experiments depends on numerous parameters including acceleration of particles, selection of target tissues, size of microparticles, and ratio of DNA to particles. Disadvantages of this method include limited control over distribution and penetration of microparticles and potential cell lethality. In addition, chromosome damage and complex multicopy vector DNA integration reduce the proportion of usable genome-edited events (Svitashev et al., 2000; Svitashev et al., 2002; Liu et al., 2019). Biolistic delivery has been employed for genome modification application in numerous species including maize (Ainley et al., 2013; Svitashev et al., 2015, 2016; Shi et al., 2017), wheat (Liang et al., 2017; Hamada et al., 2018), soybean (Jacobs et al., 2015; Li et al., 2015), and rice (Sun et al., 2016).

4.2 Electroporation

Electroporation has also been used to deliver DNA directly through the cell wall, but with limited success. This method has been used to deliver a vector containing the hygromycin-resistance gene into maize immature embryos to recover stable transgenic events (D'Halluin et al., 1992). Recently, electroporation of wheat microspores resulted in both Ds-RED expression and CRISPR/Cas9-mediated mutagenesis of an endogenous gene (Bhowmik et al., 2018). Finally, electroporation of CRE recombinase protein has been used to facilitate site-specific recombination, excising GFP and activating GUS expression in Arabidopsis suspension cells (Furuhata et al., 2019).

4.3 *Agrobacterium*

Since the early days of transformation in dicotyledonous species, *Agrobacterium* has been favored for its ability to transform an ever-increasing number of species and a range of plant tissues such as cotyledons (Nakajima et al., 2013), cotyledonary nodes (Paz et al., 2006), embryo axis (Krishnamurthy et al., 2000), stems (Moore et al., 1992), stem internodes (Maheshwari and Kovalchuk, 2016), apical meristems (Dutt et al., 2007), leaf tissue (De Block, 1988), and flower via floral-dip (Clough and Bent, 1998; Hu et al., 2019a). In contrast, for most cereals, the preferred explant is the immature embryo where the scutellar tissue is competent for both transformation and regeneration (Jones et al., 2009; Hiei et al., 2014).

Since the discovery of *Agrobacterium*-mediated plant transformation (Barton et al., 1983; Caplan et al., 1983), there has been continuous improvement in this system. Exploring different *Agrobacterium* strains led to the widespread use of such workhorse strains as LBA4404 (Ooms et al., 1981), AGL1 (Lazo et al., 1991), and the EHA strains 101 and 105 (Hood et al., 1993). Altering the expression of virulence genes (Hansen et al., 1994), or increasing the copy number of vir genes (Hiei et al., 1994; Ishida et al., 1996), resulted in further enhancements of T-DNA delivery and efficacy across an ever-increasing range of species. Such modifications to *Agrobacterium* have also helped broaden the range of genotypes accessible for genome modification (see Sardesai and Subramanyam, 2018, for review).

In its wild-type state, the virulence of *Agrobacterium* is managed through a large Ti plasmid which comprises both the T-DNA that is transferred to the host as well as virulence genes that provide the machinery to infect and deliver this T-DNA. Because of its large size (~200 kb), it is difficult to modify the Ti plasmid to reprogram the T-DNA cassette. This barrier led to the invention of a binary plasmid system in which the T-DNA component was separated from the virulence genes (Hoekema et al., 1983; Bevan et al., 1984). Further improving upon this design, Komari and colleagues introduced supplemental copies of several virulence genes in addition to the T-DNA plasmid to create what is known as the pSB1 super-binary vector (Komari et al., 1990), which has served as a system of choice for the transformation of numerous plant species (Hiei et al., 1994; Ishida et al., 1996).

Recently, Anand et al. (2018) described a ternary pVir system that simplified and improved upon pSB1 *Agrobacterium*-mediated transformation. The authors removed unnecessary sequences and split the supplemental virulence genes away from the T-DNA into a new accessory plasmid driven by a smaller and higher copy pVS1 origin of replication. In addition, they corrected deficiencies of pSB1 including restoring a functional virC operon and a full length virD2 gene and identified novel combinations of supplemental vir genes that, when added to the accessory plasmid, further improved transformation. Finally, because pSB1 was known to spontaneously develop resistance to tetracycline, the authors utilized gentamicin as an improved selectable agent. The resulting accessory system (pVir) no longer conferred spontaneous TET resistance, was smaller, and was more virulent than pSB1 (Anand et al., 2018). These modifications simplified and improved stable plant-transformation among historically recalcitrant elite maize and sorghum varieties (Anand et al., 2018; Che et al., 2018), resulting in higher frequencies of transformation and single-copy integrations (Tables 2 and 3). Importantly, moving the vir genes to a separate plasmid significantly decreased complexity and further simplified the T-DNA vector assembly process.

For both dicots and monocots, early studies focused on random DNA integration. As methods for genome modification became more sophisticated and precise, various modes of gene targeting were developed using a diverse set of tools, including CRE and FLP recombinase-mediated site-specific integration in soybean (Srivastava and Ow, 2002; Li et al., 2009) or maize (Anand et al., 2019), and homology-dependent integration using either homing endonucleases such as I-SceI (Puchta et al., 1996) or I-CreI (Gao et al., 2010), zinc-finger nucleases (Wright et al., 2005; Shukla et al., 2009), TALENs (Li et al., 2012; Haun et al., 2014), or CRISPR/Cas9 (Liang et al., 2014; Svitashev et al., 2015; Malzahn et al., 2017).

Table 2 Comparison of transformation and single-copy frequencies in the Pioneer maize inbred PH2RT, using either the helper plasmid pSB1 or the pVir helper plasmid PHP71539

Construct	Transformation frequency (%)	Single copy, no Backbone (%)
pSB1	13.7	5.1
pVir pPHP71529	31.1	9.1

Table 3 Transformation and single-copy frequencies for four different sorghum varieties using the pVIR helper plasmid PHP71539

Cultivar	Transformation frequency (%)	Single copy, no Backbone (%)
Tx430	27	16
Malisor 84-7	7.5	3.6
Tegemeo	1.5	NA
Macia	1	NA

NA, Not available.

4.4 Viral-based replicons

Plant virus-based replicons have recently become an attractive supplement to *Agrobacterium*-mediated delivery. A key advantage of a viral replicon delivery system is the prolific accumulation of nucleic acids, making the gene-editing process potentially more efficient. Thus, a transient viral platform could provide an ideal screening tool to quickly assess gRNA design and to measure desired effects. If gRNA delivery using traditional methods is a limiting factor to obtain efficient genome editing in plants, the abundant virus replication systems could potentially provide high gRNA yields for efficient, quick, and systemic editing (Cody and Scholthof, 2019).

In the past, focus on viral replicons centered mainly on production of recombinant proteins in plants (Ibrahim et al., 2019). When first attempted, plant cells were infected with a full-genome copy of the plant virus, resulting in transgene instability, limited cargo size, and raising biosafety concerns. This led to the design of a second generation of viral-based vectors that retained the replication function, supported gene expression, but lacked coat proteins and movement (Gleba et al., 2004). Removal of the mobility function necessitated delivery of the viral replicon vector to the plant cell by *Agrobacterium* or particle bombardment (Marillonnet et al., 2004). Once in the plant cell, replicon amplification increases the available template DNA, resulting in increased HDR-based editing efficiency (Baltes et al., 2014). The choice of RNA- or DNA-based viruses affects both efficacy and versatility of the expression system. RNA-based viruses have small genomes facilitating cloning and multiplexing of gRNA, and don't integrate into the plant genome. However, their cargo capacity is limited, they are unsuitable for HDR edits and have been reported to have issues with their stability and genomic integrity (Seaberg et al., 2012).

Early reports of CRISPR/Cas9-mediated gene knockouts in plants used either the RNA Tobacco Rattle Virus (TRV) (Ali et al., 2015b) or a DNA geminivirus (Baltes et al., 2014; Ali et al., 2015a). TRV has been most widely used for genome editing vector designs because of its high level of accumulation and associated gene expression in a variety of hosts. The use of TRV-based vectors has been demonstrated in *Nicotiana ssp.*, *Petunia hybrida*, and Arabidopsis (Marton et al., 2010; Honig et al., 2015; Ali et al., 2015a, 2018). Other RNA viruses used for genome editing include Tobacco Mosaic Virus (TMV) in *N. benthamiana* (Cody et al., 2017), the Pea Early Browning Virus in Arabidopsis (Ali et al., 2018), and the Beet Necrotic Yellow Vein Virus in *N. benthamiana* (Jiang et al., 2019). The TMV-based vector was shown to transiently deliver high concentrations of sgRNAs in *N. benthamiana* to obtain ~70% insertion-deletion mutations, most of which were observed within 2–3 days after *Agrobacterium* infection (Cody et al., 2017). In this manner, the authors successfully demonstrated targeting of two *Argonaute 1* (*AGO1*) paralogs with one gRNA, as well as codelivery of two adjacent gRNAs using a single construct. Recently, Barley Stripe Mosaic Virus has been reported to enable targeted genome editing in both wheat and maize (Hu et al., 2019b) expanding the use of RNA virus-derived vectors beyond model plants. DNA virus-derived vectors overcome the cargo limitations and instability associated with RNA virus-derived counterparts. DNA vectors derived from geminiviruses are most developed and implemented. The viruses in this family are characterized by a single-stranded circular DNA genome, which is converted to double-stranded DNA by plant DNA polymerases after amplification to high copy number via rolling circle replication (Hanley-Bowdoin et al., 1999). The Bean Yellow Dwarf Virus replicon system has been efficiently deployed for delivery of ZFNs, TALENs, and CRISPR/Cas9 reagents resulting in the production of calli and plants with precise DNA sequence changes (Baltes et al., 2014; Čermák et al., 2015). Similar vectors used for genome modification have been derived from other geminiviruses such as Wheat Dwarf Virus (Gil-Humanes et al., 2017; Wang et al., 2017) and Cabbage Leaf Curl Virus (Yin et al., 2015).

4.5 Nanoparticles

Nanoparticles have been explored for many years, with the earliest studies demonstrating successful delivery of either DNA (Torney et al., 2007) or proteins (Martin-Ortigosa et al., 2012) using gold-functionalized mesoporous silica delivered by particle bombardment into plant cells. Recently, the use of carbon nanotubules for delivery of siRNA (Demirer et al., 2019a) and DNA (Demirer et al., 2019b) into a range of plant species has been demonstrated for gene silencing and transgene expression, respectively. In the case of RNA delivery, the nanotubules appeared to provide protection against degradation, and for DNA they permit transient expression while precluding DNA integration into the plant genome. Interestingly, the chemistry used to pre-treat the carbon nanotubules may impact sub-cellular delivery. For example, for siRNA and DNA delivery, the carbon nanotubules were treated with polyethyleneimine which results in both nuclear and chloroplast localization. In contrast, treating with chitosan appeared to favor chloroplast delivery (Kwak et al., 2019). Irrespective of the final subcellular localization, both chemistries appeared to support transient gene expression while protecting the DNA against degradation. Efficacy of nanoparticle delivery has been demonstrated for CRISPR/Cas9-mediated editing in mammalian cells (Durfee et al., 2016). Based on the encouraging observations that nanotubule-delivered siRNA and DNA are transiently functional in plant cells, Wang et al. (2019) suggested that nanotubule delivery could potentially be developed for CRISPR/Cas9-mediated production of transgene-free genome edited plants, an outcome similar to that already demonstrated using particle delivery of RNPs (Svitashev et al., 2016).

5 Alternatives to DNA delivery

In most genome editing experiments, the gRNA as well as the Cas9 and selectable marker genes, have been delivered into plant cells using either T-DNA (*Agrobacterium* infection) or plasmid DNA (particle bombardment). In both cases, the delivered DNA can integrate into the genome leading to various side effects such as gene disruption, plant mosaicism, and potential off-site cutting (Svitashev et al., 2015; Kanchiswamy, 2016).

Direct DNA-free genome editing can be accomplished through biolistic delivery of gRNA and Cas9 either in the form of *in vitro* transcribed RNA molecules (IVTs) or as *in vitro* assembled ribonucleoprotein (RNP) complexes. Guide RNA and Cas9 mRNA have been successfully delivered to wheat immature embryos resulting in approximately 3-fold lower frequency of targeted mutagenesis in comparison to conventional DNA-based genome editing methods (Zhang et al., 2016). Lower efficiency of this approach is likely related to incomplete IVTs, partial degradation of RNA molecules during gold particles preparation, and/or RNA degradation upon delivery to the plant cells (before RNP complexes are formed).

Delivery of Cas9 and gRNA in the form of RNP complexes not only mitigates many issues associated with DNA vectors but also has several additional advantages over RNA delivery. First, the preassembled complexes are considerably more stable and not subjected to the same level of degradation as RNA molecules. We have recently demonstrated that replacement of *in vitro* transcribed gRNA with chemically synthesized molecules further increased activity of RNPs resulting in a 2–4-fold increase in targeted mutagenesis (Svitashev et al., unpublished data). Second, particle bombardment allows simultaneous delivery of thousands of RNP molecules into a single cell resulting in highly efficient target site cleavage. Third, the complex is active immediately upon delivery, eliminating the time required for transcription (i.e. for DNA) and translation of Cas9 protein (both DNA and RNA). The last two considerations are especially important for gene-editing applications relying on homology-based repair of DSB using oligonucleotides as donor templates, which are subjected to rapid degradation by cellular nucleases.

Efficient genome editing using RNP was first demonstrated in cultured human cells using electroporation-mediated delivery (Kim et al., 2014; Lin et al., 2014). Later, plant protoplasts were successfully used for RNP delivery in a variety of plants such as tobacco, Arabidopsis, lettuce, rice, and Petunia (Woo et al., 2015; Subburaj et al., 2016; Murovec et al., 2018; Park et al., 2019). However, for most monocot species, including major crops such as maize, wheat, rice, barley, and sorghum, regeneration of plants from protoplasts remains either unattainable or inefficient (Davey et al., 2005).

Delivery of RNP complexes and efficient gene editing has been recently demonstrated in maize and wheat using particle bombardment of immature embryo cells (Svitashev et al., 2016; Liang et al., 2017). The regenerated plants contained specifically targeted gene mutations at frequencies comparable to conventional DNA-based delivery methods. Editing efficiency using this approach has allowed for recovery of completely transgene-free plants with mutated alleles at high frequencies without selection. This approach has also been used to edit the maize *Acetolactate Synthase (ALS)* gene conferring resistance to the herbicide chlorsulfuron (Svitashev et al., 2016).

Considering the current limitations in transformation and regeneration for many plant species (especially those that are asexually propagated), the ability to deliver RNP complexes significantly broadens options to modify plant genomes while eliminating unwanted transgene integrations. This provides new opportunities to advance agricultural breeding practices for any plant species amenable to biolistic delivery.

6 Morphogenic genes increase transformation efficiency and extend genotype range

Genes that control the organized spatiotemporal development of plant embryos, meristems, tissue, and organs (the process of morphogenesis) have been cloned and characterized since the late 1990s. In short order, genes that regulate embryo formation and meristem maintenance began to intrigue plant transformation researchers, as a steady stream of reports demonstrated that mis-expression could result in ectopic formation of embryo- and/or meristem-like structures, with early examples including genes such as *LEC1* (Lotan et al., 1998), *LEC2* (Stone et al., 2001), *ESR1* (Banno et al., 2001), *WUS* (Zuo et al., 2002), and *BBM* (Boutilier et al., 2002). Since these early reports in Arabidopsis, new observations using orthologous genes, or newly characterized genes such as *SERK1* (Pérez-Pascual et al., 2018), *AGL15* (Harding et al., 2003; Thakare et al., 2008; Yang et al., 2014), or *STM* (Elhiti et al., 2010) were reported to either enhance preexisting somatic embryogenesis in culture, or to elicit ectopic formation of somatic embryos or meristems from differentiated tissues (for review, see Gordon-Kamm et al., 2019). While such reports were exciting, these early observations relied on constitutive expression of these morphogenic genes, which typically prevented the regeneration of normal fertile plants. Thus, use of pleiotropic genes for improved transformation has relied on recombinase-mediated excision to facilitate recovery of normal, fertile T0 plants after using either hormone biosynthesis or morphogenic genes (Ebinuma et al., 1997; or Lowe et al., 2016, respectively).

More recently, such studies have been refined, with controlled expression of morphogenic genes being used to restrict growth stimulation, enabling the subsequent recovery of transgenic plants in such diverse species as *Capsicum annuum* (Heidmann et al., 2011), *Arabidopsis thaliana* (Lutz et al., 2015; Wang et al., 2009), *Theobroma cacao* (Shires et al., 2017), *Nicotiana tabacum* (Kyo et al., 2018), *Populus tomentosa* (Deng et al., 2009), *Zea mays* (Lowe et al., 2016, 2018; Mookkan et al., 2017; Du et al., 2019), *Sorghum bicolor* (Lowe et al., 2016; Mookkan et al., 2017), and *Oryza sativa* ssp Indica (Lowe et al., 2016).

A third alternative for using morphogenic genes has been co-transformation with two *Agrobacterium* strains, one delivering a T-DNA for strong *Wus2* expression in one cell, and the second strain delivering a trait-containing T-DNA into a neighboring cell. Through either transient expression of *Wus2* (with no integration) and/or strong expression to produce WUS protein that can move to the trait gene-containing cells and stimulate embryogenesis, the trait cells form somatic embryos that regenerate without any integration of the *Wus2* T-DNA (Hoerster et al., 2020).

New morphogenic and/or growth-stimulating tools will continue to add new options to this toolbox. For example, Debernardi et al. (2020) recently report that delivery of a constitutively expressed fusion-protein composed of *Growth-Stimulating Factor4* (*Grf4*) and *Grf-Interacting Factor1* (*Gif1*) improve transformation in wheat and citrus, and naturally occurring miR396 down-regulation of *Grf4* transcript in the resultant T0 plants appears to mitigate pleiotropic problems. Similarly, *Grf5* has been used to increase transformation efficiency in sugar beet and maize (Kong et al., 2020).

In maize, the use of *Wus2* and *Bbm* has improved transformation frequency in recalcitrant inbreds by 10-fold or greater (Lowe et al., 2016), which has played an important role in rapid implementation of CRISPR/Cas9 genome editing applications at Corteva Agriscience (formerly DuPont Pioneer). In addition to simply increase the number of recovered transgenic events, there may be an equally important, although less obvious, benefit of using *Wus2/Bbm* for genome editing. Although WUS2 and BBM proteins don't directly interact with cell cycle components, cell division is stimulated (including replication during S-phase) and may be providing an HDR-conducive cellular environment. Likewise, it has been suggested that pleiotropic effects of Rep/RepA expression in plant cells harboring viral replicons may be favorable to DSB-initiated HDR-based genome editing (Baltes et al., 2014).

The regeneration bottleneck for many species can be alleviated using morphogenic regulators, facilitating genome editing. *Wus2* and *Bbm* have been used for transient CRISPR/Cas9-mediated genome editing through particle bombardment (Svitashev et al., 2015), for RNP-mediated editing (Svitashev et al., 2016), and for FLP-mediated site-specific recombination (Bett et al., 2019). This has facilitated a range of genome modifications, from simple mutagenesis to oligo-based edits and targeted gene insertion (Svitashev et al., 2015, 2016).

Additionally, they have been deployed to facilitate a wide range of CRISPR/Cas9-mediated changes in difficult maize inbreds (Chilcoat et al., 2017; Zhang et al., 2019). Similarly, *Wus2* in conjunction with other morphogenic genes has been demonstrated to be effective for genome editing in *N. benthamiana* (Maher et al., 2020).

Two examples in maize highlight the benefit of using *Wus2/Bbm* for genome editing using immature embryos as the target explant. The first focused on using CRISPR/Cas9-mediated excision of the endogenous *WAXY* allele in 11 commercially important Pioneer inbreds that were otherwise recalcitrant to transformation (Gao et al., 2020). Both Stiff-Stalk and Non-Stiff-Stalk inbreds, which hadn't been previously transformed, were chosen based on their genetics as parents in commercial hybrids. When these inbreds were transformed using *Wus2/Bbm* plus Cas9 and gRNA expression cassettes, transformation frequencies (based on numbers of regenerated T0 plants relative to the number of starting embryos) for the 11 inbreds ranged from approximately 1% to 50% (Fig. 1a). Successful *WAXY* deletions were recovered in all genotypes at frequencies ranging from 6% to 16% (Fig. 1b), relative to the number of T0 plants analyzed. These results illustrate two important points. First, the use of *Wus2/Bbm* permitted the recovery of transgenic events in all 11 inbreds, although a 50-fold range of transformation frequencies was observed. Second, the Cas9/gRNA-mediated deletions were observed to span a much narrower range of frequencies (from 6% to 16%) independent of the transformation rate. For example, PH25KM and PH1VST were at opposite ends of the transformation spectrum (49.7% and 1.7%) but had similar deletion frequencies of 11% and 12.3%, respectively. This illustrates that there was no correlation between transformation efficiency and edit frequency (Pearson's Correlation Coefficient, r = −0.21). Importantly, optimization of media

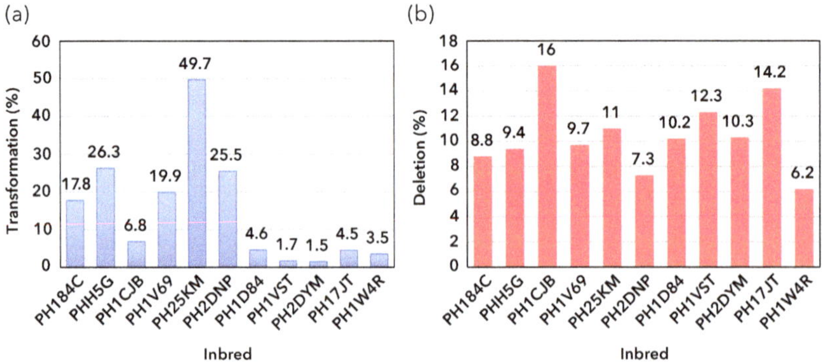

Figure 1 Transformation frequency (a) and *WAXY* deletion frequency (b) for plants recovered after delivery of a Ubiquitin-driven Cas9 and two U6-driven gRNAs, Axig1::WUS2, and PLTP::BBM (Gao et al., 2020).

Published by Burleigh Dodds Science Publishing Limited, 2021.

composition and selection (while still using the same *Wus2/Bbm* expression cassettes) has continued to improve transformation frequencies for inbred PH1VST from 1.7% to greater than 70% (Gordon-Kamm et al., unpublished result).

The second example used CRISPR/Cas9 and *Wus2/Bbm* along with a donor template to replace the endogenous ARGOS8 promoter with a constitutive maize promoter (GOS2) in a Pioneer maize inbred. As predicted from earlier transgenic studies (Habben et al., 2014; Shi et al., 2015), precise exchange of the promoter resulted in enhanced grain yield under drought conditions (Shi et al., 2017). As with the deletion of an endogenous gene in 11 inbreds, this cisgenic edit was successful because of the optimization and combination of two complementary technologies (CRISPR/Cas9-mediated editing and *Wus2/Bbm*-assisted transformation).

7 Morphogenic genes permit transformation in non-traditional explants

For many years, the requirement of using immature embryos has rendered maize transformation unattainable for most academic labs (Altpeter et al., 2016), because maintaining a consistent supply of immature embryos is both expensive and labor intensive (Que et al., 2014). Recently a viable alternative has been developed, since Lowe et al. (2016) demonstrated that *Agrobacterium*-mediated delivery of constitutively-expressed *Wus2* and *Bbm* allows transformation of both mature embryo slices and seedling-derived leaf segments to efficiently produce fertile transgenic events.

These alternative explants can also be used for genome editing. For example, a transgenic Pioneer Stiff-Stalk inbred that contained inducible *Wus2/Bbm* expression cassettes has been generated. When *Wus2* and *Bbm* were induced by the addition of ethametsulfuron, somatic embryogenesis was stimulated in leaf tissue. Using this inducible *Wus2/Bbm* germplasm as the starting point for a new experiment, seedling-derived leaf tissue was then used as the target explant for particle bombardment. To further enhance morphogenesis (beyond that provided by inducible expression), plasmids containing constitutive *Wus2* and *Bbm* expression cassettes were co-delivered with Cas9 and gRNA, as well as the template DNA (containing a *NPTII* expression cassette). It should be noted that due to high levels of *Wus2* and *Bbm* expression (inducible plus constitutive), selection using *NPTII* and G418 became less efficient, resulting in escape (wild type) plants being regenerated. However, three integration events were recovered from a total of 142 T0 plants that were regenerated and analyzed (Table 4). This result clearly shows that when *Wus2/Bbm* are used to aid the process, CRISPR/Cas9-mediated genome editing can be accomplished via leaf transformation.

Table 4 Recovery of G418-resistant T0 plants using four different levels of antibiotic selection. After PCR analysis of the total number of T0 plants for each treatment, the number of Homology-Dependent Repair (HDR) events was determined first by a positive PCR result across both the upstream and downstream flanking recombination junctions (No. HDR), and subsequently using long-PCR across the entire insertion (No. Perfect HDR)

No. seedlings	G418 (mg/l)	No. T0 Plants	No. HDR (PCR)	No. Perfect HDR
8	150	46	1	0
8	200	34	0	0
9	250	38	**4**	**3**
9	300	24	0	0

A recent variation on using morphogenic genes to assist in gene editing has also been reported by Maher et al. (2020). In this study, the authors introduced a geminiviral-based replicon directly into the wounded leaf axil after removal of the petiole in *N. benthamiana*. Replicons containing both *Wus2* and *Ipt* (isopentenyl transferase from *Agrobacterium*) expression cassettes stimulated growth of new shoots from the treated axils. The introduction of combinations of *Wus2*, *STM* (*Shootmeristemless*) and *Ipt* into the axillary region also stimulated the formation of transgenic meristem-like structures in tomato, and transgenic shoots arose from aseptically grown potato and grape cuttings. Furthermore, delivery of Cas9 and a gRNA targeting the phytoene desaturase gene (*PDS*) along with *Wus2* and *Ipt*, generated uniformly white shoots characteristic of *PDS* mutagenesis, which in one case was transmitted to progeny. While still early in development, this study clearly shows the potential of combining morphogenic genes with CRISPR/Cas9 for further development of *in planta* genome editing. This is particularly encouraging, considering that axillary meristems have been an attractive transformation target (even when morphogenic genes are not used) for such diverse species as pear (Matsuda et al., 2005), grape (Fujita et al., 2009), crambe (Qi et al., 2014), soybean (Olhoft et al., 2007), and sugarcane (Mayavan et al., 2015).

8 Future trends

Progress in plant genome editing can currently only be described as explosive, fueled by advances in many contributing areas. New CRISPR/Cas nucleases such as Cas12-based proteins with different PAM solutions are greatly increasing the potential flexibility of the system (e.g. Fonfara et al., 2014). Cas12-based type V and the Cas13-based type VI systems may also add to this versatility (Murugan et al., 2017). Recent development of the prime editing system, which relies on a Cas9 fusion with reverse transcriptase and a priming gRNA appears to be a potentially versatile and efficient genome modification system (Anzalone et al., 2019). In addition to genome modification *per se*, both type II and type I systems continue to be optimized as transcriptional activators or repressors (Cheng et al., 2013; Gilbert et al., 2013; Farzadfard et al., 2013; Rath et al., 2015; Pickar-Oliver et al., 2019).

Robust transformation/regeneration methods remain a major bottleneck in genome editing processes, not only for agronomic crops in general (Altpeter et al., 2016), but also for forestry (Chang et al., 2018), temperate fruit/nut crops (Song et al., 2019), and tropical fruits (Gómez-Lim and Litz, 2004). Difficulties in genome editing applications can also be exacerbated in crops with added genomic complexity, whether through self-incompatibility and genomic heterozygosity (e.g., in pineapple, grape and apple), or due to polyploidy as in wheat, canola, and as an extreme example, sugarcane (Botella, 2019). Rapid advances in genomics such as Next-Generation Sequencing and BioNano (Mak et al., 2016) continue to help resolve the complexity of the genomes, making targeted genome modifications more feasible.

Across species, *Agrobacterium* and physical delivery via particle bombardment are the current mainstays for crop genome editing, depending on the cargo and intended outcome. *Agrobacterium* is widely used for crops in which stably integrated T-DNA can later be segregated away. However, in cases of asexually or vegetatively propagated species where no transgene introduction is desired, particle gun delivery of RNPs has been successfully used in diverse species (Metje-Sprink et al., 2019). A resurgent interest in nanoparticle delivery may promise yet another rapid change in this landscape, permitting transient delivery of CRISPR/Cas9 components to recover non-transgenic genome-edited plants (Wang et al., 2019). Using *Agrobacterium*-mediated co-transformation for supplying morphogenic gene activity adds another tool for genome modification without integration of genes such as *Wus2* (Hoerster et al., 2020).

Changes in plant transformation and regeneration methods are also accelerating. For an increasing number of historically recalcitrant dicots, the successful use of the mature embryonic axis in soybean has been emulated in other crops, for example in grape (Dutt et al., 2007), pigeon pea (Rao et al., 2008), Camelia (Sitther et al., 2018), chickpea (Das Bhowmik et al., 2019), and cowpea (Che et al., 2019). For crops where the apical dome remains difficult to work with, the use of morphogenic genes is receiving increased attention and will continue to broaden the spectrum of plants that can successfully be edited. Using hormone biosynthesis or morphogenic genes to facilitate transformation has been reported by many groups (see Gordon-Kamm et al., 2019 for review). Recently, morphogenic genes have been seeing a resurgence of interest, and in addition to their use in producing transgenic plants (Lowe et al., 2016; Mookkan et al., 2017; Lowe et al., 2018), they are facilitating genome editing in both monocots (Svitashev et al., 2015, 2016; Zhang et al., 2019) and dicots (Maher et al., 2020).

In addition, the use of morphogenic genes can contribute to extending transformation/editing methods to an increasing range of difficult plant species. However, looking at how different species respond to morphogenic genes such as *Wus2*, *Kn1*, and *Bbm* (Gordon-Kamm et al., 2019) it becomes clear that there will probably be no universal single solution across crops. Instead, when trying to extend the use of morphogenic genes to a new crop, we would advise screening a handful of genes proven most promising in the literature (i.e. *Wus2*, *Bbm*, *Kn1/STM*, and *Ipt*), paying attention to select the most closely related ortholog possible (for plant genes), while also using optimal codon usage and promoter choices for the crop of interest. Once the optimal T-DNA has been constructed containing morphogenic gene(s) along with both a fluorescent and a selectable marker gene, this T-DNA can be used to optimize *Agrobacterium* infection and co-cultivation parameters (OD, temperature, media components, and duration) based on relative transient fluorescence levels, and then finally used to test different selection levels and timing.

As an alternative to reduce the complexity that accompanies *in vitro* organogenic or embryogenic culture methods, combining *in planta* *Agrobacterium*-mediated T-DNA delivery with morphogenic gene growth stimulation has recently been demonstrated for successful genome editing in tobacco (Maher et al., 2020) and has been proposed for more difficult species such as forestry crops (Nagle et al., 2019). Expanding on the concept of *in planta* transformation (or editing), combining this strategy with a non-integrating *Wus2* method (Hoerster et al., 2020) may prove crucial to *in planta* genome modification in long-life-cycle or asexually propagated crops. Continued development of such *in planta* strategies have great potential for simplifying and accelerating genome editing and may prove useful to a wide range of crops.

Plant genome editing is complex, requiring the effective coordination of interdependent processes and mechanisms such as delivery method, cargo, target explant, and, if appropriate, the use of morphogenic genes to aid in the recovery of fertile plants. Recent progress across all technologies is providing exciting new options. How these options fit together in a successful protocol ultimately depends on the requirements of the crop species.

9 Where to look for further information

CRISPR/Cas9-mediated genome editing in plants has become a widely publicized field, with far-ranging potential for the future of agriculture. However, genome editing *per se* represents only the tip of the technological iceberg. Below the surface, gene editing relies on the fundamentals and techniques of plant transformation, which is true for all plants. The component methods that have been combined to realize successful maize transformation thus

Published by Burleigh Dodds Science Publishing Limited, 2021.

represent the often-underappreciated foundation of current successes. These component areas include establishing reliable culture systems, optimizing DNA delivery, selecting transgenic events, and finally regenerating plants that contain the desired edits. Some of the seminal supporting technologies have been developed across a range of species but nonetheless have ended up contributing to the current state of maize transformation. Understanding these foundational technologies can be very helpful when trying to comprehend how the field has developed and its current status.

For further reading, see the following articles on using *Agrobacterium* for DNA delivery (Gelvin, 2017), old and new alternatives for delivering DNA (Demirer and Landry, 2017), development of chemical and/or visible markers for selection (or screening) of transgenic events (Rosellini, 2012), technological advances that have contributed to maize transformation (Yadava et al., 2017), utilizing CRISPR/Cas9 in maize (Chilcoat et al., 2017), and development of commercial agronomic traits in maize (Que et al., 2014).

- Chilcoat, D., Liu, Z. B. and Sander, J. (2017). Use of CRISPR/Cas9 for crop improvement in maize and soybean. In *Progress in Molecular Biology and Translational Science* (Vol. 149, pp. 27–46). Academic Press.
- Demirer, G. S. and Landry, M. P. (2017). Delivering genes to plants. *Chemical Engineering Progress*, *113*(4) 40–45.
- Gelvin, S. B. (2017). Integration of Agrobacterium T-DNA into the plant genome. *Annual Review of Genetics*, *51*, 195–217.
- Que, Q., Elumalai, S., Li, X., Zhong, H., Nalapalli, S., Schweiner, M., Fei, X., Nuccio, M., Kelliher, T., Gu, W. and Chen, Z. (2014). Maize transformation technology development for commercial event generation. *Frontiers in Plant Science* 5, 379.
- Rosellini, D. (2012). Selectable markers and reporter genes: a well furnished toolbox for plant science and genetic engineering. *Critical Reviews in Plant Sciences*, *31*(5), 401–453.
- Yadava, P., Abhishek, A., Singh, R., Singh, I., Kaul, T., Pattanayak, A. and Agrawal, P. K. (2017). Advances in maize transformation technologies and development of transgenic maize. *Frontiers in Plant Science*, *7*, 1949.

10 References

Ainley, W. M., Sastry-Dent, L., Welter, M. E., Murray, M. G., Zeitler, B., Amora, R., Corbin, D. R., Miles, R. R., Arnold, N. L., Strange, T. L., Simpson, M. A., Cao, Z., Carroll, C., Pawelczak, K. S., Blue, R., West, K., Rowland, L. M., Perkins, D., Samuel, P., Dewes, C. M., Shen, L., Sriram, S., Evans, S. L., Rebar, E. J., Zhang, L., Gregory, P. D., Urnov, F. D., Webb, S. R. and Petolino, J. F. (2013). Trait stacking via targeted genome editing. *Plant Biotechnology Journal* 11(9), 1126–1134.

Ali, Z., Abul-faraj, A., Li, L., Ghosh, N., Piatek, M., Mahjoub, A., Aouida, M., Piatek, A., Baltes, N., Voytas, D., Dinesh-Kumar, S. and Mahfouz, M. (2015a). Efficient virus-mediated genome editing in plants using the CRISPR/Cas9 uystem. *Molecular Plant* 8(8), 1288–1291.

Ali, Z., Abul-Faraj, A., Piatek, M. and Mahfouz, M. M. (2015b). Activity and specificity of TRV-mediated gene editing in plants. *Plant Signaling and Behavior* 10(10), e1044191.

Ali, Z., Eid, A., Ali, S. and Mahfouz, M. M. (2018). Pea early-browning virus-mediated genome editing via the CRISPR/Cas9 system in Nicotiana benthamiana and Arabidopsis. *Virus Research* 244, 333–337.

Altpeter, F., Baisakh, N., Beachy, R., Bock, R., Capell, T., Christou, P., Daniell, H., Datta, K., Datta, S., Dix, P. J., Fauquet, C., Huang, N., Kohli, A., Mooibroek, H., Nicholson, L., Nguyen, T. T., Nugent, G., Raemakers, K., Romano, A., Somers, D. A., Stoger, E., Taylor, N. and Visser, R. (2005). Particle bombardment and the genetic enhancement of crops: myths and realities. *Molecular Breeding* 15(3), 305–327.

Altpeter, F., Springer, N. M., Bartley, L. E., Blechl, A. E., Brutnell, T. P., Citovsky, V., Conrad, L. J., Gelvin, S. B., Jackson, D. P., Kausch, A. P., Lemaux, P. G., Medford, J. I., Orozco-Cárdenas, M. L., Tricoli, D. M., Van Eck, J., Voytas, D. F., Walbot, V., Wang, K., Zhang, Z. J. and Stewart, C. N. (2016). Advancing crop transformation in the era of genome editing. *Plant Cell* 28(7), 1510–1520.

Anand, A., Bass, S. H., Wu, E., Wang, N., McBride, K. E., Annaluru, N., Miller, M., Hua, M. and Jones, T. J. (2018). An improved ternary vector system for Agrobacterium-mediated rapid maize transformation. *Plant Molecular Biology* 97(1–2), 187–200.

Anand, A., Wu, E., Li, Z., TeRonde, S., Arling, M., Lenderts, B., Mutti, J. S., Gordon-Kamm, W., Jones, T. J. and Chilcoat, N. D. (2019). High efficiency *Agrobacterium*-mediated site-specific gene integration in maize utilizing the FLP-FRT recombination system. *Plant Biotechnology Journal* 17(8), 1636–1645.

Anzalone, A. V., Randolph, P. B., Davis, J. R., Sousa, A. A., Koblan, L. W., Levy, J. M., Chen, P. J., Wilson, C., Newby, G. A., Raguram, A. and Liu, D. R. (2019). Search-and-replace genome editing without double-strand breaks or donor DNA. *Nature* 576(7785), 149–157.

Baltes, N. J., Gil-Humanes, J., Cermak, T., Atkins, P. A. and Voytas, D. F. (2014). DNA replicons for plant genome engineering. *The Plant Cell* 26(1), 151–163.

Banno, H., Ikeda, Y., Niu, Q. W. and Chua, N. H. (2001). Overexpression of Arabidopsis ESR1 induces initiation of shoot regeneration. *The Plant Cell* 13(12), 2609–2618.

Barton, K. A., Binns, A. N., Matzke, A. J. M. and Chilton, M. (1983). Regeneration of intact tobacco plants containing full length copies of genetically engineered T-DNA, and transmission of T-DNA to R1 progeny. *Cell* 32(4), 1033–1043.

Baset, A., Finch, R. P. and Cocking, E. C. (1991). Plant regeneration from protoplasts of wild rice (Oryza rufipogon Griff.). *Plant Cell Reports* 10(4), 200–203.

Benson, E. E. and Roubelakis-Angelakis, K. A. (1994). Oxidative stress in recalcitrant tissue cultures of grapevine. *Free Radical Biology and Medicine* 16(3), 355–362.

Bett, B., Gollasch, S., Moore, A., James, W., Armstrong, J., Walsh, T., Harding, R. and Higgins, T. J. (2017). Transgenic cowpeas (*Vigna unguiculata* L. Walp) expressing Bacillus thuringiensis Vip3Ba protein are protected against the Maruca pod borer (*Maruca vitrata*). *Plant Cell, Tissue and Organ Culture (PCTOC)* 131(2), 335–345.

Bett, B., Gollasch, S., Moore, A., Harding, R. and Higgins, T. J. V. (2019). An improved transformation system for cowpea (Vigna unguiculata L. Walp) via sonication and a kanamycin-geneticin selection regime. *Frontiers in Plant Science* 10, 219.

Bevan, M. (1984). Binary Agrobacterium vectors for plant transformation. *Nucleic Acids Research* 12, 8711-8721.

Bhowmik, P., Ellison, E., Polley, B., Bollina, V., Kulkarni, M., Ghanbarnia, K., Song, H., Gao, C., Voytas, D. F. and Kagale. S. (2018). Targeted mutagenesis in wheat microspores using CRISPR/Cas9. *Scientific Reports* 8(1), 1-10.

Binding, H., Nehls, R., Schieder, O., Sopory, S. K. and Wenzel, G. (1978). Regeneration of mesophyll protoplasts isolated from dihaploid clones of *Solanum tuberosum*. *Physiologia Plantarum* 43(1), 52-54.

Bittencourt, P. A. L., Csányi, Á. and Jenes, B. (1995). Evaluation of different parameters and their influence on the PEG (polyethylene glycol) mediated gene transfer into rice (*Oryza sativa* L.) protoplasts. *Cereal Research Communications* 23, 359-365.

Botella, J. R. (2019). Now for the hard ones: is there a limit on CRISPR genome editing in crops? *Journal of Experimental Botany* 70(3), 734-737.

Boutilier, K., Offringa, R., Sharma, V. K., Kieft, H., Ouellet, T., Zhang, L., Hattori, J., Liu, C. M., van Lammeren, A. A., Miki, B.L. and Custers, J. B. (2002). Ectopic expression of BABY BOOM triggers a conversion from vegetative to embryonic growth. *The Plant Cell* 14(8), 1737-1749.

Brew-Appiah, R. A., Ankrah, N., Liu, W., Konzak, C. F., von Wettstein, D. and Rustgi, S. (2013). Generation of doubled haploid transgenic wheat lines by microspore transformation. *PLoS ONE* 8(11), e80155.

Butler, N. M., Baltes, N. J., Voytas, D. F. and Douches, D. S. (2016). Geminivirus-mediated genome editing in potato (Solanum tuberosum L.) using sequence-specific nucleases. *Frontiers in Plant Science* 7, 1045, doi.org/10.3389/fpls.2016.01045.

Caplan, A., Herrera-Estrella, L., Inzé, D., Van Haute, E., Van Montagu, M., Schell, J. and Zambryski, P. (1983). Introduction of genetic material into plant cells. *Science* 222(4625), 815-821.

Cardoza, V. and Stewart, C. N. (2003). Increased Agrobacterium-mediated transformation and rooting efficiencies in canola (*Brassica napus* L.) from hypocotyl segment explants. *Plant Cell Reports* 21(6), 599-604.

Čermák, T., Baltes, N. J., Čegan, R., Zhang, Y. and Voytas, D. F. (2015). High-frequency, precise modification of the tomato genome. *Genome Biology* 16(1), 232, doi.org/10.1186/s13059-015-0796-9.

Chang, S., Mahon, E. L., MacKay, H. A., Rottmann, W. H., Strauss, S. H., Pijut, P. M., Powell, W. A., Coffey, V., Lu, H., Mansfield, S.D and Jones, T. J. (2018). Genetic engineering of trees: progress and new horizons. *In Vitro Cellular & Developmental Biology-Plant* 54(4), 341-376.

Chaudhury, D., Madanpotra, S., Jaiwal, R., Saini, R., Kumar, P. A. and Jaiwal, P. K. (2007). *Agrobacterium tumefaciens*-mediated high frequency genetic transformation of an Indian cowpea (*Vigna unguiculata* L. Walp.) cultivar and transmission of transgenes into progeny. *Plant Science* 172(4), 692-700.

Che, P., Anand, A., Wu, E., Sander, J. D., Simon, M. K., Zhu, W., Sigmund, A. L., Zastrow-Hayes, G., Miller, M., Liu, D., Lawit, S. J., Zhao, Z. Y., Albertsen, M. C. and Jones, T. J. (2018). Developing a flexible, high-efficiency Agrobacterium-mediated sorghum transformation system with broad application. *Plant Biotechnology Journal* 16(7), 1388-1395. https://doi.org/10.1111/pbi.12879

Che, P., Chang, S., Simon, M. K., Zhang, Z., Shaharyar, A., Ourada, J., O'Neill, D., Torres-Mendoza, M., Guo, Y., Marasigan, K. M. and Vielle-Calzada, J. P. (2019). Developing a rapid and highly efficient cowpea regeneration and transformation system using embryonic axis explants. *BioRxiv*, 738971.

Chen, K., Wang, Y., Zhang, R., Zhang, H. and Gao, C. (2019). CRISPR/Cas genome editing and precision plant breeding in agriculture. *Annual Review of Plant Biology* 70, 667-697.

Cheng, A. W., Wang, H., Yang, H., Shi, L., Katz, Y., Theunissen, T. W., Rangarajan, S., Shivalila, C. S., Dadon, D. B. and Jaenisch, R. (2013). Multiplexed activation of endogenous genes by CRISPR-on, an RNA-guided transcriptional activator system. *Cell Research* 23(10), 1163-1171.

Chilcoat, D., Liu, Z. B. and Sander, J. (2017). Use of CRISPR/Cas9 for crop improvement in maize and soybean. In *Progress in Molecular Biology and Translational Science* (Vol. 149, pp. 27-46). Academic Press.

Chupeau, M. -C., Granier, F., Pichon, O., Renou, J. -P., Gaudin, V. and Chupeau, Y. (2013). Characterization of the early events leading to totipotency in an Arabidopsis protoplast liquid culture by temporal transcript profiling. *The Plant Cell* 25(7), 2444-2463.

Clough, S. J. and Bent, A. F. (1998). Floral dip: a simplified method for *Agrobacterium*-mediated transformation of *Arabidopsis thaliana*. *The Plant Journal* 16(6), 735-743.

Cocking, E. C. (1960). A method for the isolation of plant protoplasts and vacuoles. *Nature* 187(4741), 962-963.

Cody, W. B. and Scholthof, H. B. (2019). Plant virus vectors 3.0: transitioning into synthetic genomics. *Annual Review of Phytopathology* 57(1), 211-230.

Cody, W. B., Scholthof, H. B. and Mirkov, T. E. (2017). Multiplexed gene editing and protein overexpression using a tobacco mosaic virus viral vector. *Plant Physiology* 175(1), 23-35.

Côte, F. X., Teisson, C. and Perrier, X. (2001). Somaclonal variation rate evolution in plant tissue culture: contribution to understanding through a statistical approach. *In Vitro Cellular & Developmental Biology-Plant* 37(5), 539-542.

Curtin, S. J., Zhang, F., Sander, J. D., Haun, W. J., Starker, C., Baltes, N. J., Reyon, D., Dahlborg, E. J., Goodwin, M. J., Coffman, A. P., Dobbs, D., Joung, K. J., Voytas, D. F. and Stupar, R. M. (2011). Targeted mutagenesis of duplicated genes in soybean with zinc-finger nucleases. *Plant physiology* 156(2), 466-473.

Das Bhowmik, S. S., Cheng, A. Y., Long, H., Tan, G. Z. H., Hoang, T. M. L., Karbaschi, M. R., William. B., Higgins, T.J. and Mundree, S. G. (2019). Robust genetic transformation system to obtain non-chimeric transgenic chickpea. *Frontiers in Plant Science* 10, 524.

Davey, M. R., Anthony, P., Power, J. B. and Lowe, K. C. (2005). Plant protoplasts: status and biotechnological perspectives. *Biotechnology Advances* 23(2), 131-171.

Debernardi, J. M., Tricoli, D. M., Ercoli, M. F., Hayta, S., Ronald, P., Palatnik, J. F. and Dubcovsky, J. (2020). A chimera including a GROWTH-REGULATING FACTOR (GRF) and its cofactor GRF-INTERACTING FACTOR (GIF) increases transgenic plant regeneration efficiency. *bioRxiv*. doi.org/10.1101/2020.08.23.263905

De Block, M. (1988). Genotype-independent leaf disc transformation of potato (*Solanum tuberosum*) using *Agrobacterium tumefaciens*. *Theoretical and Applied Genetics* 76(5), 767-774.

de Marco, A. and Roubelakis-Angelakis, K. A. (1996). The complexity of enzymic control of hydrogen peroxide concentration may affect the regeneration potential of plant protoplasts. *Plant Physiology* 110(1), 137-145.

de Pater, S., Neuteboom, L. W., Pinas, J. E., Hooykaas, P. J. and van der Zaal, B. J. (2009). ZFN-induced mutagenesis and gene-targeting in Arabidopsis through Agrobacterium-mediated floral dip transformation. *Plant Biotechnology Journal* 7(8), 821–835.

de Pater, S., Pinas, J. E., Hooykaas, P. J. and van der Zaal, B. J. (2013). ZFN-mediated gene targeting of the Arabidopsis protoporphyrinogen oxidase gene through Agrobacterium-mediated floral dip transformation. *Plant Biotechnology Journal* 11(4), 510–515.

Delporte, F., Jacquemin, J. M., Masson, P. and Watillon, B. (2012). Insights into the regenerative property of plant cells and their receptivity to transgenesis: wheat as a research case study. *Plant Signaling & Behavior* 7(12), 1608–1620.

Demirer, G., Zhang, H., Goh, N., Chang, R. and Landry, M. (2019a). Nanotubes effectively deliver siRNA to intact plant cells and protect siRNA against nuclease degradation. *BioRxiv*. https://doi.org/10.1101/564427.

Demirer, G. S., Zhang, H., Matos, J. L., Goh, N. S., Cunningham, F. J., Sung, Y., Chang, R., Aditham, A. J., Chio, L., Cho, M., Staskawicz, B. and Landry, M. P. (2019b). High aspect ratio nanomaterials enable delivery of functional genetic material without DNA integration in mature plants. *Nature Nanotechnology* 14(5), 456–464.

Demorest, Z. L., Coffman, A., Baltes, N. J., Stoddard, T. J., Clasen, B. M., Luo, S., Retterath, A., Yabandith, A., Gamo, M. E., Bissen, J., Mathis, L., Voytas, D. F. and Zhang, F. (2016). Direct stacking of sequence-specific nuclease-induced mutations to produce high oleic and low linolenic soybean oil. *BMC Plant Biology* 16(1):225.

Deng, W., Luo, K., Li, Z. and Yang, Y. (2009). A novel method for induction of plant regeneration via somatic embryogenesis. *Plant Science* 177(1), 43–48.

D'Halluin, K., Bonne, E., Bossut, M. and Leemans, J. (1992). Transgenic maize plants by tissue electroporation. *The Plant Cell* 4(12), 1495–1505.

Dlugosz, E. M., Lenaghan, S. C. and Stewart, C. N., Jr. (2016). A robotic platform for high-throughput protoplast isolation and transformation. *Journal of Visualized Experiments* 115, e54300.

Du, X., Fang, T., Liui, Y., Huang, L., Zang, M., Wang, G., Liu, Y. and Fu, J. (2019). Transcriptome profiling predicts new genes to promoter maize callus formation and transformation. *Frontiers in Plant Science* 10, 1633, doi:10.3389/fpls.2019.01633.

Durfee, P. N., Lin, Y. S., Dunphy, D. R., Muñiz, A. J., Butler, K. S., Humphrey, K. R., Lokke, A.J., Agola, J.O., Chou, S. S., Chen, I.M. and Wharton, W. (2016). Mesoporous silica nanoparticle-supported lipid bilayers (protocells) for active targeting and delivery to individual leukemia cells. *ACS Nano* 10(9), 8325–8345.

Dutt, M., Li, Z. T., Dhekney, S. A. and Gray, D. J. (2007). Transgenic plants from shoot apical meristems of *Vitis vinifera* L."Thompson Seedless" via *Agrobacterium*-mediated transformation. *Plant Cell Reports* 26(12), 2101–2110.

Ebinuma, H., Sugita, K., Matsunaga, E. and Yamakado, M. (1997). Selection of marker-free transgenic plants using the isopentenyl transferase gene. *Proceedings of the National Academy of Sciences* 94(6), 2117–2121.

Eeckhaut, T., Lakshmanan, P. S., Deryckere, D., Van Bockstaele, E. and Van Huylenbroeck, J. (2013). Progress in plant protoplast research. *Planta* 238(6), 991–1003.

Elhiti, M., Tahir, M., Gulden, R. H., Khamiss, K. and Stasolla, C. (2010). Modulation of embryo-forming capacity in culture through the expression of Brassica genes involved in the regulation of the shoot apical meristem. *Journal of Experimental Botany* 61(14), 4069–4085.

Farzadfard, F., Perli, S. D. and Lu, T. K. (2013). Tunable and multifunctional eukaryotic transcription factors based on CRISPR/Cas. *ACS Synthetic Biology* 2(10), 604–613.

Feng, Z., Zhang, B., Ding, W., Liu, X., Yang, D., Wei, P., Cao, F., Zhu, S., Zhang, F., Mao, Y. and Zhu, J. (2013). Efficient genome editing in plants using a CRISPR/Cas system. *Cell Research* 23(10), 1229–1232.

Fisk, H. J. and Dandekar, A. M. (2004). Electroporation: introduction and expression of transgenes in plant protoplasts. In: Peña, L. (Ed.) *Transgenic Plants: Methods and Protocols*. Totowa, NJ: Humana Press, 79–90.

Fonfara, I., Le Rhun, A., Chylinski, K., Makarova, K. S., Lécrivain, A., Bzdrenga, J., Koonin, E. V. and Charpentier, E. (2014). Phylogeny of Cas9 determines functional exchangeability of dual-RNA and Cas9 among orthologous type II CRISPR-Cas systems. *Nucleic Acids Research* 42(4), 2577–2590.

Forner, J., Pfeiffer, A., Langenecker, T., Manavella, P. and Lohmann, J. U. (2015). Germline-transmitted genome editing in Arabidopsis thaliana using TAL-effector-nucleases. *PLoS ONE* 10(3), e0121056.

Fromm, M., Taylor, L. P. and Walbot, V. (1985). Expression of genes transferred into monocot and dicot plant cells by electroporation. *Proceedings of the National Academy of Sciences* 82(17), 5824–5828.

Fujita, K., Matsuoka, T., Suzuki, S. and Takayanagi, T. (2009). *In planta* transformation technique for grapevines (*Vitis vinifera* L) using dormant buds. *Journal of Plant Biochemistry and Biotechnology* 18(2), 161–167.

Furuhata, Y., Sakai, A., Murakami, T., Morikawa, M., Nakamura, C., Yoshizumi, T., Fujikura, U., Nishida, K. and Kato, Y. (2019). A method using electroporation for the protein delivery of Cre recombinase into cultured Arabidopsis cells with an intact cell wall. *Scientific Reports* 9(1), 2163.

Gao, H., Gadlage, M. J., Lafitte, H. R., Lenderts, B., Yang, M., Schroder, M., Farrell, J., Snopek, K., Peterson, D., Feigenbutz, L., Jones, S., St Clair, G., Rahe, M., Sanyour-Doyel, N., Peng, C., Wang, L., Young, J. K., Beatty, M., Dahlke, B., Hazebroek, J., Greene, T. W., Cigan, A. M., Chilcoat, N. D. and Meeley, R. B. (2020). Superior field performance of waxy corn engineered using CRISPR-Cas9. *Nature Biotechnology* 38(5), 579–581.

Gao, H., Smith, J., Yang, M., Jones, S., Djukanovic, V., Nicholson, M. G., West, A., Bidney, D., Falco, S. C., Jantz, D. and Lyznik, L. A. (2010). Heritable targeted mutagenesis in maize using a designed endonuclease. *The Plant Journal: For Cell and Molecular Biology* 61(1), 176–187.

Gao, J., Wang, G., Ma, S., Xie, X., Wu, X., Zhang, X., Wu, Y., Zhao, P. and Xia, Q. (2015). CRISPR/Cas9-mediated targeted mutagenesis in Nicotiana tabacum. *Plant Molecular Biology* 87(1–2), 99–110.

Gelvin, S. B. (2003). Agrobacterium-mediated plant transformation: the biology behind the "gene-jockeying" tool. *Microbiology and Molecular Biology Reviews* 67(1), 16–37.

Gelvin, S. B. (2017). Integration of Agrobacterium T-DNA into the plant genome. *Annual Review of Genetics* 51, 195–217.

Gilbert, L. A., Larson, M. H., Morsut, L., Liu, Z., Brar, G. A., Torres, S. E., Stern-Ginossar, N., Brandman, O., Whitehead, E. H., Doudna, J. A., Lim, W. A., Weissman, J. S. and Qi, L. S. (2013). CRISPR-mediated modular RNA-guided regulation of transcription in eukaryotes. *Cell* 154(2), 442–451.

Gil-Humanes, J., Wang, Y., Liang, Z., Shan, Q., Ozuna, C. V., Sánchez-León, S., Baltes, N. J., Starker, C., Barro, F., Gao, C. and Voytas, D. F. (2017). High-efficiency gene targeting in hexaploid wheat using DNA replicons and CRISPR/Cas9. *The Plant Journal* 89(6), 1251-1262.

Gleba, Y., Marillonnet, S. and Klimyuk, V. (2004). Engineering viral expression vectors for plants: the 'full virus' and the 'deconstructed virus' strategies. *Current Opinion in Plant Biology* 7(2), 182-188.

Gómez-Lim, M. A. and Litz, R. E. (2004). Genetic transformation of perennial tropical fruits. *In Vitro Cellular and Developmental Biology–Plant* 40(5), 442-449.

Gordon-Kamm, B., Sardesai, N., Arling, M., Lowe, K., Hoerster, G., Betts, S., & Jones, T. (2019). Using morphogenic genes to improve recovery and regeneration of transgenic plants. *Plants* 8(2), 38, doi.org/10.3390/plants8020038.

Grimsley, N., Hohn, T., Davies, J. W. and Hohn, B. (1987). Agrobacterium-mediated delivery of infectious maize streak virus into maize plants. *Nature* 325(6100), 177-179.

Grzebelus, E., Szklarczyk, M., Greń, J., Śniegowska, K., Jopek, M., Kacińska, I. and Mrożek, K. (2012). Phytosulfokine stimulates cell divisions in sugar beet (Beta vulgaris L.) mesophyll protoplast cultures. *Plant Growth Regulation* 67(1), 93-100.

Habben, J. E., Bao, X., Bate, N. J., DeBruin, J. L., Dolan, D., Hasegawa, D., Helentjaris, T. G., Lafitte, R. H., Lovan, N., Mo, H., Reimann, K. and Schussler, J. R. (2014). Transgenic alteration of ethylene biosynthesis increases grain yield in maize under field drought-stress conditions. *Plant Biotechnology Journal* 12(6), 685-693.

Hamada, H., Liu, Y., Nagira, Y., Miki, R., Taoka, N. and Imai, R. (2018). Biolistic-delivery-based transient CRISPR/Cas9 expression enables in planta genome editing in wheat. *Scientific Reports* 8(1), 14422.

Hanley-Bowdoin, L., Settlage, S. B., Orozco, B. M., Nagar, S and Robertson, D. (1999). Geminiviruses: models for plant DNA replication, transcription, and cell cycle regulation. *Critical Reviews in Plant Sciences* 18(1), 71-106.

Hansen, G., Das, A. and Chilton, M. D. (1994). Constitutive expression of the virulence genes improves the efficiency of plant transformation by Agrobacterium. *Proceedings of the National Academy of Sciences of the United States of America* 91(16), 7603-7607.

Harding, E. W., Tang, W., Nichols, K. W., Fernandez, D. E. and Perry, S. E. (2003). Expression and maintenance of embryogenic potential is enhanced through constitutive expression of AGAMOUS-LIKE15. *Plant Physiology* 133(2), 653-663.

Hatanaka, T., Choi, Y. E., Kusano, T. and Sano, H. (1999). Transgenic plants of coffee *Coffea canephora* from embryogenic callus via *Agrobacterium tumefaciens*-mediated transformation. *Plant Cell Reports* 19(2), 106-110.

Haun, W., Coffman, A., Clasen, B. M., Demorest, Z. L., Lowy, A., Ray, E., Retterath, A., Stoddard, T., Juillerat, A., Cedrone, F., Mathis, L., Voytas, D. F. and Zhang, F. (2014). Improved soybean oil quality by targeted mutagenesis of the fatty acid desaturase 2 gene family. *Plant Biotechnology Journal* 12(7), 934-940.

Hauptmann, R. M., Ozias-Akins, P., Vasil, V., Tabaeizadeh, Z., Rogers, S. G., Horsch, R. B., Vasil, I. K. and Fraley, R. T. (1987). Transient expression of electroporated DNA in monocotyledonous and dicotyledonous species. *Plant Cell Reports* 6(4), 265-270.

Heidmann, I., de Lange, B., Lambalk, J., Angenent, G. C. and Boutilier, K. (2011). Efficient sweet pepper transformation mediated by the BABY BOOM transcription factor. *Plant Cell Reports* 30(6), 1107-1115.

Hiei, Y., Ohta, S., Komari, T. and Kumashiro, T. (1994). Efficient transformation of rice (*Oryza sativa* L.) mediated by *Agrobacterium* and sequence analysis of the boundaries of the T-DNA. *The Plant Journal: For Cell and Molecular Biology* 6(2), 271-282.

Hiei, Y., Ishida, Y. and Komari, T. (2014). Progress of cereal transformation technology mediated by *Agrobacterium tumefaciens*. *Frontiers in Plant Science*, 5.

Hoekema, A., Hirsch, P. R., Hooykaas, P. J. and Schilperoort, R. A. (1983). A binary plant vector strategy based on separation of vir-and T-region of the *Agrobacterium tumefaciens* Ti-plasmid. *Nature* 303(5913), 179–180.

Hoerster, G., Wang, N., Ryan, L., Wu, E., Anand, A., McBride, K., Lowe, K., Jones, T. and Gordon-Kamm, B. (2020). Use of non-integrating *Zm-Wus2* vectors to enhance maize transformation. *In Vitro Cellular and Developmental Biology–Plant* 56(3), 265–279.

Hong, H. P., Zhang, H., Olhoft, P., Hill, S., Wiley, H., Toren, E., Hillebrand, H., Jones, T. and Cheng, M. (2007). Organogenic callus as the target for plant regeneration and transformation via Agrobacterium in soybean (*Glycine max* (L.) Merr.). *In Vitro Cellular & Developmental Biology-Plant* 43(6), 558–568.

Honig, A., Marton, I., Rosenthal, M., Smith, J., Nicholson, M., Jantz, D., Zuker, A. and Vainstein, A. (2015). Transient expression of virally delivered meganuclease *in planta* generates inherited genomic deletions. *Molecular Plant* 8(8), 1292–1294.

Hood, E. E., Gelvin, S. B., Melchers, L. S. and Hoekema, A. (1993). New *Agrobacterium* helper plasmids for gene transfer to plants. *Transgenic Research* 2(4), 208–218.

Hu, D., Bent, A. F., Hou, X. and Li, Y. (2019a). *Agrobacterium*-mediated vacuum infiltration and floral dip transformation of rapid-cycling Brassica rapa. *BMC Plant Biology* 19(1), 246.

Hu, J., Li, S., Li, Z., Li, H., Song, W., Zhao, H., Lai, J., Xia, L., Li, D. and Zhang, Y. (2019b). A barley stripe mosaic virus-based guide RNA delivery system for targeted mutagenesis in wheat and maize. *Molecular Plant Pathology* 20(10), 1463–1474.

Hu, Q., Andersen, S. B. and Hansen, L. N. (1999). Plant regeneration capacity of mesophyll protoplasts from Brassica napus and related species. *Plant Cell, Tissue and Organ Culture* 59(3), 189–196.

Ibrahim, A., Odon, V. and Kormelink, R. (2019). Plant viruses in plant molecular pharming: toward the use of enveloped viruses. *Frontiers in Plant Science* 10, 803, doi:10.3389/fpls.2019.00803.

Ishida, Y., Saito, H., Ohta, S., Hiei, Y., Komari, T. and Kumashiro, T. (1996). High efficiency transformation of maize (*Zea mays* L.) mediated by *Agrobacterium tumefaciens*. *Nature Biotechnology* 14(6), 745–750.

Ishii, S. (1988). Factors influencing protoplast viability of suspension-cultured rice cells during isolation process. *Plant Physiology* 88(1), 26–29.

Jacobs, T. B., LaFayette, P. R., Schmitz, R. J. and Parrott, W. A. (2015). Targeted genome modifications in soybean with CRISPR/Cas9. *BMC Biotechnology* 15(1), 1–10.

Jiang, N., Zhang, C., Liu, J.-Y., Guo, Z., Zhang, Z., Han, C. and Wang, Y. (2019). Development of beet necrotic yellow vein virus-based vectors for multiple-gene expression and guide RNA delivery in plant genome editing. *Plant Biotechnology Journal* 17(7), 1302–1315.

Jones, T., Lowe, K., Hoerster, G., Anand, A., Wu, E., Wang, N., Arling, M., Lenderts, B. and Gordon-Kamm, W. (2019). Maize transformation using the morphogenic genes Baby Boom and Wuschel2. In *Transgenic Plants* (pp. 81–93). Humana Press, New York, NY.

Jones, T. J. (2009). Maize tissue culture and transformation: the first 20 years. In *Molecular Genetic Approaches to Maize Improvement* (pp. 7–27). Springer, Berlin, Heidelberg.

Kanchiswamy, C. N. (2016). DNA-free genome editing methods for targeted crop improvement. *Plant Cell Reports* 35(7), 1469–1474.

Kao, K. N. and Michayluk, M. R. (1980). Plant regeneration from mesophyll protoplasts of alfalfa. *Zeitschrift für Pflanzenphysiologie* 96(2), 135–141.

Kiełkowska, A. and Adamus, A. (2012). An alginate-layer technique for culture of Brassica oleracea L. protoplasts. *In Vitro Cellular and Developmental Biology–Plant* 48(2), 265–273.

Kim, S., Kim, D., Cho, S. W., Kim, J. and Kim, J. -S. (2014). Highly efficient RNA-guided genome editing in human cells via delivery of purified Cas9 ribonucleoproteins. *Genome Research* 24(6), 1012–1019.

Klein, T. M., Wolf, E. D., Wu, R. and Sanford, J. C. (1987). High-velocity microprojectiles for delivering nucleic acids into living cells. *Nature* 327(6117), 70–73.

Kofer, W., Koop, H. -U., Wanner, G. and Steinmüller, K. (1998). Mutagenesis of the genes encoding subunits A, C, H, I, J and K of the plastid NAD(P)H plastoquinone-oxidoreductase in tobacco by polyethylene glycol-mediated plastome transformation. *Molecular and General Genetics MGG* 258(1–2), 166–173.

Kong, J., Martin-Ortigosa, S., Finer, J., Orchard, N., Gunadi, A., Batts, L. A., Thakare, D., Rush, B., Schmitz, O., Stuiver, M. and Olhoft, P. and Pacheco-Villalobos, D. (2020). Overexpression of the transcription factor GROWTH-REGULATING FACTOR5 improves transformation of dicot and monocot species. *Frontiers in plant science*, 11, 1389.

Komari, T. (1990). Transformation of cultured cells of *Chenopodium quinoa* by binary vectors that carry a fragment of DNA from the virulence region of pTiBo542. *Plant Cell Reports* 9, 303–306.

Koornneef, M., Hanhart, C. J. and Martinelli, L. (1987). A genetic analysis of cell culture traits in tomato. *Theoretical and Applied Genetics* 74(5), 633–641.

Krens, F. A., Molendijk, L., Wullems, G. J. and Schilperoort, R. A. (1982). In vitro transformation of plant protoplasts with Ti-plasmid DNA. *Nature* 296(5852), 72–74.

Krishnamurthy, K. V., Suhasini, K., Sagare, A. P., Meixner, M., de Kathen, A., Pickardt, T. and Schieder, O. (2000). *Agrobacterium* mediated transformation of chickpea (*Cicer arietinum* L.) embryo axes. *Plant Cell Reports* 19(3), 235–240.

Kyo, M., Maida, K., Nishioka, Y. and Matsui, K. (2018). Coexpression of WUSCHEL related homeobox (WOX) 2 with WOX8 or WOX9 promotes regeneration from leaf segments and free cells in *Nicotiana tabacum* L. *Plant Biotechnology* 35(1), 23–30.

Kwak, S. Y., Lew, T. T. S., Sweeney, C. J., Koman, V. B., Wong, M. H., Bohmert-Tatarev, K., Snell, K. D., Seo, J. S., Chua, N. H. and Strano, M. S. (2019). Chloroplast-selective gene delivery and expression in planta using chitosan-complexed single-walled carbon nanotube carriers. *Nature Nanotechnology* 14(5), 447–455.

Lazo, G. R., Stein, P. A. and Ludwig, R. A. (1991). A DNA transformation-competent Arabidopsis genomic library in *Agrobacterium*. *Bio/Technology* 9(10), 963–967.

Li, J. F., Norville, J. E., Aach, J., McCormack, M., Zhang, D., Bush, J., Church, G.M. and Sheen, J. (2013). Multiplex and homologous recombination-mediated genome editing in Arabidopsis and *Nicotiana benthamiana* using guide RNA and Cas9. *Nature Biotechnology 31*(8), 688–691.

Li, T., Liu, B., Spalding, M. H., Weeks, D. P. and Yang, B. (2012). High-efficiency TALEN-based gene editing produces disease-resistant rice. *Nature Biotechnology* 30(5), 390–392.

Li, Z., Burow, M. D. and Murai, N. (1990). High frequency generation of fertile transgenic rice plants after PEG-mediated protoplast transformation. *Plant Molecular Biology Reporter* 8(4), 276–291.

Li, Z., Liu, Z. B., Xing, A., Moon, B. P., Koellhoffer, J. P., Huang, L., Ward, R. T., Clifton, E., Falco, S. C. and Cigan, A. M. (2015). Cas9-guide RNA directed genome editing in soybean. *Plant Physiology* 169(2), 960–970.

Li, Z., Xing, A., Moon, B. P., McCardell, R. P., Mills, K. and Falco, S. C. (2009). Site-specific integration of transgenes in soybean via recombinase-mediated DNA cassette exchange. *Plant Physiology* 151(3), 1087–1095.

Liang, Z., Chen, K., Li, T., Zhang, Y., Wang, Y., Zhao, Q., Liu, J., Zhang, H., Liu, C., Ran, Y. and Gao, C. (2017). Efficient DNA-free genome editing of bread wheat using CRISPR/Cas9 ribonucleoprotein complexes. *Nature Communications* 8(1), 14261.

Liang, Z., Zhang, K., Chen, K. and Gao, C. (2014). Targeted mutagenesis in *Zea mays* using TALENs and the CRISPR/Cas system. *Journal of Genetics and Genomics = Yi Chuan Xue Bao* 41(2), 63–68.

Lin, C. S., Hsu, C. T., Yang, L. H., Lee, L. Y., Fu, J. Y., Cheng, Q. W., Wu, F. H., Hsiao, H. C., Zhang, Y., Zhang, R., Chang, W. J., Yu, C. T., Wang, W., Liao, L. J., Gelvin, S. B. and Shih, M. C. (2018). Application of protoplast technology to CRISPR/Cas9 mutagenesis: from single-cell mutation detection to mutant plant regeneration. *Plant Biotechnology Journal* 16(7), 1295–1310.

Lin, S., Staahl, B. T., Alla, R. K. and Doudna, J. A. (2014). Enhanced homology-directed human genome engineering by controlled timing of CRISPR/Cas9 delivery. *eLife* 3, e04766, doi:10.7554/eLife.04766.

Liu, J., Nannas, N. J., Fu, F. F., Shi, J., Aspinwall, B., Parrott, W. A. and Dawe, R. K. (2019). Genome-scale sequence disruption following biolistic transformation in rice and maize. *The Plant Cell* 31(2), 368–383.

Lloyd, A., Plaisier, C. L., Carroll, D. and Drews, G. N. (2005). Targeted mutagenesis using zinc-finger nucleases in Arabidopsis. *Proceedings of the National Academy of Sciences* 102(6), 2232–2237.

Lotan, T., Ohto, M. A., Yee, K. M., West, M. A., Lo, R., Kwong, R. W., Yamagishi, K., Fischer, R.L., Goldberg, R.B. and Harada, J. J. (1998). Arabidopsis LEAFY COTYLEDON1 is sufficient to induce embryo development in vegetative cells. *Cell* 93(7), 1195 -1205.

Lowe, K., La Rota, M., Hoerster, G., Hastings, C., Wang, N., Chamberlin, M., Wu, E., Jones, T. and Gordon-Kamm, W. (2018). Rapid genotype "independent" *Zea mays* L. (maize) transformation via direct somatic embryogenesis. *In Vitro Cellular and Developmental Biology–Plant* 54(3), 240–252.

Lowe, K., Wu, E., Wang, N., Hoerster, G., Hastings, C., Cho, M., Scelonge, C., Lenderts, B., Chamberlin, M., Cushatt, J., Wang, L., Ryan, L., Khan, T., Chow-Yiu, J., Hua, W., Yu, M., Banh, J., Bao, Z., Brink, K., Igo, E., Rudrappa, B., Shamseer, P., Bruce, W., Newman, L., Shen, B., Zheng, P., Bidney, D., Falco, C., Register, J., Zhao, Z., Xu, D., Jones, T. and Gordon-Kamm, W. (2016). Morphogenic regulators *Baby boom* and *Wuschel* improve monocot transformation. *Plant Cell* 28(9), 1998–2015.

Lutz, K. A., Martin, C., Khairzada, S. and Maliga, P. (2015). Steroid-inducible BABY BOOM system for development of fertile Arabidopsis thaliana plants after prolonged tissue culture. *Plant Cell Reports* 34(10), 1849–1856.

Maas, C. and Werr, W. (1989). Mechanism and optimized conditions for PEG mediated DNA transfection into plant protoplasts. *Plant Cell Reports* 8(3), 148–151.

Maher, M. F., Nasti, R. A., Vollbrecht, M., Starker, C. G., Clark, M. D. and Voytas, D. F. (2020). Plant gene editing through *de novo* induction of meristems. *Nature Biotechnology* 38(1), 84–89.

Maheshwari, P. and Kovalchuk, I. (2016). *Agrobacterium*-mediated stable genetic transformation of *Populus angustifolia* and *Populus balsamifera*. *Frontiers in Plant Science* 7, 296, doi:10.3389/fpls.2016.00296.

Mak, A. C., Lai, Y. Y., Lam, E. T., Kwok, T. P., Leung, A. K., Poon, A., Mostovoy, Y., Hastie, A. R., Stedman, W., Anantharaman, T., Andrews, W., Zhou, X., Pang, A. W., Dai, H., Chu, C., Lin, C., Wu, J. J., Li, C. M., Li, J. W., Yim, A. K., Chan, S., Sibert, J., Džakula, Ž, Cao, H., Yiu, S. M., Chan, T. F., Yip, K. Y., Xiao, M. and Kwok, P. Y. (2016). Genome-wide structural variation detection by genome mapping on nanochannel arrays. *Genetics* 202(1), 351–362.

Malzahn, A., Lowder, L. and Qi, Y. (2017). Plant genome editing with TALEN and CRISPR. *Cell and Bioscience* 7(1), 21.

Marillonnet, S., Giritch, A., Gils, M., Kandzia, R., Klimyuk, V. and Gleba, Y. (2004). *In planta* engineering of viral RNA replicons: efficient assembly by recombination of DNA modules delivered by *Agrobacterium*. *Proceedings of the National Academy of Sciences of the United States of America* 101(18), 6852–6857.

Martin-Ortigosa, S., Valenstein, J. S., Lin, V. S.-Y., Trewyn, B. G. and Wang, K. (2012). Gold functionalized mesoporous silica nanoparticle mediated protein and DNA codelivery to plant cells via the biolistic method. *Advanced Functional Materials* 22(17), 3576–3582.

Marton, I., Zuker, A., Shklarman, E., Zeevi, V., Tovkach, A., Roffe, S., Ovadis, M., Tzfira, T. and Vainstein, A. (2010). Nontransgenic genome modification in plant cells. *Plant Physiology* 154(3), 1079–1087.

Matsuda, N., Gao, M., Isuzugawa, K., Takashina, T. and Nishimura, K. (2005). Development of an Agrobacterium-mediated transformation method for pear (*Pyrus communis* L.) with leaf-section and axillary shoot-meristem explants. *Plant Cell Reports* 24(1), 45–51.

Mayavan, S., Subramanyam, K., Jaganath, B., Sathish, D., Manickavasagam, M. and Ganapathi, A. (2015). *Agrobacterium*-mediated in planta genetic transformation of sugarcane setts. *Plant Cell Reports* 34(10), 1835–1848.

Metje-Sprink, J., Menz, J., Modrzejewski, D. and Sprink, T. (2019). DNA-free genome editing: past, present and future. *Frontiers in Plant Science* 9, 1957, doi:10.3389/fpls.2018.01957.

Meurer, C. A., Dinkins. R. D., and Collins, G. B. (1998). Factors affecting soybean cotyledonary node transformation. *Plant Cell Reports* 18(3–4), 180–186.

Mookkan, M., Nelson-Vasilchik, K., Hague, J., Zhang, Z. J. and Kausch, A. P. (2017). Selectable marker independent transformation of recalcitrant maize inbred B73 and sorghum P898012 mediated by morphogenic regulators BABY BOOM and WUSCHEL2. *Plant Cell Reports* 36(9), 1477–1491.

Moore, G. A., Jacono, C. C., Neidigh, J. L., Lawrence, S. D. and Cline, K. (1992). *Agrobacterium*-mediated transformation of citrus stem segments and regeneration of transgenic plants. *Plant Cell Reports* 11-11(5–6), 238–242.

Mórocz, S., Donn, G., Nérneth, J. and Dudits, D. (1990). An improved system to obtain fertile regenerants via maize protoplasts isolated from a highly embryogenic suspension culture. *TAG. Theoretical and Applied Genetics. Theoretische und Angewandte Genetik* 80(6), 721–726.

Murovec, J., Guček, K., Bohanec, B., Avbelj, M. and Jerala, R. (2018). DNA-Free genome editing of *Brassica oleracea* and *B. rapa* protoplasts using CRISPR-Cas9 ribonucleoprotein complexes. *Frontiers in Plant Science* 9, 1594, doi:10.3389/fpls.2018.01594.

Murugan, K., Babu, K., Sundaresan, R., Rajan, R. and Sashital, D. G. (2017). The revolution continues: newly discovered systems expand the CRISPR-Cas toolkit. *Molecular Cell* 68(1), 15–25.

Nagle, M., Déjardin, A., Pilate, G. and Strauss, S. H. (2019). Opportunities for innovation in genetic transformation of forest trees. *Frontiers in Plant Science* 9, 1443, doi:10.3389/fpls.2018.01443

Nakajima, I., Sato, Y., Saito, T., Moriguchi, T. and Yamamoto, T. (2013). *Agrobacterium*-mediated genetic transformation using cotyledons in Japanese pear (*Pyrus pyrifolia*). *Breeding Science* 63(3), 275–283.

Niedz, R. P. (2006). Regeneration of somatic embryos from sweet orange (*C. sinensis*) protoplasts using semi-permeable membranes. *Plant Cell, Tissue and Organ Culture* 84(3), 353–357.

Niedz, R. P., McKendree, W. L. and Shatters, R. C. (2003). Electroporation of embryogenic protoplasts of sweet orange (Citrus sinensis (L.) Osbeck) and regeneration of transformed plants. *In Vitro Cellular and Developmental Biology–Plant* 39(6), 586–594.

Olhoft, P. M., Bernal, L. M., Grist, L. B., Hill, D. S., Mankin, S. L., Shen, Y., Kalogerakis, M., Wiley, H., Toren, E., Song, H., Hillebrand, H. and Jones, T. (2007). A novel *Agrobacterium rhizogenes*-mediated transformation method of soybean [*Glycine max* (L.) Merrill] using primary-node explants from seedlings. *In Vitro Cellular and Developmental Biology–Plant* 43(6), 536–549.

Ooms, G., Hooykaas, P. J., Moolenaar, G. and Schilperoort, R. A. (1981). Crown gall plant tumors of abnormal morphology, induced by *Agrobacterium tumefaciens* carrying mutated octopine Ti plasmids; analysis of T-DNA functions. *Gene* 14(1–2), 33–50.

Papadakis, A. K., Paschalidis, K. A. and Roubelakis-Angelakis, K. A. (2005). Biosynthesis profile and endogenous titers of polyamines differ in totipotent and recalcitrant plant protoplasts. *Physiologia Plantarum* 125(1), 10–20.

Papadakis, A. K., Siminis, C. I. and Roubelakis-Angelakis, K. A. (2001). Reduced activity of antioxidant machinery is correlated with suppression of totipotency in plant protoplasts. *Plant Physiology* 126(1), 434–444.

Park, J., Choi, S., Park, S., Yoon, J., Park, A. Y. and Choe, S. (2019). DNA-free genome editing via ribonucleoprotein (RNP) delivery of CRISPR/Cas in lettuce. In: Qi, Y. (Ed.) *Plant Genome Editing with CRISPR Systems: Methods and Protocols*. New York, NY: Springer, 337–354.

Parrott, W. A., Hoffman, L. M., Hildebrand, D. F., Williams, E. G. and Collins, G. B. (1989). Recovery of primary transformants of soybean. *Plant Cell Reports* 7(8), 615–617.

Paszkowski, J., Shillito, R. D., Saul, M., Mandák, V., Hohn, T., Hohn, B. and Potrykus, I. (1984). Direct gene transfer to plants. *The EMBO Journal* 3(12), 2717–2722.

Pati, P. K., Sharma, M. and Ahuja, P. S. (2005). Extra thin alginate film: an efficient technique for protoplast culture. *Protoplasma* 226(3–4), 217–221.

Paz, M. M., Martinez, J. C., Kalvig, A. B., Fonger, T. M. and Wang, K. (2006). Improved cotyledonary node method using an alternative explant derived from mature seed for efficient *Agrobacterium*-mediated soybean transformation. *Plant Cell Reports* 25(3), 206–213.

Pérez-Pascual, D., Jiménez-Guillen, D., Villanueva-Alonzo, H., Souza-Perera, R., Godoy-Hernández, G. and Zúñiga-Aguilar, J. J. (2018). Ectopic expression of the *Coffea canephora* SERK1 homolog-induced differential transcription of genes involved in auxin metabolism and in the developmental control of embryogenesis. *Physiologia Plantarum* 163(4), 530–551.

Pickar-Oliver, A., Black, J. B., Lewis, M. M., Mutchnick, K. J., Klann, T. S., Gilcrest, K. A., Sitton, M. J., Nelson, C. E., Barrera, A., Bartelt, L. C., Reddy, T. E., Beisel, C. L., Barrangou, R. and Gersbach, C. A. (2019). Targeted transcriptional modulation with type I CRISPR-Cas systems in human cells. *Nature Biotechnology* 37(12), 1493-1501.

Pigeaire, A., Abernethy, D., Smith, P. M., Simpson, K., Fletcher, N., Lu, C. Y., Atkins, C. A. and Cornish, E. (1997). Transformation of a grain legume (*Lupinus angustifolius* L.) via *Agrobacterium tumefaciens*-mediated gene transfer to shoot apices. *Molecular Breeding* 3(5), 341-349.

Popelka, J. C., Gollasch, S., Moore, A., Molvig, L. and Higgins, T. J. (2006). Genetic transformation of cowpea (Vigna unguiculata L.) and stable transmission of the transgenes to progeny. *Plant Cell Reports* 25(4), 304-312.

Power, J. B. and Cocking, E. C. (1970). Isolation of leaf protoplasts: macromolecule uptake and growth substance response. *Journal of Experimental Botany* 21(1), 64-70.

Puchta, H., Dujon, B. and Hohn, B. (1996). Two different but related mechanisms are used in plants for the repair of genomic double-strand breaks by homologous recombination. *Proceedings of the National Academy of Sciences* 93(10), 5055-5560.

Qi, W., Tinnenbroek-Capel, I. E., Schaart, J. G., Huang, B., Cheng, J., Visser, R. G., Van Loo, E. N. and Krens, F. A. (2014). Regeneration and transformation of *Crambe abyssinica*. *BMC plant biology 14*(1), 235.

Que, Q., Elumalai, S., Li, X., Zhong, H., Nalapalli, S., Schweiner, M., Fei, X., Nuccio, M., Kelliher, T., Gu, W., Chen, Z. and Chilton, M. D. (2014). Maize transformation technology development for commercial event generation. *Frontiers in plant science 5*, 379.

Que, Q., Chilton, M. D. M., Elumalai, S., Zhong, H., Dong, S. and Shi, L. (2019). Repurposing macromolecule delivery tools for plant genetic modification in the era of precision genome engineering. In: Kumar, S., Barone, P. and Smith, M. (Eds) *Transgenic Plants: Methods and Protocols*. New York, NY: Springer, 3-18.

Rao, K. V., Rathore, K. S. and Hodges, T. K. (1995). Physical, chemical and physiological parameters for electroporation-mediated gene delivery into rice protoplasts. *Transgenic Research* 4(6), 361-368.

Rao, K. S., Sreevathsa, R., Sharma, P. D., Keshamma, E. and Kumar, M. U. (2008). In planta transformation of pigeon pea: a method to overcome recalcitrancy of the crop to regeneration in vitro. *Physiology and Molecular Biology of Plants* 14(4), 321 -328.

Rath, D., Amlinger, L., Hoekzema, M., Devulapally, P. R. and Lundgren, M. (2015). Efficient programmable gene silencing by Cascade. *Nucleic Acids Research* 43(1), 237-246.

Raveendar, S. and Ignacimuthu, S. (2010). Improved Agrobacterium Mediated Transformation in Cowpea *Vigna unguiculata* L. Walp. *Asian Journal of Plant Sciences* 9(5), 256.

Rhodes, C. A., Lowe, K. S. and Ruby, K. L. (1988). Plant regeneration from protoplasts isolated from embryogenic maize cell cultures. *Nature Biotechnology* 6(1), 56-60.

Rouan, D., Montané, M. -H., Alibert, G. and Teissiè, J. (1991). Relationship between protoplast size and critical field strength in protoplast electropulsing and application to reliable DNA uptake in *Brassica*. *Plant Cell Reports* 10(3), 139-143.

Sanford, J. C. (2000). The development of the biolistic process. *In Vitro Cellular and Developmental Biology–Plant* 36(5), 303-308.

Sardesai, N. and Subramanyam, S. (2018). Agrobacterium: a genome-editing tool-delivery system. In: Gelvin, S. B. (Ed.) *Agrobacterium Biology*. Cham: Springer, 463–488.

Saunders, J. A., Lin, C. H., Hou, B. H., Cheng, J., Tsengwa, N., Lin, J. J., Smith, C. R., McIntosh, M. S. and Van Wert, S. (1995). Rapid optimization of electroporation conditions for plant cells, protoplasts, and pollen. *Molecular Biotechnology* 3(3), 181–190.

Seaberg, B. L., Hsieh, Y. C., Scholthof, K. B. G. and Scholthof, H. B. (2012). Host impact on the stability of a plant virus gene vector as measured by a new fluorescent local lesion passaging assay. *Journal of Virological Methods* 179(2), 289–294.

Shepard, J. F. and Totten, R. E. (1977). Mesophyll cell protoplasts of potato: isolation, proliferation, and plant regeneration. *Plant Physiology* 60(2), 313–316.

Shi, J., Gao, H., Wang, H., Lafitte, H. R., Archibald, R. L., Yang, M., Hakimi, S. M., Mo, H. and Habben, J. E. (2017). ARGOS8 variants generated by CRISPR-Cas9 improve maize grain yield under field drought stress conditions. *Plant Biotechnology Journal* 15(2), 207–216.

Shi, J., Habben, J. E., Archibald, R. L., Drummond, B. J., Chamberlin, M. A., Williams, R. W., Lafitte, H. R. and Weers, B. P. (2015). Overexpression of ARGOS genes modifies plant sensitivity to ethylene, leading to improved drought tolerance in both Arabidopsis and maize. *Plant Physiology* 169(1), 266–282.

Shillito, R. D., Saul, M. W., Paszkowski, J., Müller, M. and Potrykus, I. (1985). High efficiency direct gene transfer to plants. *Nature Biotechnology* 3(12), 1099–1103.

Shires, M. E., Florez, S. L., Lai, T. S. and Curtis, W. R. (2017). Inducible somatic embryogenesis in *Theobroma cacao* achieved using the DEX-activatable transcription factor-glucocorticoid receptor fusion. *Biotechnology Letters* 39(11), 1747–1755.

Shukla, V. K., Doyon, Y., Miller, J. C., DeKelver, R. C., Moehle, E. A., Worden, S. E., Mitchell, J. C., Arnold, N. L., Gopalan, S., Meng, X., Choi, V. M., Rock, J. M., Wu, Y., Katibah, G. E., Zhifang, G., McCaskill, D., Simpson, M. A., Blakeslee, B., Greenwalt, S. A., Butler, H. J., Hinkley, S. J., Zhang, L., Rebar, E. J., Gregory, P. D. and Urnov, F. D. (2009). Precise genome modification in the crop species Zea mays using zinc-finger nucleases. *Nature* 459(7245), 437–441.

Sidorov, V., Gilbertson, L., Addae, P. and Duncan, D. (2006). Agrobacterium-mediated transformation of seedling-derived maize callus. *Plant Cell Reports* 25(4), 320–328.

Sitther, V., Tabatabai, B., Enitan, O. and Dhekney, S. (2018). Agrobacterium-mediated transformation of *Camelina sativa* for production of transgenic plants. *Journal of Biological Methods* 5(1).

Song, G. Q., Prieto, H. and Orbovic, V. (2019). *Agrobacterium*-mediated transformation of tree fruit crops: methods, progress, and challenges. *Frontiers in Plant Science* 10, 226, doi:10.3389/fpls.2019.00226.

Srivastava, V. and Ow, D. W. (2002). Biolistic mediated site-specific integration in rice. *Molecular Breeding* 8(4), 345–349.

Stone, S. L., Kwong, L. W., Yee, K. M., Pelletier, J., Lepiniec, L., Fischer, R. L., Goldberg, R. B. and Harada, J. J. (2001). LEAFY COTYLEDON2 encodes a B3 domain transcription factor that induces embryo development. *Proceedings of the National Academy of Sciences* 98(20), 11806–11811.

Subburaj, S., Chung, S. J., Lee, C., Ryu, S. M., Kim, D. H., Kim, J. S., Bae, S. and Lee, G. J. (2016). Site-directed mutagenesis in Petunia× hybrida protoplast system using direct delivery of purified recombinant Cas9 ribonucleoproteins. *Plant Cell Reports* 35(7), 1535–1544.

Sun, Y., Zhang, X., Wu, C., He, Y., Ma, Y., Hou, H., Guo, X., Du, W., Zhao, Y. and Xia, L. (2016). Engineering herbicide-resistant rice plants through CRISPR/Cas9-mediated homologous recombination of acetolactate synthase. *Molecular Plant* 9(4), 628–631.

Svitashev, S., Ananiev, E., Pawlowski, W. P. and Somers, D. A. (2000). Association of transgene integration sites with chromosome rearrangements in hexaploid oat. *Theoretical and Applied Genetics* 100(6), 872–880.

Svitashev, S., Schwartz, C., Lenderts, B., Young, J. K. and Cigan, A. M. (2016). Genome editing in maize directed by CRISPR-Cas9 ribonucleoprotein complexes. *Nature Communications* 7(1), 1–7.

Svitashev, S., Young, J. K., Schwartz, C., Gao, H., Falco, S. C. and Cigan, A. M. (2015). Targeted mutagenesis, precise gene editing, and site-specific gene insertion in maize using Cas9 and guide RNA. *Plant Physiology* 169(2), 931–945.

Svitashev, S. K., Pawlowski, W. P., Makarevitch, I., Plank, D. W. and Somers, D. A. (2002). Complex transgene locus structures implicate multiple mechanisms for plant transgene rearrangement. *The Plant Journal* 32(4), 433–445.

Takebe, I., Labib, G. and Melchers, G. (1971). Regeneration of whole plants from isolated mesophyll protoplasts of tobacco. *Naturwissenschaften* 58(6), 318–320.

Takebe, I., Otsuki, Y. and Aoki, S. (1968). Isolation of tobacco mesophyll cells in intact and active state. *Plant and Cell Physiology* 9(1), 115–124.

Thakare, D., Tang, W., Hill, K. and Perry, S. E. (2008). The MADS-domain transcriptional regulator AGAMOUS-LIKE15 promotes somatic embryo development in Arabidopsis and soybean. *Plant Physiology* 146(4), 1663–1672.

Toda, E., Koiso, N., Takebayashi, A., Ichikawa, M., Kiba, T., Osakabe, K., Osakabe, Y., Sakakibara, H., Kato, N. and Okamoto, T. (2019). An efficient DNA-and selectable-marker-free genome-editing system using zygotes in rice. *Nature Plants* 5(4), 363.

Torney, F., Trewyn, B. G., Lin, V. S. -Y. and Wang, K. (2007). Mesoporous silica nanoparticles deliver DNA and chemicals into plants. *Nature Nanotechnology* 2(5), 295–300.

Uchimiya, H. and Murashige, T. (1974). Evaluation of parameters in the isolation of viable protoplasts from cultured tobacco cells. *Plant Physiology* 54(6), 936–944.

Von Klercker, J. (1892). A method for isolating living protoplasts. *Ofvers Vetensk. Akad. Forh. Stockholm* 49, 463.

Wan, Y., Widholm, J. M. and Lemaux, P. G. (1995). Type I callus as a bombardment target for generating fertile transgenic maize (*Zea mays* L.). *Planta* 196(1), 7–14.

Wang, J. W., Grandio, E. G., Newkirk, G. M., Demirer, G. S., Butrus, S., Giraldo, J. P. and Landry, M. P. (2019). Nanoparticle-mediated genetic engineering of plants. *Molecular Plant* 12(8), 1037–1040.

Wang, M., Lu, Y., Botella, J. R., Mao, Y., Hua, K. and Zhu, J. (2017). Gene targeting by homology-directed repair in rice using a geminivirus-based CRISPR/Cas9 system. *Molecular Plant* 10(7), 1007–1010.

Wang, X., Niu, Q. W., Teng, C., Li, C., Mu, J., Chua, N. H. and Zuo, J. (2009). Overexpression of PGA37/MYB118 and MYB115 promotes vegetative-to-embryonic transition in Arabidopsis. *Cell Research* 19(2), 224–235.

Woo, J. W., Kim, J., Kwon, S. I., Corvalán, C., Cho, S. W., Kim, H., Kim, S., Kim, S., Choe, S. and Kim, J. (2015). DNA-free genome editing in plants with preassembled CRISPR-Cas9 ribonucleoproteins. *Nature Biotechnology* 33(11), 1162–1164.

Wright, D. A., Townsend, J. A., Winfrey Jr., R. J., Irwin, P. A., Rajagopal, J., Lonosky, P. M., Hall, B. D., Jondle, M. D. and Voytas, D. F. (2005). High-frequency homologous recombination in plants mediated by zinc-finger nucleases. *The Plant Journal* 44(4), 693–705.

Xie, K. and Yang, Y. (2013). RNA-guided genome editing in plants using a CRISPR-Cas system. *Molecular Plant* 6(6), 1975–1983.

Xu, R.-F., Li, H., Qin, R.-Y., Li, J., Qiu, C., Yang, Y., Ma, H., Li, L., Wei, P. and Yang, J. (2015). Generation of inheritable and "transgene clean" targeted genome-modified rice in later generations using the CRISPR/Cas9 system. *Scientific Reports* 5(1), 11491, doi:10.1038/srep11491.

Xu, Z.-H. and Xue, H.-W. (1999). Plant regeneration from cultured protoplasts. In: Soh, W.-Y. and Bhojwani, S. S. (Eds) *Morphogenesis in Plant Tissue Cultures*. Dordrecht, Netherlands: Springer, 37–70.

Yan, B., Reddy, M. S., Collins, G. B. and Dinkins, R. D. (2000). *Agrobacterium tumefaciens*-mediated transformation of soybean [*Glycine max* (L.) Merrill.] using immature zygotic cotyledon explants. *Plant Cell Reports* 19(11), 1090–1097.

Yang, Z., Li, C., Wang, Y., Zhang, C., Wu, Z., Zhang, X., Liu, C. and Li, F. (2014). GhAGL15s, preferentially expressed during somatic embryogenesis, promote embryogenic callus formation in cotton (Gossypiumhirsutum L.). *Molecular Genetics and Genomics: MGG* 289(5), 873–883.

Yin, K., Han, T., Liu, G., Chen, T., Wang, Y., Yu, A. Y. and Liu, Y. (2015). A geminivirus-based guide RNA delivery system for CRISPR/Cas9 mediated plant genome editing. *Scientific Reports* 5, 14926.

Zhang, F., Maeder, M. L., Unger-Wallace, E., Hoshaw, J. P., Reyon, D., Christian, M., Li, X., Pierick, C. J., Dobbs, D., Peterson, T., Joung, J. K. and Voytas, D. F. (2010). High frequency targeted mutagenesis in *Arabidopsis thaliana* using zinc finger nucleases. *Proceedings of the National Academy of Sciences* 107(26), 12028–12033.

Zhang, Q., Zhang, Y., Lu, M. H., Chai, Y., Jiang, Y., Zhou, Y., Wang, X. and Chen, Q. (2019). A novel ternary vector system united with morphogenic genes enhances CRISPR/Cas delivery in maize. *Plant Physiology* 181(4), 1441–1448.

Zhang, Y., Liang, Z., Zong, Y., Wang, Y., Liu, J., Chen, K., Qiu, J. L. and Gao, C. (2016). Efficient and transgene-free genome editing in wheat through transient expression of CRISPR/Cas9 DNA or RNA. *Nature Communications* 7, 12617.

Zhao, T., Wang, Z., Su, L., Sun, X., Cheng, J., Zhang, L., Karunga, S. K., Han, Y., Li, S. and Xin, H. (2017). An efficient method for transgenic callus induction from *Vitis amurensis* petiole. *PLoS ONE* 12(6), e0179730.

Zuo, J., Niu, Q. W., Frugis, G. and Chua, N. H. (2002). The WUSCHEL gene promotes vegetative-to-embryonic transition in Arabidopsis. *The Plant Journal* 30(3), 349–359.

Chapter 9

Advances in screening plants for edits and off-targets

Chun Wang and Kejian Wang, China National Rice Research Institute, China

1 Introduction

With the widespread use of gene-editing technology, especially the clustered regularly interspaced short palindromic repeats (CRISPR)/CRISPR-associated proteins (Cas) system and base editors, an increasing number of mutations need to be genotyped. In plants, the genotypes produced after gene editing include heterozygous, biallelic, homozygous, chimeric (with ≥3 edited allelic events within a single individual) and wild-type. It has become a challenge to identify genome-edited plants at the early stages from the massive number of plants produced from the transformation process. Several molecular methods have been developed in recent years to identify genome-edited plants for different purposes. In plant research, this includes evaluating the editing efficiency of new target sites or genome-editing tools in protoplasts or calli and obtaining homozygous or biallelic mutants for further research, and different identification strategies can be selected for different goals.

http://dx.doi.org/10.19103/AS.2020.0082.14

In this chapter, we introduce some of the most commonly used, simple and efficient methods for identifying genome-edited mutations. All of these methods require polymerase chain reaction (PCR) assays to amplify the sequence around the target sites as the first step, followed by enzyme digestion, signal acquisition or sequencing. The reader can choose the appropriate method for identification according to their experimental purposes. For example, four outcomes can be obtained when targeting diploid plants (like rice and maize): heterozygous mutants, biallelic mutants, homozygous mutants and mosaic mutants. The methods of PCR/restriction enzyme (RE), mismatch cleavage assay, PCR/ribonucleoprotein (RNP), high resolution melting (HRM) analysis, annealing at critical temperature (ACT-PCR), Sanger sequencing chromatogram analysis, next-generation sequencing (NGS) (like high-throughput tracking of mutations (Hi-TOM)) can be used to identify biallelic and homozygous mutants in these diploid outcomes. When targeting polyploid plants, such as wheat, the results are exceedingly complex. The methods of mismatch cleavage assay, HRM analysis and Sanger sequencing chromatogram analysis are unsuitable for identifying these polyploid plants.

Each section compares the advantages and disadvantages of each method as well as the scope of application.

2 PCR/RE assay

The PCR/RE assay is relatively simple and sensitive for screening genome-edited mutants. A prerequisite of this strategy is that the target sites (the sites of double-stranded breaks, DSBs) must include an RE site. If a mutation has occurred, the RE site would be disrupted, and the mutant amplicons would therefore be resistant to RE digestion, showing as uncleaved bands following electrophoresis (Shan et al., 2013). There are three outcomes: (1) the PCR band is completely cleaved, indicating that the plant is not mutated; (2) the band is partially cleaved, indicating that the plant is heterozygous or a chimaera; and (3) the band is uncleaved, indicating that the plant is homozygous or biallelic. Compared with other detection methods, PCR/RE has the following advantages: (1) it can detect all types of mutations, including single-nucleotide polymorphisms (SNPs), insertions and deletions; (2) it is highly sensitive for the detection of mutation efficiency and (3) the process is simple, and has a low cost. These advantages make the PCR/RE become one of the most widely used method in plant genome editing and detection, especially in polyploid plants. The limitation of this method is that it is constrained by the availability of RE sites near the target sequences.

3 Mismatch cleavage assay

In cases where the PCR/RE assay cannot be applied to a target locus because there is no appropriate RE site, the mismatch cleavage assay can be used.

The mutation is detected using enzymes that cleave DNA heteroduplexes at mismatch sites formed by single or multiple nucleotides, thereby overcoming the limitation of PCR/RE (Shan et al., 2014). Available mismatch-specific endonucleases include T7 endonuclease 1 (T7E1) and Surveyor nuclease (Vouillot et al., 2015). T7E1 outperforms the Surveyor nuclease in terms of sensitivity with deletion substrates, whereas Surveyor is better for detecting single nucleotide changes. So T7E1 is the preferred enzyme to scan mutations triggered by engineered nucleases, and Surveyor is the preferred enzyme to scan mutations triggered by base editor. The PCR product is designed to amplify 300–600 bp surrounding the genomic target site. To avoid missing detection of homozygous mutants, PCR products of WT plant and genome-edited plant were mixed in equal quantities, and the amplicons are denatured and renatured to form heteroduplexes. Then, the reaction products are digested with mismatch-specific endonuclease and analysed by agarose gel electrophoresis. The annealed heteroduplexes are cleaved by mismatch-specific endonucleases, whereas the WT homoduplexes remain intact. Moreover, the indel mutation frequency can be estimated from the band intensity. The mismatch cleavage assay is simple and cost-effective and can be performed using standard laboratory techniques and equipment. Furthermore, it can detect polymorphisms, including single base-pair mismatches, insertions and deletions, at low frequencies. Thus, this method is widely employed in the context of mutagenesis projects using engineered nucleases. However, if the mutation rate is less than 5%, it cannot be detected even with T7E1.

4 PCR/RNP

Because the CRISPR/Cas system also recognises and cleaves the target sequence in vitro, the ribonucleoprotein complexes (RNP) of Cas protein and guide RNA (gRNA) can be used to detect mutation at the selected targets. The CRISPR nuclease RNPs can digest PCR products complementary to the guide RNA sequences (WT) but fail to digest PCR products with mutated sequences (mutants). Liang et al. (2018) described a PCR/RNP method based on the RNA-guided endonucleases SpCas9 or FnCpf1 for identifying indels generated by genome editing in both polyploid and diploid plants, without the requirement for a RE site. Briefly, proteins (SpCas9, AsCpf1 or FnCpf1) were expressed in *Escherichia coli* and purified in vitro, and the corresponding target gRNAs (sgRNA for SpCas9, crRNA for FnCpf1 and AsCpf1) were synthesised using the HiScribe T7 In Vitro Transcription Kit (New England Biolabs). Then, the RNP was prepared by incubating the protein together with the gRNA. The purified CRISPR RNP complexes were applied to cleave the PCR products for genome-edited mutation detection in hexaploid wheat and diploid rice, and heterozygous or homozygous/biallelic mutations were successfully identified. Compared with the mismatch cleavage assay, mutation detection with RNPs is unaffected by background noise (SNPs) around the target site using the PCR/RNP method, which has great significance for polyploid plants, such as wheat. The PCR/RNP method is suitable for applications in polyploid plants, especially when a mutation target needs to be detected from a large number of samples. However, this method is relatively expensive because the gRNA synthesis is a costly process.

5 High-resolution melting (HRM) analysis

High-resolution melting (HRM) curve analysis is a fluorescence-based technique that measures the amount of double-stranded DNA at different temperatures, thereby revealing the melting temperature (T_m) of a particular PCR product. While a homoduplex product generated from a homozygous DNA sample will have a particular T_m, a heteroduplex product generated from a heterozygous individual will have an additional T_m, which is generally a much lower T_m signature. Thus, the mutation could be identified by the T_m signals. HRM analysis is a powerful technique used in molecular biology to detect mutations and polymorphisms differences in double-stranded DNA samples (Montgomery et al., 2007).

Because amplicon sizes >150 bp decrease sensitivity in HRM analysis, the amplicon lengths should be chosen between 80 bp and 95 bp, and GC contents of the amplicons should be ranged from 32% to 47%. The HRM protocol requires no additional manipulations (no digestion steps or electrophoresis) after the PCR assay (Thomas et al., 2014). Instead, a 5-min HRM scan and *in silico* analysis is performed automatically by the dedicated HRM instrument (e.g. LightScanner). This melting data can be analysed to detect sequence variations, such as SNPs, small insertions and deletions. However, HRM analysis did not distinguish WT and homozygous 1-bp insertion reliably, and this shortcoming can be overcome by performing another PCR reaction with mixing genomic DNA of WT and potential homozygous mutant at 1:1 ratio. HRM is highly efficient for mutation detection across multiple genome-editing platforms, making it particularly suitable for mutation detection in a large number of samples. The disadvantage of this method is that it requires expensive equipment. Denbow et al. (2017) reported that the melt curve analysis also could be obtained by using a regular RT-PCR machine, even though the sensitivity is decreased compared to a HRM instrument and the experiment takes longer. This result showed that detection of mutation using the HRM method might not require a dedicated hardware.

6 ACT-PCR

Hua et al. (2017) described a method annealing at critical temperature PCR (ACT-PCR) for identifying mutants, which is a cost-effective and sensitive screening technique based on conventional PCR. A conventional PCR cycle comprises three steps: denaturation, annealing and extension. The annealing temperature is critical for primer-template pairing for a successful PCR assay. An optimal temperature suppresses mismatched annealing, thereby reducing the generation of non-specific products. Based on this theory, ACT-PCR is developed by designing specific primers for the target sites and determining the critical annealing temperature by performing a preliminary gradient PCR (see Fig. 1). The forward primer, named the DSB site-specific (DS) primer, flanks the DSB site, with its 3′ end containing a 4-bp overhang relative to the DSB site to ensure specificity and sensitivity for WT gene binding and PCR amplification. The reverse primer is located outside the DSB site and has a higher T_m value than that of the DS primer to ensure DNA template binding at the critical annealing temperature. After performing a preliminary gradient PCR assay with WT DNA as the template to determine the critical annealing temperature, conventional PCR can be performed at the critical annealing temperature. If a biallelic or homozygous mutant is present, the DS primer will not bind to the mutated sequence, and no amplicons will be obtained. ACT-PCR has been used to accurately distinguish CRISPR/Cas9-induced mutants from WT in both rice and zebrafish (Hua et al., 2017). Moreover, the method can be adapted to accurately determine the mutation frequency in cultured cells with real-time PCR.

In contrast to other methods for mutant identification, ACT-PCR is less restricted and does not need expensive equipment. Compared with the PCR/RE assay, ACT-PCR can be utilised for detection regardless of the available RE sites. Moreover, ACT-PCR is highly sensitive for identifying CRISPR/Cas9-induced mutants, even 1-bp indels. The main disadvantage of this method is that heterozygotes cannot be identified. The simplicity of ACT-PCR makes it particularly suitable for rapid, large-scale screening of CRISPR/Cas9-induced mutants in plants, especially for screening homozygous mutants in the second generation of transformed plants.

7 Analysis of Sanger sequencing chromatograms

Although the methods described above can identify mutants, none of them can obtain the specific mutated sequences of the targeted alleles. Most of the time, it is necessary to determine the mutated sequences to ascertain the changes in the protein sequences following gene editing. Sanger sequencing (direct sequencing) is a commonly used method to know the sequence of mutations. In the conventional procedure, the targeted genomic region is amplified by PCR, the individual DNA molecules are cloned into a bacterial vector and multiple clones are sequenced for each targeted editing site. This approach can distinguish biallelic, heterozygous and chimeric mutations in the

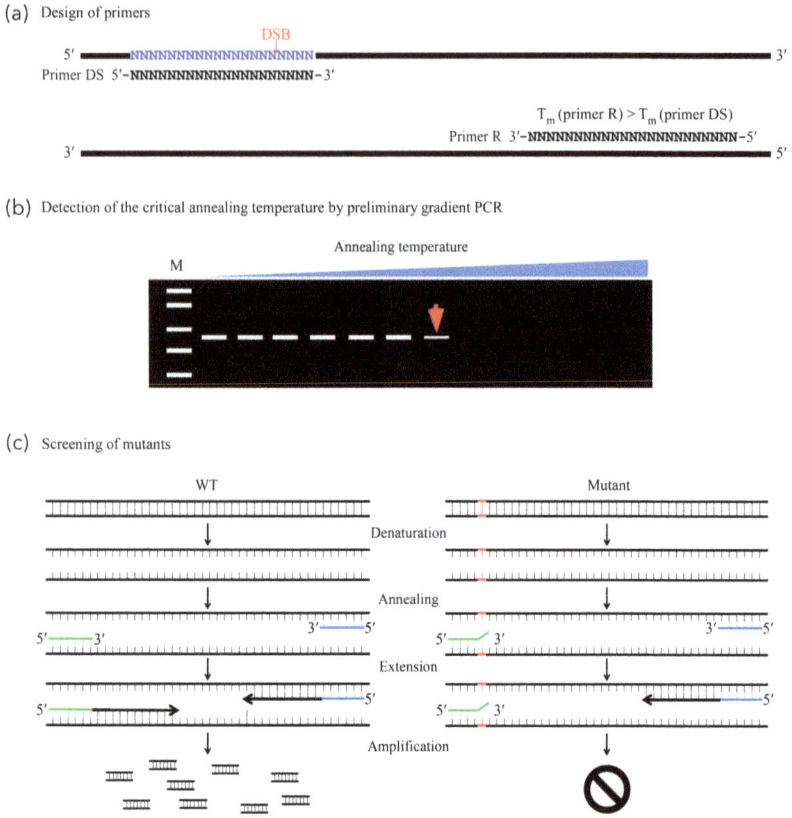

Figure 1 Schematic illustration of ACT-PCR. (a) Design of primers. The forward primer, double-strand break site-specific (DS) primer, is designed with a 4-bp overhang relative to the DSB site. The reverse primer, primer R, is designed with a higher melting temperature (Tm) value than the DS primer. (b) Detection of the critical annealing temperature by preliminary gradient PCR. The critical annealing temperature (red arrow) is defined as the highest temperature at which PCR products could be detected. (c) Screening of mutants. The primer DS, primer R and mutation site are labelled in green, blue and red, respectively. At the critical annealing temperature, amplicons could be obtained from the wild-type (WT) gene but not from the CRISPR/Cas9-induced mutant because of the introduction of mismatches at the target site. (Reproduced from [Hua et al., 2017] with permission from the *Journal of Genetics and Genomics*).

first transgenic generation and can determine the frequency of the desired editing event in the pool of protoplasts by an increase in the number of clones (Shan et al., 2013). However, this approach is labour intensive, time-consuming and relatively costly.

The PCR products of the selected targets can also be sequenced directly, providing an easy and quick method of identification. However, direct sequencing of PCR products containing heterozygous and biallelic mutations results in superimposed peaks in the sequencing chromatograms, which is hard to analyse. Brinkman et al. (2014) and Ma et al. (2015) developed the methods of tracking of indels by decomposition (TIDE) and degenerate sequence decoding (DSD), respectively, to decode superimposed sequencing chromatograms of PCR products.

The TIDE method requires two parallel PCR reactions (one uses the WT as the template) followed by a pair of standard Sanger sequencing analyses, and the raw sequencing traces are decoded using the TIDE web tool (http://tide.nki.nl). The algorithm accurately reconstructs the spectrum of indels from the sequence traces, and the web tool reports the identity of the detected indels and their frequencies. However, TIDE can only detect overall indel frequencies and no nucleotide substitutions or specifically designed indels. TIDER (Tracking of Insertions, Deletions and Recombination events), a redesigned version of TIDE, were developed to overcome the limitation of TIDE (Brinkman et al., 2018). TIDER can estimate the incorporation frequency of any type of template-directed mutations (including point mutations) and separate it from the background spectrum of additional indels. TIDER is suitable for analysing a pool of protoplast with WT DNA, indels and the designed mutation, much more quantitative and informative than the mismatch-specific endonucleases assays. It is important to note that the accuracy of TIDER relies on the quality of the PCR products and the sequence reads, because the TIDER algorithm analyses individual peak heights in the input sequence traces.

The DSD method decodes superimposed sequencing chromatograms as follows. First, starting from the first overlapping-peak position on the chromatogram, a short degenerate sequence is manually generated that is adjacent to the anchor sequence, which sits upstream of the first overlapping-peak. Then, the degenerate sequence is queried twice against the intact reference sequence using a sequence analysis programme to detect the matched sequence. Finally, the anchor sequence is linked with the query-matched sequences to generate the allele sequences or, if detecting only one matching hit, the second allele sequence is generated by subtracting the allele 1 nucleotides from the degenerate bases. Liu et al. (2015) also developed a web-based tool (http://skl.scau.edu.cn/dsdecode/), DSDecode, to automatically decode superimposed and regular sequencing chromatograms derived from PCR amplicons containing various types of mutations.

These user-friendly web tools can be used to genotype various types of uniform mutations and to significantly accelerate the analysis of mutant sequences. This method accounts for a high proportion of the current research on plant genome editing. However, it is unsuitable for detecting polyploids that present as complex superimposed sequencing chromatograms, and is expensive to carry out large-scale screening.

8 Next generation sequencing (NGS)

Next generation sequencing (NGS), also known as high-throughput sequencing, is the catch-all term used to describe a DNA sequencing technology that has revolutionised genomic research. The critical difference between Sanger sequencing and NGS is the sequencing volume. While the Sanger method only sequences a single DNA fragment at a time, NGS involves massively parallel sequencing and is capable of simultaneously sequencing thousands to millions of fragments per run. Although NGS is at a disadvantage in terms of sequencing length, normally it can only measure 150 bp, which is far less than the length of Sanger sequencing, NGS of PCR amplicon around the induced break site still become a powerful tool to analyse the mutation spectrum. With NGS platforms becoming available for most research institutions, NGS is widely used for measuring the frequencies of mutations and off-target changes induced by gene-editing tools. Amplicon sequencing is the most accurate method to detect genome editing efficiency, especially in the analysis of experimental results in protoplasts. Except for large fragment deletions, mutation types of SNPs, small insertions and deletions at specific location can be accurately detected.

However, the analysis of raw NGS data is difficult for many researchers. Thus, several web-based tools have been developed to analyse the NGS data containing genome editing-induced mutations by pooling samples. These can achieve a reliable sensitivity of 0.01%, such as CRISPR-GA (Guell et al., 2014), AGEseq (Xue and Tsai, 2015), Cas-Analyzer (Park et al., 2017) and CRISPResso (Pinello et al., 2016). These software packages aid researchers to analyse NGS data to strictly assess the quality of a genome-editing experiment by evaluating mutation frequencies and types. If not added barcode and adaptor sequences to the PCR product during the amplification process, the disadvantage of these methods is that they can only analyse one sample at a time, which is inefficient and data wasteful.

Liu et al. (2019) established a simplified and streamlined workflow based on NGS to analyse multiple samples simultaneously. This involved the construction of an NGS library using fixed-bridge sequences and fixed-barcoding primers, and the development an online tool, Hi-TOM (high-throughput tracking of mutations, available at http://www.hi-tom.net/hi-tom/) to track the mutations, with a precise percentage, for multiple samples and multiple target sites (see Fig. 2). This platform can facilitate the application of NGS sequencing in large samples where mutation identification is required, particularly for laboratories that do not have researchers with NGS or bioinformatics skills.

The library construction for Hi-TOM includes two rounds of PCR. In the first round, the initial PCR primers include a target-specific sequence with common bridging sequences (5′-ggagtgagtacggtgtgc-3′ and 5′-gagttggatgctggatgg-3′) added at the 5′ end. The forward or reverse target-specific primer is designed

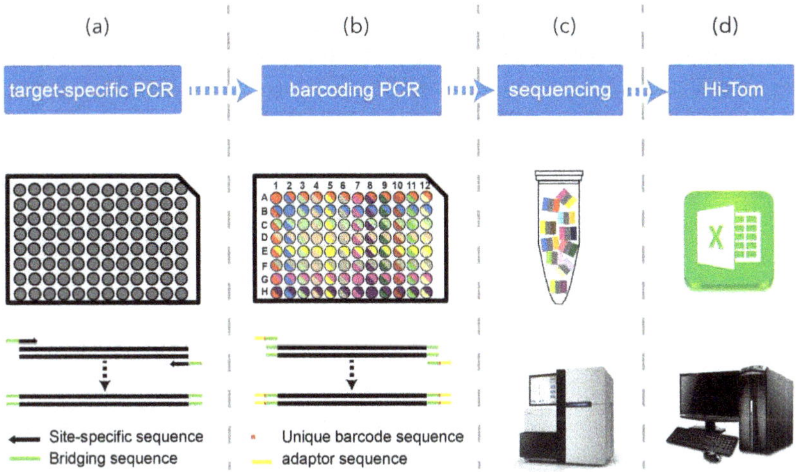

Figure 2 Schematic illustration of the workflow of Hi-TOM. (a) The samples are amplified using site-specific primers (target-specific PCR). (b) The products of the first-round PCR are used as templates for the second-round PCR (barcoding PCR) in the 96-hole plate kit. By barcoding PCR, the products of each sample are barcoded. (c) All products of the second-round PCR are pooled in equimolar amounts in a single tube and sent for NGS. (d) Hi-TOM analyses the data sample-by-sample and exports the results in Excel format. (Reproduced from [Liu et al., 2019] with permission from the *Science China Life Sciences*).

within 10–100 nucleotides of the targeted sites to ensure that the mutation can be detected by NGS. The samples are amplified using the target-specific primers by PCR reaction. Then, the first-round PCR products are labelled with fixed barcodes and bridging sequences by fixed primers during the second-round PCR. These fixed primers contain 12 forward and 8 reverse primers with four nucleic acid bases (barcodes) that contain a difference of at least two nucleotides with each other. Thus, it is possible to create a uniquely barcoded amplicon for up to 96 samples. Upon completion of the second round of PCR, all PCR products can be mixed as a single sample and sequenced using an Illumina HiSeq platform. In most cases, 1 Gb sequencing data, the size of which is approximately 100 Mb after gzip file compression, is sufficient for mutation analysis of 96 samples about 30000 reads of every sample. A corresponding online Hi-TOM platform (http://www.hi-tom.net/hi-tom/) has been developed for high-throughput mutation sequence decoding. The users need only to upload the raw gzip-compressed data from NGS sequencing and the reference sequence of the targets. Then, the Hi-TOM platform automatically decodes and aligns the sequences based on the barcodes and the reference sequence and integrates the results of 96 samples into a Microsoft Excel spreadsheet. The mutation types and positions of each sample are listed in detail in the Excel spreadsheet, including the number of reads, ratio, mutation type, mutated bases and DNA sequence.

Using the Hi-TOM strategy, Liu et al. (2019) analysed the samples from rice, hexaploid wheat and human cells. The results revealed that the Hi-TOM tool had a high level of reliability and sensitivity in tracking various mutations, especially the complex chimeric mutations that are frequently induced by genome editing. Hi-TOM does not require the design of special barcode primers, cumbersome parameter configuration or additional data analysis. Thus, the streamlined NGS library construction and comprehensive result output make Hi-TOM particularly suitable for the high-throughput identification of all types of mutations induced by CRISPR/Cas systems. Of course, NGS is also expensive and requires at least several days at current stage. With the development of high-throughput sequencing technology, the speed of sequencing will increase, and the price will drop. Thus, Hi-TOM is expected to be the best choice to identify most genome-editing mutations.

9 Detection of off-target effects

Genome-editing technologies may cause non-specific and unintended genetic modifications due to incorrect recognition and the cleavage of predetermined DNA sequences with one to a few nucleotide mismatches, termed off-target effects. Off-target cleavage can produce unintended effects that can confound the analysis of a phenotype of interest. In particular, off-target effects can impose cellular toxicity or disrupt functional genomic regions. The avoidance of off-target effects is an important challenge for therapeutic purposes. However, in agricultural applications, off-target editing is not such a major concern. Because plant breeding is ultimately based on a trait or phenotype selection even with off-target mutation, and most off-target mutations can be segregated away by genetic recombination, the genomic background variations matter less. A basic reference of genome-wide sequence scan as well as some software and web-based tools have been developed to identify all protospacer-adjacent motif-restricted off-target sites for CRISPR/Cas systems *in silico*, such as Cas-OFFinder (Bae et al., 2014), Off-Spotter (Pliatsika and Rigoutsos, 2015) and FlashFry (McKenna and Shendure, 2018). Further, the CRISPRitz tool (Cancellieri et al., 2019) can presume off-target sites, including supporting genomic searching with DNA/RNA bulges and mismatches of arbitrary size, as specified by the user, as well as considering genetic variants (variant-aware). Researchers can use these software packages to predict off-target sites and then perform experiments in the same manner as used for the detection of target mutations to determine whether the off-target change actually occurs.

Needless to say, the most rigorous detection of off-target effects should be achieved through whole-genome sequencing (WGS). Tang et al. (2018) performed a WGS analysis of 34 plants edited by Cas9 and 15 plants edited by Cpf1 in T_0 and T_1 generations along with 20 diverse control plants in rice.

The results showed that most mutations in the edited plants were random and created by the tissue-culture process. Among 12 Cas9 sgRNAs and three Cpf1 CRISPR RNAs (crRNAs) assessed by WGS, only one Cas9 sgRNA resulted in off-target mutations in T_0 lines at the sites predicted by the computer programs. The comprehensive and rigorous analysis of WGS data across multiple sample types suggests that both Cas9 and Cpf1 nucleases are highly specific in generating targeted DNA modifications and that changes due to off-targeting can be avoided by designing gRNAs with high specificity.

Jin et al. (2019) performed WGS sequencing analysis of rice plants treated with a third-generation cytosine base editor (BE3), high-fidelity cytosine base editor (HF1-BE3) or adenine base editor (ABE). It was revealed that BE3 and HF1-BE3, but not ABE, induced substantial genome-wide off-target mutations, which were mostly C→T type single-nucleotide variants (SNVs) that appeared to be enriched in genic regions. Notably, the treatment of rice with BE3 or HF1-BE3 in the absence of sgRNA also increased the number of genome-wide SNVs.

10 Conclusion and future trends

Except these methods described in the previous sections, there are many other methods that have been developed to identify mutants, such as PAGE-based genotyping assay (Zhu et al., 2014) and fluorescent PCR capillary gel electrophoresis (Ramlee et al., 2015), but these methods still have the disadvantage that they are either labour-intensive or expensive. In some experiments, researchers generated mutants with a large chromosomal segment deletion by designing two nearby sgRNAs, so that the genotypes can be determined by the length of the PCR fragment simply (Zhou et al., 2014). However, due to the different cutting efficiency of sgRNA, not all two sgRNAs together will necessarily produce large fragment deletions.

Most of the time, multiple methods will be used to screen the mutants from genome-edited plants (see Table 1). The genotypes of the first-generation plants produced by gene editing are often heterozygous, biallelic and chimeric. In this case, it is recommended to use these methods of PCR/RE, mismatch cleavage assay, PCR/RNP or HRM for preliminary screening to exclude unedited plants. Then Sanger sequencing could be performed to ensure the mutated sequences to obtain the desired mutants. Using such a strategy, researchers can not only get mutants accurately, but also save time and money. When a large number of multiple mutants need to be identified at the next-generation gene-edited plants, ACT-PCR and Hi-TOM should be the good choice. In summary, these methods mentioned above could be flexibly selected according to laboratory conditions and experimental purposes for mutant identification rapidly and accurately.

Table 1 Overview of methods for genome-edited mutations detection

Methods	Determination of mutation type	Accuracy	Time	Labour	Cost	Throughput	Limitations	Suitable for application
PCR/RE	No	High	3-4 h	Moderate	Low	Moderate	Availability of restriction site	Protoplast mixture, T0 and T1 plants
Mismatch cleavage assay	No	Moderate	3-4 h	Moderate	Moderate	Moderate	Misses substitutions	Protoplast mixture, T0 plants
PCR/RNP	No	Moderate	3-4 h	High	High	Moderate	Labour intensive	Protoplast mixture, T0 and T1 plants
HRM	No	Moderate	2-3 h	Low	Moderate	High	Misses large indels	T0 plants
ACT-PCR	No	Moderate	2-3 h	Low	Low	High	Unable to detect heterozygotes	T1 plants
Sanger sequencing	Yes	High	1 day	Moderate	Moderate	Low	Unable to detect polyploid plants	T0 plants
NGS	Yes	High	>2 days	Low	Low~high[a]	High	Misses large indels, too few samples will be expensive	Protoplast mixture, T0 and T1 plants

[a] The price is inversely proportional to the number of samples.

In the long term, the NGS technology will be the mainstream mutation detection method in the future because of outstanding advantages of high accuracy, high throughput and high sensitivity. If there are enough samples to be tested, the price of detection through high-throughput sequencing will decrease and the detection speed will increase. Based on the Hi-TOM platform, our laboratory provides services of high-throughput sequencing and data analysis, which can do $1.75 per sample and provide mutation detection results within 7 days. Due to the advantages of convenience, accuracy and low cost, we believe that Hi-TOM will become the most widely used detection method in the future, applicable to different species and materials.

11 Where to look for further information

There are some current reviews that summarize the methods for detection of mutations generated by genome editing, such as:

- Detection of on-target and off-target mutations generated by CRISPR/Cas9 and other sequence-specific nucleases. (https://www.sciencedirect.com/science/article/pii/S0734975016301586).
- How to start your monocot CRISPR/Cas project: plasmid design, efficiency detection and offspring analysis. (https://thericejournal.springeropen.com/articles/10.1186/s12284-019-0354-2).
- The latest detection methods can be tracked by searching PubMed literature database (https://pubmed.ncbi.nlm.nih.gov/) through keywords such as 'genome editing, CRISPR, detection'.

12 References

Bae, S., Park, J. and Kim, J. S. (2014). Cas-OFFinder: a fast and versatile algorithm that searches for potential off-target sites of Cas9 RNA-guided endonucleases, *Bioinformatics* 30(10), 1473–5.

Brinkman, E. K., Chen, T., Amendola, M. and Van Steensel, B. (2014). Easy quantitative assessment of genome editing by sequence trace decomposition, *Nucleic Acids Res.* 42(22), e168.

Brinkman, E. K., Kousholt, A. N., Harmsen, T., Leemans, C., Chen, T., Jonkers, J. and Van Steensel, B. (2018). Easy quantification of template-directed CRISPR/Cas9 editing, *Nucleic Acids Res.* 46(10), e58.

Cancellieri, S., Canver, M. C., Bombieri, N., Giugno, R. and Pinello, L. (2019). CRISPRitz: rapid, high-throughput, and variant-aware in silico off-target site identification for CRISPR genome editing, *Bioinformatics* 36(7), 2001–8.

Denbow, C. J., Lapins, S., Dietz, N., Scherer, R., Nimchuk, Z. L. and Okumoto, S. (2017). Gateway-compatible CRISPR-Cas9 vectors and a rapid detection by high-resolution melting curve analysis, *Front. Plant Sci.* 8, 1171.

Guell, M., Yang, L. and Church, G. M. (2014). Genome editing assessment using CRISPR Genome Analyzer (CRISPR-GA), *Bioinformatics* 30(20), 2968–70.

Hua, Y., Wang, C., Huang, J. and Wang, K. (2017). A simple and efficient method for CRISPR/Cas9-induced mutant screening, *J. Genet. Genomics* 44(4), 207–13.

Jin, S., Zong, Y., Gao, Q., Zhu, Z., Wang, Y., Qin, P., Liang, C., Wang, D., Qiu, J. L., Zhang, F. and Gao, C. (2019). Cytosine, but not adenine, base editors induce genome-wide off-target mutations in rice, *Science* 364(6437), 292–5.

Liang, Z., Chen, K., Yan, Y., Zhang, Y. and Gao, C. (2018). Genotyping genome-edited mutations in plants using CRISPR ribonucleoprotein complexes, *Plant Biotechnol. J.* 16(12), 2053–62.

Liu, Q., Wang, C., Jiao, X., Zhang, H., Song, L., Li, Y., Gao, C. and Wang, K. (2019). Hi-TOM: a platform for high-throughput tracking of mutations induced by CRISPR/Cas systems, *Sci. China Life Sci.* 62(1), 1–7.

Liu, W., Xie, X., Ma, X., Li, J., Chen, J. and Liu, Y. G. (2015). DSDecode: a web-based tool for decoding of sequencing chromatograms for genotyping of targeted mutations, *Mol. Plant* 8(9), 1431–3.

Ma, X., Chen, L., Zhu, Q., Chen, Y. and Liu, Y. G. (2015). Rapid decoding of sequence-specific nuclease-induced heterozygous and biallelic mutations by direct sequencing of PCR products, *Mol. Plant* 8(8), 1285–7.

McKenna, A. and Shendure, J. (2018). FlashFry: a fast and flexible tool for large-scale CRISPR target design, *BMC Biol.* 16(1), 74.

Montgomery, J., Wittwer, C. T., Palais, R. and Zhou, L. (2007). Simultaneous mutation scanning and genotyping by high-resolution DNA melting analysis, *Nat. Protoc.* 2(1), 59–66.

Park, J., Lim, K., Kim, J. S. and Bae, S. (2017). Cas-analyzer: an online tool for assessing genome editing results using NGS data, *Bioinformatics* 33(2), 286–8.

Pinello, L., Canver, M. C., Hoban, M. D., Orkin, S. H., Kohn, D. B., Bauer, D. E. and Yuan, G. (2016). Analyzing CRISPR genome-editing experiments with CRISPResso, *Nat. Biotechnol.* 34(7), 695–7.

Pliatsika, V. and Rigoutsos, I. (2015). "Off-Spotter": very fast and exhaustive enumeration of genomic lookalikes for designing CRISPR/Cas guide RNAs, *Biol. Direct* 10, 4.

Ramlee, M. K., Yan, T., Cheung, A. M., Chuah, C. T. and Li, S. (2015). High-throughput genotyping of CRISPR/Cas9-mediated mutants using fluorescent PCR-capillary gel electrophoresis, *Sci. Rep.* 5, 15587.

Shan, Q., Wang, Y., Li, J. and Gao, C. (2014). Genome editing in rice and wheat using the CRISPR/Cas system, *Nat. Protoc.* 9(10), 2395–410.

Shan, Q., Wang, Y., Li, J., Zhang, Y., Chen, K., Liang, Z., Zhang, K., Liu, J., Xi, J. J., Qiu, J. L. and Gao, C. (2013). Targeted genome modification of crop plants using a CRISPR-Cas system, *Nat. Biotechnol.* 31(8), 686–8.

Tang, X., Liu, G., Zhou, J., Ren, Q., You, Q., Tian, L., Xin, X., Zhong, Z., Liu, B., Zheng, X., Zhang, D., Malzahn, A., Gong, Z., Qi, Y., Zhang, T. and Zhang, Y. (2018). A large-scale whole-genome sequencing analysis reveals highly specific genome editing by both Cas9 and Cpf1 (Cas12a) nucleases in rice, *Genome Biol.* 19(1), 84.

Thomas, H. R., Percival, S. M., Yoder, B. K. and Parant, J. M. (2014). High-throughput genome editing and phenotyping facilitated by high resolution melting curve analysis, *PLoS ONE* 9(12), e114632.

Vouillot, L., Thelie, A. and Pollet, N. (2015). Comparison of T7E1 and surveyor mismatch cleavage assays to detect mutations triggered by engineered nucleases, *G3 (Bethesda)* 5(3), 407–15.

Xue, L. J. and Tsai, C. J. (2015). AGEseq: analysis of genome editing by sequencing, *Mol. Plant* 8(9), 1428–30.

Zhou, H., Liu, B., Weeks, D. P., Spalding, M. H. and Yang, B. (2014). Large chromosomal deletions and heritable small genetic changes induced by CRISPR/Cas9 in rice, *Nucleic Acids Res.* 42(17), 10903–14.

Zhu, X., Xu, Y., Yu, S., Lu, L., Ding, M., Cheng, J., Song, G., Gao, X., Yao, L., Fan, D., Meng, S., Zhang, X., Hu, S. and Tian, Y. (2014). An efficient genotyping method for genome-modified animals and human cells generated with CRISPR/Cas9 system, *Sci. Rep.* 4, 6420.

Chapter 10

Targeted modification of promoters

Andika Gunadi and Ning Zhang, Boyce Thompson Institute, USA; and John J. Finer, The Ohio State University, USA

1 Introduction

Plant growth and development depends on sets of genes expressed in various tissues during different developmental phases and under changing environmental conditions. Gene expression in plants is regulated through changes in mRNA levels, achieved by alterations to local chromatin structure, transcriptional activity, mRNA stability and translation (de Nadal et al., 2011). Of the factors influencing gene expression, transcriptional regulation, mediated by the activity of the promoter, is arguably the single most important factor affecting gene expression. The promoter largely determines when, where and how a gene is transcribed. Precise editing of the promoter region would therefore allow all imaginable changes possible to the regulation of native genes.

Generally, a promoter is defined as the DNA sequence directly upstream of the gene coding sequence (see Fig. 1) and is the main component responsible for initiation and regulation of gene transcription. But the promoter can be more loosely defined to include more distal regions which can influence gene expression in some way. If a distal region of DNA folds back to bring its associated proteins in proximity to the proximal promoter region, and DNA expression is altered, it can be argued that the distal regions are part of the promoter. However, for routine analysis of the promoter, sequences of 1–2.5 kb upstream of various gene coding regions are typically isolated, allowing for the characterization of functional promoter sequence motifs, termed *cis*-acting

http://dx.doi.org/10.19103/AS.2020.0082.29

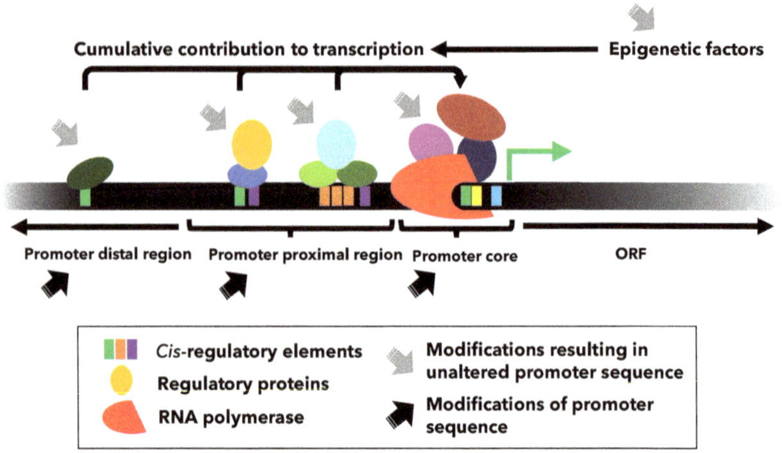

Figure 1 Anatomy of plant promoter and potential targets for modification.

regulatory elements (*cis*-elements). The most well-known elements, the TATA box and the initiator element, enable the recruitment of a set of general transcription factors that subsequently form a pre-initiation complex recognized by RNA polymerases (Maston et al., 2006). Other critical *cis*-elements are usually located in the proximal region of a promoter, within 500 bp of the transcription start site (TSS). The unique placement, sequence variability and combination of *cis*-elements within a given promoter affect the transcriptional regulation of the associated gene, specifying qualities such as the timing, expression intensity, inducibility and tissue specificity of gene expression (Hernandez-Garcia and Finer, 2014).

In plants, natural variations within the same promoter lead to changes in gene expression, which understandably contribute to differences in phenotypes (Swinnen et al., 2016), such as root morphology in maize (*Zea mays*) (Li et al., 2019), grain size in rice (*Oryza sativa*) (Duan et al., 2017), and flowering time and maturity in soybean (*Glycine max*) and *Arabidopsis* (Zhao et al., 2018; Liu et al., 2014a). Studies aimed at comparing natural variation in promoter sequences can offer important information for crop improvement. However, such studies are often limited by the available diversity of plant germplasm and the laborious process of identifying the associated naturally occurring mutations. Complementing the discovery of naturally occurring promoter variants, plant promoters can be intentionally modified and expressed *in planta* to determine the function of their *cis*-elements, develop customized promoter variants for transgenic crop improvements and make precise changes in native gene expression through promoter genome editing.

Through genome editing technologies, the targeted modification of a promoter in its native context within the genome allows for the study of promoter function with precision never before achievable, enables fine modulations of native gene expression, and can accelerate the generation of useful promoter variants for crop improvement. To date, genome editing technologies have been most widely used for gene functional analysis through the editing of gene coding regions, although several studies have shown precise gene regulation by tethering activators or repressors to the targeted promoter regions (Lowder et al., 2015; Piatek et al., 2015; Mahfouz et al., 2012; Gupta et al., 2012; Guan et al., 2002). However, few studies have reported promoter changes by modifying specific *cis*-elements within the native promoter sequences in plants.

This chapter covers the budding field of targeted modifications in the upstream gene regulatory regions of plants, as well as its potential for crop improvement. The first portion of the chapter provides the context and background information on plant promoters and leading introns, which are often associated with promoter regions. Use of transgenic approaches for promoter analysis is then presented as a means of generating and analyzing promoter variants, which leads to a better understating of promoter functionality. The remainder of the chapter covers the principles of promoter editing, followed by the approaches for promoter editing based on the recent advancements in genome editing technology.

2 Plant promoters, *cis*-acting regulatory elements and leading introns

Studies of plant promoters are important for understanding some of the basic mechanisms of gene regulation, and they are utilized for precisely controlling transgene expression in transgenic plants. Based on their regulatory patterns, promoters (see Fig. 1) are typically classified into three types: constitutive, spatiotemporal and inducible (Hernandez-Garcia and Finer, 2014). To date, many plant promoters have been isolated and characterized, among which several have been used in the production of transgenic plants (Christensen et al., 1992; Zhang et al., 1991). Although not often used in commercial transgenics, the most widely used promoter in plant transformation is still the constitutive Cauliflower Mosaic Virus 35S (CaMV35S) promoter (Odell et al., 1985), which originated from a plant virus. The CaMV35S promoter has been extensively characterized, but as it is constitutive and of viral origin, it is not an adequate representation of plant promoters and offers a limited range of specificity and intensity of gene regulation in plants.

2.1 Defining the promoter region

Next-generation sequencing has provided a tremendous amount of genomic sequence information, allowing relatively simple access to all components of the genome, including the promoter region. Although gene coding regions can be identified with relative ease, annotation of the regulatory regions of genes is more challenging (Morey et al., 2011). Bioinformatics methods have been developed to identify regulatory regions across a genome, by simply assuming that functional regulatory sequences that primarily control transcription initiation are usually located immediately upstream of the TSSs and house conserved regulatory elements (Maston et al., 2006). While the TSS is clearly the start of transcription, in promoter-based studies, this sequence is the end of the promoter, with regulatory sequences radiating back into the promoter region. The start or 5′ end of the promoter is difficult or impossible to determine, unless another gene is located upstream, and it is unlikely that the promoter extends into the coding region of the gene.

Defining the promoter region can be accomplished using three complementary approaches: sequence analysis, DNA structural features analysis and promoter dissection. Sequence analysis of the promoter relies on the hypothesis that the promoter core usually contains highly conserved *cis*-elements, such as the TATA box and the initiator element. However, short signature sequences, such as the relatively short consensus TATA box sequence (TATATA/T), typically appear every 250 bp in genomic sequences and are thus unlikely to be exclusive to promoter regions (Wasserman and Sandelin, 2004). Therefore, a TATA-box-based promoter prediction tends to produce a large number of false positives across the whole genome (Wasserman and Sandelin, 2004). On the other hand, the variability and even absence of the consensus sequences compromise the ability to detect true promoters (Morey et al., 2011). Finally, the fact that a growing number of genes have multiple transcription initiation sites greatly complicates the use of TSSs in determining the regulatory regions of a gene (Wasserman and Sandelin, 2004).

DNA structural features, such as bendability, curvature, free energy and GC skew, seem to show characteristic patterns in the promoter region proximal to the TSS (Rangannan and Bansal, 2009; Florquin et al., 2005; Fujimori et al., 2005; Kanhere and Bansal, 2005). For instance, an *in silico* program PromPredict was developed to predict promoters across the entire genome on the basis of unique free energy patterns in the adjoining regions of TSSs (Morey et al., 2011). Notably, the regions upstream of TSSs are presumed to be less stable than the downstream regions, so there is a significant difference of free energy between the upstream and downstream regions of the TSS, typically with a peak of free energy just upstream of the TSS. PromPredict has shown improved prediction capability of promoters in *Arabidopsis* and rice genomes (Morey et al., 2011), but it is based only on free energy of DNA sequences and is biased toward unstable regions such as AT regions. Integration of other DNA features into future versions of this program will likely give better performance of promoter prediction.

Although analyses of sequence and structural features provide much information on where a promoter may be located and how it may function, approaches in defining the promoter region should always be complemented by tools that allow stepwise dissection or modification of the promoter. Promoter dissection is typically performed through simple promoter truncation (Zhang et al., 2015), where the promoter region is physically reduced in a stepwise fashion so that the measured promoter activity declines when an essential component of the promoter is removed. This classical approach allows some crude determination of the upstream start of a promoter but the precision of genome editing in allowing targeted modification of promoter sequences will allow the generation of more detailed information for defining the limits and functionality of the promoter. Up to this point, however, most promoter analysis studies utilize transgenic approaches of some type, where the promoter is cloned, modified and reintroduced.

2.2 Roles of cis-acting regulatory elements in gene regulation

Regulatory properties of promoters are mainly determined by their contributing *cis*-elements (see Fig. 1). Each promoter typically contains clusters of regulatory elements that function independently or in combination to recruit transcription factors and other proteins to regulate gene expression (Spitz and Furlong, 2012). How transcription factors recognize their short DNA elements across the genome and the knowledge of the additional factors that influence the functionality of DNA elements are critical in understanding the mechanisms of promoter regulation in gene expression. In general, known *cis*-elements can be categorized into those conferring constitutive or spatiotemporal high-levels of gene expression, and those rendering the cognate promoter inducible by various endogenous and exogenous signals (see Table 1). Many elements may exhibit dual features; elements conferring high gene expression could also be induced by physical or chemical signals, and *vice versa*. And, since most *cis*-elements are located and may cluster in the proximal promoter region, those elements may themselves contain core recognition sequences with distal modifiers, adding to the complexity of promoter-mediated gene expression.

Few studies have specifically focused on the identification and characterization of *cis*-elements contributing to high gene expression (see Table 1). The well-known G-box element, first identified as a light-responsive element (Giuliano et al., 1988), was also later demonstrated to contribute to gene expression enhancement (Zhang et al., 2019; Ishige et al., 1999; McKendree and Ferl, 1992). G-box elements may be composed of a strong core element sequence of 6–8 nt with modifiers contributing to specificity. Other stimulatory elements have also been identified from strong constitutive plant promoters. For instance, the TEF1 box (AGGGGCANNNNNGTAANT) and

Table 1 Exemplary categories of plant *cis*-elements

Categories of plant *cis*-regulatory elements	Examples	References
Constitutive or spatial high-level expression	AG-box, G-box, TEO1 box, TEF1 box, PY-repeat, RY-repeat	Zhang et al. (2019), Ishige et al. (1999), Bobb et al. (1997), Daraselia et al. (1996), Curie et al. (1993), McKendree and Ferl (1992),Giuliano et al. (1988)
Light responsive	AAG-box, AE box, box II, GAP box, GT-1	Thum et al. (2001), Zhou (1999), Park et al. (1996), Gillmartin et al. (1992)
Drought responsive	CRT, DRE, MYC-like motif	Agarwal et al. (2017), Tran et al. (2004), Baker et al. (1994)
Pathogen responsive	D-box, GST, JASE1 box, JERE, PR-1 motif, S-box, SARE box, W1-box, W-2 box	Koschmann et al. (2012), Liu et al. (2011), Rushton et al. (2002), He and Gan (2001), Lebel et al. (1998), Shah and Klessig (1996)
Wound responsive	DRE, GCC box, GST, JERE, W1-box, W2-box	Hernandez-Garcia and Finer (2016), Rushton et al. (2002)
Ethylene responsive	GCC box	Agarwal et al. (2017), Fujimoto et al. (2000)
Abscisic acid responsive	ABRE	Shen and Ho (1995), Baker et al. (1994), Guiltinan et al. (1990)
Auxin responsive	AuxRE	Ulmasov et al. (1995)

Teo1 box (CCCTAA), contributing to high gene expression in rapidly growing tissues, were first identified from a promoter driving an *Arabidopsis* elongation factor 1A gene (Curie et al., 1993). Stimulating elements can be found in both regulatory DNA sequences and in other genomic regions including coding regions and intronic sequences. For instance, the PY-rich stretch (TTTCTTCTCT) in the 5′ untranslated region (5′ UTR) contributes to the high transcription levels of the native gene (Daraselia et al., 1996). Interestingly, elements contributing to high gene expression can also give rise to a spatiotemporal expression pattern, most of which drove very high-level expression in young and rapidly growing tissues such as root and shoot meristems, young embryos, and vascular tissues (Zhang et al., 2015; Curie et al., 1993). This is not surprising, as cells with high metabolic rates require high transcription rates.

Most of the elements identified as being responsive to environmental stimuli (see Table 1) have been isolated and characterized from inducible promoters. Inducible promoters rapidly activate the expression of cognate genes once exposed to external stimuli, and afterward return to a basal level after removal of the stimulus (Weake and Workman, 2010). *Cis*-elements are mainly responsible for inducing those expression changes, and studies of these elements present in stress-inducible gene promoters provide critical information for gene regulation and plant signaling under stress conditions. Pathogen-inducible promoters and their contributing elements have received increasing attention due to their importance in agriculture. In nature, plants have developed native defense mechanisms for protection against pathogen attack. Specific regulatory elements in pathogen-inducible promoters contribute to the increased transcription levels of defense genes under pathogen attack and either have low or no activity when the pathogen is absent. The sensitivity and inducibility of many of these elements have been validated using synthetic promoters (Rushton et al., 2002; Mazarei et al., 2008), providing critical information for better understanding plant-pathogen interactions and for engineering crops with enhanced disease resistance.

2.3 Identification of cis-acting regulatory elements in promoters

Most methods for predicting transcription factor-binding sites (TFBSs) assume that each transcription factor binds independently to its target sequence. However, in reality, transcription factor binding is also influenced by the context of flanking sequences and by the presence of additional proteins including other transcription factors or co-factors (Slattery et al., 2011; Hollenhorst et al., 2009), as combinatorial interactions of proteins binding to different DNA targets are essential for gene regulation (Lloyd et al., 2001). The simplest and most commonly used (but notably inaccurate) method of identifying potential DNA regulatory elements in promoters is the use of experimentally annotated DNA element databases, such as plant transcription factor-binding site databases PLACE (Higo et al., 1999) and PlantCARE (Lescot et al., 2002). Motifs in PLACE are from vascular plants, while PlantCARE not only collects DNA motifs from plant genes, but also provides a portal of tools for *in silico* promoter analysis. Another database TRANSFAC (Matys et al., 2003) focuses on eukaryotic gene regulation and provides data on transcription factors, experimentally proven TFBSs, and consensus binding sequences. However, these database resources, especially PLACE and PlantCARE, typically reveal an excessive number of motifs for a queried promoter sequence (Gunadi et al., 2016), and further experimental validation of those predicted elements is necessary (Hernandez-Garcia and Finer, 2014).

To find similarly regulated promoters or promoter elements, gene expression data is typically first used to identify co-regulated genes. Given that a set of genes are co-expressed, it is generally assumed that those genes are co-regulated by the same family of transcription factors (Hudson and Quail, 2003). By finding conserved sequences among the promoters of co-regulated genes, functional motifs may be identified without any prior knowledge of TFBSs. Although alignment techniques can be used to discover conserved DNA motifs to some extent, with a large number of promoters, it is difficult to distinguish element sequences that are highly similar and bound by related proteins but with divergent functions (Hudson and Quail, 2003). In comparison, the detection of over-represented motifs (enumerative methods) among those promoters is more productive and widely used (Jensen and Knudsen, 2000; Vanet et al., 2000). The enumerative method searches for small sequence motifs enriched to statistically significant levels in a set of given promoters. Hudson and Quail (2003) modified the enumerative method for the identification of over-represented sequences by comparing the tested co-regulated promoters to the rest of the promoters in the genome, which may largely reduce false positives. Other notable tools for the identification of over-represented motifs in sets of biological sequences include MEME (Bailey et al., 2006), Wordseeker (Lichtenberg et al., 2010) and HOMER (Heinz et al., 2010).

Several high-throughput technologies, such as comparative genomic sequence analysis (Margulies and Birney, 2008), DNA microarray (Hudson and Quail, 2003), chromatin immunoprecipitation (ChIP; Wang et al., 2002), ChIP-chip (Horak and Snyder, 2002), ChIP-seq (Valouev et al., 2008) and the yeast-one hybrid screening system (Zeng et al., 2008) have been developed to rapidly and efficiently identify sets of genomic targets for DNA-binding proteins. For instance, quantitative characterization of DNA-binding site preferences for zinc-finger proteins was performed using DNA microarray-based methods (Bulyk et al., 2001), facilitating the genome-wide identification of putative TFBSs. Other early attempts to study genome-wide distribution of TFBSs were performed using ChIP with a modification of serial analysis of gene expression or followed by hybridization on a DNA microarray (Impey et al., 2004; Horak and Snyder, 2002). Next-generation sequencing further revolutionized this field and several highly accurate studies on the identification of genome-wide TFBSs were conducted using ChIP-Seq (Valouev et al., 2008; Johnson et al., 2007; Robertson et al., 2007). ChIP-exo, an improved modification of ChIP-seq, integrates an additional DNase digestion step that was proven to aid in reducing false-positive binding interactions and further improved the predictive resolution of TFBSs (Rhee and Pugh, 2011). Compared to low-throughput experimental technologies for TFBS identification in the pre-genomic age, *in vitro* and *in vivo* high-throughput methods could provide large volumes of data on TFBSs and improve the understanding of genome-wide DNA-protein interactions in space and time. Overall, these data of predicted TFBSs are more accurate than those obtained from motif prediction. However, the definitive function of TFBSs can only be obtained through functional analysis of how these sequences impact the promoter activity.

2.4 Measuring promoter activity using transgenic approaches

For measurement of promoter activity, reporter gene assays provide a robust means for the evaluation of spatial and temporal regulation of putative promoters, through the production of reporter proteins which are easy to visualize and quantify. A series of fluorescent proteins such as the green fluorescent protein (GFP) permit highly sensitive and nondestructive monitoring of gene expression using both transient expression and expression in stably transformed tissues. Transient expression of a reporter gene can be visualized and measured as soon as 1.5 h after gene introduction (Ponappa et al., 1999) but transient expression may not accurately reflect all of the regulatory properties of promoters of interest. For expression in stably transformed tissue, the reporter gene must be stably introduced into the genome and express in tissues derived from a single targeted cell. This is certainly much more time-consuming but stable expression usually provides more comprehensive information on promoter activity. As nearly all transgenic approaches display promoter functionality outside of the native context of the promoter in the genome, native epigenomic effects on promoter activity are likely altered, and this must be considered when interpreting promoter function using different approaches.

The rapid results of transient expression assays with promoters regulating reporter genes allow fast quantitative analysis of promoter strength and inducibility (Hernandez-Garcia and Finer, 2014). Previously, we have established a transient expression assay by introducing promoter constructs into lima bean cotyledons via particle bombardment and subsequently tracking GFP expression using an automated image collection and analysis system (Gunadi et al., 2019; Hernandez-Garcia et al., 2010b; Chiera et al., 2007). For the discovery and validation of stress-inducible promoters or contributing *cis*-elements, *Agrobacterium*-mediated transient transformation in tobacco leaves (Agroinfiltration) have also been utilized to rapidly assess the inducibility of promoters or synthetic promoter constructs under various inductive treatments (Liu et al., 2011; Zahur et al., 2009). Since the DNA introduction process itself can cause certain types of stresses, interpretation of promoter induction needs to be thoughtfully evaluated with both of these rapid DNA introduction approaches. Although transient expression assays are effective for rapid evaluation of promoter strength, transient expression results may not accurately reflect *in planta* gene regulation as with stably transformed plants.

Characterization of promoters using stably transformed transgenic plants is often more desirable for measuring promoter activity, allowing fine analysis of promoter activity at various tissues and at different developmental stages. However, the generation of transgenic plants, especially transgenic crops, requires an extended time period, and it is still challenging to generate transgenic plants for some plant species. Since evaluation of large numbers of promoter variants in stably transformed plants is not always practical in some target plant species of interest, *Arabidopsis* and tobacco have been widely used as model systems for promoter studies as they are so easily transformed. However, the use of heterologous transgenic systems for promoter evaluation may lead to altered regulatory activity of the promoter, if substantive differences in types of transcription factors between species exist. For instance, *VR-ACS1* from mung bean is an auxin-inducible ACC synthase gene, but its promoter exhibited a high-level constitutive activity in transgenic *Arabidopsis* and tobacco (Cazzonelli et al., 2005).

Although promoter analysis in stably transformed tissues can provide some information on the regulatory machinery, the following considerations are needed in the determination of promoter activity in transgenic plants: (1) promoter evaluation in transgenic plants may not reflect their native regulatory properties especially in heterologous systems due to the absence of certain regulatory proteins (Hernandez-Garcia and Finer, 2014); (2) random insertion of promoter sequences into the plant genome often leads to variations in promoter strength and specificity; (3) transgene copy number can also influence gene expression and promoter evaluation among independent transgenic events; (4) promoter size or 'completeness' has an important role in promoter strength or specificity, as critical regulatory elements or enhancers may exist in far upstream promoter region that is not always included in the promoter constructs. As promoter analysis using transient and stable expression usually reduced or eliminated the influence of native epigenetic regulation, this approach can lead to somewhat unexpected gene regulatory patterns. Therefore, the use of genome editing tools for manipulation of promoter sequences in their native genomic context will undoubtedly provide more relevant information on native gene regulatory properties.

2.5 Synthetic promoters

Synthetic promoters contain engineered combinations of a core promoter with various dissected *cis*-element motifs (see Fig. 2). Synthetic promoters are used to help identify core promoters and *cis*-elements, and study the spatial and temporal activity of those elements. A typical synthetic promoter consists of *cis*-elements from plant promoters, with varied *cis*-element architectures (the position, spacing, copy number, specific combination of element sequences), which can recreate native and novel regulatory sequences to precisely modulate gene expression (Liu et al., 2013). Synthetic promoters are most commonly composed of a core promoter with upstream tetrameric repeats of the *cis*-element under evaluation (see Fig. 2) (Rushton et al., 2002; Hernandez-Garcia and Finer, 2016). Use of isolated *cis*-element tetramers with a core promoter sequence leads to an enhanced sensitivity for detection of inductive properties of that *cis*-element. Elements can be evaluated as monomers, dimers, timers, tetramers or even in more copies; however, a higher copy number of elements does not necessarily give rise to higher activity (Rushton et al., 2002). The core sequence from the CaMV35S promoter has been widely used for element

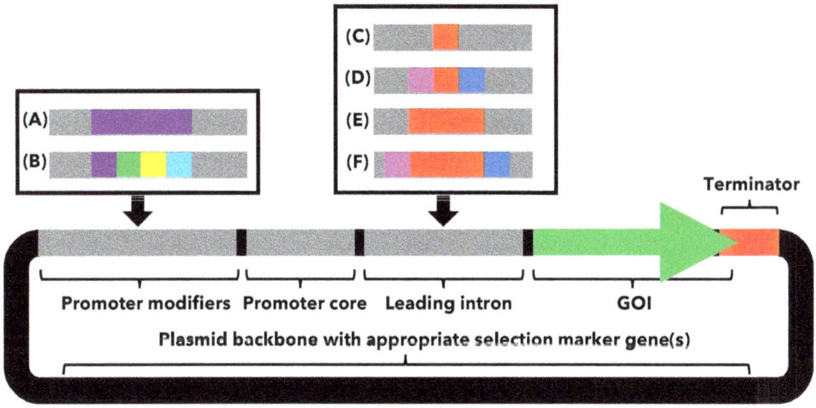

Figure 2 Representative construct of synthetic promoter and leading intron regulating gene of interest (GOI). Marker genes such as GFP may be used to measure changes in gene expression. Color-coded boxes indicate representative motifs within the synthetic promoter and intron, including tetramer of the same *cis*-element (A), tetramer of different *cis*-elements (B), and combinations of stimulatory intron motifs (C, D, E, F). Synthetic promoter and leading intron variations can be compared to construct with unmodified promoter and/or unmodified leading intron (positive controls), as well as only promoter core and/or no leading intron (negative controls).

evaluation due to its low background and efficient transcription initiation (Hernandez-Garcia and Finer, 2014).

Several in-depth studies using synthetic promoters elucidated the regulatory functions of pathogen-inducible *cis*-motifs for pathogen inducibility and plant defense (Koschmann et al., 2012; Liu et al., 2011; Rushton et al., 2002). In a pioneering study (Rushton et al., 2002), candidate elements (either W1, W2, GCC, JERE, S, Gst1 or D boxes) were evaluated in synthetic promoters and characterized as pathogen-responsive and/or defense-signal responsive, and the mechanism of pathogen/defense responsiveness was conserved in plant species at the *cis*-element level. In another study, an *Agrobacterium*-mediated transient expression assay in *Nicotiana benthamiana* was used to rapidly evaluate pathogen inducibility of synthetic promoters harboring putative pathogen-inducible elements (Liu et al., 2011). Liu et al. (2013) further advanced the phytosensing system by analyzing synthetic promoters containing candidate *cis*-elements in stably transformed tobacco and *Arabidopsis* plants. Combined with bioinformatics tools and high-throughput gene expression profiles, the use of synthetic promoters to experimentally evaluate novel *cis*-sequences responsible for specific stimuli holds great potential for the study of transcriptional activation and defense signaling in plants at the promoter level (Koschmann et al., 2012; Venter, 2007). It is not likely that genome editing of single *cis*-elements in native promoters to generate *cis*-element variants will lead to the same large variations in gene expression as is currently observed with *cis*-element tetramers in synthetic promoters. Therefore, it is beneficial to use synthetic promoters to first map *cis*-element variants and identify suitable targets, which can then be targeted for genome editing (Zhang et al., 2019).

2.6 Leading introns and their roles in gene regulation

Introns are DNA sequences within transcribed regions of genes that are processed and spliced during RNA maturation. In plants, more than 80% of genes contain at least one intron, with an average density of four introns per gene (Morello and Breviario, 2008). Interestingly, housekeeping genes or highly expressed genes tend to contain leading introns in their 5'UTRs (Hernandez-Garcia et al., 2010a), most of which were found to significantly enhance gene expression (Rose, 2019). In some cases, introns exhibited tissue- or development-specific patterns of gene regulation (Jeong et al., 2006; Jeon et al., 2000; Fu et al., 1995), possibly by providing binding sites for spatial- or temporal-specific regulatory factors in living cells. Promoters containing a leading intron (see Fig. 2) can greatly increase levels of gene expression and can be used in transgenic plants to elevate transgene expression. In commercial Roundup-Ready 2 Yield soybean (Fincher et al., 2006), part of the *Arabidopsis* elongation factor 1A intron was used, along with its native promoter (Axelos et al., 1989), although the reason for use of the intron was not specified in this patent.

Similar to promoters, introns may contain multiple regulatory sequences such as intronic enhancers or repressors that can provide binding sites for splicing factors or transcription factors and other transcriptional activators (Morello and Breviario, 2008). For instance, an intronic enhancer was identified and characterized from the leading intron of a soybean polyubiquitin (Gmubi) promoter, which played an essential role for the high activity of the promoter (Grant et al., 2017; De La Torre and Finer, 2015). This type of stimulating intron usually can boost gene expression independent of its orientation and position, similar to classical transcriptional enhancer sequences typically located far upstream of the transcriptional start site of the gene. In some cases, introns were also demonstrated to function as promoters, capable of driving a weak but consistent gene expression (Qi et al., 2007; Morello et al., 2002, 2006; Salgueiro et al., 2000). However, most stimulating introns identified so far do not contain recognized regulatory elements or enhancers contributing to the increased gene expression, and they can only augment gene expression when they are placed within the transcribed region in a correct orientation. This type of intron enhancement is defined as intron-mediated enhancement (IME) (Mascarenhas et al., 1990), and may regulate gene expression directly or indirectly during the intron splicing process, even though splicing is not sufficient for IME.

At the molecular level, intron-containing genes likely enhance transcriptional efficiency either through interactions between factors influencing intron splicing transcription or by modifying the chromatin structure into a transcriptionally active state through the recruitment of chromatin-remodeling proteins (Rose, 2008; Rose et al., 2008; Le Hir et al., 2003; Mattick and Gagen, 2001; Liu et al., 1995). Other studies also revealed that splicing signals can influence multiple steps of mRNA processing, including 5′ capping (Lewis and Izaurralde, 1997), 3′ polyadenylation (Proudfoot et al., 2002), mRNA nuclear export (Luo and Reed, 1999) and mRNA decay (Kertesz et al., 2006). Whether at the DNA or RNA level, the presence of an enhancing intron in genes has been proposed to make the transcriptional or translational machinery more processive (Rose, 2004; Parra et al., 2011).

Not all efficiently spliced introns display IME, indicating specific sequences or unique structures within the stimulatory introns play a critical role in enhancing gene expression (Gallegos and Rose, 2015). To identify and characterize contributing introns and intron regulatory elements, synthetic introns (see Fig. 2) can be generated to evaluate the enhancement of gene expression. In a similar approach to synthetic promoters, candidate intron sequences can be manipulated in different orientation, spacing and copy numbers within synthetic introns to determine the specific effects of intronic sequences on gene expression enhancement (Grant et al., 2017; Zhang et al., 2016; De La Torre and Finer, 2015).

3 Principles of targeted promoter modification

Genome editing technologies have been widely used in plants for gene functional analysis by generating gene mutations, deletions, insertions or gene substitution (Baltes and Voytas, 2015). Nevertheless, sequence-specific nucleases can also be utilized as robust tools for modifying gene regulation (see Tables 2 and 3). Catalytically inactive sequence-specific nucleases fused with transcriptional activators, repressors or epigenetic modifiers could be designed to bind to specific promoter or coding regions of genes of interest to modulate gene expression without altering the target sequences (Papikian et al., 2019; Liu et al., 2014b). Complementarily, sequence-specific nucleases can also be used to modify critical *cis*-elements in the promoter sequence of target genes to precisely control the level or specificity of gene expression by direct promoter editing.

Several principles for promoter editing are feasible based on our understanding of plant gene regulatory sequences and the *cis*-elements that contribute to their functionality. *Cis*-elements are typically composed of a conserved core element motif and flanking sequences, making the modulation of gene expression possible by precise DNA sequence modification within these

Table 2 General outcomes of targeted promoter modification strategies

Approaches using sequence-specific nucleases	Alter promoter activity	Maintain promoter sequence	Modify existing element	Generate new element	Insert or replace promoter sequence larger than 44 bp
Transcriptional modification	✓	✓			
Epigenetic modification	✓	✓			
Small indel	✓		✓	✓	
Large deletion	✓		✓		
Base editing	✓		✓	✓	
Targeted integration	✓		✓	✓	✓
Prime editing	✓		✓	✓	

Table 3 Reports of targeted promoter modification in plants

Promoter modification	Approach	Plant	Reference
Transcriptional silencing and activation	CRISPRi and CRISPRa	*Arabidopsis thaliana*, *Nicotiana benthamiana*, rice	Papikian et al. (2019), Lowder et al. (2015, 2018), Piatek et al. (2015)
Epigenetic regulation	Targeted methylation	*Arabidopsis thaliana*	Papikian et al. (2019)
Induced promoter variation	Targeted indel and large deletion	Grapefruit, orange, rice, tomato	Peng et al. (2017), Rodríguez-Leal et al. (2017), Duan et al. (2016), Jia et al. (2016), Li et al. (2012)
Promoter substitution	Targeted integration	Cassava, tomato	Hummel et al. (2018), Čermák et al. (2015)

sequences. Change of one or a few nucleotides in the appropriate promoter sequence could create a new functional motif or conversely, destroy a known functional motif, leading to the change of the promoter intensity or specificity (Zhang et al., 2019). Similarly, modification of the flanking sequences of a *cis*-regulatory element can also significantly increase or decrease the expression of the target genes. For instance, modification of the 2 bp flanking sequence of the G-box core element in a synthetic promoter led to an over 10-fold increase in gene expression in transiently transformed lima bean cotyledons and a two-fold increase in stably transformed soybean tissues (Zhang et al., 2019), suggesting that genome editing of the promoter by similar subtle changes to the core elements or their flanking nucleotides can result in predictable and desirable changes in native gene expression. Promoter editing in *cis*-elements or their flanks may be achieved through NHEJ-mediated random indels, large deletions, targeted integration of donor sequence for large insertion or substitution (Hanh et al., 2018), base editing (Shan and Voytas, 2018) or the emerging prime editing (Anzalone et al., 2019) approaches (see Table 2). However, precise modification of regulatory motifs in native promoters by genome editing is still at an early stage, and successful editing requires a better understanding and/or experimental manipulation of regulatory element composition and structure within the promoter.

Precise promoter editing can be more challenging than editing within the coding region due to minimal information available on gene regulatory regions compared to the gene coding sequence, as well as the higher likelihood of encountering repetitive elements within the noncoding regions. However, compared to transgenics, promoter editing can potentially result in superior crop improvement due to the maintenance of native genetic background, leading to potentially less off-target effects. Additionally, small targeted modifications can mimic naturally occurring mutations and can result in a gradient of phenotypes (Rodríguez-Leal et al., 2017). Large modifications to the promoter region, depending on the design, may or may not mimic naturally occurring mutations and may therefore be considered as transgenic (Shew et al., 2018).

In light of these considerations, we have formulated some general guidelines for targeted modification of promoter sequence in plants (see Fig. 3). A recommended approach is to first generate and evaluate promoter *cis*-element variants using synthetic promoter approaches and classical gene introduction (Zhang et al., 2019), so that the most desirable promoter modification can be determined (see Fig. 4). For the more well-studied promoters and promoter families, some *cis*-elements may have already been identified and functionally validated. However, targeted modification of less well-studied promoter sequences will require more effort to identify their pattern of transcriptional regulation, as well as their key motifs. The best targets and strategies for targeted promoter modification can be gleaned once key motifs within the promoters are functionally validated (see Figs 3 and 4) and the appropriate promoter variants are constructed and evaluated. For production of commercial promoter edited plants, a predetermination of optimal outcomes for promoter modification seems straightforward and useful. These approaches for targeted promoter modification (see Table 2) and some examples (see Table 3) are further discussed below.

4 Approaches for targeted promoter modification

Promoters that regulate genes involved in plant growth and development, as well as abiotic and biotic stress responses, and nutrient acquisition are among the key targets for crop improvement (Hernandez-Garcia and Finer,

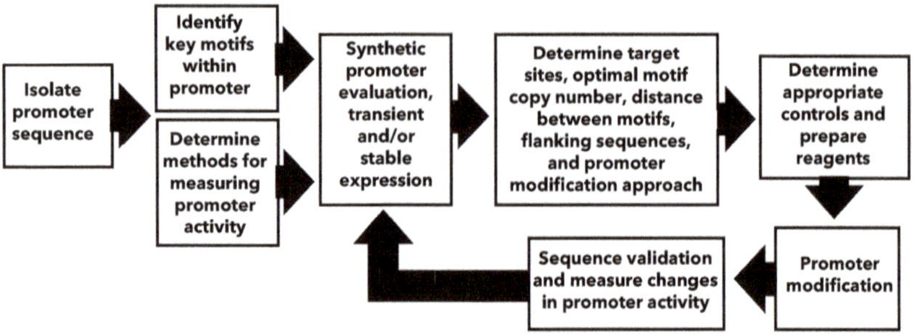

Figure 3 Guideline for targeted modification of upstream regulatory sequences in plants.

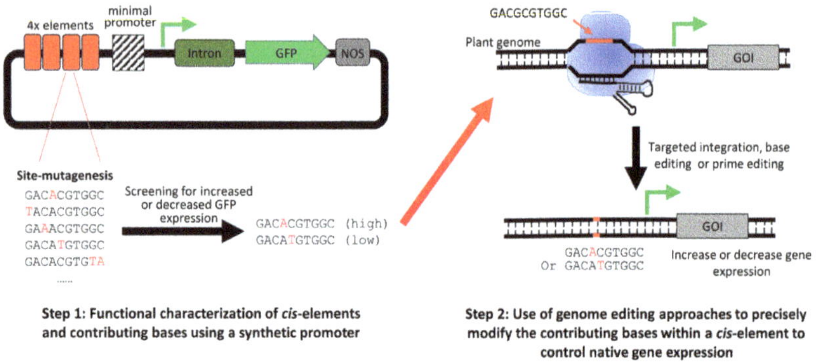

Figure 4 Schematic of the use of synthetic promoters to expedite targeted promoter editing.

2014; Potenza et al., 2004). The growing toolbox of available genome editing approaches is especially useful for the targeted modifications of these promoter types and other plant gene regulatory regions, including leading introns. As highlighted in the previous section, these approaches fall under two general categories: (1) targeted transcriptional modification resulting in unaltered promoter sequence and (2) targeted alteration of the promoter sequence (see Fig. 1; see Table 2).

4.1 Targeted transcriptional modification

The use of catalytically inactive sequence-specific nucleases, such as CRISPR interference (CRISPRi) and CRISPR activation (CRISPRa) target the promoter sequence, altering the promoter activity without editing the DNA. This indirect transcriptional regulation is achievable through the use of either deactivated nuclease alone (e.g. dCas9) or the deactivated nuclease fused with transcriptional repressors or activators (Lowder et al., 2015, 2018; Piatek et al.,

2015; Qi et al., 2013). The binding of dCas to functional regions of the promoter, such as to the transcription start site or to TFBSs, sterically prevents the binding and/or activity of corresponding proteins involved in the gene transcription (Qi et al., 2013). In comparison, the use of dCas fused with transcriptional repressors or activators not only sterically hinder endogenous proteins involved in transcription, but also recruit endogenous proteins that repress (CRISPRi) or activate (CRISPRa) transcription, enabling an altered regulation of transcriptional activity (La Russa and Qi, 2015). Unlike genome editing approaches that change the promoter sequence permanently, the activity of CRISPRi and CRISPRa are reversible (Larson et al., 2013), depending on the expression of the modified dCas protein and the sgRNA. Hence, the regulated expression of the modified Cas protein and sgRNA are useful approaches for limiting (CRISPRi) or increasing (CRISPRa) the strength and spatiotemporal activity of the targeted promoter. The use of CRISPRi and CRISPRa have been reported in several plant species (see Table 3).

4.2 Chromatin modification and targeted methylation

The accessibility of plant promoter sequences to transcription factors and DNA polymerase binding are regulated in part by elaborate epigenetic factors that include chromatin structure modifiers and promoter methylation (see Fig. 1; To et al., 2015; Pikaard and Scheid, 2014). Thus far, most studies in plants that characterize the effects of epigenetic factors on gene expression patterns rely on correlations between preexisting epigenetic marks and phenotypic changes. Recent modifications to the CRISPR machinery, however, can better study this relationship through enabling targeted epigenetic changes (Adli, 2018). Similar in concept to CRISPRi and CRISPRa, the dCas protein can also be fused to DNA methyltransferases (Vojta et al., 2016), histone modifiers (Hilton et al., 2015; Kearns et al., 2015), and to SunTag peptides that can recruit multiple endogenous copies of these factors to induce targeted epigenetic changes (Pflueger et al., 2018; Tanenbaum et al., 2014). Utilizing the SunTag systems, a recent study in *Arabidopsis* has successfully modified the epigenome of a floral development gene, leading to the activation and suppression of its native promoter activity (Papikian et al., 2019).

4.3 Targeted mutagenesis and deletion

Although conceptually one of the simplest applications of genome editing, targeted mutagenesis and deletions are powerful approaches for discovering novel promoter variants. Targeted mutagenesis typically generates small insertions or deletions at a targeted double-stranded break site. In comparison, targeted large deletions usually require at least two double-stranded break sites flanking the deletion site. Using these two approaches, obtaining promoter variants with altered transcriptional activity is achievable through either targeting a functional *cis*-element and its proximal sequences, or alternatively through creating a novel functional *cis*-element (Zhang et al., 2019). These strategies for modifications of native promoters with previously known gene functions were successfully implemented to generate genome-edited tomatoes (*Solanum lycopersicum* and *Solanum pimpinellifolium*) having a phenotypic gradient of fruit sizes, floral organ numbers and plant architecture (Rodríguez-Leal et al., 2017). The involvement of a GT-1 *cis*-element in salt stress in rice (*Oryza sativa*) was confirmed using CRISPR mutagenesis of that element in the promoter (Duan et al., 2016). In grapefruit (*Citrus paradisi*) and orange (*Citrus sinensis*), a targeted mutation in the CsLOBP promoter led to increased resistance to the plant bacterial pathogen *Xanthomonas citri* (Peng et al., 2017; Jia et al., 2016). All of these promoter edits using classical CRISPR mutagenesis resulted in the generation of targeted but random insertions or deletions.

4.4 Base editing

Specific base substitutions within the plant genome are achievable through the use of modified nucleases. The Cas9 nuclease was altered to function as a nickase that cuts only one strand of the DNA, and fused to catalytic enzymes that perform the base substitutions (Tan et al., 2019; Kang et al., 2018; Shan and Voytas, 2018). As a promoter modification strategy, base editing has the potential to introduce mutations to modify functional *cis*-elements or render active previously non-functional motifs at greater precision than indels or large deletion strategies. Although base editing shows tremendous potential for promoter editing in plants, this approach is yet to be appropriately pursued.

4.5 Targeted integration

Homology-dependent repair (HDR) and non-homologous end joining (NHEJ)-based strategies are useful for insertion of donor template sequence into a targeted location in the genome of various plants (Bonawitz et al., 2019; Hanh et al., 2018; Schmidt et al., 2019; Begemann et al., 2017). Targeted integration or knock-in also has a vast potential for insertion or replacement of whole or parts of genes and regulatory sequences (up to several thousand bp in length). However, targeted integration approaches are presently limited by low efficiencies across plant species (0–10%), requiring ongoing optimization efforts (Collonnier et al., 2017).

Using targeted integration approaches, whole or partial promoter sequences can potentially be replaced. This targeted replacement can be achieved using at least two double stranded breaks and the addition of donor promoter sequence. Additionally, one to several *cis*-elements within a promoter can be replaced with a donor template containing any multiple or combination of *cis*-element. For example, a single element can be replaced by a tetramer of the same element to enhance transcription factor-binding affinity (Rushton et al., 2002). In addition, novel combinations of *cis*-elements can also be introduced into the promoter using targeted integration approaches. To date, targeted swapping of plant promoter and/or promoter parts have yet to be reported, although the targeted insertion of the viral-based CaMV35S promoter to regulate native plant genes has been reported (Hummel et al., 2018; Čermák et al., 2015). Using CRISPR/Cas, Hummel et al. (2018) inserted a donor sequence containing a double CaMV35S promoter and partially modified coding sequence into the native *EPSPS* gene of cassava, resulting in glyphosate-resistant plants. Using both TALENs and CRISPR/Cas, Čermák et al. (2015) inserted donor sequence containing the NPTII gene construct and a CaMV35S promoter upstream of the native tomato gene controlling anthocyanin biosynthesis, resulting in anthocyanin-rich purple-colored plants.

4.6 Prime editing

Prime editing (Anzalone et al., 2019) offers several unique advantages for targeted promoter modification (see Table 2). Using prime editing guide RNA (pegRNA) as both guide and template for the editing of DNA sequence, prime editing circumvents the reliance on donor DNA template and DNA double-stranded break, with an additional advantage of less off-targets compared to other genome editing strategies (Anzalone et al., 2019). Moreover, with the flexibility of the pegRNA design, prime editing approach is also less constrained by the location of protospacer adjacent motif (PAM) sites of the targeted DNA sequence. So far, prime editing is capable of single-base editing, targeted deletion of up to 80 bp, and targeted insertion of up to 44 bp (Anzalone et al., 2019). Since the length of most known plant *cis*-elements is below 44 bp, prime editing is likely suitable for specific modification of a wide array of plant gene regulatory sequences along with their flanking sequences. In human cells, the efficiency of prime editing (20–50%) is generally higher than targeted integration using HDR (below 10%) (Anzalone et al., 2019), making it a viable genome editing approach with a potential for further improvements. To date, the breadth of capabilities for prime editing have not yet been reported in plants.

5 Conclusion and future trends

Although substantial concepts and tools for targeted promoter modification in plants are available (see Table 2), very few studies on genome editing of promoter regions have been reported to date (see Table 3). Considering the importance of modifying gene regulatory sequences to induce beneficial traits for agriculture, along with the rapid progress of genome editing tools, we anticipate a robust development in this field as means of complementing mainstream breeding strategies. Several limitations, however, need to be overcome for effective targeted modification of plant promoters.

Notably, plant promoters and non-coding sequences commonly contain repetitive sequences, which may have gene regulatory functions. Hence, for experimental purposes and for crop improvements, the possibility of off-target effects (Wolt et al., 2016; Kuscu et al., 2014; Fu et al., 2013) must be considered for all targeted promoter modification approaches. Depending on the promoter modification approach and the type of sequence-specific nuclease used, finding appropriate targets that are unique within the plant genome is ideal (Baltes and Voytas, 2015), yet if the aim is to target specific recurring sequences such as *cis*-elements, targeting some repetitive sequences may be unavoidable. Several strategies are available to overcome off-targets in promoter modification. One solution is to use selective breeding to preserve the edited promoter sequence while removing off-target edits (e.g. via backcrossing), but this process is laborious. Alternatively, specific promoter modifications can first be tested in a relatively short timeframe using transient expression systems (Gunadi et al., 2019; De la Torre and Finer, 2015), and once a desirable edit is identified, a large region of the promoter can be replaced using targeted integration between unique target flanking sequences.

Another limitation for promoter targeting that is inherent to the use of CRISPR technology is the availability of a Protospacer-adjacent motif (PAM) site within the promoter sequence. Regardless of the Cas nuclease, CRISPR ribonucleic protein complexes are unable to bind to the desired sequence without a compatible PAM site. This limitation is gradually being alleviated with the increasing discovery and characterization of natural variants of Cas nuclease, and engineering Cas nucleases (as well as their compatible sgRNAs) with variable sequence specificities (Raitskin et al., 2019; Cebrian-Serrano and Davies, 2017). Alternatively, PAM site restrictions within the promoter can be overcome through targeted replacement of a large region in or beyond the promoter sequence using unique target flanking sequences. As the repertoire for CRISPR/Cas and other genome editing tools continues to rapidly evolve, we are hopeful for these technical limitations for promoter editing to be resolved, leaving only limitations of the creativity in applying these tools.

6 Where to look for further information

6.1 Databases of plant promoters

- DNAshape (http://rohslab.usc.edu/DNAshape/serverBackend.php; Zhou et al., 2013).
- Phytozome (http://.phytozome.net; Goodstein et al., 2012).
- PLAZA (http://bioinformatics.psb.ugent.be/plaza/; Proost et al., 2009).
- PromPredict (http://nucleix.mbu.iisc.ernet.in/prompredict/prompredict.html; Morey et al., 2011).
- TFBSshape (http://rohslab.usc.edu/TFBSshape/; Yang et al., 2013).

6.2 Databases of plant promoter motifs

- PlantCARE (http://bioinformatics.psb.ugent.be/webtools/plantcare/html/; Lescot et al., 2002).
- New PLACE (http:// dna.affrc.go.jp/PLACE/?action=newplace; Higo et al., 1999).
- JASPAR (http://jaspar.genereg.net; Khan et al., 2018).
- TRANSFAC (http://gene-regulation.com/pub/databases.html#transfac; Matys et al., 2003).
- PlantPAN3.0 (http://plantpan.itps.ncku.edu.tw; Chow et al., 2019).
- PlantTFDB (http://planttfdb.cbi.pku.edu.cn/; Jin et al., 2017).

6.3 Tools for identification of motifs within promoters and leading introns

- MEME (http://meme-suite.org; Bailey et al., 2006).
- Wordseeker (http://omictools.com/wordseeker-tool; Lichtenberg et al., 2010).
- HOMER (http://homer.ucsd.edu/homer/; Heinz et al., 2010).

6.4 General references for genome editing approaches and materials

- http://blog.addgene.org.
- http://addgene.org.

7 References

Adli, M. (2018). The CRISPR tool kit for genome editing and beyond, *Nat. Commun.* 9(1), 1911.

Agarwal, P. K., Gupta, K., Lopato, S. and Agarwal, P. (2017). Dehydration responsive element binding transcription factors and their applications for the engineering of stress tolerance, *J. Exp. Bot.* 68(9), 2135–2148.

Anzalone, A. V., Randolph, P. B., Davis, J. R., Sousa, A. A., Koblan, L. W., Levy, J. M., Chen, P. J., Wilson, C., Newby, G. A., Raguram, A. and Liu, D. R. (2019). Search-and-replace genome editing without double-strand breaks or donor DNA, *Nature* 576(7785), 149–157.

Axelos, M., Bardet, C., Liboz, T., Le Van Thai, A., Curie, C. and Lescure, B. (1989). The gene family encoding the *Arabidopsis thaliana* translation elongation factor eEF1A: molecular cloning, characterization and expression, *Mol. Gen. Genet.* 219(1–2), 106–112.

Bailey, T. L., Williams, N., Misleh, C. and Li, W. W. (2006). MEME: discovering and analyzing DNA and protein sequence motifs, *Nucleic Acids Res.* 34, W369–W373.

Baker, S. S., Wilhelm, K. S. and Thomashow, M. S. (1994). The 5′-region of Arabidopsis thaliana corl5a has cis-acting elements that confer cold-, drought- and ABA-regulated gene expression, *Plant Mol. Biol.* 24, 701–713.

Baltes, N. J. and Voytas, D. F. (2015). Enabling plant synthetic biology through genome engineering, *Trends Biotechnol.* 33(2), 120–131.

Begemann, M. B., Gray, B. N., January, E., Gordon, G. C., He, Y., Liu, H., Wu, X., Brutnell, T. P., Mockler, T. C. and Oufattole, M. (2017). Precision insertion and guided editing of higher plant genomes using Cpf1 CRISPR nucleases, *Sci. Rep.* 7, 11606.

Bobb, A. J., Chern, M. S. and Bustos, M. M. (1997). Conserved RY-repeats mediate transactivation of seed-specific promoters by the developmental regulator PvALF, *Nucleic Acids Res.* 25(3), 641–647.

Bonawitz, N. D., Ainley, W. M., Itaya, A., Chennareddy, S. R., Cicak, T., Effinger, K., Jiang, K., Mall, T. K., Marri, P. R., Samuel, J. P., Sardesai, N., Simpson, M., Folkerts, O., Sarria, R., Webb, S. R., Gonzalez, D. O., Simmonds, D. H. and Pareddy, D. R. (2019). Zinc finger nuclease-mediated targeting of multiple transgenes to an endogenous soybean genomic locus via non-homologous end joining, *Plant Biotechnol. J.* 17(4), 750–761.

Bulyk, M. L., Huang, X., Choo, Y. and Church, G. M. (2001). Exploring the DNA-binding specificities of zinc fingers with DNA microarrays, *Proc. Natl. Acad. Sci. U.S.A.* 98(13), 7158–7163.

Cazzonelli, C. I., McCallum, E. J., Lee, R. and Botella, J. R. (2005). Characterization of a strong, constitutive mung bean (*Vigna radiata* L.) promoter with a complex mode of regulation *in planta*, *Transgen. Res.* 14(6), 941–967.

Cebrian-Serrano, A. and Davies, B. (2017). CRISPR-Cas orthologues and variants: optimizing the repertoire, specificity and delivery of genome engineering tools, *Mamm. Genome* 28(7–8), 247–261.

Čermák, T., Baltes, N. J., Čegan, R., Zhang, Y. and Voytas, D. F. (2015). High-frequency, precise modification of the tomato genome, *Genome Biol.* 16(1), 232.

Chiera, J. M., Bouchard, R. A., Dorsey, S. L., Park, E., Buenrostro-Nava, M. T., Ling, P. P. and Finer, J. J. (2007). Isolation of two highly active soybean (*Glycine max* (L.) Merr.) promoters and their characterization using a new automated image collection and analysis system, *Plant Cell Rep.* 26(9), 1501–1509.

Chow, C. N., Lee, T. Y., Hung, Y. C., Li, G. Z., Tseng, K. C., Liu, Y. H., Kuo, P. L., Zheng, H. Q. and Chang, W. C. (2019). Plant PAN3.0: a new and updated resource for reconstructing transcriptional regulatory networks from ChIP-seq experiments in plants, *Nucleic Acids Res.* 47(D1), D1155–D1163.

Christensen, A. H., Sharrock, R. A. and Quail, P. H. (1992). Maize polyubiquitin genes: structure, thermal perturbation of expression and transcript splicing, and promoter activity following transfer to protoplasts by electroporation, *Plant Mol. Biol.* 18(4), 675–689.

Collonnier, C., Guyon-Debast, A., Maclot, F., Mara, K., Charlot, F. and Nogué, F. (2017). Towards mastering CRISPR-induced gene knock-in in plants: survey of key features and focus on the model *Physcomitrella patens*, *Methods* 121–122, 103–117.

Curie, C., Axelos, M., Bardet, C., Atanassova, R., Chaubet, N. and Lescure, B. (1993). Modular organization and developmental activity of an *Arabidopsis thaliana* EF-1α gene promoter, *Mol. Gen. Genet.* 238(3), 428–436.

Daraselia, N. D., Tarchevskaya, S. and Narita, J. O. (1996). The promoter for tomato 3-hydroxy-3-methylglutaryl coenzyme A reductase gene 2 has unusual regulatory elements that direct high-level expression, *Plant Physiol.* 112(2), 727–733.

De La Torre, C. M. and Finer, J. J. (2015). The intron and 5′ distal region of the soybean Gmubi promoter contribute to very high levels of gene expression in transiently and stably transformed tissues, *Plant Cell Rep.* 34(1), 111–120.

de Nadal, E., Ammerer, G., and Posas, F. (2011). Controlling gene expression in response to stress. *Nat. Rev. Genet.*, 12(12), 833–845.

Duan, P., Xu, J., Zeng, D., Zhang, B., Geng, M., Zhang, G., Huang, K., Huang, L., Xu, R., Ge, S., Qian, Q. and Li, Y. (2017). Natural variation in the promoter of GSE5 contributes to grain size diversity in rice, *Mol. Plant* 10(5), 685–694.

Duan, Y. B., Li, J., Qin, R. Y., Xu, R. F., Li, H., Yang, Y. C., Ma, H., Li, L., Wei, P. C. and Yang, J. B. (2016). Identification of a regulatory element responsible for salt induction of rice OsRAV2 through *ex situ* and *in situ* promoter analysis, *Plant Mol. Biol.* 90(1–2), 49–62.

Fincher, K. L., Flasinski, S. and Wilkinson, J. Q. (2006). Glyphosate resistant plants using hybrid promoter constructs, U.S. Patent No. 7,141,722 Nov 28, 2006, Washington, DC: U.S. Patent and Trademark Office.

Florquin, K., Saeys, Y., Degroeve, S., Rouze, P. and Van de Peer, Y. (2005). Large-scale structural analysis of the core promoter in mammalian and plant genomes, *Nucleic Acids Res.* 33(13), 4255–4264.

Fu, H., Kim, S. Y. and Park, W. D. (1995). A potato Sus3 sucrose synthase gene contains a context-dependent 3′ element and a leader intron with both positive and negative tissue-specific effects, *Plant Cell* 7(9), 1395–1403.

Fu, Y., Foden, J. A., Khayter, C., Maeder, M. L., Reyon, D., Joung, J. K. and Sander, J. D. (2013). High-frequency off-target mutagenesis induced by CRISPR-Cas nucleases in human cells, *Nat. Biotechnol.* 31(9), 822–826.

Fujimori, S., Washio, T. and Tomita, M. (2005). GC-compositional strand bias around transcription start sites in plants and fungi, *BMC Genomics* 6, 26.

Fujimoto, S. Y., Ohta, M., Usui, A., Shinshi, H. and Ohme-Takagi, M. (2000). *Arabidopsis* ethylene-responsive element binding factor act as transcriptional activators or repressors of GCC box-mediated gene expression, *Plant Cell* 12(3), 393–404.

Gallegos, J. E. and Rose, A. B. (2015). The enduring mystery of intron-mediated enhancement, *Plant Sci.* 237, 8–15.

Gillmartin, P. M., Memelink, J., Hiratsuka, K., Kay, S. A. and Chua, N. H. (1992). Characterization of a gene encoding a DNA binding protein with specificity for a light-responsive element, *Plant Cell* 4(7), 839–849.

Giuliano, G., Pichersky, E., Malik, V. S., Timko, M. P., Scolnik, P. A. and Cashmore, A. R. (1988). An evolutionarily conserved protein binding sequence upstream of a plant light-regulated gene, *Proc. Natl. Acad. Sci. U.S.A.* 85(19), 7089–7093.

Goodstein, D. M., Shu, S., Howson, R., Neupane, R., Hayes, R. D., Fazo, J., Mitros, T., Dirks, W., Hellsten, U., Putnam, N. and Rokhsar, D. S. (2012). Phytozome: a comparative platform for green plant genomics, *Nucleic Acids Res.* 40(Database issue), D1178–D1186.

Grant, T. N., De La Torre, C. M., Zhang, N. and Finer, J. J. (2017). Synthetic introns help identify sequences in the 5′ UTR intron of the *Glycine max* polyubiquitin (Gmubi) promoter that give increased promoter activity, *Planta* 245(4), 849–860.

Guan, X., Stege, J., Kim, M., Dahmani, Z., Fan, N., Heifetz, P., Barbas, C. F. and Briggs, S. P. (2002). Heritable endogenous gene regulation in plants with designed polydactyl zinc finger transcription factors, *Proc. Natl. Acad. Sci. U.S.A.* 99(20), 13296–13301.

Guiltinan, M. J., Marcotte Jr., W. R. and Quatrano, R. S. (1990). A plant leucine zipper protein that recognizes an abscisic acid response element, *Science* 250(4978), 267–271.

Gunadi, A., Dean, E. A. and Finer, J. J. (2019). Transient transformation using particle bombardment for gene expression analysis. *Methods Mol. Biol.* 1864, 67–79.

Gunadi, A., Rushton, P. J., McHale, L. K., Gutek, A. H. and Finer, J. J. (2016). Characterization of 40 soybean (*Glycine max*) promoters, isolated from across 5 thematic gene groups, *Plant Cell. Tissue Organ Cult.* 127(1), 145–160.

Gupta, M., DeKelver, R. C., Palta, A., Clifford, C., Gopalan, S., Miller, J. C., Novak, S., Desloover, D., Gachotte, D., Connell, J., Flook, J., Patterson, T., Robbins, K., Rebar, E. J., Gregory, P. D., Urnov, F. D. and Petolino, J. F. (2012). Transcriptional activation of Brassica napus β-ketoacyl-ACP synthase II with an engineered zinc finger protein transcription factor, *Plant Biotechnol. J.* 10(7), 783–791.

Hanh, F., Eisenhut, M., Mantegazza, O. and Weber, A. P. M. (2018). Homology-directed repair of a defective *Glabrous* gene in *Arabidopsis* with Cas9-based gene targeting, *Front. Plant Sci.* 9, 424.

He, Y. and Gan, S. (2001). Identical promoter elements are involved in regulation of the OPR1 gene by senescence and jasmonic acid in *Arabidopsis*, *Plant Mol. Biol.* 47(5), 595–605.

Heinz, S., Benner, C., Spann, N., Bertolino, E., Lin, Y. C., Laslo, P., Cheng, J. X., Murre, C., Singh, H. and Glass, C. K. (2010). Simple combinations of lineage-determining transcription factors prime *cis*-regulatory elements required for macrophage and B cell identities, *Mol. Cell* 38(4), 576–589.

Hernandez-Garcia, C. M., Bouchard, R. A., Rushton, P. J., Jones, M. L., Chen, X., Timko, M. P. and Finer, J. J. (2010a). High level transgenic expression of soybean (*Glycine max*) GmERF and Gmubi gene promoters isolated by a novel promoter analysis pipeline, *BMC Plant Biol.* 10, 237.

Hernandez-Garcia, C. M., Chiera, J. M. and Finer, J. J. (2010b). Robotics and dynamic image analysis for studies of gene expression in plant tissues, *J. Vis. Exp.* 39(39), 1733.

Hernandez-Garcia, C. M. and Finer, J. J. (2014). Identification and validation of promoters and *cis*-acting regulatory elements, *Plant Sci.* 17, 109–119.

Hernandez-Garcia, C. M. and Finer, J. J. (2016). A novel *cis*-acting element in the GmERF3 promoter contributes to inducible gene expression in soybean and tobacco after wounding, *Plant Cell Rep.* 35(2), 303–316.

Higo, K., Ugawa, Y., Iwamoto, M. and Korenaga, T. (1999). Plant *cis*-acting regulatory DNA elements (PLACE) database: 1999, *Nucleic Acids Res.* 27(1), 297–300.

Hilton, I. B., D'Ippolito, A. M., Vockley, C. M., Thakore, P. I., Crawford, G. E., Reddy, T. E. and Gersbach, C. A. (2015). Epigenome editing by a CRISPR-Cas9-based acetyltransferase activates genes from promoters and enhancers, *Nat. Biotechnol.* 33(5), 510–517.

Hollenhorst, P. C., Chandler, K. J., Poulsen, R. L., Johnson, W. E., Speck, N. A. and Graves, B. J. (2009). DNA specificity determinants associate with distinct transcription factor functions, *PLoS Genet.* 5(12), e1000778.

Horak, C. E. and Snyder, M. (2002). ChIP-chip: a genomic approach for identifying transcription factor binding sites, *Methods Enzymol.* 350, 469–483.

Hudson, M. E. and Quail, P. H. (2003). Identification of promoter motifs involved in the network of phytochrome A-regulated gene expression by combined analysis of genomic sequence and microarray data, *Plant Physiol.* 133(4), 1605–1616.

Hummel, A. W., Chauhan, R. D., Cermak, T., Mutka, A. M., Vijayaraghavan, A., Boyher, A., Starker, C. G., Bart, R., Voytas, D. F. and Taylor, N. J. (2018). Allele exchange at the EPSPS locus confers glyphosate tolerance in cassava, *Plant Biotechnol. J.* 16(7), 1275–1282.

Impey, S., McCorkle, S. R., Cha-Molstad, H., Dwyer, J. M., Yochum, G. S., Boss, J. M., McWeeney, S., Dunn, J. J., Mandel, G. and Goodman, R. H. (2004). Defining the CREB regulon: a genome-wide analysis of transcription factor regulatory regions', *Cell* 119(7), 1041–1054.

Ishige, F., Takaichi, M., Foster, R., Chua, N. and Oeda, K. (1999). A G-box motif (GCCACGTGCC) tetramer confers high-level constitutive expression in dicot and monocot plants, *Plant J.* 18(4), 443–448.

Jensen, L. J. and Knudsen, S. (2000). Automatic discovery of regulatory patterns in promoter regions based on whole cell expression data and functional annotation, *Bioinformatics* 16(4), 326–333.

Jeon, J. S., Lee, S., Jung, K. H., Jun, S. H., Kim, C. and An, G. (2000). Tissue-preferential expression of a rice alpha-tubulin gene, OsTubA1, mediated by the first intron, *Plant Physiol.* 123(3), 1005–1014.

Jeong, Y. M., Mun, J. H., Lee, I., Woo, J. C., Hong, C. B. and Kim, S. G. (2006). Distinct roles of the first introns on the expression of *Arabidopsis* profilin gene family members, *Plant Physiol.* 140(1), 196–209.

Jia, H., Orbovic, V., Jones, J. B. and Wang, N. (2016). Modification of the PthA4 effector binding elements in type I CsLOB1 promoter using Cas9/crRNA to produce transgenic Duncan grapefruit alleviating XccΔpthA4:dCsLOB1.3 infection, *Plant Biotechnol. J.* 14(5), 1291–1301.

Jin, J. P., Tian, F., Yang, D. C., Meng, Y. Q., Kong, L., Luo, J. and Gao, G. (2017). PlantTFDB 4.0: toward a central hub for transcription factors and regulatory interactions in plants, *Nucleic Acids Res.* 45(D1), D1040–D1045.

Johnson, D. S., Mortazavi, A., Myers, R. M. and Wold, B. (2007). Genome-wide mapping of *in vivo* protein-DNA interactions, *Science* 316(5830), 1497–1502.

Kang, B. C., Yun, J. Y., Kim, S. T., Shin, Y., Ryu, J., Choi, M., Woo, J. W. and Kim, J. S. (2018). Precision genome engineering through adenine base editing in plants, *Nat. Plants* 4(7), 427–431.

Kanhere, A. and Bansal, M. (2005). A novel method for prokaryotic promoter prediction based on DNA stability, *BMC Bioinformatics* 6, 1.

Kearns, N. A., Pham, H., Tabak, B., Genga, R. M., Silverstein, N. J., Garber, M. and Maehr, R. (2015). Functional annotation of native enhancers with a Cas9-histone demethylase fusion, *Nat. Methods* 12(5), 401–403.

Kertesz, S., Kerenyi, Z., Merai, Z., Bartos, I., Palfy, T., Barta, E. and Silhavy, D. (2006). Both introns and long 3′-UTRs operate as *cis*-acting elements to trigger nonsense-mediated decay in plants, *Nucleic Acids Res.* 34(21), 6147–6157.

Khan, A., Fornes, O., Stigliani, A., Gheorghe, M., Castro-Mondragon, J. A., van der Lee, R., Bessy, A., Chèneby, J., Kulkarni, S. R., Tan, G., Baranasic, D., Arenillas, D. J., Sandelin, A., Vandepoele, K., Lenhard, B., Ballester, B., Wasserman, W. W., Parcy, F. and Mathelier, A. (2018). JASPAR 2018: update of the open-access database of transcription factor binding profiles and its web framework, *Nucleic Acids Res.* 46, D260–D266.

Koschmann, J., Machens, F., Becker, M., Niemeyer, J., Schulze, J., Bülow, L., Stahl, D. J. and Hehl, R. (2012). Integration of bioinformatics and synthetic promoters leads to the discovery of novel elicitor-responsive *cis*-regulatory sequences in *Arabidopsis*, *Plant Physiol.* 160(1), 178–191.

Kuscu, C., Arslan, S., Singh, R., Thorpe, J. and Adli, M. (2014). Genome-wide analysis reveals characteristics of off-target sites bound by the Cas9 endonuclease, *Nat. Biotechnol.* 32(7), 677–683.

La Russa, M. F. and Qi, L. S. (2015). The new state of the art: Cas9 for gene activation and repression, *Mol. Cell. Biol.* 35(22), 3800–3809.

Larson, M. H., Gilbert, L. A., Wang, X., Lim, W. A., Weissman, J. S. and Qi, L. S. (2013). CRISPR interference (CRISPRi) for sequence-specific control of gene expression, *Nat. Protoc.* 8(11), 2180–2196.

Le Hir, H., Nott, A. and Moore, M. J. (2003). How introns influence and enhance eukaryotic gene expression, *Trends Biochem. Sci.* 28(4), 215–220.

Lebel, E., Heifetz, P., Thorne, L., Uknes, S., Ryals, J. and Ward, E. (1998). Functional analysis of regulatory sequences controlling PR-1 gene expression in *Arabidopsis*, *Plant J.* 16(2), 223–233.

Lescot, M., Dehais, P., Thijs, G., Marchal, K., Moreau, Y., Van de Peer, Y., Rouzé, P. and Rombauts, S. (2002). PlantCARE, a database of plant *cis*-acting regulatory elements and a portal to tools for *in silico* analysis of promoter sequences, *Nucleic Acids Res.* 30(1), 325–327.

Lewis, J. D. and Izaurralde, E. (1997). The role of the cap structure in RNA processing and nuclear export, *Euro. J. Biochem.* 247(2), 461–469.

Li, P., Pan, T., Wang, H., Wei, J., Chen, M., Hu, X., Zhao, Y., Yang, X., Yin, S., Xu, Y., Fang, H., Liu, J., Xu, C. and Yang, Z. (2019). Natural variation of ZmHKT1 affects root morphology in maize at the seedling stage, *Planta* 249(3), 879–889.

Li, T., Liu, B., Spalding, M. H., Weeks, D. P. and Yang, B. (2012). High-efficiency TALEN-based gene editing produces disease-resistant rice, *Nat. Biotechnol.* 30(5), 390–392.

Lichtenberg, J., Kurz, K., Liang, X., Al-ouran, R., Neiman, L., Nau, L. J., Welch, J. D., Jacox, E., Bitterman, T., Ecker, K., Elnitski, L., Drews, F., Lee, S. S. and Welch, L. R. (2010). Wordseeker: concurrent bioinformatics software for discovering genome-wide patterns and word-based genomic signatures, *BMC Bioinformatics* 11 (Suppl. 12), S6.

Liu, K., Sandgren, E. P., Palmiter, R. D. and Stein, A. (1995). Rat growth hormone gene introns stimulate nucleosome alignment in vitro and in transgenic mice, *Proc. Natl. Acad. Sci. U.S.A.* 92(17), 7724–7728.

Liu, L., Adrian, J., Pankin, A., Hu, J., Dong, X., von Korff, M. and Turck, F. (2014a). Induced and natural variation of promoter length modulates the photoperiodic response of FLOWERING LOCUS T, *Nat. Commun.* 5, 4558.

Liu, W., Rudis, M. R., Peng, Y., Mazarei, M., Millwood, R. J., Yang, J. P., Xu, W., Chesnut, J. D. and Stewart, C. N. (2014b). Synthetic TAL effectors for targeted enhancement of transgene expression in plants, *Plant Biotechnol. J.* 12(4), 436–446.

Liu, W., Mazarei, M., Rudis, M. R., Fethe, M. H. and Stewart Jr., C. N. (2011). Rapid in vivo analysis of synthetic promoters for plant pathogen phytosensing, *BMC Biotechnol.* 11, 108.

Liu, W., Yuan, J. S. and Stewart Jr., C. N. (2013). Advanced genetic tools for plant biotechnology, *Nat. Rev. Genet.* 14(11), 781–793.

Lloyd, G., Landini, P. and Busby, S. (2001). Activation and repression of transcription initiation in bacteria, *Essays Biochem.* 37, 17–31.

Lowder, L. G., Zhang, D., Baltes, N. J., Paul, J. W., Tang, X., Zheng, X., Voytas, D. F., Hsieh, T. F., Zhang, Y. and Qi, Y. (2015). A CRISPR/Cas9 toolbox for multiplexed plant genome editing and transcriptional regulation, *Plant Physiol.* 169(2), 971–985.

Lowder, L. G., Zhou, J., Zhang, Y., Malzahn, A., Zhong, Z., Hsieh, T. F., Voytas, D. F., Zhang, Y. and Qi, Y. (2018). Robust transcriptional activation in plants using multiplexed CRISPR-Act2.0 and mTALE-Act systems, *Mol. Plant* 11(2), 245–256.

Luo, M. J. and Reed, R. (1999). Splicing is required for rapid and efficient mRNA export in metazoans, *Proc. Natl. Acad. Sci. U.S.A.* 96(26), 14937–14942.

Mahfouz, M. M., Li, L., Piatek, M., Fang, X., Mansour, H., Bangarusamy, D. K. and Zhu, J. (2012). Targeted transcriptional repression using a chimeric TALE-SRDX repressor protein, *Plant Mol. Biol.* 78(3), 311–321.

Margulies, E. H. and Birney, E. (2008). Approaches to comparative sequence analysis: towards a functional view of vertebrate genomes, *Nat. Rev. Genet.* 9, 303–313.

Mascarenhas, D., Mettler, I. J., Pierce, D. A. and Lowe, H. W. (1990) Intron-mediated enhancement of heterologous gene expression in maize, *Plant Mol. Biol.* 15(6), 913–920.

Maston, G. A., Evans, S. K. and Green, M. R. (2006). Transcriptional regulatory elements in the human genome, *Annu. Rev. Genom. Hum. Genet.* 7(1), 29–59.

Mattick, J. S. and Gagen, M. J. (2001). The evolution of controlled multitasked gene networks: the role of introns and other noncoding RNAs in the development of complex organisms, *Mol. Biol. Evol.* 18(9), 1611–1630.

Matys, V., Fricke, E., Geffers, R., Gossling, E., Haubrock, M., Hehl, R., Hornischer, K., Karas, D., Kel, A. E., Kel-Margoulis, O. V., Kloos, D. U., Land, S., Lewicki-Potapov, B., Michael, H., Münch, R., Reuter, I., Rotert, S., Saxel, H., Scheer, M., Thiele, S. and Wingender, E. (2003). TRANSFAC: transcriptional regulation, from patterns to profiles, *Nucleic Acids Res.* 31(1), 374–378.

Mazarei, M., Teplova, I., Hajimorad, M. R. and Stewart Jr., C. N. (2008). Pathogen phytosensing: plants to report plant pathogens, *Sensors* 8(4), 2628–2641.

McKendree Jr., W. L. and Ferl, R. J. (1992). Functional elements of the *Arabidopsis* Adh promoter include the G-box, *Plant Mol. Biol.* 19(5), 859–862.

Morello, L., Bardini, M., Cricrì, M., Sala, F. and Breviario, D. (2006). Functional analysis of DNA sequences controlling the expression of the rice OsCDPK2 gene, *Planta* 223(3), 479–491.

Morello, L., Bardini, M., Sala, F. and Breviario, D. (2002). A long leader intron of the Ostub16 rice β-tubulin gene is required for high-level gene expression and can autonomously promote transcription both *in vivo* and *in vitro*, *Plant J.* 29(1), 33–44.

Morello, L. and Breviario, D. (2008). Plant spliceosomal introns: not only cut and paste, *Curr. Genomics* 9(4), 227–238.

Morey, C., Mookherjee, S., Rajasekaran, G. and Bansal, M. (2011). DNA free energy-based promoter prediction and comparative analysis of *Arabidopsis* and rice genomes, *Plant Physiol.* 156(3), 1300–1315.

Odell, J. T., Nagy, F. and Chua, N. H. (1985). Identification of DNA sequences required for activity of the cauliflower mosaic virus 35S promoter, *Nature* 313(6005), 810–812.

Papikian, A., Liu, W., Gallego-Bartolomé, J. and Jacobsen, S. E. (2019). Site-specific manipulation of *Arabidopsis* loci using CRISPR SunTag systems, *Nat. Commun.* 10(1), 729.

Park, S. C., Kwon, H. B. and Shih, M. C. (1996). *cis*-acting elements essential for light regulation of the nuclear gene encoding the A subunit of chloroplast glyceraldehyde 3-phosphate dehydrogenase in *Arabidopsis thaliana*, *Plant Physiol.* 112(4), 1563–1571.

Parra, G., Bradnam, K., Rose, A. B. and Korf, I. (2011). Comparative and functional analysis of intron-mediated enhancement signals reveals conserved features among plants, *Nucleic Acids Res.* 39(13), 5328–5337.

Peng, A., Chen, S., Lei, T., Xu, L., He, Y., Wu, L., Yao, L. and Zou, X. (2017). Engineering canker resistant plants through CRISPR/Cas9-targeted editing of the susceptibility gene CsLOB1 promoter in citrus, *Plant Biotechnol. J.* 10, 1011–1013.

Pflueger, C., Tan, D., Swain, T., Nguyen, T., Pflueger, J., Nefzger, C., Polo, J. M., Ford, E. and Lister, R. (2018). A modular dCas9-SunTag DNMT3A epigenome editing system overcomes pervasive off-target activity of direct fusion dCas9-DNMT3A constructs, *Genome Res.* 28(8), 1193–1206.

Piatek, A., Ali, Z., Baazim, H., Li, L., Abulfaraj, A., Al-Shareef, S., Aouida, M. and Mahfouz, M. M. (2015). RNA-guided transcriptional regulation *in planta* via synthetic dCas9-based transcription factors, *Plant Biotechnol. J.* 13(4), 578–589.

Pikaard, C. S. and Scheid, O. M. (2014). Epigenetic regulation in plants, *Cold Spring Harb. Perspect. Biol.* 6(12), a019315.

Ponappa, T., Brzozowski, A. E. and Finer, J. J. (1999). Transient expression and stable transformation of soybean using the jellyfish green fluorescent protein (GFP), *Plant Cell Rep.* 19(1), 6–12.

Potenza, C., Aleman, L. and Sengupta-Gopalan, C. (2004). Targeting transgene expression in research, agricultural, and environmental applications: promoters used in plant transformation, *In Vitro Cell. Dev. Biol. - Plant* 40(1), 1–22.

Proost, S., Van Bel, M., Sterck, L., Billiau, K., Van Parys, T., Van de Peer, Y. and Vandepoele, K. (2009). Plaza: a comparative genomics resource to study gene and genome evolution in plants, *Plant Cell* 21(12), 3718–3731.

Proudfoot, N. J., Furger, A. and Dye, M. J. (2002). Integrating mRNA processing with transcription, *Cell* 108(4), 501–512.

Qi, L. S., Larson, M. H., Gilbert, L. A., Doudna, J. A., Weissman, J. S., Arkin, A. and Lim, W. (2013). Repurposing CRISPR as an RNA-guided platform for sequence-specific control of gene expression, *Cell* 152(5), 1173–1183.

Qi, X. T., Zhang, Y. X. and Chai, T. Y. (2007). The bean PvSR2 gene produces two transcripts by alternative promoter usage, *Biochem. Biophys. Res. Commun.* 356(1), 273–278.

Raitskin, O., Schudoma, C., West, A. and Patron, N. J. (2019). Comparison of efficiency and specificity of CRISPR-associated (Cas) nucleases in plants: an expanded toolkit for precision genome engineering, *PLoS ONE* 14(2), e0211598.

Rangannan, V. and Bansal, M. (2009). Relative stability of DNA as a generic criterion for promoter prediction: whole genome annotation of microbial genomes with varying nucleotide base composition, *Mol. Biosyst.* 5(12), 1758–1769.

Rhee, H. S. and Pugh, B. F. (2011). Comprehensive genome-wide protein-DNA interactions detected at single-nucleotide resolution, *Cell* 147(6), 1408–1419.

Robertson, G., Hirst, M., Bainbridge, M., Bilenky, M., Zhao, Y., Zeng, T., Euskirchen, G., Bernier, B., Varhol, R., Delaney, A., Thiessen, N., Griffith, O. L., He, A., Marra, M., Snyder, M. and Jones, S. (2007). Genome-wide profiles of STAT1 DNA association using chromatin immunoprecipitation and massively parallel sequencing, *Nat. Methods* 4(8), 651–657.

Rodríguez-Leal, D., Lemmon, Z. H., Man, J., Bartlett, M. E. and Lippman, Z. B. (2017). Engineering quantitative trait variation for crop improvement by genome editing, *Cell* 171(2), 470–480.e8.

Rose, A. B. (2004). The effect of intron location on intron-mediated enhancement of gene expression in *Arabidopsis*, *Plant J.* 40(5), 744–751.

Rose, A. B. (2008). Intron-mediated regulation of gene expression. In: Reddy, A. S. N., Golovkin, M. (Eds) *Nuclear Pre-mRNA Processing in Plants, Current Topics in Microbiology and Immunology*, vol. 326. Berlin, Heidelberg: Springer, pp. 277–290.

Rose, A. B. (2019). Introns as gene regulators: a brick on the accelerator, *Front. Genet.* 9, 672.

Rose, A. B., Elfersi, T., Parra, G. and Korf, I. (2008). Promoter-proximal introns in *Arabidopsis thaliana* are enriched in dispersed signals that elevate gene expression, *Plant Cell* 20(3), 543–551.

Rushton, P. J., Reinstädler, A., Lipka, V., Lippok, B. and Somssich, I. E. (2002). Synthetic plant promoters containing defined regulatory elements provide novel insights into pathogen- and wound-induced signaling, *Plant Cell* 14(4), 749–762.

Salgueiro, S., Pignocchi, C. and Parry, M. A. (2000). Intron-mediated gusA expression in tritordeum and wheat resulting from particle bombardment, *Plant Mol. Biol.* 42(4), 615–622.

Schmidt, C., Pacher, M. and Puchta, H. (2019). DNA break repair in plants and its application for genome engineering. In: Kumar, S., Barone, P., and Smith, M. (Eds), *Transgenic Plants: Methods and Protocols*, New York: Humana Press, vol. 1864, pp. 67–80.

Shah, J. and Klessig, D. F. (1996). Identification of a salicylic acid-responsive element in the promoter of the tobacco pathogenesis-related β-1,3-glucanase gene, PR-2d, *Plant J.* 10(6), 1089–1101.

Shan, Q. and Voytas, D. F. (2018). Editing plant genes one base at a time, *Nat. Plants* 4(7), 412–413.

Shen, Q. and Ho, T. H. (1995). Functional dissection of an abscisic acid (ABA)-inducible gene reveals two independent ABA-responsive complexes each containing a G-box and a novel *cis*-acting element, *Plant Cell* 7(3), 295–307.

Shew, A. M., Nalley, L. L., Snell, H. A., Nayga Jr., R. M. and Dixon, B. L. (2018). CRISPR versus GMOs: public acceptance and valuation, *Glob. Food Sec.* 19, 71–80.

Slattery, M., Riley, T., Liu, P., Abe, N., Gomez-Alcala, P., Dror, I., Zhou, T., Rohs, R., Honig, B., Bussemaker, H. and Mann, R. (2011). Cofactor binding evokes latent differences in DNA binding specificity between Hox proteins, *Cell* 147(6), 1270–1282.

Spitz, F. and Furlong, E. E. (2012). Transcription factors: from enhancer binding to developmental control, *Nat. Rev. Genet.* 13(9), 613–626.

Swinnen, G., Goossens, A. and Pauwels, L. (2016). Lessons from domestication: targeting *cis*-regulatory elements for crop improvement, *Trends Plant Sci.* 21(6), 506–515.
Tan, J., Zhang, F., Karcher, D. and Bock, R. (2019). Engineering of high-precision base editors for site-specific single nucleotide replacement, *Nat. Commun.* 10(1), 439.
Tanenbaum, M. E., Gilbert, L. A., Qi, L. S., Weissman, J. S. and Vale, R. D. (2014). A protein-tagging system for signal amplification in gene expression and fluorescence imaging, *Cell* 159(3), 635–646.
Thum, K. E., Kim, M., Morishige, D. T., Eibl, C., Koop, H. U. and Mullet, J. E. (2001). Analysis of barley chloroplast psbD light-responsive promoter elements in transplastomic tobacco, *Plant Mol. Biol.* 47(3), 353–366.
To, T. K., Saze, H. and Kakutani, T. (2015). DNA methylation within transcribed regions, *Plant Physiol.* 168(4), 1219–1225.
Tran, L. S., Nakashima, K., Sakuma, Y., Simpson, S. D., Fujita, Y., Maruyama, K., Fujita, M., Seki, M., Shinozaki, K. and Yamaguchi-Shinozaki, K. (2004). Isolation and functional analysis of *Arabidopsis* stress-inducible NAC transcription factors that bind to a drought-responsive cis-element in the early responsive to dehydration stress 1 promoter, *Plant Cell* 16(9), 2481–2498.
Ulmasov, T., Liu, Z. B., Hagen, G. and Guilfoyle, T. J. (1995). Composite structure of auxin responsive elements, *Plant Cell* 7(10), 1611–1623.
Valouev, A., Johnson, D. S., Sundquist, A., Medina, C., Anton, E., Batzoglou, S., Myers, R. M. and Sidow, A. (2008). Genome-wide analysis of transcription factor binding sites based on ChIP-Seq data, *Nat. Methods* 5(9), 829–834.
Vanet, A., Marsan, L., Labigne, A. and Sagot, M. F. (2000). Inferring regulatory elements from a whole genome: an analysis of *Helicobacter pyloriσ* 80 family of promoter signals, *J. Mol. Biol.* 297(2), 335–353.
Venter, M. (2007). Synthetic promoters: genetic control through *cis* engineering, *Trends Plant Sci.* 12(3), 118–124.
Vojta, A., Dobrinić, P., Tadić, V., Bočkor, L., Korać, P., Julg, B., Klasić, M. and Zoldoš, V. (2016). Repurposing the CRISPR-Cas9 system for targeted DNA methylation, *Nucleic Acids Res.* 44(12), 5615–5628.
Wang, H., Tang, W., Zhu, C. and Perry, S. E. (2002). A chromatin immunoprecipitation (ChIP) approach to isolate genes regulated by AGL15, a MADS domain protein that preferentially accumulates in embryos, *Plant J.* 32(5), 831–843.
Wasserman, W. W. and Sandelin, A. (2004). Applied bioinformatics for the identification of regulatory elements, *Nat. Rev. Genet.* 5(4), 276–287.
Weake, V. M. and Workman, J. L. (2010). Inducible gene expression: diverse regulatory mechanisms, *Nat. Rev. Genet.* 11(6), 426–437.
Wolt, J. D., Wang, K., Sashital, D. and Lawrence-Dill, C. J. (2016). Achieving plant CRISPR targeting that limits off-target effects, *Plant Genome* 9(3), 1–8.
Yang, L., Zhou, T., Dror, I., Mathelier, A., Wasserman, W. W., Gordân, R. and Rohs, R. (2013). TFBSshape: a motif database for DNA shape features of transcription factor binding sites, *Nucleic Acids Res.* 42, D148–D155.
Zahur, M., Maqbool, A., Irfan, M., Barozai, M. Y. K., Qaiser, U., Rashid, B., Husnain, T. and Riazuddin, S. (2009). Functional analysis of cotton small heat shock protein promoter region in response to abiotic stresses in tobacco using Agrobacterium-mediated transient assay, *Mol. Biol. Rep.* 36(7), 1915–1921.
Zeng, J., Yan, J., Wang, T., Mosbrook-Davis, D., Dolan, K. T., Christensen, R., Stormo, G. D., Haussler, D., Lathrop, R. H., Brachmann, R. K. and Burgess, S. M. (2008). Genome

wide screens in yeast to identify potential binding sites and target genes of DNA-binding proteins, *Nucleic Acids Res.* 36(1), e8.

Zhang, N., McHale, L. K. and Finer, J. J. (2015). Isolation and characterization of "GmScream" promoters that regulate highly expressing soybean (*Glycine max* Merr.) genes, *Plant Sci.* 241, 189–198.

Zhang, N., McHale, L. K. and Finer, J. J. (2016). A leader intron of a soybean elongation factor 1A (eEF1A) gene interacts with proximal promoter elements to regulate gene expression in synthetic promoters, *PLoS ONE* 11(11), e0166074.

Zhang, N., McHale, L. K. and Finer, J. J. (2019). Changes to the core and flanking sequences of G-box elements lead to increases and decreases in gene expression in both native and synthetic soybean promoters, *Plant Biotechnol. J.* 17(4), 724–735.

Zhang, W., McElroy, D. and Wu, R. (1991). Analysis of rice Act1 5′ region activity in transgenic rice plants, *Plant Cell* 3(11), 1155–1165.

Zhao, L., Li, M., Xu, C., Yang, X., Li, D., Zhao, X., Wang, K., Li, Y., Zhang, X., Liu, L., Ding, F., Du, H., Wang, C., Sun, J. and Li, W. (2018). Natural variation in GmGBP1 promoter affects photoperiod control of flowering time and maturity in soybean, *Plant J.* 96(1), 147–162.

Zhou, D. X. (1999). Regulatory mechanism of plant gene transcription by GT-elements and GT-factors, *Tends Plant Sci.* 4(6), 210–214.

Zhou, T., Yang, L., Lu, Y., Dror, I., Dantas Machado, A. C., Ghane, T., Di Felice, R. and Rohs, R. (2013). DNAshape: a method for the high-throughput prediction of DNA structural features on a genomic scale, *Nucleic Acids Res.* 41, W56–W62.

Chapter 11

The regulation of genome-edited crops

Gregory Jaffe, Center for Science in the Public Interest, USA

1 Introduction

Scientists are applying genome editing to crops with the hope and expectation that some of those new crops will provide benefits when released as commercial products. National governments need to determine whether, and if so, which of those products require regulation because they are made using genome-editing technologies. It is the job of regulators to ensure that only products that are safe for humans, animals, and the environment enter the market. Should they apply existing national laws that regulate genetically modified organisms (GMOs)? Should they regulate genome-edited products in the same fashion as they regulate conventionally bred crops? Or should they develop new regulatory procedures and requirements that are uniquely tailored to genome-edited products? How to regulate genome-edited crops is complicated by the fact that there are many different genome-editing techniques that produce crops with a full spectrum of different types of genomic changes -- including but not limited to deletions, additions, and substitutions. Products of genome editing have different genotypic changes and phenotypic traits. This chapter will provide background on genome editing and the debate over oversight of genome-edited crops. Then, it will describe how the Cartagena Biosafety Protocol might apply to genome-edited products as well as the regulatory oversight adopted by different countries. Finally, it will analyze the different regulatory systems of countries and describe issues that arise from those regulatory oversight policies and procedures.

http://dx.doi.org/10.19103/AS.2020.0082.16

2 Background on genome editing and its regulation

Before discussing the regulation of genome-edited crops, it is important to provide some information about genome editing as well as some information about the international discussions that have occurred in the past few years on the issues surrounding the regulation of genome-edited products. This section will start by describing what is meant by genome editing and the range of changes that can be made to an organism's genome. Next, there will be a discussion about regulation, including how regulation of genome-edited organisms fits into broader regulations of crops, food, and products from agricultural biotechnology. Finally, the section will touch on the complexities of the discussion among government officials and different interested stakeholders on how to appropriately regulate genome-edited products.

2.1 What is genome editing?

Genome-editing (also called 'gene editing') techniques allow scientists to make precise changes to an organism's genetic material at a pre-programmed location. Those techniques can be used to turn genes on or off, modify the regulation and function of a gene, and introduce genes that confer new traits. A neutral definition that encompasses the many different types of genome editing is the definition from a recent US National Academy of Sciences report: 'Specific modification of the DNA of an organism to create mutations or introduce new alleles or new genes' (NASEM, 2017).

To make desired changes in an organism's genome, scientists use genome editing to recognize a specific location in the organism's genome and cut the DNA at that location. Currently, the most widely used genome-editing technique is called clustered regularly interspaced short palindromic repeats (CRISPRs), although transcription activator-like effector nucleases (TALENs) and zinc finger nucleases (ZFNs) are also used at a lower frequency. New editing technologies will most likely be discovered and employed in the coming years.

Once a cell's DNA is cut, it is normally repaired perfectly by its own DNA repair mechanisms. At times, however, the repair mechanisms can make mistakes that create desirable edits. For example, the cell can repair the cut in a somewhat random fashion, resulting in an addition or deletion as small as a single base-pair or as large as hundreds of base-pairs. A random repair can also mistakenly result in small base-pair substitutions. Alternatively, the cell can repair the break in its DNA using a repair template that is designed by the scientist and introduced into the cell along with the genome-editing components. The template can be designed to result in a predetermined deletion of a section of the cell's existing DNA. It can also be designed to result in a substitution of one or a few targeted base-pairs to correct or modify a mutant gene. The template can also include one or more genes from the same or a different species to be inserted into the organism's genome at the location of the cut or to swap one allele of a gene for a different allele. For example, an additional copy of an existing gene could be added to increase production of the product that the gene codes for. Finally, some genome-editing techniques only introduce into the cell the specific guide RNA molecule (without its corresponding DNA) to locate and cut the genome in a particular spot. (Metje-Sprink et al., 2019). The different outcomes from genome editing a crop are summarized in Fig. 1.

The term 'genome editing' in this chapter does not have the same meaning as the commonly used terms 'genetic engineering', 'genetically modified', or 'genetically modified organism' (typically referred to as 'GMOs'). While both the scientific and legal definitions of 'genetic engineering' or 'GMO' vary throughout the world, in this chapter, 'genetic engineering' of plants will refer to a process whereby scientists extract a gene with a useful trait from an organism and insert it into the genome of a different organism most often using either an *Agrobacterium* or the 'gene gun'. There is no control over where this insertion happens in the genome. Genome editing can also insert a new gene into an organism's genome, but it does so more precisely, because breaks are made in the DNA at specific targeted locations and a template with the new additional gene is added to the cell so it is copied into the DNA when repairing the break. Therefore, some genome-edited products can mimic genomic changes that occur through traditional genetic engineering, such as the addition of a single gene from a different species (often referred to as a 'foreign' gene or a transgene). In other instances, the changes caused by genome editing can mimic

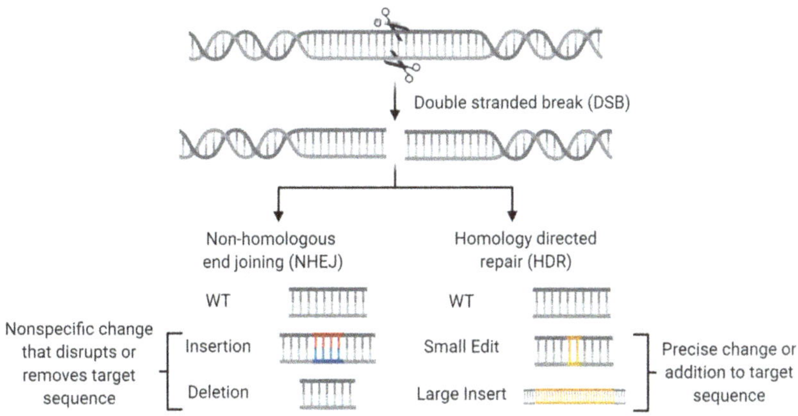

Figure 1 Diagram of most common genome editing and the different genomic outcomes. After the double-stranded break (DSB), non-homologous end joining can result in a return to the wild type (WT) or a random deletion or insertion. Homology directed repair involves a template which introduces specific edits (usually small or large additions). Image courtesy of Morgan Carter and Biorender.

those produced through conventional breeding techniques, such as chemical mutagenesis or irradiation (which result in random changes to an organism's DNA as the result of inaccurate DNA repair). Understanding how genome-edited organisms compare to organisms produced with other breeding techniques has been an important part of the discussion on how those products are regulated.

In addition to making a single change in an organism's genome, genome-editing techniques can produce targeted genomic changes - deletions, substitutions, or additions - at one or more locations in an organism at the same time (Putcha, 2016). A scientist also can stack multiple genome-edited changes into a single organism over several generations (which can also be done with traditional genetic engineering). Therefore, the degree of change in a genome-edited organism's genome can vary tremendously from a single gene edit to multiple edits. The type and degree of change is important both in determining potential risks from the organism as well as how it might be regulated. As stated in a report on the bioeconomy by the World Economic Forum (2018):

> [D]epending on the type of graduated change in an organism's genome, gene editing could produce a product identical to one created with 'traditional' mutagenesis or a product identical to a transgenic GMO...No longer are there clear distinctions for 'transgenic methods,' conventional breeding, biotechnology, and other methods and techniques to develop improved crops and animals. (WEF, 2018)

Thus, genome editing does not lead to monothetic products, but instead can achieve many different outcomes. It is just a tool that can be utilized to make a wide variety of genomic changes. This acknowledgement of varied scientific outcomes is critical for any discussion of regulation because policymakers and regulators need to decide whether to treat all products of genome editing identically or differentiate between them based on the wide-ranging changes and outcomes possible.

Finally, for genome editing, while modification can be achieved at a precise location, modifications, referred to as 'off-targets' can inadvertently occur at other places in the organism's DNA, especially at places which have a similar base-pair sequence as the targeted location. Those off-target changes often have no phenotypic impact on the organism, but investigation usually is required to confirm there is no inadvertent adverse impact. Predicting and analyzing the impact of off-targets is an issue that has been raised by scientists carrying out genome editing as well as stakeholders arguing for and against the need for genome-editing specific regulation.

2.2 What is regulation and why focus on genome-editing specific regulation?

A government's legal authority to regulate comes from laws passed by the Legislative branch of government and enacted by the Executive branch.

Regulatory obligations are either set forth in those laws or in regulations, which are formal orders or rules usually from a government agency in the Executive branch meant to oversee either a product or the conduct of an individual. There are many reasons governments may regulate a product or activity, but this analysis will focus on regulations established to ensure an independent check of a product's or activity's safety to humans, animals, and/or the environment.[1]

Government regulation can take many forms. Many people associate safety regulation with a process in which the government conducts a risk assessment and decides on whether to allow a product to enter commerce based on scientific evidence. However, there are many other forms of government oversight, some that are less stringent and others that are more stringent. Regulation can involve providing information so that the government and the public are informed about products and potential risks. The U.S. Surgeon General's warnings on tobacco and alcohol products are examples of this type of regulation. Other regulations involve the government establishing general conditions for safe conduct (such as research standards) and then requiring individuals and institutions to demonstrate compliance with those standards. Pharmaceutical companies are subject to these types of regulations, which ensure drugs have been adequately tested for safety and efficacy before they can be prescribed. At the other extreme, regulation can prohibit or greatly restrict certain products or conduct. For example, the U.S. Environmental Protection Agency bans certain chemicals, such as polychlorinated biphenyls (PCBs), if they demonstrate excessive risks to human or environmental health.

Food and agricultural products derived from genome-edited crops, like food and agricultural products produced using other techniques, will be subjected to various general laws and regulations. Most countries have laws and regulations to ensure the safety of food that enters commerce. These types of regulations make no distinction between foods produced with genome editing, conventional breeding, or any other method of producing a crop or animal. Similarly, seeds produced using genome editing would be regulated as all new seed varieties are regulated, using a country's varietal protection laws, plant protection laws, or other laws involving the safety, purity, and transport of seed.

This chapter does not discuss laws that will generally apply equally to genome-edited products and other similar products in a product category, such as food, plants, animals, seeds, and so on. Instead, it will focus on regulations that are tailored specifically to genome-edited products, meaning

1 Different countries use different terminology for legislation enacted by the legislative branch, and rules or orders set forth by the executive branch. For example, some countries refer to 'Act' to describe a law while others call similar legal instruments 'Directives.' Legal documents that qualify as regulations might be called regulations in some countries and 'resolutions' in others. In this document, legal documents will be referred to as they are identified by each country. When not discussing a specific document from a country, the terms for legal documents in the United States will be used in the text.

the regulatory requirements only apply because the product was genome edited. The most obvious example of laws and/or regulations that apply to a subcategory of products produced in a specific way is biosafety laws that regulate GMOs. As discussed below, these laws in some countries, but not others, encompass some genome-edited products. Another example of specific regulation of a sub-category of products is the regulation of pesticides. Pesticides are chemicals that could be regulated under a general chemical law. However, many countries often have special regulations for pesticidal chemicals because of their unique properties and purpose, which is to kill a pest.

The issue discussed in this chapter with respect to genome-edited products is not whether they should be subjected to any regulation, but whether they should be subjected to a country's current GMO regulations or have some other genome-editing special regulatory process due to the fact that the product was made with genome editing. A reason for genome-edited products to get special regulation would be that those products and/or the process by which they are made pose potential risks that might not be adequately covered by generally applicable laws. It is outside the scope of this chapter to include a discussion of the potential risks from genome editing and genome-edited products and whether those potential risks are different or similar to potential risks from other breeding methods and the products made from them. However, the next section does briefly describe the different perspectives on how genome-edited products should be regulated.

2.3 International debate about regulating genome-edited agricultural products

In the global discussion over whether and how to regulate genome-edited agricultural products, the question dominating the discussion has been whether and in which cases genome-edited products should be considered 'GMOs' and fall within a country's national biosafety regulatory systems for GMOs (WEF, 2018; Eckerstorfer et al., 2019). If they are not 'GMOs', then the question is whether they should be regulated as conventionally bred products or whether a country should establish new regulations specifically for genome-edited products.

Many scientists and seed developers, and some governments, have argued that if a molecular genome-editing technique is used to alter a crop's existing genome but does not introduce DNA from a different species ('foreign' DNA), the resulting product is no different from a crop with its genome altered using other traditional breeding methods. Chemical mutagenesis and ionizing radiation, which cause random changes (additions and deletions) in an organism's genome, are currently exempt from product-specific regulation in most countries around the world. Some argue that molecular genome-editing techniques that create similar products (but with greater precision) should also be exempt. This position was articulated by the American Seed Trade Association, which wrote: 'Plant varieties developed through the latest breeding methods should not be differentially regulated if they are similar to or indistinguishable from varieties that could have been produced through earlier breeding methods' (ASTA, 2020). For advocates who argue that many genome-edited products are not GMOs, the criteria they advocate using to show they are not GMOs include: (1) whether there is 'foreign' DNA in the final product; (2) whether the genome edit could have been 'found in nature'; or (3) whether the edit could have been produced through 'conventional' or 'traditional' breeding methods, including chemical mutagenesis and irradiation breeding (WEF, 2018).

On the other side of the debate, some scientists, consumer and environmental groups, and other stakeholders assert that genome-editing techniques are a category of 'genetic engineering' that could result in harmful products and should, therefore, meet the substantive legal obligations required for GMOs. As stated by the Non-GMO Project: 'Regardless of whether foreign DNA is used, though, any process where nucleic acid is engineered in the laboratory is genetic engineering, and the resulting products are GMOs' (NGMO, 2020). In fact, some of these stakeholders even argue that genome editing could be more dangerous than genetic engineering and should have more stringent regulatory oversight. A Friends of the Earth report claimed that genome editing had 'the potential for significant unintended consequences' and that CRISPR might cause 'genetic havoc' (FOE, 2018).

There is also a middle ground in the debate. That position acknowledges that genome-edited products may need some sort of specific regulation but that the laws and regulations that have been put in place around the world for GMOs are not appropriate. As stated by the World Economic Forum's Bio-Innovation Report:

> Appropriate science-based and proportionate oversight could provide predictable and transparent rules that enable developers to bring products to market and an independent safety assessment that consumers can trust. Alternatively, overly stringent and costly regulation could prevent many beneficial products from entering the marketplace, while no regulation could cause consumers to question their safety. (WEF, 2018)

In this middle ground, regulation could be established that is tailored to the potential risks, including the level of novelty of the techniques and products, and any other issues that arise from the various genome-editing techniques and the resulting products.

3 Whether the Cartagena Biosafety Protocol applies to genome-edited crops

While most countries in the world have not yet determined what type of regulatory oversight, if any, is or will be required for genome-edited products, many countries are discussing the issue and several countries already have established regulatory policies. This section analyzes whether the Cartagena Biosafety Protocol (CBP) - an international agreement that aims to ensure the safe use and trade of living modified organisms - applies to genome-edited organisms. Many countries around the world have biosafety regulatory systems that implement the requirements of the CBP and directly adopt many of its substantive provisions (including key definitions). If the scope of that international agreement includes some or all genome-edited products, that will determine how many countries regulate such edited products.

The Cartagena Biosafety Protocol, which is a sub-agreement to the United Nations' 1992 Convention on Biological Diversity (CBD), is an international treaty enacted to ensure 'an adequate level of protection in the field of the safe transfer, handling, and use of living modified organisms (LMO)'[2] (CBP, 2000). While the obligation to establish national frameworks for biosafety stems from Article 8g of the CBD, for many countries, the CBP has been the starting point for their GMO oversight (CBD, 1992). The CBP establishes a regulatory framework for carrying out scientifically sound risk assessments to identify and evaluate potential adverse effects of LMOs on biodiversity in the context of transboundary movement of LMOs. It was adopted in 2000 and came into force in 2003 when 50 countries had ratified it. As of the end of 2019, 172 countries and the European Union are parties to the CBP, but the United States is not (CBP, 2020).

Many countries have adopted laws and regulations that implement the CBP's provisions and those national laws and regulations utilize many of the terms and definitions from the CBP. With the development of genome editing, there have been discussions both at international meetings involving the CBP and within countries about to what extent the CBP covers organisms resulting from the different types of genome-editing techniques. Analyzing whether the CBP includes organisms developed through genome editing can help determine whether countries whose biosafety laws mirror provisions from the CBP likely will include organisms developed through genome editing. Examples of countries whose biosafety regulatory systems are based largely on the CBP include Argentina and Kenya.

2 The Cartagena Protocol uses the term 'living modified organism'. Almost all countries implementing the Cartagena Protocol use the term 'genetically modified organism' as well as most other international discussions about this topic. In virtually all instances, there is no scientific or substantive distinction between an LMO and a GMO. Therefore, except when specifically referring to language in the Cartagena Protocol, this article uses GMOs to be consistent with international best practice.

To determine whether the CBP's provisions incorporate genome-edited organisms, one must analyze whether those organisms fall within the CBP's definition of an LMO, which is defined as 'any living organism that possesses a novel combination of genetic material obtained through the use of modern biotechnology'. The two critical phrases in this definition are 'novel combination of genetic material and 'obtained through the use of modern biotechnology'. 'Novel combination of genetic material' is not defined in the CBP or the CBD and so that phrase is let up to individual countries to interpret its scope.

The CBP defines 'modern biotechnology' as the application of:

a In vitro nucleic acid techniques, including recombinant deoxyribonucleic acid (DNA) and direct injection of nucleic acid into cells or organelles, or
b Fusion of cells beyond the taxonomic family,

that overcome natural physiological reproductive or recombination barriers and that are not techniques used in traditional breeding and selection (CBP Article 3, 2000).

Analyzing the definition of 'modern biotechnology', it is difficult to argue that genome editing is not included. Most genome-editing techniques use '*in vitro* nucleic acid techniques' or 'recombinant DNA' so they satisfy part (a) of that definition. Also, it is hard to argue that they are 'traditional breeding and selection techniques'. In addition, they overcome natural recombination barriers because they make specific targeted genomic changes in an organism's genome and those changes can be done at a time in the cell's life cycle when natural recombination does not occur. Therefore, whether a genome-edited organism is subjected to regulation under the CBP as an LMO will hinge on the interpretation of the phrase 'a novel combination of genetic material' in the definition of 'living modified organism'.

To determine if a genome-edited crop is 'novel', one could argue that unless the genome-edited change is identical to an existing allele that has been scientifically documented in an individual organism, it is a novel combination of genetic material. This would make most genome-edited organisms qualify as LMOs, even ones caused by random DNA repair unless the random change results in an allele whose existence already has been documented in that species. Alternatively, one could argue that if the change could arise as a mutation naturally because the genetic alteration is a limited small change to the organism's existing genome without incorporating foreign DNA, it is not a novel combination of genetic material (even if the altered allele has not yet been documented). That interpretation would exempt from the definition of LMOs most genome-edited organisms with insertions or deletions produced during natural DNA repair as well as alleles that already exist within the species. Only recombination that crosses species barriers or involves foreign DNA would qualify as an LMO. However, this argument becomes more difficult to make when multiple genes are edited in an organism, because as more edited genes are placed (or 'stacked') into one organism, the likelihood of that organism's genomic changes occurring naturally decreases. Of course, the issue arises of whether you look at each genomic change individually or the cumulative change in the organism's genome to determine novelty.

Therefore, the CBP definitions do not provide a clear answer about the extent to which they capture gene-edited organisms because 'novel combination of genetic material' could have varied interpretations, some more and some less inclusive. Novelty could also vary with time because genomes are constantly changing. Until there is a decision by the collective membership of all the countries who are parties to the CBP about whether, and to what extent, to include genome-edited organisms, countries will have to choose the interpretation they believe is correct and most appropriate for their situation.

For all genome-edited organisms which do qualify as LMOs, the CBP requires scientific risk assessments and advanced informed consent by a country before the LMO is released into the environment. It also requires that the identity of those organisms and the decision made to approve them be made publicly available. The CBP sets up a Biosafety Clearinghouse mechanism which contains that information and most regulators also make that information available to the public through public notifications, public dockets, and/or websites (CBP Article 20, 2000). Therefore, for countries that signed onto the CBP, if a genome-edited product satisfies the definition of an LMO, the approval of that product will be transparent and available to the public. However, if a genome-edited product is not an LMO and does not fall within the scope of the CBP, the individual country's regulatory system will determine the degree of transparency as discussed in the country case studies below.

4 Country case studies demonstrating regulation of genome-edited crops

This section describes the regulatory policies for genome-edited products of several 'leader' countries.[3] The countries were chosen to represent the range of regulatory systems currently adopted for genome-edited products around the world. Those country systems are likely to become models that other countries around the world look to and potentially replicate when establishing their own regulatory policy.

3 While the focus of this chapter is on the regulation of genome edited crops, the CBP and most country regulatory systems apply equally to crops and animals. Where a country has regulations or other documents that address genome edited animal oversight, they are included briefly in the discussion in this section. If no reference is made to animals, it can be assumed that the regulatory system being discussed treats crops and animals in an identical manner.

Before reading about the different national regulatory systems set forth below, it is important to understand that each country cannot start from scratch and simply ask the question 'What regulation, if any, is needed for genome-edited agricultural products?' Countries already have enacted laws and regulations that may apply to genome-edited products. A country with a biosafety law that defines GMOs broadly may have no choice but to regulate genome-edited products as GMOs (at least until that law is amended or repealed). Another country may have a narrow definition of GMOs in their biosafety law and could determine whether to extend the law to cover genome-edited products or to regulate genome-edited products de novo. A third country's biosafety law may not include genome editing, but a different law, such as a law regulating new technologies, may apply to genome editing. Therefore, when reading about the different national regulatory systems set forth below, keep in mind that how a country decides to regulate genome-edited products must be viewed within the context of that country's existing laws and regulations. At the end of the section, Table 1 summarizes key aspects of the regulatory systems for genome-edited crops in the different countries.

4.1 Argentina – mandatory consultation to determine which products are genetically modified organisms (GMOs)

Argentina has a functional biosafety regulatory system that regulates GMOs and Argentina has been growing and exporting GMOs for many years. That biosafety regulatory system involves multiple laws and regulations implemented by several agencies (Burachik, 2012), including Resolution 763/2011 issued by the Ministry of Agriculture. Argentina's regulatory system defines a GMO using the CBP definitions of 'living modified organism' and 'modern biotechnology'. To obtain a permit to commercialize a GMO, the government conducts and weighs the results of three assessments: (1) an assessment on potential impacts on the agroecosystem; (2) an assessment of food safety; and (3) an evaluation of potential impacts on Argentina's international trade (CRS, 2015).

Argentina was one of the first countries to set forth a regulation establishing how it would regulate genome-edited plants. After several years of internal government discussion and external consultation with stakeholders, Argentina decided that every genome-editing technique utilizes in vitro recombinant DNA and fulfills that portion of the 'modern biotechnology' definition, but not every genome-edited product necessarily contains a 'novel combination of genetic material'. They then established 2015 Resolution no. 173/15, a regulation applicable to Innovative Biotechnology/New Breeding Techniques ('NBT') for plants. (AMALF, 2015), which sets forth a regulatory procedure for genome-edited crops different from that for GMOs. In addition to relying on the definition of a GMO to justify their new regulatory process, they also relied on the definition of an 'event', which is 'the insertion into the plant genome in a stable and joint form, of one or more genes or DNA sequences that are part of a defined genetic construction' (AMALF, 2015). In 2017, Argentina updated its animal biotechnology regulatory system with Resolution no. 79-E/17, which established the same regulatory oversight procedures for animals produced using NBTs (AMALF, 2017).

Resolution no. 173/15 sets up a 60-day consultation process with the National Advisory Commission on Agricultural Biotechnology (CONABIA) to determine if the NBT product has a new combination of genetic material. The developer submits a detailed application with data on (1) breeding methodology; (2) the introduced trait or characteristic; and (3) evidence of genetic change in the final product (Article 2). The regulator then looks at this data and determines whether there is a stable insertion of one or more genes or DNA sequences that were part of a template inserted into the genome with the genome-editing components. The developer must also show that no component molecules used to genome edit the organism remain in the DNA of the final plant product (such as a temporary transgene) (Article 3). The regulation assumes that if the plant is a GMO for any stage of its development, it remains a GMO until there is scientific evidence to show the introduced DNA is gone (Article 3) (AMALF, 2015). Finally, the regulation does not define NBTs, which allows for flexibility to encompass future techniques and arguably prevents the regulation from becoming outdated as the science develops (Whelan and Lema, 2015). Examples of current NBTs that are not genome edited, but are included in the consultation process are grafting, reverse breeding, and agro-infiltration (Whelan and Lema, 2015). The consultation process set up by Resolution no. 173/15 is summarized in Fig. 2.

To submit a consultation package, an applicant must be registered under a National Registry of Operators with Genetically Modified Plant Organisms (Resolution no. 46/04, Article 6). An applicant can submit an inquiry about a hypothetical product for a preliminary ruling, but must then submit a full consultation with scientific data for a final ruling before a product may enter commerce (Article 7) (AMALF, 2015). Argentina's regulatory body established a consultative form in a 2019 Administrative Order to standardize the information to be submitted. That form requires the expected molecular biology information to show that no foreign DNA has been added but it also requires phenotypic information, such as the organism's new phenotype, examples of products on the market with the same phenotype, analysis of any other potential impacts beyond the anticipated phenotype, and any expected changes to manage the resulting organism (AMALF, 2019a). This information is helpful to the regulator because they have the responsibility, even if the genome-edited crop is not a GMO, to recommend, based on scientific grounds, that the Ministry of

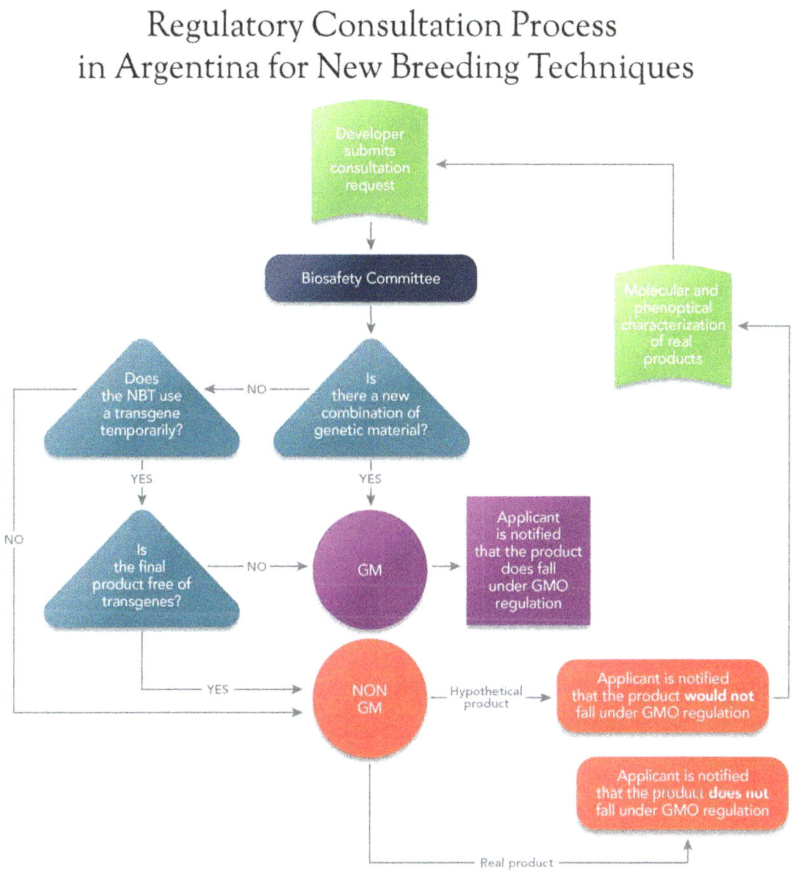

Figure 2 Regulatory consultation process in Argentina for new breeding techniques. Source: Adapted from Whelan, A. I., and Lema, M. A. (2019). Regulation of Genome Editing in Plant Biotechnology: Argentina. In 'Regulation of Genome Editing in Plant Biotechnology' (pp.19-62), Springer, Cham.

Agriculture follow up on the crop where its traits and/or novelty make it worth monitoring. (AMALF, 2019b).

Applying the resolution's language to the different types of changes possible to a crop's genome from genome editing, it is likely that: (1) deletions of any size in an organism's genome will not be considered GMOs; (2) small additions may or may not be considered GMOs depending on whether they were with the addition of an introduced DNA template; and (3) introductions of one or more genes will be considered GMOs, unless the introduced gene can be shown to be a '"perfect allelic replacement" completely indistinguishable from the outcome of a sexual crossing' (Lema, 2019). The genome-edited crops that are determined to be a GMO require government approval after completing the three assessments discussed above.

It is impossible to review how this system has operated in practice because the Argentina regulatory system does not provide detailed information about products reviewed under Resolution no. 173/15. The public does not have access to the product submissions nor the decisions by the regulator. Therefore, it is impossible for the public to know which products are or are not determined to be GMOs and what scientific evidence informed those determinations. It is unclear how the regulator has treated products with multiple edits, or how many base-pairs can be added before the DNA sequence is considered 'novel'. Similarly, one cannot determine what analysis was made about off-targets and any impact they might have. Finally, the lack of transparency makes it impossible to know which products are entering the market and could cause trade problems if those products end up in markets where they have not yet satisfied regulatory requirements. Therefore, while Argentina has a functional and operational case-by-case process for determining oversight of genome editing and has reviewed a number of products, the lack of transparency makes it impossible to judge how well the system is working and how decisions are being made.

4.2 Australia – some products are exempt and others are genetically modified organisms (GMOs)

Australia regulates GMOs under the Gene Technology Act of 2000 and has approved GMO varieties of cotton and canola that are grown in the country. That statute regulates organisms modified by 'gene technology', which is defined as 'any technique for the modification of genes or other genetic material' (GTA, 2000). A plain reading of that definition would include not only genetic engineering, but also genome editing and even some traditional breeding techniques such as mutagenesis. However, in the Gene Technology Regulations of 2001, certain techniques, such as radiation-induced and chemical-induced mutagenesis, are exempt (GTT, 2001).

In 2019, the Gene Technology Regulations were amended to address organisms modified with genome editing (GTA, 2019). Those regulations now exempt two types of organisms: (1) a 'mutant organism in which the mutational event did not involve the introduction of foreign nucleic acid'; and (2) 'an organism modified by repair of single-stranded or double-stranded breaks of genomic DNA induced by a site-directed nuclease, if a nucleic acid template was not added to guide homology-directed repair' (GTA, 2019). This means that when site-directed nuclease techniques are used to modify an organism's genome without any nucleic acid template, the organism is not a GMO. However, for this to apply, the site-directed nuclease technique must involve a protein and not the introduction of a transgene or its expressed product. An organism produced using a transgene is regulated as a GMO if the transgene is integrated into the final product or if it is transiently present. When the transgene and its product are degraded, then the organism is no longer a GMO. So, if the Cas9 and guide RNA come from DNA that is either integrated in the genome or is translated in the cell, it is a GMO. However, if the organism's DNA break is caused by the introduction into the cell of a Cas9 protein and a guide RNA without the DNA that codes for the guide RNA, it is not a GMO. Similarly, using TALENs/ZFNs would be exempt if the proteins are delivered directly into the cell without DNA. With genome editing involving a nucleic acid template, the organism is a GMO and whether the template contains a native or foreign segment of DNA is immaterial.

The 2019 amendments to the regulations also state that an organism is a GMO if the 'organism is modified using oligonucleotide-directed mutagenesis' or if the 'organism is modified with site-directed nuclease techniques involving templates to guide the repair action' (Australia, 2019). As the Australian regulatory office stated:

> In each case, the method used to modify the organism is central to determining whether or not the organism is a GMO. The number of resulting nucleotide changes, whether insertions or deletions, or whether the resulting nucleotide sequence may be found in sexually compatible species is not a deciding factor. (OGTR, 2019)

Therefore, Australia amended its regulations to address genome editing by determining that GMO status will hinge on whether a template has been introduced to change the organism's DNA. It does not depend on whether the introduced DNA is native or foreign. It also specifies which genome editing techniques produce organisms that are exempt from oversight and which ones are treated as GMOs. Those genome-edited crops treated as GMOs under the regulations require submission of an application and the granting of a license by the Office of the Gene Regulator before conducting any activities with the product.

4.3 The European Union – regulates genome-edited products the same as GMOs

The European Union has an elaborate biosafety regulatory system that regulates GMOs using several different Directives and Regulations, including 'EU Directive 2001/18 on the Deliberate Release into the Environment of Genetically Modified Organisms' (EU, 2001) and 'EU Regulation 1829/2003 on Genetically Modified Food and Feed' (EU, 2003). The EU defines GMO as 'an organism, . . . in which the genetic material has been altered in a way that does not occur naturally by mating and/or natural recombination' (EU, 2001). There is mandatory pre-market approval for all GMOs based on detailed science-based risk assessments. In addition, there is an exclusion from the substantive provisions of Direction 2001/18 for organisms produced using mutagenesis if they do not involve the use of recombinant nucleic acid molecules (EU, 2001).

European regulators' first discussion on how to apply these directives to genome-edited products started in 2007 in the New Techniques Working Group (NTWG), established by the European Commission. Parallel to the discussions in the NTWG, the European Food Safety Agency (EFSA) issued a technical report in 2012 about the scientific safety assessment needed for plants developed with the addition of a gene using ZFNs. Their conclusion was that the environmental and food safety of those products was appropriately addressed with the risk assessment requirements set forth for GMOs (EFSA, 2012).

Before European Union regulators could determine administratively whether they would apply the same oversight for genome-edited products that they conduct for GMOs, nine French NGOs brought a court case in France in 2016 that asked, among other things: (1) whether EU Directive 2001/18 applies to organisms developed through mutagenesis; (2) what the scope of the mutagenesis exemption in the Directive is; and (3) how these provisions apply to genome-editing techniques.

The case was elevated to the European Court of Justice (ECJ), who issued a decision in July 2018 (ECJ, 2018). The question before the ECJ was whether the exemption from being required to comply with the substantive requirements of the Directive for certain mutagenesis techniques applies to mutagenesis from modern breeding techniques, such as genome editing. The ECJ confirmed that organisms developed though mutagenesis are GMOs but exempt from any substantive oversight. It also decided that all genome-edited products were GMOs that did not qualify for the mutagenesis exemption (ECJ, 2018). Therefore, genome-edited organisms are GMOs and are required to comply with all the obligations required for GMOs, including the requirements to obtain pre-market approval based on a risk assessment.

To reach this decision, the ECJ found the following. First, the Directive includes as GMOs organisms made by mutagenesis methods and so all organisms developed with mutagenic techniques, including those made from chemical-induced or radiation-induced mutagenesis, are GMOs. Then, the ECJ said that the Directive exempts a subset of mutagenesis techniques 'which have been conventionally used in a number of applications and have a long safety record' from the Directive's substantive requirements (i.e. they are still GMOs but don't need any oversight). That exemption applies to conventional mutagenesis breeding methods, such as chemical mutagenesis, because the Parliament determined they had a long safety record. However, the ECJ found that some mutagenesis techniques, including genome-editing techniques, are new, do not have a long safety record, and thus do not qualify for the exemption. Therefore, they are GMOs under the Directive and must meet all the substantive requirements (ECJ, 2018). This means that the 'mutagenesis exemption' is frozen in time and new techniques cannot qualify for it (at least arguably not until they have a long history of safe use).

As the highest Court in the EU, the ECJ decision is the law and the situation can only be changed if the EU Parliament either amends the Directive or passes a new Directive specifically addressing genome-edited products. Numerous stakeholders in Europe who believe the ECJ decision was wrongly decided are mobilizing to change the regulatory system, but it is unclear whether they will be successful (EUSAGE, 2019). In November 2019, the EU Council of Ministers asked the European Commission to conduct a study on the status under Union Law of organisms developed through genomic techniques (EU, 2020).

The European Commission began consultations in 2020 on how it should regulate organisms developed with genome editing. Any legislative change would take several years to adopt and implement. EU regulators could more quickly make administrative changes that would reduce the regulatory requirements for categories of genome-edited products as a quicker interim measure (Wasmer, 2019). However, until changes are made legislatively or administratively, all genome-edited products will require pre-market approval and be regulated with the same procedures and requirements as GMOs. Those procedures require the submission of an application with detailed information about the new variety, the completion of a detailed risk assessment, an opportunity for public review and comment on the submission, and then a decision by EU officials.

4.4 Japan - mandatory notifications to determine whether a product is a GMO

Japan is a party to the CBP and has three agencies - Ministry of Environment (MOE), Ministry of Agriculture, Forestry and Fisheries (MAFF), and Ministry of Health, Labor and Welfare (MHLW) -- that regulate GMOs. In 2004, it established its Cartagena Act, which defines an LMO as 'living organisms having nucleic acids obtained by utilizing a technique for processing nucleic acids outside the cell for the purpose of transferring or replicating the nucleic acids by transferring them into a cell, virus, or viroid' (Tsuda et al., 2019). Japan does not grow any GMOs, but it has approved them to be imported and to be used in food and feed.

A plain reading of Japan's definition of an LMO suggests that most, if not all, genome-edited plants are GMOs. However, in 2019 and 2020, the three agencies each established policies for genome-edited products. Their policies each require a case-by-case review of genome-edited products to see if they qualify as a GMO, but each agency policy contains different details on what information and analysis the agency will conduct.

The MOE, which implements the CBP in Japan, set forth a policy in 2019 entitled, 'About the Handling of Organisms Produced by the Use of Genome

Editing Technology that Do Not Meet the Definition of "Genetically Modified Organism" in the Cartagena Act' (Tsuda et al., 2019). According to the policy, 'any organism that has inserted extracellularly nucleic acid (including RNA) is an LMO' unless there is 'complete removal of the inserted nucleic acid (including RNA), or its replicated product' (MOE, 2019). Japan interprets this to mean that if the genome is cut with an artificial nuclease without any foreign nucleic acid, and if the repair is random without a template being introduced into the cell, the resulting organism is not a GMO. However, if the product is developed using an artificial nuclease with foreign DNA, even if the DNA repair occurs naturally, it is a GMO. Similarly, if any inserted nucleic acid is incorporated into the final product, it is a GMO regardless of whether the artificial nuclease cutting the nucleic acid has foreign nucleic acid. There is no minimum size requirement for the extracellularly processed nucleic acid (Tsuda et al., 2019).

For genome-edited products not considered as GMOs, the MOE still has some regulatory requirements. If the genome-edited product will be used in a contained facility, it must follow the same containment measures as GMOs. If the genome-edited product is to be released outside the laboratory, then the developer must submit the following information and wait for approval to proceed from MOE:

a the fact that the organism does not possess remnants of extracellularly processed nucleic acid or its replicated product,
b taxonomical species of the modified organism,
c method of genome editing used for the modification,
d modified gene and its function,
e change in traits added by the modification,
f whether there is a change of traits other than those described in (e),
g usage of the organism, and
h discussion on possible influences on biological diversity when the organism is used (MOE, 2019).

If the organism could have any risk to biodiversity, 'the competent national authority will require additional information from the user; then, necessary measures can be taken'. MOE will post information about each organism after it has been reviewed by MOE on the Japan Biosafety Clearinghouse (it will make public items b, e, g, and h in the above information list) (MOE, 2019). The scientific data to be submitted to MOE to establish that there is no introduced nucleic acids and no impacts from off-targets are yet to be determined by MOE. The MOE's regulatory pathway for determining the regulatory status of a genome-edited crop is summarized in Fig. 3.

The Ministry of Agriculture, Forestry and Fisheries (MAFF), which has primary jurisdiction over agriculture, forests, and fish products, released its procedures for the handling of genome-edited organisms in October 2019 (Notification No. 2743) in compliance with the Japan's Cartagena Act (FAS, 2019a). That policy closely tracks the MOE policy and requires notification with the same information that is submitted to the MOE, which is used by the MAFF to decide whether there is no extracellularly processed nucleic acid or a replicated product in the genome-edited organism. The policy also reiterates that containment measures from Article 12 of the Japan's Cartagena Act are required, even for gene-edited organisms that do not qualify as GMOs. Finally, the policy states that if new breeding technologies are developed in the future to modify organisms, the MAFF should be contacted to determine what information is needed (FAS, 2019a).

In 2019, Japan's Ministry of Health, Labour, and Welfare (MHLW) released its final policy titled, 'Food Sanitation Guidelines on the Handling of Genome-edited Food and Food Additives.' Under that policy, 'genome-editing technology' is defined as 'technology that modifies a specific site of a base sequence on a chromosome in order to add a specific function' (FAS, 2019b, 2020). In addition, if a foreign gene or fragments of such genes are contained in the end product, it shall fall under the recombinant DNA techniques and is treated identical to a GMO. If a genome-edited food has no added foreign DNA but has 'deletion of base(s), substitution, and insertion of a few bases from failure to repair at the cleavage site', it is not subject to GMO oversight but

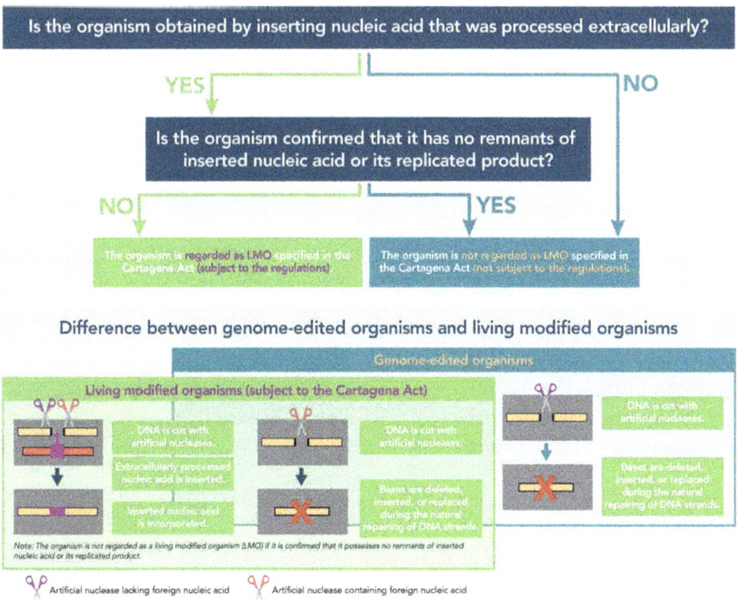

Figure 3 Process for determining regulatory oversight of genome-edited crops in Japan.

instead to a pre-market notification process. The pre-market notification, which is also required for imported products, includes:

1 crop and variety name, summary of the developed food,
2 method of the genome-editing technology applied and details of the modification,
3 information regarding confirmation of absence of introduced foreign gene or residual fragments,
4 information regarding confirmation that the identified changes in the DNA do not produce allergens or enhance known inherent toxic substances, and do not result in adverse effects on human health,
5 in case modifications have been made to change a metabolic pathway to enhance or reduce specific components, information regarding changes in the main components in the targeted metabolic pathway, and
6 the month/year of launch (FAS, 2019b, 2020).

The MHLW then makes its decision to approve the notification and makes some of the information submitted in the notification available to the public.[4] While the notification process is clearly more streamlined and does not require a detailed risk assessment, it does require similar information about the organism – such as allergenicity and toxicity – that is comparable to what is required of a GMO (FAS, 2019b, 2020).

4.5 New Zealand – GMOs and genome-edited products are regulated the same

In New Zealand, the Hazardous Substances and New Organism Act of 1996 regulates the introduction of GMOs into the environment. A GMO is defined as 'any organism in which any of the genes or other genetic material (a) has been modified by in vitro techniques' (New Zealand, 1996). Regulations under the Act identify which organisms are not considered genetically modified. Section 3(1)(b) identifies as not genetically modified:

> Organisms that are regenerated from organs, tissues, or cell cultures, including those produced through selection and propagation of somaclonal variants, embryo rescue and cell fusion (including protoplast fusion or chemical or radiation treatments that cause changes in chromosome number or cause chromosome rearrangements). (New Zealand, 1996)

4 Japan's MAFF established a policy entitled 'Feed Safety Guidelines on the Handling of Genome Edited Feed and Feed Additives" (FAS, 2020). In addition, Japan's Consumer Affairs Agency (CAA) established a policy that genome-edited foods that do not contain foreign DNA are not subject to the Food Labeling Standards (FAS, 2019c). Both these policies apply the same triggers and have similar provisions to the other policies discussed in the body of this article.

In 2013, the Environmental Protection Agency (EPA) determined that products resulting from modifications by ZFNs and TALENS were GMOs because they were modified by in vitro techniques but qualified as exempt under Section 3(1)(b) (NZ EPA, 2013). That decision was appealed to the High Court of New Zealand by the Sustainability Council of New Zealand Trust.

In 2014, the High Court determined that modifications by ZFNs and TALENS did not qualify for the exemption from being treated as GMOs. The Court found that the only thing in common with all the techniques identified in 3(1)(b) is that they were in common use at the time of the regulation and that it made sense to exempt them because they were 'well understood and established' and there was less likelihood of 'scientific and technical uncertainty about the effects of such techniques' (NZHC, 2014). For ZFNs and TALENs, the Court found that these technologies are 'relatively new' and their environmental effects were not certain. It therefore found that the exclusion only applied to the identified techniques and even if it applied to 'similar techniques', these were not sufficiently similar. Adding new techniques that are exempt is a decision for the legislation, and not for an administrative agency (NZHC, 2014). Therefore, in New Zealand, like in the EU, all genome-edited products will be treated as GMOs for the purposes of environmental safety oversight because they use *in vitro* methods to alter DNA and do not qualify for any exemption.[5] As of July 2020, there is research on GMOs in New Zealand but GMOs manufactured in New Zealand are not commercially available, so treating genome-edited crops as GMOs significantly reduces the likelihood of their commercialization. (NZ, 2020)

4.6 Nigeria – amended biosafety law to include genome editing

The National Biosafety Management Agency Act of 2015 establishes the National Biosafety Management Agency (NBMA) to carry out biosafety regulation in Nigeria. The Act requires NBMA to approve all activities involving GMOs, which are defined with the same definition found in the CBP ('novel combination of genetic material obtained through modern biotechnology') (NBMA, 2015). The NBMA is one of the few biosafety regulatory agencies in Africa to have approved GMOs for use by farmers (e.g. Bt cotton and Bt cowpea).

Nigeria enacted an Amendment to the NBMA Act in 2019 giving the NBMA the authority to approve all activities involving gene editing (as well as gene drives and synthetic biology). That Amendment defines 'gene editing' as 'a type of genetic engineering in which DNA is inserted, deleted, modified or replaced in the genome of a living organism' (NBMA, 2019). A plain reading of that

5 Australia and New Zealand jointly regulate food safety through Food Standards Australia New Zealand (FSANZ). They issued a report in December 2019 calling for a revision in their definitions related to gene technology to address genome editing. The report acknowledges "divergent views" on the acceptability and risk of genome edited foods and how to regulate them but does not set forth what will eventually be proposed or adopted. (FSANZ, 2019).

definition suggests a broad legal scope that covers deletions, replacements, and insertions, independent of whether the introduced DNA or gene is from a sexually compatible species or not (i.e. foreign DNA is irrelevant). So, this Amendment makes it clear that genome-edited products are different from GMOs but that they require approval under the biosafety law. It is anticipated that the NBMA will establish regulations or guidelines that set forth how they will implement this new legal authority and whether they will set up similar or different oversight procedures for different types of genome-edited techniques and/or products.

4.7 United States – generally treats GMOs and genome-edited products the same but exempts certain genome-edited crops

The United States' biosafety regulatory system was described in a policy titled 'Coordinated Framework on Biotechnology' in 1986 (OSTP, 1986). That policy states that products of biotechnology are to be regulated using existing laws and regulations by three primary agencies – the U.S. Department of Agriculture, (USDA), U.S. Environmental Protection Agency (EPA), and the Food and Drug Administration (FDA). Thus, agencies regulate genetically engineered and gene-edited products using laws and regulations that cover a broad range of products.

Plants produced using genetic engineering and the products made from those plants can be regulated by one, two, or all three agencies. USDA regulates genetically engineered plants utilizing the Plant Protection Act and requires permits for interstate movement, importation, and the growing of engineered plants outdoors, both for confined field trials and unconfined releases, unless the engineered plant has been granted non-regulated status (USDA, 2010). However, the USDA only regulates products it considers to be potential 'plant pests'. Until the revised regulations issued in 2020 (discussed below), this has included engineered plants produced using *Agrobacterium* in the transformation process because *Agrobacterium* is a 'plant pest'. However, USDA has exempted from oversight genetically engineered plants produced using the gene gun or other techniques that directly introduce DNA if there is no genetic material from organisms classified as plant pests. Using this same rationale, USDA has also exempted numerous genome-edited products from oversight when no organisms that are plant pests are involved. The USDA has found that the genome-editing components of CRISPR-Cas 9, TALENs, and ZFNs are not themselves plant pests (USDA, 2020a). Unlike distinctions made by regulators in Argentina, Australia, and Japan, the USDA's decisions to regulate through June 2020 have nothing to do with the presence or use of 'foreign' DNA. The USDA has exempted from regulation transgenic plants with foreign DNA, and the USDA would have exempted from oversight a plant where CRISPR was used to introduce a gene from non-plant-pest organism if such an application had been submitted to it.

In 2020, the USDA published new regulations for genetically engineered organisms with significant changes on how it regulates genetically modified organisms with specific provisions for genome-edited plants (called the Sustainable, Ecological, Consistent, Uniform, Responsible, Efficient, (SECURE) Rule) (USDA, 2020b). First, the new regulations define 'genetic engineering' to include 'techniques that use recombinant, synthesized, or amplified nucleic acids to modify or create a genome' (USDA, 2020b). This broad definition includes genetic engineering that adds one or more new genes to an organism and gene-editing techniques which edit an organism's existing genome. From the USDA's perspective, genome editing is just one type of genetic engineering.

The USDA sets forth several exemptions that eliminate any oversight for some genome-edited plants. The USDA exempts products with a single deletion of any size, a single base-pair substitution, or a single addition of a known sequence of any size (if the sequence is from the plant's gene pool). The USDA's rationale for these exemptions is that since a single deletion, substitution, or addition produces a plant that could be achieved by conventional breeding methods, and because conventionally bred plants historically have not raised plant pest risks, gene-edited plants that are the same as products that could be achieved through conventional breeding also will not pose plant pest risks (USDA, 2020b). For this reason, the USDA limits the exemption for additions to known DNA sequences that currently exist in the plant's gene pool. If an existing gene is introduced into the plant, it is exempt, but if that gene is changed in any manner (e.g. corrected or optimized), then it is not exempt. By contrast, deletions and substitutions where a random change is made are exempt, whether or not the resulting genomic sequence already exists in the plant's gene pool. The USDA rule allows developers to self-determine whether they qualify for these exemptions. It provides the developer with a voluntary option to seek confirmation of their self-determination from the USDA. There is no requirement to inform USDA when the developer self-determines their product's exempt status (USDA, 2020b).

The SECURE rule does not exempt genome-edited crops that have more than one edit to the crops' genome (USDA, 2020b). The USDA determined that, while it is possible to make more than one edit in a crop's genome through conventional breeding methods, they could not reach that conclusion for all plant species and for all types of edits, especially linked mutations or multiple mutations in a single gene. Therefore, they limited the exemption to single edits, which are 'genetic changes that could practically be achieved by conventional breeding methods in any plant' (USDA, 2020b). The USDA also clarified that for polyploid crops, a change in a gene on one chromosome is exempt but if the change is made on other chromosomes, it is considered a multi-edited crop and not exempt. This means that making a homozygously edited tomato (a deletion) would be exempt but a homozygously edited wheat with a gene deleted in all six copies would not be exempt. The USDA rule does establish a process to add exemptions for multi-edited crops, but it requires an application with scientific data that shows the multiple edits could be achieved through conventional breeding (USDA, 2020b).

For multiple edits and additions of a DNA sequence that cannot currently be found in a plant's gene pool, those genome-edited plants require some review by the USDA and possibly oversight by the USDA as a potential plant pest. The developer either voluntarily agrees to substantive oversight by seeking a permit to carry out activities with the edited plant (which includes interstate movement, importation, and outdoor activities - research and commercial plantings) or they submit an application to the USDA to conduct a Regulatory Status Review (RSR). The RSR starts with an initial 180-day process where the USDA determines if the GE plant has any plausible plant pest risks. That initial review does not require any plant-specific 'laboratory or field-test data'. If the USDA determines there are no plausible plant pest risks, the USDA sends a letter to the developer stating the plant is not regulated and publishes the letter on its website. If the USDA cannot conclude there are no plausible risks, then the USDA conducts the second part of the RSR, which is a detailed evaluation of potential plant pest risks (which can take up to 15 months). That analysis will result either in a determination that the plant is regulated and grown under permit or not regulated because it does not have any plant pest risks (USDA, 2020b). The USDA's new regulatory system set forth in the SECURE rule is summarized in Fig. 4.

The EPA regulates products that are pesticides under the Federal Fungicide, Insecticide and Rodenticide Act (FIFRA), including any pesticidal substances that have been engineered into plants. Those plant-incorporated protectants (PIPs) need to be registered, as do biological and chemical pesticides (EPA, 2020). The EPA has registered the pesticidal substance in genetically engineered Bt corn and cotton varieties as well as in crops engineered to be virus-resistant (squash and papaya). The agency has not publicly stated how it will regulate genome-edited plants. However, if a genome-edited plant has pesticidal characteristics, it is reasonable to assume that it will need to be registered as a pesticide and go through the same registration process as PIPs, especially if the pesticidal properties come from the presence of an added transgene.

The FDA regulates most food, including food from plants, under its authority in the Federal Food, Drug and Cosmetic Act (FFDCA). The FDA established a voluntary consultation process for foods derived from engineered plants in the mid-1990s (FDA, 1997). The agency has not stated that the same voluntary regulatory procedures apply to foods derived from genome-edited plants. However, the first genome-edited food crop that has been grown commercially in the United States (Calyxt's high-oleic soybeans) did submit a voluntary

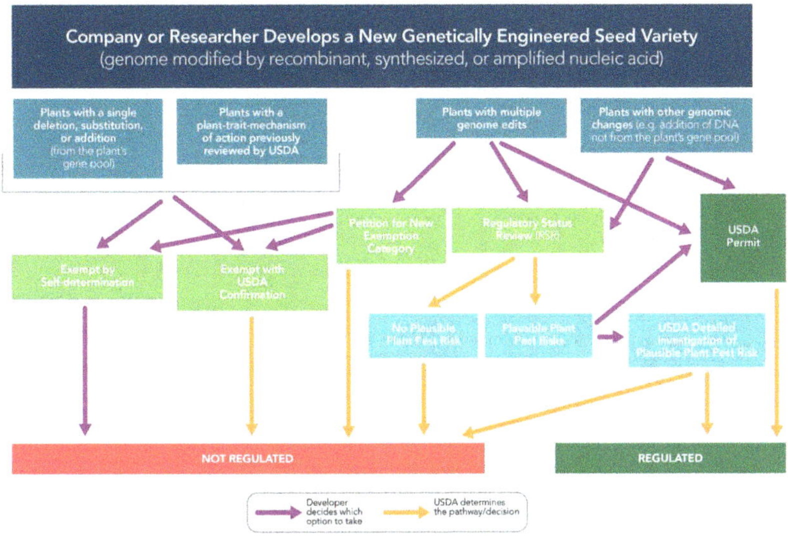

Figure 4 USDA genetically engineered plant oversight under new regulations. Source: Center for Science in the Public Interest.

consultation package and the FDA reviewed the submission in the same manner as it reviews submissions for genetically engineered plants (FDA, 2019).

The FDA also regulates genetically engineered animals and is the only federal agency with specific oversight of those animals. The FDA regulates them under the 'new animal drug' provision of FFDCA. A 'new animal drug' is anything that alters the structure or function of an animal and the FDA issued a guidance on its oversight in 2008 (Guidance #187). FDA reissued a revised version of Guidance #187 in 2017 entitled 'Regulation of Intentionally Altered Genomic DNA in Animals' (FDA, 2017). That Guidance makes clear that the same regulatory oversight that applies to genetically engineered animals applies to genome-edited animals, including single-base-pair deletions or additions (FDA, 2017).

In summary, the United States is still determining how different agencies will regulate genome-edited products. Most genome-edited products are or will be regulated under the same laws, regulations, and policies as GMOs. When those regulations are applied to genome-edited products, the result is the same oversight for animals but little or no mandatory oversight for certain genome-edited plants, depending on the number of edits and whether additions come from the plant's gene pool. For genome-edited plants, if a developer submits information to either the FDA or the USDA voluntarily for any regulatory process (such as the FDA voluntary consultation and the USDA's confirmation process), that information will be available to the public. However,

Table 1 Summary of regulation of genome-edited crops in different countries

Country	National biosafety law and/or regulation	Definition of key terms in biosafety law/regulations	Which genome-edited crops are exempt from oversight?	Regulatory oversight procedures for genome-edited products	Which genome-edited products treated as GMOs?	Information required for regulatory process for genome-edited crop
Argentina	The regulatory process in Argentina is established by various resolutions including, Resolution 763/2011 issued by the Ministry of Agriculture	**'GMO'** is a 'novel combination of genetic material created with modern biotechnology'[a]; **'event'** is 'the insertion into the plant genome in a stable and joint form, of one or more genes or DNA sequences that are part of a defined genetic construction'	No genome edited products are exempt (every product must go through a consultative process)	**Mandatory case-by case review** of products to determine if novel DNA has been added and they are a GMO	Genome edited products with foreign DNA from a template or with genes from a sexually compatible species (unless they are a perfect allelic swap)	Information on gene editing method and changes made; confirmation of no addition of foreign DNA; phenotype and potential impact of the phenotype
Australia	Gene Technology Act of 2000; Gene Technology Regulations of 2019	**gene technology** is 'any technique for the modification of genes or other genetic material'	Genome edited products modified through a mutational event without introduction of any foreign DNA and those that did not use a template to guide homologous repair	**Regulated as GMOs** based on the genome editing process utilized	Genome edited organisms modified with oligonucleotide-directed mutagenesis, site-directed nuclease techniques involving templates to guide the repair action, or any other introduction of any new DNA (even transiently)	Same information as required for GMOs

European Union	EU Directive 2001/18 on the Deliberate Release into the Environment of Genetically Modified Organisms; EU Regulation 1829/2003 on Genetically Modified Food and Feed	**GMO** is 'an organism, . . . in which the genetic material has been altered in a way that does not occur naturally by mating and/or natural recombination'	No genome-edited products are exempt; (only crops modified through mutagenesis if it is not done using recombinant DNA and the technique has a long history of safe use)	**Regulated as GMOs** with full risk assessment and pre-market approval	All genome edited crops	Same information as required for GMOs (In 2020, EFSA is reviewing which aspects of risk assessment are needed for deletions and substitutions)
Japan	Law Concerning the Conservation and Sustainable Use of Biological Diversity through Regulations on the Use of Living Modified Organisms (Law No. 97 of 2003)	**LMO** is 'living organisms having nucleic acids obtained by utilizing a technique for processing nucleic acids outside the cell for the purpose of transferring or replicating the nucleic acids by transferring them into a cell, virus, or viroid'. **genome-editing technology** is 'technology that modifies a specific site of a base sequence on a chromosome in order to add a specific function'	No genome edited products are exempt	**Mandatory notification** for all genome edited crops. Some will be regulated as GMOs. Products not found to be GMOs are required to conduct research under GMO containment conditions and submit pre-market notifications for clearance	Genome edited crops that have added nucleic acid using a template or if they have a replicated product from any inserted nucleic acid	Products that are not GMOs are still required to submit phenotype information for added traits, information on possible influences on biological diversity, and information on food safety changes (allergens, toxicity, etc.)

(*Continued*)

Table 1 (*Continued*)

Country	National biosafety law and/or regulation	Definition of key terms in biosafety law/regulations	Which genome-edited crops are exempt from oversight?	Regulatory oversight procedures for genome-edited products	Which genome-edited products treated as GMOs?	Information required for regulatory process for genome-edited crop
New Zealand	Hazardous Substances and New Organism Act of 1996	**GMO** is 'any organism in which any of the genes or other genetic material – (a) have been modified by in vitro techniques; or (b) are inherited or otherwise derived, through any number of replications, from any genes or other genetic material which has been modified by in vitro techniques'	No genome edited products are exempt (exemptions in the regulations do not apply because genome editing techniques are not identified as exempt and are not sufficiently like the exempt techniques)	**Regulated as GMOs**	All genome edited crops	Same information as required for GMOs
Nigeria	The National Biosafety Management Agency Act of 2015; 2019 Amendments to the National Biosafety Management Agency Act.	**GMO** is 'any organism . . . that possesses a novel combination of genetic material obtained through the use of modern biotechnology';[b] **gene editing** is 'a type of genetic engineering in which DNA is inserted, deleted, modified or replaced in the genome of	No genome edited products are exempt	**Regulated under the Biosafety Law** (details to be determined in forthcoming regulations/ guidance)	All genome-edited crops regulated under National Biosafety Management Agency Act	To be determined

United States Department of Agriculture[c]	The Plant Protection Act, 7 U.S.C. 7701; Sustainable, Ecological, Consistent, Uniform, Responsible, Efficient, (SECURE) Rule of 2020, 7 CFR Part 340	**genetic engineering** is 'techniques that use recombinant, synthesized, or amplified nucleic acids to modify or create a genome'	Products with a single deletion, substitution, or addition (if it is from the same plant's gene pool); multiple edits may be exempt in the future	**Regulated as GMOs** (i.e. products undergo review as potential plant pests); a streamlined regulatory status review process is used to determine if crop requires oversight	Genome edited crops with plausible plant pest risks	No laboratory or field data is needed to complete initial regulatory status review; if the more detailed 15-month review is required, then specific data needed on the crop and trait being assessed and potential to be a plant pest

[a] This definition is identical to the definition in the Cartagena Biosafety Protocol.
[b] This is the same definition as in the Cartagena Biosafety Protocol.
[c] The U.S. Department of Agriculture (USDA) is the only US agency that has established different policies and regulations for genome edited crops. The other two agencies - the U.S. Environmental Protection Agency (EPA) and the U.S. Food and Drug Administration (FDA) - currently apply their existing regulatory process for GMOs to genome-edited crops.

if a developer chooses to market their genome-edited plant without submitting information to a regulatory agency, only the developer and whomever the developer provides information to will know about the product.

5 Analysis of regulations in different countries

When one reviews how different countries regulate genome-edited crops, one finds both similarities and differences among the different regulatory systems. The following analysis highlights some key issues that result from analyzing and comparing those regulatory systems.

5.1 Most genome-edited products will be subject to specific regulation in almost all countries

Many genome editing advocates argue for no special genome editing oversight for at least some products and want those products to be treated identically to conventionally bred varieties. However, virtually all the regulatory systems elaborated above, which are representative of the regulation that will likely occur throughout the world, have some level of genome-editing specific regulation. If these regulatory systems are used as models for countries to adopt, then most countries will have some regulatory oversight that uniquely applies to genome-edited products.

One can view specific regulation for genome-edited products on a continuum, with a high level of oversight and inclusivity at one end and exemptions from oversight for some or all genome-edited products at the other end. The European Union and New Zealand could be found at the end of the continuum where there is a high degree of oversight. Both countries currently treat all genome-edited products identically to GMOs, which require pre-market approval after completing a detailed risk assessment. Nigeria's recent amendment to their National Biosafety Management Agency Act also explicitly puts oversight of genome-edited products under a pre-market approval requirement. However, the details of whether the process and scientific evidence for approval of genome-edited products in Nigeria will be the same as for GMOs have yet to be decided.

Argentina and Japan - which both have 'case-by-case' oversight - are in the middle of the continuum, as they do not exempt any genome-edited products from all oversight, but products vary in the information needed by the regulators and the procedures that the regulators will follow. In those regulatory systems, the degree of oversight depends on the types of edits and the product's final composition. However, in each of those countries, there is a mandatory requirement to consult and/or submit an application for every genome-edited product, and the product cannot proceed to market unless they receive a decision by the government agency based on the application and information they submitted. As stated by an Argentinian regulator, the Argentinian consultation process for genome-edited plants and animals results in a government-regulatory determination that the genome-edited product is or is not a GMO (Lema, 2019).

Australia's system is similar to the system in Argentina and Japan in that it separates genome-edited products into different categories based on the techniques used in the development of the products as well as its final characteristics. However, it does exempt a small subset of genome-edited products from any oversight under their Gene Technology Act (if the change is the result of random mistake caused when the cell repairs its broken DNA). Therefore, Australia and other countries (such as the United States) which exempt certain genome-edited products from any special oversight inhabit the other end of the continuum as it currently is today – where at least some genome-edited products require no specific oversight. Interestingly, the United States exemption could be perceived as narrow or large depending on whether one believes most genome-edited products will be single edits, which are exempt, or multiple edits, which require some oversight.

The conclusion from reviewing these varied oversight systems is that most countries will require some specific regulatory oversight for many genome-edited products. For many products, it will likely be less comprehensive and less burdensome than that country's oversight of GMOs. This may be adequate for some genome-edited products that have simple changes to one or a couple of genes since those products may be safer than a product with an introduced transgene. However, for some genome-edited products that most closely resemble GMOs, those products will be required to complete a detailed review and approval process with government regulators. Only rarely (e.g. for some products in Australia or the United States) will genome-edited products be treated identical to conventionally bred products and require no specific safety oversight.

5.2 The details matter – the criteria for case-by-case review vary in different countries

For countries that have decided how genome-edited crops will be regulated, the most common regulatory system involves some sort of case-by-case assessment where some products are treated identical to GMOs and others get a less thorough review. In addition to Argentina, Australia, and Japan, other countries that have adopted similar systems include Brazil, Colombia, and Chile (Schmidt et al., 2020). However, a genome-edited product could receive different degrees of oversight and require different information to be submitted in countries arguably having similar case-by-case regulatory systems. Therefore, the details around the genome-editing technique utilized, the composition of the final product, and the requirements of each country's regulatory system will determine the degree and kind of oversight in most regulatory systems. The result is that a genome-edited product's oversight could vary significantly from country to country.

Two examples show how, depending on the details of the editing process and the final product, the same product could receive different levels of oversight in different countries. The first example is a genome-edited crop where a gene from a sexually compatible species is introduced into the organism in place of the original copy. In Argentina, this product would not be considered a GMO and only needs to complete the consultation process. The developer need only show that the introduced gene is a '"perfect allelic replacement" completely indistinguishable from the outcome of a sexual crossing'. The same treatment would occur for that addition in the United States for USDA's regulatory oversight. However, in Australia, that product would be regulated as a GMO because a 'template' was used to introduce the gene - whether the gene is from a sexually compatible species is irrelevant. A similar result would occur in Japan as they consider a genome-edited product a GMO if any 'extracellular DNA is inserted, independent of whether it comes from a sexually compatible species'. In almost all country regulatory systems, if foreign DNA is introduced through genome editing and that foreign DNA remains, the resulting product is regulated as a GMO.

The second example is a genome-edited product where a gene has been knocked out through a small deletion. In Argentina, that product would only require a consultation in which the developer needs to prove that no 'foreign DNA' is present in the final organism. Whether the deletion was random or occurred from a DNA template is immaterial if the DNA template is not integrated into the final genome-edited product. In Australia, the degree of regulatory oversight depends on whether the deletion occurred randomly or was the result of a temporarily introduced DNA template. The former would be exempt from any special oversight while the latter would be treated as a GMO. That is also the case in Japan. However the regulations in Japan also distinguish for edits that are repaired randomly between edits where the DNA is cut using an artificial nuclease containing or lacking 'foreign nucleic acid'. If the artificial nuclease used to cut the DNA includes 'foreign nucleic acid', it is a GMO, and if it lacks 'foreign nucleic acid', it is not. Finally, the USDA SECURE rule would exempt that product with a single deletion because it is comparable to a product that could be developed with conventional breeding but does not exempt organisms with multiple deletions.

The amount of information in the consultation also can vary by country. For the second example above involving a simple deletion, Argentina is primarily interested in obtaining information about the techniques used and the final product to prove there is no foreign DNA in the final product. They do require an analysis of off-targets, but it is unclear from the regulations what this entails. Japan, on the other hand, requires not only information about the techniques used and the characteristics of the final product but also information on its usage, its impacts on biological diversity, as well as confirmation that it does not produce allergens or enhance toxins. Therefore, while Japan does not treat a genome-edited product with a deletion as a GMO, it still requires information similar to the information that would be required if a developer conducted a GMO risk assessment.

5.3 Many provisions of the regulatory systems are not science-based

None of the countries that have established regulatory systems for genome-edited products seem to have succeeded in making regulatory procedures and requirements that are based primarily or exclusively on scientific evidence. Instead, they integrate into their regulatory systems distinctions - sometimes arbitrary - without a scientific basis. For example, several countries, including Argentina and Japan, decide on the degree of oversight based on whether the genome-edited product has introduced foreign DNA. Anyone who has taken high school or college biology knows that DNA is the same molecule in all species and that many species have significant overlaps of common DNA (e.g. humans and other primates share over 90% of their DNA) (Cummins, 2018). There is evidence that many organisms that are considered safe contain introduced DNA from another species that occurred long ago (e.g. sweet potato with genes from *Agrobacterium*) (Kyndt et al., 2015). In addition, there is no evidence that 'foreign' DNA itself is inherently risky and needs special oversight - instead, it would depend on the introduced gene, what the gene does, the molecule it encodes, and the resulting introduced trait. Regulatory systems that establish arbitrary classifications perpetuate a view that introducing 'foreign' DNA into an organism automatically increases its potential risk, and that all genomic changes within an organism's own DNA are safe, yet neither of those two propositions is supported by scientific evidence.

Regulations in several countries also do not distinguish between different types of random deletions. In some countries (e.g. Argentina, Australia, Japan, and the United States), a single base-pair deletion that affects one gene is treated the same as a deletion of thousands of base-pairs, which might impact multiple genes and multiple cellular pathways. Most scientists would not categorize the potential phenotypic and genotypic impact of these two examples as the same, yet these regulatory systems do.

In addition, some regulatory systems assume that if a genome edit is similar to what could arise through conventional breeding, it is not risky enough to require oversight. The rationale is that conventional crop varieties have not had oversight in the past so they must not have significant risk and the same treatment should be given for genome-edited crops that could arise through conventional breeding. The USDA SECURE rule specifically exempts single-sequence deletions and substitutions because the same genotype and phenotype could arise from conventional breeding (such as chemical mutagenesis or irradiation breeding methods). Those exemptions continue 'process-based' regulation, which exempts products based on their method of development. A science-based regulatory system would not regulate based solely on how the product was made but instead determine whether a crop's introduced trait or phenotype has any potential risks based on the product's characteristics along with some consideration of how the phenotype was developed.

Could a national regulatory system for genome-edited products base its level of oversight on scientific evidence? Yes, if the oversight conducted was influenced by both the final product and the process used to achieve that product. Such a regulatory system would likely determine the degree of oversight and whether regulation is needed at all by considering criteria such as: (1) the type of edit(s) made; (2) the process for making the edit(s); (3) the resulting genotype of the organism; and (4) the organism's resulting phenotype. Some regulatory systems, such as in Argentina and Australia, consider the first three of these criteria but not the resulting phenotype, which is a key factor in determining any potential risks to humans or the environment.

5.4 Harmonization of regulatory oversight by different countries is unlikely (especially in the short run)

Many stakeholders advocate for countries to adopt the same regulatory procedures and requirements. Harmonization, if it were possible, would have numerous benefits, including cost savings and making international trade and collaborative research significantly easier. However, based on the review of the range of regulatory systems in the countries highlighted in this chapter, it is unlikely that harmonization will occur, especially in the short run. The regulatory systems in the United States and the European Union treat genome-edited products differently and harmonizing those two systems would require legislative action. Enacting and implementing legislation would take years. In addition, the regulatory systems that are considered 'case-by-case', treating some genome-edited products as GMOs and others with less oversight, may at a superficial level already seem harmonized. However, when one reviews the details, they treat identical products differently and require different levels of information, making them much less-harmonized than one might expect. Therefore, for the immediate future, global regulatory harmony for genome-edited products is a distant dream. This also has been the case for biosafety oversight of GMOs. However, it is possible that there could be regional harmonization or subregional harmonization between neighboring countries. Several Latin American countries have used the Argentinian mandatory consultation process as a model for determining how they might regulate genome-edited products (Schmidt et al., 2020; Metje-Sprink et al., 2019).

5.5 Transparency is critical but is inconsistent among regulatory systems

In today's world, information – both accurate and inaccurate – is available on the internet 24 hours a day around the world. Some individuals and stakeholders will want to know whether a product has been genome-edited. How will that farmer, grain handler, retailer, consumer, or government official find factually correct information? If that information is not available or is perceived by some people to be hidden, will that lack of information translate into concern that genome-edited products are unsafe and avoidance of them in the marketplace? Transparency about which products have been genome-edited and where they are in commerce will be important for ensuring that: (1) consumers who believe this information is important to them can find it; (2) the market can meet consumer expectations by providing non-genome-edited products for those who want them; (3) regulators know which products qualify for oversight; and (4) international commerce can occur without trade disruptions.

While the primary purpose of biosafety regulatory systems is to ensure safety, a corollary benefit is that those regulatory systems provide transparency about emerging biotech products to the public, other countries, and the marketplace. Regulatory oversight depends on transparency to identify regulated products. Information regarding regulatory decisions on GMOs has been made available throughout the world through international and national mechanisms, and this information has been critical to smooth international trade disputes and to enable stakeholders in the food chain to address the demand for information by some consumers. In contrast, the degree of transparency around regulated genome-edited products seems to vary depending on the country. Japan's regulatory system for genome-edited products is relatively transparent. The different regulatory agencies publicly provide information on genome-edited products that do not fall under the GMO regulations, including information about each organism and its modified traits. At the other extreme, Argentina conducts consultations for all genome-edited organisms, but does not release any information about those applications nor its decisions that the organism does not qualify as a GMO. In the United States, if developers self-determine that their genome-edited crops qualify for an exemption, they do not need to notify the USDA or the public, so there is no transparency. However, if a genome-edited crop developer makes a submission to the USDA to confirm its self-determination, obtains a permit, or goes through the Regulatory Status Review process (as would be required for genome-edited plants with multiple edits), the USDA will make its determinations publicly available. In other countries, such as Australia, it is unclear what information will be provided publicly on organisms that are not regulated as GMOs. For many regulators and other stakeholders, transparency regarding the regulatory status of genome-edited products is important (Eckerstorfer et al., 2019). When governments are not providing consistent and accurate information, third parties will generate their own lists of genome-edited products that vary in their accuracy. Such a result could lead to misinformation and hinder genome-edited product acceptance. Some stakeholders have called for national or international registries that identify all commercial genome-edited products in agriculture and food (Jaffe, 2019; Eckerstorfer et al., 2019).

5.6 Addressing organisms with multiple or complex genomic changes

Countries that have established regulatory policies for genome-edited products generally explain how their systems operate using simple editing examples - for example, a deletion to knock out a gene, a base-pair change to correct a gene, or the addition of a new gene. However, scientists have been successful editing multiple genes at one time in an organism. In addition, products could be developed where edited genes are stacked into a single organism to create a trait from the cumulative or synergistic effect of multiple gene changes, such as changing a metabolic pathway or causing the organism to produce a new molecule. How complex genome-edited products will be regulated in most countries is unclear.

As stated earlier, one of the arguments for exempting genome-edited products from regulation is that the same genomic allele or point mutation could be achieved from conventional breeding and should not be considered a 'novel' organism that might pose risks that need oversight. That scientific argument seldom applies to complex genome-edited products because some of them could not occur naturally or be developed with conventional breeding. It may be relatively easy to argue and prove scientifically that a single-point mutation likely could occur naturally and is not novel, but it is harder to scientifically support the same argument when multiple edits are involved, either done at the same time or stacked by crossing plants over several generations (Custers et al., 2019). The USDA SECURE rule does not exempt crops with multiple edits for just this reason - that from a practical standpoint, multiple edits are not universally achievable in all crop varieties using conventional breeding methods. Also, it is hard to argue that the potential risks of making one precise edit in one location versus multiple edits (such as making two dozen precise edits at different locations) are the same. Multiple edits increase the possibility of off-targets. The potential risk, not just the novelty, should determine the degree and type of regulatory oversight for safety.

How products with complex genome editing will be regulated in most countries will only be determined when those products move toward commercial marketing and those developers approach regulators asking what oversight is required. The exception is the USDA SECURE rule. The USDA does not approach this issue using 'novelty' as the regulatory trigger. Instead it has exempted products that could practically occur using conventional breeding and finds that it can only make that scientific conclusion for single edits within a genome.

A plain reading of the regulations in Argentina or Australia suggests that if each one of the multiple edits in a crop individually is the result of template-less repair and does not introduce new DNA, then the whole organism with multiple edits would not be considered a GMO. However, if the result of the edits is a new molecule or elevated levels of existing molecules, it is possible that the analysis of whether to regulate the organism could be different. Such a product could be more like a GMO than a conventionally bred product.

As stated earlier, many countries will look to whether these complex genome changes are GMOs under the Cartagena Protocol. While one may be able to scientifically establish that knocking out one gene or correcting a single base-pair mutation 'could occur in nature' and is not a 'novel combination of genetic material', - the trigger for inclusion under the Cartagena Protocol -- those arguments are much more difficult when many genes are altered in an organism or multiple genes are edited to change a metabolic pathway. Clearly, the more edits that are made in an organism, the harder it is to say that the organism's genomic changes are not 'novel'. Therefore, in many countries, including those with established regulatory procedures for genome-edited products, there remains uncertainty about the level of oversight for complex genome-edited products.

6 Conclusion and future trends

The science around genome editing is moving rapidly and many different crop varieties will be developed that could benefit farmers and consumers around the world. However, before those crops can be grown by farmers, they will need to comply with each country's applicable laws and regulations that ensure those products are safe. In all countries, genome-edited crops will need to comply with the same laws and regulations applicable to crop varieties developed using conventional breeding techniques. In some countries and for some genome-edited crops, compliance with those laws and regulations will be sufficient to market the genome-edited crop. For other countries, some or all genome-edited crops will need to comply with additional laws and regulations that only apply to genome-edited products (or that equally apply to GMO crops developed with genetic engineering). For most countries, whether to require specific regulations for genome-edited crops has not yet been determined.

The regulatory systems that have been established in several 'leader' countries around the world – Argentina, Australia, the European Union, Japan, New Zealand, Nigeria, and the United States – vary tremendously in their approach to regulate genome-edited crops. If one views the degree of oversight on a continuum from minimal to stringent, the EU and New Zealand would be at the stringent extreme, regulating each genome-edited product in the same manner as it regulates GMOs – requiring mandatory pre-market approval after conducting a detailed scientific risk assessment. At the other end of the continuum is the United States and Australia, which exempt certain genome-edited products (such as deletions that occur spontaneously after a DNA break) from any regulation other than what would be required for conventionally bred crops. In the middle of the continuum would be Argentina and Japan, which require mandatory government consultations for all genome-edited crops with limited scientific information and then determine if the genome-edited crop qualifies as a GMO and requires more stringent oversight.

The conclusions that can be reached from reviewing those regulatory systems are as follows:

- All countries will regulate some genome-edited products using the same regulations that are required for GMOs (primarily genome edited products that add DNA or a foreign gene).
- Most countries will require some regulation of all genome-edited products solely because they were genome edited, although that oversight could be a streamlined consultation process set up to ensure that the genome-edited crop is not a GMO.
- The regulatory systems for genome-edited products in different countries will differ in their detail, meaning that identical genome-edited crops will be treated differently in different countries (e.g. a genome-edited crop may be considered a GMO in one country and not in another).
- Many provisions in the regulatory systems are not necessarily based on science, such as treating changes made with a template different from those made randomly or using 'foreign' DNA as a trigger for oversight independent of the phenotype introduced by the added DNA.
- Global harmonization of regulations for genome-edited crops is not likely in the immediate future.
- The degree of transparency about which genome-edited products have been reviewed by regulators and which ones are commercial products will vary tremendously by country.

- The regulatory pathway for complex genome-edited products with multiple edits is much less certain than the pathway for simple edits.

A regulatory system that ensures the safety of a product should be transparent and predictable, treat products proportionally based on their potential risk, and use scientific evidence and risk assessments to ensure safety before a product is released into the market. The policy debate over how to regulate genome-edited crops has primarily focused on whether a genome-edited crop should be treated as a GMO or treated as a conventionally bred crop. Some genome-edited crops may be less risky than GMOs while other genome-edited products may be riskier than conventionally bred crops. Genome editing of crops will produce many different types of products, each of which has its own risk profile. Any regulatory system needs to consider both the genotype and phenotype of the final product as well as the process by which it was created. It should treat different categories of genome-edited products (e.g. deletions, additions, multiple edits) proportionately based on their potential risk. Instead of trying to pigeonhole new products with a range of potential risk profiles into the existing regulatory systems for GMOs or conventionally bred products, countries should establish regulatory oversight procedures that ensure safety in a proportionate, science-based and risk-based manner.

7 References

AMALF (2015). Argentina Ministry of Agriculture, Livestock and Fisheries: Resolution 173/15, Available at: http://servicios.infoleg.gob.ar/infolegInternet/anexos/245000-249999/246978/norma.htm (Accessed June 8, 2020).

AMALF (2017). Argentina Ministry of Agriculture, Livestock and Fisheries: Resolution 79-E/17/, Available at: http://servicios.infoleg.gob.ar/infolegInternet/anexos/285000-289999/287337/norma.htm (Accessed June 8, 2020).

AMALF (2019a). Argentina Ministry of Agriculture, Livestock and Fisheries: Administrative Order 1/2019. DI-2019-1-APN-DB#MPYT.

AMALF (2019b). Argentina Ministry of Agriculture, Livestock and Fisheries: Resolution 36/2019. RESOL-2019-36-APN-SAYBI#MPYT.

ASTA (2020). American Seed Trade Association Frequently Asked Questions on Plant Breeding, Available at: https://www.betterseed.org/wp-content/uploads/ASTA-Resource_NBT_Updates.pdf (Accessed July 21, 2020).

Burachik, M. (2012). Regulation of GM Crops in Argentina, *GM Crops and Food*, 3(1), 48–51.

CBD (1992). Convention on Biological Diversity (CBD), Available at: https://www.cbd.int/convention/text/ (Accessed July 20, 2020).

CBP (2000). Cartagena Protocol on Biosafety to the Convention on Biological Diversity (CBP), Available at: https://www.cbd.int/doc/legal/cartagena-protocol-en.pdf (Accessed July 10, 2020).

CBP (2020). Cartagena Protocol on Biosafety Official Website, Available at: https://bch.cbd.int/protocol/ (Accessed June 7, 2020).

CRS (2015). Congressional Research Service (CRS – USA): "Restrictions on Genetically Modified Organisms: Argentina", Available at: https://www.loc.gov/law/help/restrictions-on-gmos/argentina.php (Accessed April 23, 2020).

Cummins, E. (2018). Bananas: Your Cousin, Maybe?, *Popular Science*, August 14, 2018, Available at: https://www.popsci.com/humans-genetically-linked-to-bananas/ (Accessed June 8, 2020).

Custers, R., Casacuberta, J. M., Eriksson, D., Sági, L. and Schiemann, J. (Jan 2019). Genetic Alterations that do or do not occur naturally; Consequences for genome-edited organisms in the context of regulatory oversight, *Frontiers in Bioengineering and Biotechnology* 6, Article 213.

ECJ (2018). European Court of Justice (ECJ). In: Case C-528/16, July 25, 2018, Available at: http://curia.europa.eu/juris/document/document.jsf;jsessionid=93215E86AD2241BFA04AEEAFD64E00D7?text=&docid=204387&pageIndex=0&doclang=en&mode=req&dir=&occ=first&part=1&cid=3070137 (Accessed June 7, 2020).

Eckerstorfer, M. F., Engelhard, M., Heissenberger, A., Simon, S. and Teichmann, H. (Feb 2019). Plants Developed by New Genetic Modification Technologies – Comparison of Existing Regulatory Frameworks in the EU and non-EU Countries, *Frontiers in Bioengineering and Biotechnology* 7, Article 26.

EFSA (2012). European Food Safety Authority (EFSA): Scientific Opinion Addressing the Safety Assessment of Plants Developed with Zinc Finger Nuclease 3 and Other Site-Directed Nucleases with Similar Function, *EFSA Journal* 10(10), 2943.

EPA (2020). United States Environmental Protection Agency (EPA): 40 CFR 174. Procedures and Requirements for Plant-Incorporated Protectants, Available at: https://ecfr.io/Title-40/pt40.26.174 (Accessed June 7, 2020).

EU (2001). Directive 2001/18/EC of the European Parliament and of the Council of 12 March 2001 on the Deliberate Release into the Environment of Genetically Modified Organisms and Repealing Council Directive 90/220/EEC – Commission Declaration, Available at: https://eur-lex.europa.eu/legal-content/EN/TXT/?uri=CELEX:32001L0018 (Assessed July 31, 2020).

EU (2003). Regulation (EC) No 1829/2003 of the European Parliament and of the Council of 22 September 2003 on genetically modified food and feed, Available at: https://eur-lex.europa.eu/legal-content/EN/ALL/?uri=CELEX:32003R1829 (Accessed July 31, 2020).

EU (2020). European Commission Study on New Genomic Techniques, Available at: https://ec.europa.eu/food/plant/gmo/modern_biotech/new-genomic-techniques_en (Accessed July 20, 2020).

EUSAGE (2019). Open Statement of the European Union Sustainable Agriculture through Genome Editing, Available at: https://www.uu.nl/sites/default/files/Open%20Statement%20EU-SAGE%20July%202020_EN.pdf (Assessed July 31, 2020).

FAS (2019a). United States Foreign Agriculture Service (FAS) GAIN Report: "MAFF Guidance for the Handling of Genome Edited Organisms under the Cartagena Act," November 19, 2019.

FAS (2019b). United States Foreign Agriculture Service (FAS): GAIN Report: "Japan Invites Comments on Genome-Edited Food Handling Procedures," July 5, 2019.

FAS (2019c). United States Foreign Agriculture Service (FAS): Gain Report, "Labeling Guidance for Genome Edited Food Products," October 8, 2019.

FAS (2020). United States Foreign Agriculture Service (FAS): GAIN Report: "Japan Finalizes Handling Procedures for Genome-Edited Feed and Feed Additives," February 14, 2020.

FDA (1997). United States Food and Drug Administration (FDA): Consultation Procedures Under FDA's 1992 Statement of Policy for Food Derived from New Plant Varieties, Available at: https://www.fda.gov/food/ingredients-additives-gras-packaging-guidance-documents-regulatory-information/consultation-procedures-under-fdas-1992-statement-policy-foods-derived-new-plant-varieties (Accessed June 7 2020).

FDA (2017). United States Food and Drug Administration (FDA): Guidance #187 entitled "Regulation of Intentionally Altered Genomic DNA in Animals", https://www.fda.gov/media/74614/download (Accessed June 7, 2020).

FDA (2019). United States Food and Drug Administration (FDA): Letter to Calyxt from FDA Regarding High Oleic Soybean Dated 2/26/19, Available at: https://www.fda.gov/media/120707/download (Accessed June 7, 2020).

Friends of the Earth (2018). Gene-Edited Organisms in Agriculture: Risks and Unexpected Consequences, Available at: https://1bps6437gg8c169i0y1drtgz-wpengine.netdna-ssl.com/wp-content/uploads/2018/09/FOE_GenomeEditingAgReport_final.pdf (Accessed June 8, 2020).

FSANZ (2019). Food Standards Australia New Zealand (FSANZ) Final Report "Review of Food Derived Using New Breeding Techniques," December 2019, Available at: https://www.foodstandards.gov.au/consumer/gmfood/Pages/Review-of-new-breeding-technologies-.aspx (Accessed July 28, 2020).

GTA (2000). Gene Technology Act (Australia), Available at: https://www.legislation.gov.au/Details/C2016C00792 (Accessed July 10, 2020).

GTA (2019). Gene Technology Amendment (2019 Measures No. 1) Regulations 2019 (Australia), Available at: https://www.legislation.gov.au/Details/F2019L00573/Explanatory%20Statement/Text (Accessed July 10, 2020).

GTT (2001). Gene Technology Regulations (2001) (Australia), Available at: https://www.legislation.gov.au/Details/F2016C00615 (Accessed July 10, 2020).

Jaffe, G. (2019). Biotech Blog: A Call for a National Registry of Gene-Edited Crops April 22, 2019, Available at: https://www.pnas.org/content/112/18/5844 (Accessed June 8, 2020).

Kyndt, T., Quispe, D., Zhai, H., Jarret, R., Ghislain, M., Liu, Q., Gheysen, G. and Kreuze, J. F. (2015). The Genome of Cultivated Sweet Potato Contains Agrobacterium T-DNAs with Expressed Genes: An Example of a Naturally Transgenic Food Crop, *Proceedings of the National Academy of Sciences* 112(18), 5844–5849.

Lema, M. A. (2019). Regulatory Aspects of Gene Editing in Argentina, *Transgenic Research* 28(S2), 147–150.

Metje-Sprink, J., Menz, J., Modrzejewski, D. and Sprink, T. (Jan 2019). DNA Free Genome Editing: Past, Present and Future, *Frontiers in Plant Science*, 9, Article 1957.

MOE (2019). Ministry of Environment (MOE) (Japan): "To Genome Editing Technologies Users", Available at: https://www.env.go.jp/press/2_2_%20genome%20editing_En.pdf (Accessed August 25, 2020).

NASEM (2017). National Academies of Science, Engineering and Medicine (NASEM): Preparing for Future Products of Biotechnology, Available at: https://www.nap.edu/catalog/24605/preparing-for-future-products-of-biotechnology (Accessed June 8, 2020).

NBMA (2019). Nigeria National Biosafety Management Act Amendments, Available at: https://nbma.gov.ng/wp-content/uploads/2017/06/Amendment-of-NBMA-Act-2019.pdf (Accessed June 7, 2020).

NBMA (2015). Nigeria National Biosafety Management Act (2015), Available at: https://www.nbma.gov.ng/wp-content/uploads/2017/06/National_Biosafety_Management_Agency__Act_2015-signed.pdf (Accessed June 7, 2020).

New Zealand (1996). Hazardous Substance and New Organisms Act, Available at: http://www.legislation.govt.nz/act/public/1996/0030/latest/whole.html#DLM381222 (Accessed June 7, 2020).

NGMO (2020). Non-GMO Project Website, Available at: https://www.nongmoproject.org/blog/emerging-techniques-in-biotechnology-pose-new-risks-to-the-non-gmo-supply-chain/ (Accessed June 8, 2020).

NZ (2020). New Zealand Ministry for the Environment webpage on "about genetic modification in New Zealand,", Available at: https://www.mfe.govt.nz/more/hazards/new-organisms/genetic-modification-new-zealand/about-gm-new-zealand (Accessed July 31, 2020).

NZ EPA (2013). New Zealand Environmental Protection Agency (NZ EPA): Decision Whether or not an Organism Is a New Organism under the Hazardous Substance and New Organism Act, Available at: https://epa.govt.nz/assets/FileAPI/hsno-ar/APP201381/03e845f138/APP201381-APP201381-Decision.pdf (Accessed June 7, 2020).

NZHC (2014). New Zealand High Court, Wellington Registry (NZ Court Decision). Court Decision CIV 2013-485-877 (2014) NZHC 1067. Judgement of Mallon J.

OGTR (2019). Office of the Gene Technology Regulator (OGTR) (Australia): Overview of the Amendments to the Gene Technology Regulations 2001, September 2019, Available at: http://www.ogtr.gov.au/internet/ogtr/publishing.nsf/Content/23A002A6AC8A4908CA258478000D2031/$File/Overview%20of%20amendments%20to%20the%20Gene%20Technology%20Regulations%202001.pdf (Accessed June 7, 2020).

OSTP (1986). United States Office of Science and Technology Policy (OSTP): Coordinated Framework for the Regulation of Biotechnology, *US Federal Register*, 51(123), 23302.

Putcha, H. (2016). Using CRISPR/Cas in three dimensions: towards synthetic plant genomes, transcriptomes and epigenomes, *The Plant Journal* 87, 5–15.

Schmidt, S. M., Belisle, M. and Frommer, W. B. (2020). The evolving landscape around genome editing in agriculture, *EMBO Reports*, 21(6), e50680.

Tsuda, M., Watanabe, K. N. and Ohsawa, R. (2019). Regulatory Status of Genome-edited Organisms under the Japanese Cartagena Act, *Frontiers in Bioengineering and Biotechnology* 7, Article 387.

USDA (2010). United States Department of Agriculture (USDA): 7 CFR Part 340. Regulation Entitled: Introduction of Organisms and Products Altered or Produced through Genetic Engineering Which Are Plant Pests or Which There is Reason to Believe Are Plant Pests. https://www.aphis.usda.gov/biotechnology/downloads/7_cfr_340.pdf (Accessed June 7, 2020).

USDA (2020a). United States Department of Agriculture – Biotechnology Regulatory Services (BRS): Regulated Articles Letters of Inquiry, Available at: https://www.aphis.usda.gov/aphis/ourfocus/biotechnology/am-i-regulated/regulated_article_letters_of_inquiry/regulated_article_letters_of_inquiry (Accessed June 7, 2020).

USDA (2020b). United States Department of Agriculture (USDA): Movement of Certain Genetically Engineered Organisms Final Rule, *US Federal Register*, 85(96), 29790–29838.
Wasmer, M. (June 2019). Roads Forward for European GMO Policy – Uncertainties in the Wake of ECJ Judgment Have to be Mitigated by Regulatory Reform, *Frontiers in Bioengineering and Biotechnology*, 7, Article 132.
WEF (2018). World Economic Forum: "Bioinnovation in the Food System. Toward a New Chapter in Multistakeholder Collaborations", Available at: http://www3.weforum.org/docs/WEF_Bio-Innovation_Dialogue_WP.pdf (Accessed June 8, 2020).
Whelan, A. I. and Lema, M. A. (2015). Regulatory Framework for Gene Editing and Other New Breeding Techniques (NBTs) in Argentina, *GM Crops and Food* 6(4), 253–265.

Part 2

Applications

Chapter 12

Genome editing of barley

Martin Becker, Leibniz Institute of Plant Genetics and Crop Plant Research (IPK), Germany; and Goetz Hensel, Leibniz Institute of Plant Genetics and Crop Plant Research (IPK), Germany, and Palacký University, Czech Republic

1 Introduction

Although barley is of great importance for the brewing and animal feed industries and is regarded as a model for small grain cereals, only few results on targeted gene modification using CRISPR/Cas endonuclease technology have been published to date. This may be due to the relatively long generation time and the small number of laboratories that can transform barley. Even if many results of basic research on transient systems such as biolistic bombardment of barley leaves (Budhagatapalli et al., 2016) or PEG-mediated transformation of protoplasts were obtained routineously (Gerasimova et al., 2018; Hahn et al., 2019), stable heredity in the offspring is always required for practical application. Since in barley, unlike in dicotyledons, cannot use leaves to regenerate plants and the regeneration from protoplasts only works for a few genotypes, efficient protocols are required for the production of such plants. The same holds true for the usage of DNA-free ribonucleoprotein complexes in protoplasts.

Precise genome editing (or gene targeting) using sequence-specific endonucleases becomes state of the art. It was shown for model and non-model organisms that randomly induced insertions or deletions (InDels) and

http://dx.doi.org/10.19103/AS.2020.0082.18

precise editing using homology-directed repair (HDR) can be used to achieve heritable mutations (Zhang et al., 2019).

In this chapter, the frontiers and achievements of the currently used techniques in barley genome modification will be shown and discussed.

The principle of targeted mutagenesis in plants originated from the zinc-finger nucleases and meganucleases which had their disadvantage in costs and complexity. Thus, they become uncommon in plant science. The mutation rate was too less, but both were able to mutate the desired loci as shown multiple times in the literature (Peer et al., 2015; Ran et al., 2018). Nevertheless, TAL effector-like nucleases (TALEN) and RNA-guided endonucleases (Cas-enzymes) guided their way in plant science due to their ability for easy cloning and relatively low costs which come along with a higher stability and editing availability. Zinc-finger nucleases and TALEN have in common that they use the dimerization of the FokI-endonuclease to create a cut between two specific recognition sites in the genome. In contrast, the Cas enzymes introduced a double-strand break (DSB) upstream of a specific recognition site (Protospacer adjacent motif, PAM) by its endonuclease domains. The Cas9 enzyme needs a sequence-specific RNA (single guide RNA; sgRNA) and a tracer RNA (trRNA). While the sgRNA recognizes the specific region in the genome the trRNA enables the binding in the Cas9 due to its two-dimensional structure. The complex of sgRNA and trRNA is called guide RNA (gRNA). An in vitro assembly of gRNA and Cas9 builds the ribonucleotide-protein-complex (RNP). Compared to FokI-based sequence-specific endonucleases, Cas9 can be used also as protein complex, inducing mutations at predefined target sites as a DNA-free approach. Regarding off-targets, FokI-fusion endonucleases are, in theory, much more specific due to the need of two sequence-specific motifs and therefore a longer stretch of DNA, which might be specific for a particular plant genome. However, off-targets using Cas-enzymes can be avoided by off-target predictions software (e.g. Deskgen); furthermore, it has already been reported that off-targets using Cas-enzymes should occur more often in theory, which is not the case practically (Modrzejewski et al., 2018). Another aspect often less discussed is the fact that a given guide RNA might be present several times in the plant genome, but the mandatory adjacent PAM motif is rarely available. Since functionality is only provided if both elements (gRNA and Cas protein) can bind, this dramatically reduces the number of putative off-targets.

The majority of approaches using sequence-specific endonucleases target a transgene within a transgenic plant, so-called gene targeting, for example, β-glucuronidase in *Arabidopsis thaliana* (Wolter et al., 2018). Some approaches are targeting an endogenic locus in the plant cell which is more likely a precise gene editing rather than gene targeting. The induction of a DSB in the plant cell triggers the cell-specific repair mechanism. In principle, such mutations occur often by failures in transcription or natural radiation and can be easily repaired by non-homologues end joining (NHEJ) or HDR. The single use of a binary vector carrying Cas-enzyme and gRNA or a single RNP will result in mutations done by a failure in the repair of DSB by the NHEJ. Authors call these approaches frequently as precise, knowing that they target a precise position in the genome and inducing a randomly occurring mutation in the target. It is not predictable which modification will appear in the progeny in an NHEJ-based approach. More precise is the HDR-mediated allele exchange. In HDR the cell copies the gene located on the sister-chromatid or a customized DNA fragment to repair the failure in the other DNA copy. The genes involved in NHEJ and HDR are well described in mammalian cells (Deriano and Roth, 2013), but not much is known in plant cells (Schmidt et al., 2019), especially in barley.

In addition, HDR can be used for precise allele exchange to introduce an in-frame predicted mutation at the target of choice, already shown in maize and wheat (Svitashev et al., 2016; Wang et al., 2014). The repair template should be highly accessible with flanking sites homologous to the target-sequence. Up-to-date repair template properties are not well investigated, used were single-stranded DNA oligos (ssDNA) or double-stranded DNA oligos (dsDNA). In DNA-based approaches, the repair template properties are limited to dsDNA because the repair template has to be expressed together with the RNP. Čermák et al. (2015) used a gemini-virus-based system for transient expression of the repair template with 100-fold higher gene-dose compared to other systems. The use of virus-based systems in a transgenic approach triggers regulatory problems when it comes to crops and is still under discussion (Mehta et al., 2019). Another approach is the stable expression of the repair template within the left and right border of the transferred transgene. Problems here are the uncontrolled integration of the repair template into the genome. Using a pre-assembled RNP complex co-transfected into a cell allows much higher repair-template ratios. Basically, the repair template is integrated synchronously into the cell with the RNP. This can be done by particle bombardment (Svitashev et al., 2016) or transfection of protoplast. The latter is not suitable in barley because the regeneration of plants from protoplast is hardly possible and genotype-dependent. But protoplast-assays can be used to test the functionality of a gRNA-Cas-expression cassette or an RNP complex.

Barley is a diploid small-grain model organism and has frequently been used to integrate transgenic DNA of choice. Transformation techniques have been widely described by the use of immature embryos of the cultivar (cv.) Golden Promise using *Agrobacterium tumefaciens*-mediated gene transfer or particle bombardment (Tingay et al., 1997; Bartlett et al., 2008; Hensel et al., 2008). Other cultivars, such as cv. Igri can also be used for haploid-induction in combination with *A. tumefaciens*-mediated gene transfer in pollen-cultures

(Kumlehn et al., 2006). Also, other varieties are described in the literature to be used for transformation, but in most cases with much lower transformation efficiencies (Hensel et al., 2008). The agronomic use of barley is restricted to animal food and some purposes in the human diet but, more important, it can also be seen as a model organism for more complex cereals such as wheat.

2 Expression systems used for targeted mutagenesis

Targeted mutagenesis intends in barley had been carried out using binary vectors harbouring a Cas-expression cassette together with a gRNA transformed by *A. tumefaciens* or biolistic gene transfer. In principle, two different promoters express both components necessary for RNP formation in the cell. The Cas-cassette is normally driven by a constitutive polymerase II promoter, while the gRNA is driven by a polymerase III promoter sequence. Non-codon-optimized and codon-optimized versions of Cas have been used in the literature (see Table 1). In order to efficiently induce DSB, efforts were done to optimize the expression of the Cas and gRNA within the cell. Only if both components are equally abundant in the cell it will result in heritable mutations in the progeny. This is also one of the challenges in the DNA-based use of sequence-specific endonucleases because the data in other species indicates a reduced activity of specific promoters in pollen and embryogenic tissues (Wang et al., 2015; Hamilton et al., 1998). Furthermore, a constant expression of a transgene in very high doses triggers silencing reactions and/or off-target modifications. To avoid these effects RNPs were successfully used in several organisms such as maize, tobacco and wheat (Svitashev et al., 2016; Schedel et al., 2017; Zhang et al., 2016). The RNP complex in the plant cell induces modifications in the genome without integration of a transgene. Such an approach would have been called DNA-free because the abundance of the RNP is transient and its survival is restricted to some hours or days in the cell as demonstrated in mouse cell cultures (Harrington et al., 2017).

The constructs published in barley so far contain exclusively Streptococcus pyogenes Cas9 (SpCas9) nuclease, which according to the publications have been partially codon-optimized for maize (see Table 1). Since the level of expression of the two components involved (gRNA and Cas-protein) is decisive for the mutation success, the constitutive maize *Ubiquitin 1* promoter was used for the expression of the *Cas9* endonuclease. For the expression of RNA, either the rice *U3* or the wheat *U6* promoter were used in various publications (Gerasimova et al., 2018; Holubová et al., 2018; Lawrenson et al., 2015). Only in the work on the knockout of *MORC1* (Kumar et al., 2018) was the endogenous barley polymerase III *U3* promoter used. Kapusi and colleagues (2017) used the rice *U6* promoter and, in combination with SpCas9, achieved the highest mutation efficiency of 78% in primary transgenic plants.

Table 1 Summary of site-directed mutagenesis in barley

Target (GenBank or Gene ID)	Target tissue	Method of gene transfer	Promoter driving *Cas*	Cas codon optimization	Cas variant	No of gRNAs	Promoter driving gRNA	Mutations efficiency[1] [%]	No of plants analysed	Mutant detection method	Heritability	Reference
PM19-1 (KT336449) *PM19-3* (KT336450)	IEs	*A. tumefaciens*	*ZmUbi1*	Canonical	SpCas9	1	*TaU6*	23.0 10.0	13 10	PCR/RE, Sanger seq.	Yes	Lawrenson et al. 2015
PAPhy_a promoter	IEs	*A. tumefaciens*	*ZmUbi1*	Wheat	SpCas9	1	*TaU6*	44.0	64	PCR/RE, subcloning, Sanger seq.	Yes	Holme et al. 2017
ENGase (MLOC_10039.2)	IEs Callus	*A. tumefaciens* Biolistic	*ZmUbi1*	Canonical	SpCas9	5[2] 2	*OsU6*	78.3 100.0	23 7	Subcloning, Sanger seq.	Yes	Kapusi et al. 2017
Nud1 (HORVU7Hr1G089930) *Vrs1* (HORVU2Hr1G092290)	Protoplasts	PEG	*ZmUbi1*	Maize	SpCas9	1	*OsU3*	22.0 6.0	-	Amplicon sequencing	No	Gerasimova et al. 2018
CKX1 (HORVU3Hr1Go19850)	IEs	*A. tumefaciens*	*ZmUbi1*	Maize	SpCas9	1	*OsU3*	65.2	23	PCR/RE, Sanger seq.	Yes	Holubová et al. 2018
MORC1 (HG316119.1)	IEs Callus	*A. tumefaciens* Biolistic	*ZmUbi1*	Human	SpCas9	1	*HvU3* *OsU3*	100.0 70.0	12 10	TIDE[3], Sanger seq.	Yes	Kumar et al. 2018
CKX1 (HORVU3Hr1G019850) *CKX1* + *CKX3* (HORVU1Hr1G042360) *Nud1* (HORVU7Hr1G089930)	IEs	*A. tumefaciens*	*ZmUbi1*	Maize	SpCas9	1 2 1	*TaU6*	66.0* 12.5 64.3*	71 72 28	PCR/RE, Sanger seq.	Yes	Gasparis et al. 2018
CMF7 (AK366098)	IEs	*A. tumefaciens*	*ZmUbi1*	Maize	SpCas9	1	*OsU3*	5.0	20	PCR/RE, Sanger seq.	Yes	Li et al. 2019a
CMF3 (HORVU6Hr1G021460.2)	IEs	*A. tumefaciens*	*ZmUbi1*	Maize	SpCas9	2	*OsU3*	11.8	34	PCR/RE, Sanger seq.	Yes	Li et al. 2019b
GBSS1a (AK373583) *PTST1*	IEs	*A. tumefaciens*	*ZmUbi1*	Wheat	SpCas9	1	*TaU6*	100.0 100.0	9 8	Subcloning, Sanger seq.	Yes No	Zhong et al. 2019

[1]Number of proven mutations/number of analysed plants.
[2]on individual plasmids.
[3]TIDE – Tracking of InDels by Decomposition, Brinkman et al., 2014.
*Mutation efficiency is based on PCR/RE, since just some of the mutations were confirmed by sequencing.

3 Enhancing plant development and performance

Topics worked on in barley are as varied as their cultivation and application. The question is how can the yield be secured or even increased under the changing climatic conditions. It should be noted that the yield is a multifactorial quantity that contains basic knowledge in various areas of nutrient uptake and supply, as well as plant and spike architecture. How are molecules synthesized and distributed within the plant? Which enzymes are involved in the different processes and which are the key points to shift the process in one direction or the other? Yield is often measured in grain weight per acre (e.g. tons per hectare). However, there are also data on thousand grain weight (TGW) or number of shoots/plant or number of grains/spike.

The first publication on targeted genome modification using CRISPR/Cas endonuclease technology in barley was published in 2015. The *PM19* gene, which is involved in the expression of the dormancy trait, was eliminated in stable transgenic plants (Lawrenson et al., 2015). In barley, which carries four copies of *PM19*, alleles differ only by some single-nucleotide polymorphisms (SNPs). The authors used two different gRNA constructs, which should switch off all four gene copies with small sequence deviations. Sequence analysis of the 20 primary transgenic plants showed a mutation efficiency of 10–23% of the *PM19* alleles investigated. The authors were also able to show that the mutations detected in the T0 plants are inherited in the offspring, which is a prerequisite for the application of the technology.

CRISPR/Cas endonuclease constructs for two genes involved in the formation of spikes and grains were investigated in protoplasts of the Siberian barley variety Aley (Gerasimova et al., 2018). One of the two, the *Vrs1* gene, is significantly involved in the expression of the characteristic row-type. In addition, one must know that a barley spikelet consists of flowers, of which in two-rowed barley only the middle flower is fertile and forms a grain. The two lateral florets do not form grains due to the expression of *Vrs1*, which is a repressor of flower fertility. Six-rowed barley had a mutation in the *Vrs1* gene leading to the formation of three grains per spikelet (Komatsuda et al., 2007). Approaches to reduce *Vrs1* transcription by inverted-repeat-mediated RNA interference (RNAi) have shown that the two-rowed barley model cv. Golden Promise can be converted into six-rowed barley (Sakuma et al., 2013). However, since this approach only means a reduction and not complete elimination of *Vrs1* gene function, the lateral grains were much smaller than those of the central florets. In order to select efficient gRNAs for the same target, protoplast experiments of the Siberian cv. Aley were performed. Deep sequencing of *Vrs1* amplicons in the PEG-transformed Aley protoplasts revealed that about 6% of the cells had InDels at the target sequence. In parallel, the transformation was more successful with constructs that triggered the knockout of the *Nud1* gene. This gene is involved in the fixed linkage of lemma and palea, the two petals, to the barley grain and thus ensures that there is hulled or naked barley (Taketa et al., 2008). Here guides were tested for two different positions in the *Nud1* gene. While for one target region comparable efficiencies as for the *Vrs1* gene were detected (8% vs. 6%), for the second target region nearly more than every 5th sequence showed InDels (22%) (Gerasimova et al., 2018). In another study, Gasparis and co-workers targeted the same gene and observed a high mutation efficiency (64% based on PCR/RE results) and showed the expected naked phenotype of the grains (Gasparis et al., 2018). Evidence of how differences in mutation efficiency of on- and off-targets are influenced by the appropriate selection of gRNAs is described elsewhere in this book.

Plant hormones, such as auxins, cytokinins, ABA or strigolactones, have a variety of functions and are significantly involved in various processes of plant development. Holubová et al. (2018) recently published an approach in which the genetic manipulation of the cytokinin homoeostasis plays a role. It is important to maintain a balance between synthesis and degradation of cytokinins in different tissues or developmental stages. In a first RNAi-mediated experiment, the abundance of an enzyme involved in the degradation of cytokinin (CKX1) was reduced. As a result, a 10% increase in the number of shoots was achieved in field trials, resulting in 18–39% more yield per square metre. The yield increase was accompanied by a loss of grain size which resulted in a decrease in thousand grain weight. A complete knockout showed comparable results regarding the change of the cytokinin balance, but field experiments could not yet be carried out.

Gasparis et al. (2019) followed a similar approach, but came to some contrasting conclusions. In this study the activity of Cytokinin oxidase/dehydrogenase (CKX) could be strongly reduced by targeted knockout of *CKX1* and *CKX3* genes. However, the authors describe a prolonged primary root length and increased root surface for *ckx1* mutants, which is in clear contrast to the study by Holubová et al. (2018). Furthermore, no yield increase could be observed.

An enzyme that plays an important role in grain germination for the provision of phosphorus was investigated using TALEN and CRISPR/Cas technology (Holme et al., 2017). The authors wanted to cut out cis-elements of the promoter that are responsible for the temporal and spatial expression of the *PAPhy_a* gene. Both approaches led to heritable changes in which mutations of different lengths were obtained. The subsequent analysis of the grains showed that the *PAPhy_a* gene is mainly responsible for the phytase activity of the barley grain and that only mutations that cut out the entire promoter area showed significantly reduced enzyme activity. The knockout of the *PAPhy_a* gene led to delayed germination, as the grains could not store enough phytase during their formation. Since CRISPR/Cas technology mainly produces InDels of −1/+1 nucleotides, most of the plants studied were only slightly inhibited in their activity. A small proportion of the plants contained larger deletions and were therefore useful for the desired application.

Starch is the main source of energy for human nutrition. Consequently, precise knowledge of the production and accumulation of starch in plants is of great importance. Starch consists of amylose and amylopectin and is processed in plants by a coordinated sequence of enzymatic reactions. Since relatively little was known about the activity of two main actors, namely Protein Targeting to Starch 1 (PTST1) and Granule-Bound Starch Synthase (GBSS) in the barley storage organ, in the endosperm, Zhong et al. (2019) recently published a functional analysis using overexpression, bi-molecular fluorescence microscopy and targeted knockout. They were able to show that both proteins interact directly and are responsible for starch storage. The *gbss1a* and *ptst1* mutants showed significantly altered patterns of starch storage in the grain. Since homozygous *ptst1* mutants contained almost no starch, the resulting grains could not germinate. Only heterozygous plants with at least one wild-type allele were able to germinate and could be examined in the next generation.

Chloroplasts are the power house of the cell, and thus a fundamental component of plant development. Their maturation from proplastides is caused by the joint action of nuclear and plastidic genes. Mutations in genes that impair the biogenesis of chloroplasts lead to pigment-free or striped leaves. A famous chloroplast mutant of barley, which was created in the 1950s after X-ray treatment and first described by Hagemann and Scholz (1962), is called *albostrians*. It took until 2019 for the corresponding gene to be described. Li et al. (2019a) have shown by means of TILLING and CRISPR/Cas technology that the *albostrians* gene belongs to the family of CCT (CONSTANS, CO-like and TOC1) motif family (CMF) of putative transcription factors (Cockram et al., 2012). Plants with a mutation in the gene now called *CMF7* have leaf sectors that do not contain chlorophyll, and thus complement the original *albostrians* mutant. Another study (Li et al., 2019b) showed that *CMF3* is the ohnologous gene of *CMF7*. Ohnologous genes are gene pairs caused by whole genome duplication, whereby at least one gene has to undergo neofunctionalization. Accordingly, the phenotype of the *cmf3* mutant is not green-white striped, but xantha colour. Parallel to *CMF7*, the gene function of *CMF3* was also detected by TILLING and CRISPR/Cas. Within the *cmf3* CRISPR/Cas mutants, a new protein domain important for the functionality of the *CMF3* gene could be identified.

A major issue due to the changing climatic conditions is the resistance to salination and drought. Here as well, the use of CRISPR/Cas endonuclease technology, through its precision and speed, can bring decisive progress for the development of new varieties. However, this also requires knowledge of the basic principles of the gene networks involved.

4 Molecular understanding of resistance against plant pathogens

Another big topic is of course yield losses caused by insects, fungi and other phytopathogens. A broad knowledge is sparsely published and partly contains transgenic approaches and positive examples, like the broad-spectrum fungus resistance, conferred by the *Mlo* gene in wheat. Due to increasing travel activity combined with changing climatic conditions, the spread of pests will increase. As a consequence, the understanding and manipulation of resistance mechanisms in barley plants is of great importance.

Microchidia (MORC) proteins first described in *A. thaliana* conferring resistance to the turnip crinkle virus (TCV) (Kang et al., 2008, 2010, 2012) were also studied in barley. Reports that RNAi-mediated reduction of the *HvMORC2* gene showed increased resistance to mildew (*Blumeria graminis*) were confirmed by CRISPR/Cas-mediated knockout of *MORC1* (Kumar et al., 2018). In addition to mildew resistance, increased resistance to the necrotrophic fungus (*Fusarium graminearum*) and increased genome stability were also demonstrated.

5 Targeted mutagenesis for molecular farming purposes

In addition to medicinal and aromatic plants, crops such as tobacco, maize, wheat and barley are also used for the production of pharmaceutically relevant products (Schillberg et al., 2019). While tobacco is mainly used for transient expression due to its large leaves, the grains are mainly used as a bioreactor. These are characterized by special features such as the ability to store proteins and metabolites without degradation, a comparably high protein content and the ability to process complex proteins and modify them post-translationally. In addition, the media for plant cell cultures are cheaper than human and animal systems and do not contain potential impurities or toxins that complicate purification and make it more expensive. Glycosylation of proteins contributes significantly to enzyme activity and is therefore of great importance in the plant production of recombinant proteins. Since human and plant glycosylation patterns differ, attempts have been made to influence this process using CRISPR/Cas technology. Kapusi et al. (2017) therefore selected five different gRNAs specifically for the *Endo-N-Acetyl-b-D-Glucosaminidase* (*ENGase*) gene and transferred them in pairs into barley both biolistically and by *A. tumefaciens*-mediated transformation. The resulting transgenic barley plants exhibited different mutation patterns. A functional characterization with regard to altered glycosylation patterns was not part of the manuscript and must be clarified in future experimentation.

6 Conclusion

Targeted mutagenesis using CRISPR/Cas endonuclease technology is well established in barley, but genome editing (replacement or introduction of a customized sequence at a predefined genomic position) is still challenging. In other plants, it was shown that HDR using a gemini-virus-driven repair template on one T-DNA together with the RNP formatting cassettes was working as well in tomato or tobacco (Baltes et al., 2014; Čermák et al., 2011). The use of RNP has already been demonstrated in several protoplast assays (Kim et al., 2016; Murovec et al., 2018) which were also able to introduce heritable mutations in plants regenerated from protoplasts (Lin et al., 2018) but also from immature embryos in wheat (Wang et al., 2014). Currently appearing problems are the regeneration without selective pressure due to the absence of selectable markers using RNP complexes. To overcome this problem it has already been demonstrated that the *ALS* or *MAR1* gene can be used to introduce a herbicide or kanamycin-resistance to a plant cell (Wolter et al., 2018; Veillet et al., 2019; Conte et al., 2009). Only the ALS concept could be feasible for barley tissue culture, because kanamycin selection in barley, as induced by *MAR1*, is less promising because of the high number of escapes.

Alternatively, by multiplexing several gRNA for different genes, a selectable marker can be addressed with one gRNA, while synchronously the other gRNA targets the gene of choice as already shown by Wang et al. (2014) and Gao et al. (2015) in wheat and tobacco.

The delivery of RNPs is still the major challenge. The protein has to be transferred in a native buffer to ensure its activity. By using protoplasts, it is possible to use cationic transfection reagents or PEG to easily introduce an RNP into the cell (Svitashev et al., 2016; Murovec et al., 2018). In dicotyledonous plants, totipotency is given and complete plants can be regenerated from a single protoplast which is not the case for monocotyledonous plants. Thus, this possibility can be used only for transient assays or proof-of-concept experiments. The RNP has to be transferred in another way into the cell. Using a cationic transfections reagent it is possible to air-dry physiological active RNP onto gold-nanoparticles for biolistic transformation (Svitashev et al., 2016). This method is still the state-of-the-art for the transformation of barley and wheat in many cases. The use of cationic transfection reagents limits also the use of the repair template for HDR. Unfortunately, it is not possible to combine a dsDNA- or ssDNA-repair template together with an RNP since the repair template is somehow only bound to the nanoparticle in very low amounts. To overcome this problem several approaches have to be evaluated such as the usage of a SNAP-tag-Cas endonuclease together with a biotinylated repair template (Savic et al., 2018). In this technique provided by New England Biolabs the biotinylated repair template binds to the modified Cas-enzyme. In theory this should enhance the availability of the repair template and triggers the HDR in the cell. Binding of this SNAP-RNP-repair template complex into liposomes by the cationic transfection reagent should work much more efficient and increase the transformation rate. Unfortunately, the SNAP-tag-based Cas endonuclease is rarely available and the system is quite expensive. The prime editing technique offers perspectives as an alternative gene-editing method without the introduction of a double-strand break (Anzalone et al., 2019; Marzek et al., 2020). Here a spCas9 H840A protein is fused to an engineered reverse transcriptase. The endonuclease works like a canonical Cas9 nickase, so a nick is induced at a target of choice with a PAM-sequence. The reverse transcriptase together with a 'prime editing extended guide RNA' can then rewrite the mutation of choice at the locus of choice. Prime editing does not need any template DNA and, thus, offers to be easier to handle. Furthermore, it was described that the formation of InDels decrease, while a similar or better efficiency compared to Cas9-mediated HDR systems was achieved. Another promising aspect is the use of single-walled carbon nanotubes or carbon nanodots. Single-walled carbon nanotubes were used to transfer the *GFP* gene into the leaves of several organisms (Demirer et al., 2019). Carbon nanodots were used for same targets with more or less same results. The difference between both systems is the shape of the nanoparticles and the added functional groups (Doyle et al., 2019). These systems could be used to synchronously transfer DNA, protein or siRNAs into the cell addressing genes responsible for NHEJ to enhance the HDR efficiency. Even possible would be the transient expression of a Cas9 and gRNA-harbouring plasmid which edits transiently the gDNA in the plant cell resulting in heritable mutations. The single-walled carbon nanotubes could be transferred by vacuum infiltration as already used for *A. tumefaciens*-mediated transformation of immature embryos of barley. Furthermore, the repair template could bind to the single-walled carbon nanotubes to enhance the coating of the gold nanoparticles and afterwards can be used for biolistic transformation thereby enhancing the dose and availability of the repair template.

7 Where to look for further information

The Leibniz Institute of Plant Genetics and Crop Plant Research (IPK) is one of the world's leading international institutions in the field of plant genetics and cropscience. The scientific interest of the research group Plant Reproductive Biology is focussed on plant developmental processes such as gametophyte development, fertilization as well as embryogenesis and seed development that constitute the main subjects of investigation and manipulation.

A wider source of information is another book covering the field of barley research and biotechnology:

- Kumlehn, J., Hiekel, S. and Budhagatapalli, N. (2019): Site-directed genome modification in barley and wheat. In: Ordon F, Friedt W (Eds.), *Advances in Breeding Techniques for Cereal Crops*. (Series: Burleigh Dodds Series in Agricultural Science, No. 60) Cambridge, UK: Burleigh Dodds Science Publishing Limited. ISBN 978-1-78676-244-3.

A good comprehensive review article about genetic transformation would be:

- Hensel, G. (2020): Genetic transformation of Triticeae cereals – Summary of almost three-decade's development. *Biotechnology Advances* 40: 107484. doi.org/10.1016/j.biotechadv.2019.107484.

A good comprehensive review article about CRISPR/Cas technology would be:

- Koeppel, I., Hertig, C., Hoffie, R. and Kumlehn, J. (2019): Cas endonuclease technology - a quantum leap in the advancement of barley and wheat genetic engineering. *Int. J. Mol. Sci.* 20:2647. dx.doi.org/10.3390/ijms20112647.

Since 2019 a COST Action to plant genome editing was funded and is aimed in collecting and strengthening European activities in this field. More informations can be found at: https://plantgenomeediting.eu/.

The yearly conference in plant in vitro biology: https://sivb.org/meetings/ is a recommend forum for the latest information in the plant gene editing field.

8 References

Anzalone, A. V., Randolph, P. B., Davis, J. R., Sousa, A. A., Koblan, L. W., Levy, J. M., Chen, P. J., Wilson, C., Newby, G. A., Raguram, A. and Liu, D. R. 2019. Search-and-replace genome editing without double-strand breaks or donor DNA. *Nature* 576(7785), 149–57. doi:10.1038/s41586-019-1711-4.

Baltes, N. J., Gil-Humanes, J., Cermak, T., Atkins, P. A. and Voytas, D. F. 2014. DNA replicons for plant genome engineering. *Plant Cell* 26(1), 151–63. doi:10.1105/tpc.113.119792.

Bartlett, J. G., Alves, S. C., Smedley, M., Snape, J. W. and Harwood, W. A. 2008. High-throughput *Agrobacterium*-mediated barley transformation. *Plant Methods* 4, 22. doi:10.1186/1746-4811-4-22.

Brinkman, E. K., Chen, T., Amendola, M. and van Steensel, B. 2014. Easy quantitative assessment of genome editing by sequence trace decomposition. *Nucleic Acids Research* 42(22), e168. doi:10.1093/nar/gku936.

Budhagatapalli, N., Schedel, S., Gurushidze, M., Pencs, S., Hiekel, S., Rutten, T., Kusch, S., Morbitzer, R., Lahaye, T., Panstruga, R., Kumlehn, J. and Hensel, G. 2016. A simple test for the cleavage activity of customized endonucleases in plants. *Plant Methods* 12, 18. doi:10.1186/s13007-016-0118-6.

Čermák, T., Doyle, E. L., Christian, M., Wang, L., Zhang, Y., Schmidt, C., Baller, J. A., Somia, N. V., Bogdanove, A. J. and Voytas, D. F. 2011. Efficient design and assembly of custom TALEN and other TAL effector-based constructs for DNA targeting. *Nucleic Acids Research* 39(12), e82. doi:10.1093/nar/gkr218.

Čermák, T., Baltes, N. J., Čegan, R. et al. 2015. High-frequency, precise modification of the tomato genome. *Genome Biology* 16, 232. https://doi.org/10.1186/s13059-015-0796-9.

Cockram, J., Thiel, T., Steuernagel, B., Stein, N., Taudien, S., Bailey, P. C. and O'Sullivan, D. M. 2012. Genome dynamics explain the evolution of flowering time CCT domain gene families in the Poaceae. *PLOS ONE* 7(9), e45307. doi:10.1371/journal.pone.0045307.

Conte, S., Stevenson, D., Furner, I. and Lloyd, A. 2009. Multiple antibiotic resistance in Arabidopsis is conferred by mutations in a chloroplast-localized transport protein. *Plant Physiology* 151(2), 559–73. doi:10.1104/pp.109.143487.

Demirer, G. S., Zhang, H., Matos, J. L., Goh, N. S., Cunningham, F. J., Sung, Y., Chang, R., Aditham, A. J., Chio, L., Cho, M. J., Staskawicz, B. and Landry, M. P. 2019. High aspect ratio nanomaterials enable delivery of functional genetic material without DNA integration in mature plants. *Nature Nanotechnology* 14(5), 456–64. doi:10.1038/s41565-019-0382-5.

Deriano, L. and Roth, D. B. 2013. Modernizing the nonhomologous end-joining repertoire: alternative and classical NHEJ share the stage. *Annual Review of Genetics* 47, 433–55. doi:10.1146/annurev-genet-110711-155540.

Doyle, C., Higginbottom, K., Swift, T. A., Winfield, M., Bellas, C., Benito-Alifonso, D., Fletcher, T., Galan, M. C., Edwards, K. and Whitney, H. M. 2019. A simple method for spray-on gene editing in planta. *bioRxiv*. doi:10.1101/805036805036.

Gao, J., Wang, G., Ma, S., Xie, X., Wu, X., Zhang, X., Wu, Y., Zhao, P. and Xia, Q. 2015. CRISPR/Cas9-mediated targeted mutagenesis in *Nicotiana tabacum*. *Plant Molecular Biology* 87(1–2), 99–110. doi:10.1007/s11103-014-0263-0.

Gasparis, S., Kała, M., Przyborowski, M., Łyżnik, L. A., Orczyk, W. and Nadolska-Orczyk, A. 2018. A simple and efficient CRISPR/Cas9 platform for induction of single and multiple, heritable mutations in barley (*Hordeum vulgare* L.). *Plant Methods* 14, 111. doi:10.1186/s13007-018-0382-8.

Gasparis, S., Przyborowski, M., Kała, M. and Nadolska-Orczyk, A. 2019. Knockout of the HvCKX1 or HvCKX3 gene in barley (*Hordeum vulgare* L.) by RNA-guided Cas9 nuclease affects the regulation of cytokinin metabolism and root morphology. *Cells* 8(8), 782. doi:10.3390/cells8080782.

Gerasimova, S. V., Korotkova, A. M., Hertig, C., Hiekel, S., Hoffie, R., Budhagatapalli, N., Otto, I., Hensel, G., Shumny, V. K., Kochetov, A. V., Kumlehn, J. and Khlestkina, E. K. 2018. Targeted genome modification in protoplasts of a highly regenerable Siberian barley cultivar using RNA-guided Cas9 endonuclease. *Vavilov Journal of Genetics and Breeding* 22, 1033–9. doi:10.18699/VJ18.447.

Hagemann, R. and Scholz, F. 1962. A case of gene induced mutations of the plasmotype in barley. *Theoretical and Applied Genetics* 32, 50–9. doi:10.1007/BF00709127.

Hahn, F., Korolev, A., Loures, L. S. and Nekrasov, V. 2019. A modular cloning toolkit for genome editing in plants. *bioRxiv*. doi:10.1101/738021738021

Hamilton, D. A., Schwarz, Y. H. and Mascarenhas, J. P. 1998. A monocot pollen-specific promoter contains separable pollen-specific and quantitative elements. *Plant Molecular Biology* 38(4), 663–9. doi:10.1023/A:1006083725102.

Harrington, L. B., Paez-Espino, D., Staahl, B. T., Chen, J. S., Ma, E., Kyrpides, N. C. and Doudna, J. A. 2017. A thermostable Cas9 with increased lifetime in human plasma. *Nature Communications* 8(1), 1424. doi:10.1038/s41467-017-01408-4.

Hensel, G., Valkov, V., Middlefell-Williams, J. and Kumlehn, J. 2008. Efficient generation of transgenic barley: the way forward to modulate plant-microbe interactions. *Journal of Plant Physiolog* 165(1), 71–82. doi:10.1016/j.jplph.2007.06.015.

Holme, I. B., Wendt, T., Gil-Humanes, J., Deleuran, L. C., Starker, C. G., Voytas, D. F. and Brinch-Pedersen, H. 2017. Evaluation of the mature grain phytase candidate HvPAPhy_a gene in barley (Hordeum vulgare L.) using CRISPR/Cas9 and TALENs. *Plant Molecular Biology* 95(1–2), 111–21. doi:10.1007/s11103-017-0640-6.

Holubová, K., Hensel, G., Vojta, P., Tarkowski, P., Bergougnoux, V. and Galuszka, P. 2018. Modification of barley plant productivity through regulation of cytokinin content by reverse-genetics approaches. *Frontiers in Plant Science* 9, 1676. doi:10.3389/fpls.2018.01676.

Kang, H. G., Kuhl, J. C., Kachroo, P. and Klessig, D. F. 2008. CRT1, an Arabidopsis ATPase that interacts with diverse resistance proteins and modulates disease resistance to turnip crinkle virus. *Cell Host Microbe* 3(1), 48–57. doi:10.1016/j.chom.2007.11.006.

Kang, H. G., Oh, C. S., Sato, M., Katagiri, F., Glazebrook, J., Takahashi, H., Kachroo, P., Martin, G. B. and Klessig, D. F. 2010. Endosome associated CRT1 functions early in resistance gene-mediated defense signaling in Arabidopsis and tobacco. *Plant Cell* 22(3), 918–36. doi:10.1105/tpc.109.071662.

Kang, H. G., Hyong, W. C., von Einem, S., Manosalva, P., Ehlers, K., Liu, P. P., Buxa, S. V., Moreau, M., Mang, H. G., Kachroo, P., Kogel, K. H. and Klessig, D. F. 2012. CRT1 is a nuclear-translocated MORC endonuclease that participates in multiple levels of plant immunity. *Nature Communications* 3, 1297. doi:10.1038/ncomms2279.

Kapusi, E., Corcuera-Gomez, M., Melnik, S. and Stoger, E. 2017. Heritable genomic fragment deletions and small indels in the putative ENGase gene induced by CRISPR/Cas9 in barley ENGase. *Frontiers in Plant Science* 8, 540. doi:10.3389/fpls.2017.00540.

Kim, D., Kim, J., Hur, J. K., Been, K. W., Yoon, S. H. and Kim, J. S. 2016. Genome-wide analysis reveals specificities of Cpf1 endonucleases in human cells. *Nature Biotechnology* 34(8), 863–8. doi:10.1038/nbt.3609.

Komatsuda, T., Pourkheirandish, M., He, C., Azhaguvel, P., Kanamori, H., Perovic, D., Stein, N., Graner, A., Wicker, T., Tagiri, A., Lundqvist, U., Fujimura, T., Matsuoka, M., Matsumoto, T. and Yano, M. 2007. Six-rowed barley originated from a mutation in a homeodomain-leucine zipper I-class homeobox gene. *PNAS* 104(4), 1424–9. doi:10.1073/pnas.0608580104.

Kumar, N., Galli, M., Ordon, J., Stuttmann, J., Kogel, K. H. and Imani, J. 2018. Further analysis of barley MORC1 using a highly efficient RNA-guided Cas9 gene-editing system. *Plant Biotechnology Journal* 16(11), 1892–903. doi:10.1111/pbi.12924.

Kumlehn, J., Serazetdinova, L., Hensel, G., Becker, D. and Lörz, H. 2006. Genetic transformation of barley (*Hordeum vulgare* L.) via infection of androgenetic pollen cultures with *Agrobacterium tumefaciens*. *Plant Biotechnology Journal* 4(2), 251–61. doi:10.1111/j.1467-7652.2005.00178.x.

Lawrenson, T., Shorinola, O., Stacey, N., Li, C. D., Ostergaard, L., Patron, N., Uauy, C. and Harwood, W. 2015. Induction of targeted, heritable mutations in barley and Brassica oleracea using RNA-guided Cas9 nuclease. *Genome Biology* 16, 258. doi:10.1186/s13059-015-0826-7.

Li, M., Hensel, G., Mascher, M., Melzer, M., Budhagatapalli, N., Rutten, T., Himmelbach, A., Beier, S., Korzun, V., Kumlehn, J., Borner, T. and Stein, N. 2019a. Leaf variegation and impaired chloroplast development caused by a truncated CCT domain gene in albostrians barley. *Plant Cell* 31(7), 1430–445. doi:10.1105/tpc.19.00132.

Li, M., Hensel, G., Melzer, M., Junker, A., Tschiersch, H., Arend, D., Kumlehn, J., Börner, T. and Stein, N. 2019b. Mutation of the ALBOSTRIANS ohnologous gene HvCMF3 impairs chloroplast development and thylakoid architecture in barley due to reduced plastid translation. *bioRxiv* 756833. doi:10.1101/756833.

Lin, C. S., Hsu, C. T., Yang, L. H., Lee, L. Y., Fu, J. Y., Cheng, Q. W., Wu, F. H., Hsiao, H. C., Zhang, Y., Zhang, R., Chang, W. J., Yu, C. T., Wang, W., Liao, L. J., Gelvin, S. B. and Shih, M. C. 2018. Application of protoplast technology to CRISPR/Cas9 mutagenesis: from single-cell mutation detection to mutant plant regeneration. *Plant Biotechnology Journal* 16(7), 1295–310. doi:10.1111/pbi.12870.

Marzek, M., Brąszewska-Zalewska, A. and Hensel, G. 2020. Prime editing: a new way for genome editing. *Trends in Cell Biology* 30(4), 257–9. doi:10.1016/j.tcb.2020.01.004.

Mehta, D., Stürchler, A., Anjanappa, R. B., Zaidi, S. S. E. A., Hirsch-Hoffmann, M., Gruissem, W. and Vanderschuren, H. 2019. Linking CRISPR-Cas9 interference in cassava to the evolution of editing-resistant geminiviruses. *Genome Biology* 20(1), 80. doi:10.1186/s13059-019-1678-3.

Modrzejewski, D., Hartung, F., Sprink, T., Krause, D., Kohl, C., Schiemann, J. and Wilhelm, R. 2018. What is the available evidence for the application of genome editing as a new tool for plant trait modification and the potential occurrence of associated off-target effects: a systematic map protocol. *Environmental Evidence* 7(1), 1–8. doi:10.1186/s13750-018-0130-6.

Murovec, J., Guček, K., Bohanec, B., Avbelj, M. and Jerala, R. 2018. DNA-free genome editing of brassica oleracea and B. Rapa protoplasts using CRISPR-cas9 ribonucleoprotein complexes. *Frontiers in Plant Science* 9, 1594. doi:10.3389/fpls.2018.01594.

Peer, R., Rivlin, G., Golobovitch, S., Lapidot, M., Gal-On, A., Vainstein, A., Tzfira, T. and Flaishman, M. A. 2015. Targeted mutagenesis using zinc-finger nucleases in perennial fruit trees. *Planta* 241(4), 941–51. doi:10.1007/s00425-014-2224-x.

Ran, Y., Patron, N., Kay, P., Wong, D., Buchanan, M., Cao, Y. Y., Sawbridge, T., Davies, J. P., Mason, J., Webb, S. R., Spangenberg, G., Ainley, W. M., Walsh, T. A. and Hayden, M. J. 2018. Zinc finger nuclease-mediated precision genome editing of an endogenous gene in hexaploid bread wheat (*Triticum aestivum*) using a DNA repair template. *Plant Biotechnology Journal* 16(12), 2088–101. doi:10.1111/pbi.12941.

Sakuma, S., Pourkheirandish, M., Hensel, G., Kumlehn, J., Stein, N., Tagiri, A., Yamaji, N., Ma, J. F., Sassa, H., Koba, T. and Komatsuda, T. 2013. Divergence of expression pattern contributed to neofunctionalization of duplicated HD-Zip I transcription factor in barley. *New Phytologist* 197(3), 939–48. doi:10.1111/nph.12068.

Savic, N., Ringnalda, F. C. A. S., Lindsay, H., Berk, C., Bargsten, K., Li, Y., Neri, D., Robinson, M. D., Ciaudo, C., Hall, J., Jinek, M. and Schwank, G. 2018. Covalent linkage of the

DNA repair template to the CRISPR-Cas9 nuclease enhances homology-directed repair. *eLife* 7, 1–18. doi:10.7554/eLife.33761.

Schedel, S., Pencs, S., Hensel, G., Müller, A., Rutten, T. and Kumlehn, J. 2017. RNA-guided Cas9-induced mutagenesis in tobacco followed by efficient genetic fixation in doubled haploid plants. *Frontiers in Plant Science* 7, 1–14. doi:10.3389/fpls.2016.01995.

Schillberg, S., Raven, N., Spiegel, H., Rasche, S. and Buntru, M. 2019. Critical analysis of the commercial potential of plants for the production of recombinant proteins. *Frontiers in Plant Science* 10, 720. doi:10.3389/fpls.2019.00720.

Schmidt, C., Pacher, M. and Puchta, H. 2019. Efficient induction of heritable inversions in plant genomes using the CRISPR/Cas system. *Plant Journal* 98(4), 577–89. doi:10.1111/tpj.14322.

Svitashev, S., Schwartz, C., Lenderts, B., Young, J. K. and Mark Cigan, A. 2016. Genome editing in maize directed by CRISPR-Cas9 ribonucleoprotein complexes. *Nature Communications* 7, 13274. doi:10.1038/ncomms13274.

Taketa, S., Amano, S., Tsujino, Y., Sato, T., Saisho, D., Kakeda, K., Nomura, M., Suzuki, T., Matsumoto, T., Sato, K., Kanamori, H., Kawasaki, S. and Takeda, K. 2008. Barley grain with adhering hulls is controlled by an ERF family transcription factor gene regulating a lipid biosynthesis pathway. *PNAS* 105(10), 4062–7. doi:10.1073/pnas.0711034105.

Tingay, S., McElroy, D., Kalla, R., Fieg, S., Wang, M., Thornton, S. and Brettell, R. 1997. *Agrobacterium tumefaciens*-mediated barley transformation. *Plant Journal* 11(6), 1369–76. doi:10.1046/j.1365-313X.1997.11061369.x.

Veillet, F., Perrot, L., Chauvin, L., Kermarrec, M. P., Guyon-Debast, A., Chauvin, J. E., Nogué, F. and Mazier, M. 2019. Transgene-free genome editing in tomato and potato plants using *Agrobacterium*-mediated delivery of a CRISPR/Cas9 cytidine base editor. *International Journal of Molecular Sciences* 2019, 20, 402). doi:10.3390/ijms20020402.

Wang, Y., Cheng, X., Shan, Q., Zhang, Y., Liu, J., Gao, C. and Qiu, J. L. 2014. Simultaneous editing of three homoeoalleles in hexaploid bread wheat confers heritable resistance to powdery mildew. *Nature Biotechnology* 32(9), 947–51. doi:10.1038/nbt.2969.

Wang, Z. P., Xing, H. L., Dong, L., Zhang, H. Y., Han, C. Y., Wang, X. C. and Chen, Q. J. 2015. Egg cell-specific promoter-controlled CRISPR/Cas9 efficiently generates homozygous mutants for multiple target genes in Arabidopsis in a single generation. *Genome Biology* 16, 144. doi:10.1186/s13059-015-0715-0.

Wolter, F., Klemm, J. and Puchta, H. 2018. Efficient in planta gene targeting in Arabidopsis using egg cell-specific expression of the Cas9 nuclease of Staphylococcus aureus. *Plant Journal* 94(4), 735–46. doi:10.1111/tpj.13893.

Zhang, Y., Liang, Z., Zong, Y., Wang, Y., Liu, J., Chen, K., Qiu, J. L. and Gao, C. 2016. Efficient and transgene-free genome editing in wheat through transient expression of CRISPR/Cas9 DNA or RNA. *Nature Communications* 7, 12617. doi:10.1038/ncomms12617.

Zhang, Y., Malzahn, A. A., Sretenovic, S. and Qi, Y. 2019. The emerging and uncultivated potential of CRISPR technology in plant science. *Nature Plants* 5(8), 778–94. doi:10.1038/s41477-019-0461-5.

Zhong, Y., Blennow, A., Kofoed-Enevoldsen, O., Jiang, D. and Hebelstrup, K. H. 2019. Protein targeting to starch 1 is essential for starchy endosperm development in barley. *Journal of Experimental Botany* 70(2), 485–96. doi:10.1093/jxb/ery398.

Chapter 13

Genome editing of maize

Jacob D. Zobrist, Morgan McCaw, Minjeong Kang, Alan L. Eggenberger, Keunsub Lee, and Kan Wang, Iowa State University, USA

1 Introduction

Over the course of 10 000 years, selective breeding converted a Mesoamerican grass with little agricultural value, known as teosinte, into the important commercial crop maize (*Zea mays*) (Gewin, 2003; Doebley, 2004). Historically, breeding efforts have concentrated on increasing the yield, aesthetics, harvestability, and robustness of maize (Tuxill et al., 2010; Andorf et al., 2019). Modern breeders continue to seek improvements to maize, but traditional breeding programs are constrained by long generation times, linkage drag, and the loss of genetic diversity during domestication (Nannas and Dawe, 2015; Andorf et al., 2019). Whole genome sequencing analysis, detailed genetic mapping studies, and the ability to transform maize have led to new methods to overcome some of these constraints inherent to traditional selective breeding (Gordon-Kamm et al., 1990; Schnable et al., 2009; Eichten et al., 2011; for a review see Andorf et al., 2019). More recently, precision genome editing methods have been successfully used in many major commercial crops that are important for supporting the world's growing population (Ji et al., 2013; Luo et al., 2016; Jaganathan et al., 2018). In this chapter, we will review recent advancements in maize genome editing technologies, the current state of the art breeding technology in maize and other systems.

http://dx.doi.org/10.19103/AS.2020.0082.19

Early methods to alter the maize genome utilized transposons, radiation and mutagenic chemicals to introduce random mutations (Neuffer et al., 1997). The randomly mutagenized lines were then subjected to large forward and reverse genetic screens to drive gene discovery (Neuffer et al., 1997; Settles et al., 2007). Random mutagenesis introduces both desired and undesired mutations into the genome (Nannas and Dawe, 2015). Evaluation of a mutant requires multiple generations of backcrossing to separate from other mutations, a process that is costly and time-consuming.

Transposon tagging allows reverse genetic screening in maize, but requires massive populations and extensive screening, making it a difficult task for many labs. To facilitate transposon tagging, the Uniform Mu program was developed as a community resource for the genetic study of maize (Settles et al., 2007; Liu et al., 2009). Despite having these tools for genetic discovery, precision genome editing is more desirable because it allows for the maize genome to be modified in a directed and controlled manner (Andorf et al., 2019).

Transgenic approaches to introduce genetic modifications into the maize genome involve random integration of DNAs through *Agrobacterium*-mediated transformation or direct DNA transfer methods, for example, biolistic transformation (Chen et al., 2019). These methods have generated a number of agriculturally useful products and proved to be valuable research tools (ISAAA, 2017). However, due to the lack of control of where transgenes integrate into the maize genome, transgene expression levels can be unpredictable. This phenomenon, known as 'position effect', results from the effect of adjacent chromosomal DNA on transgene expression. In addition, random integration of transgenes can disrupt endogenous genes (Brandle et al., 1995; Ni et al., 1995; Elmayan and Vaucheret, 1996; Mlynárová et al., 1996; Stewart et al., 2003; Nagaya et al., 2005).

To circumvent the position effect and insertion-site complexity in transgenic plants, researchers have been searching for precision engineering tools since the early days of biotechnology. Recombinase-based systems such as Cre/*lox* from the bacteriophage P1 (Sauer and Henderson, 1988), FLP/*FRT* from yeast (Anand et al., 2019), and INT from the *Streptomyces* phage phiC31 were hot pursuits in the turn of the century (Ow, 2005, 2016; Srivastava and Thomson, 2016). These systems allow for the insertion of a precise single-copy DNA fragment into the target site of the maize genome (Ow, 2016; Anand et al., 2019). It was also shown that delivery of the recombinase Cre as protein into a transgenic maize line carrying the *loxP* sites led to precise gene excision with a frequency of 20% (Martin-Ortigosa et al., 2014). However, the main limitation of the recombinase-based system is that it requires the establishment of a 'founder' line, which carries target sites in well-characterized genome locations that permit transgene insertion.

It was not until the sequence-specific nucleases (SSNs) were developed as genome editing tools that precision genome engineering became widely applicable (Silva et al., 2011; Gaj et al., 2013; Chen et al., 2019). This chapter focuses on recent advancements in maize genome engineering using SSNs. SSNs such as meganucleases, zinc finger nucleases (ZFNs), transcription activator-like effector nucleases (TALENs), and the clustered regularly interspaced short palindromic repeat (CRISPR)/CRISPR-associated proteins (e.g., Cas9 and Cas12a) have been shown to efficiently create site-specific double-strand DNA breaks (DSBs) in various organisms including maize. We review here their applications in maize genome editing (Table 1).

2 Site-specific engineering in maize: an overview

Genome editing is the process by which a specific site in the genome is targeted for manipulation. Precisely targeted genome editing in maize has been achieved by inducing site-specific DSBs resulting in sequence modifications during the repair process (Belhaj et al., 2013; Gaj et al., 2013). These targeted DSBs are mainly repaired by two distinct pathways in plants: non-homologous end-joining (NHEJ) and homology-directed repair (HDR) (Manova and Gruszka, 2015). The plant cell's dominant method of repair, NHEJ, does not require a repair template and can result in short insertions or deletions (indels) that can disrupt gene function by creating frameshift mutations or by inserting or deleting amino acids in the encoded protein (Gorbunova and Levy, 1999; Najera et al., 2019). HDR is a less common method of DNA repair in plant cells, and relies on a donor repair template that must share homology with the borders of the break site. This template can come from a sister chromatid or a duplicate gene, or it can be exogenously supplied into the cell along with the SSN. HDR can be used to introduce point mutations or insert a desired engineered sequence into the plant genome (Voytas and Gao, 2014).

2.1 Meganucleases

The earliest method used to create precise genome modifications in maize utilized meganucleases, also known as homing endonucleases, which creates DSBs at rare 14–40 bp recognition sites (Edgell, 2009). D'Halluin et al. (2008) tested the feasibility of the I-*Sce*I endonuclease for genome editing in maize through two rounds of transformation. They first produced a transgenic maize line carrying a target DNA fragment containing an I-*Sce*I recognition site adjacent to a promoter-less herbicide-resistance gene (*bar*) cassette. This line was then transformed with a plasmid containing a maize-optimized I-*Sce*I gene and a repair template consisting of a cauliflower mosaic virus (CaMV) 35S promoter between sequences flanking the target I-*Sce*I site. Correct insertion of the repair template would result in the CaMV 35S promoter driving *bar* gene

Table 1 List of publications on maize genome editing using site-specific nucleases

Time	First author (journal)	Title	Innovation	Transformation methods (type of explant)	Type of Genome editing	Genes
2008	D'Halluin et al. (Plant Biotechnology Journal)	Homologous recombination: a basis for targeted genome optimization in crop species such as maize	Use of meganuclease I-*Sce*I to precisely introduce promoter to activate integrated *bar* gene	PEG transfection, *Agrobacterium*-mediated, particle gun bombardment (protoplast, embryogenic cell culture)	Meganuclease	*bar,* I-*Sce*I
2009	Shukla et al. (Plant Biotechnology Journal)	Precise genome modification in the crop species *Zea mays* using zinc-finger nucleases	Zinc finger FokI precision genome edits created INDELS within the *IPK1* gene through NHEJ	Silicon carbide whisker (maize callus)	ZFN	*IPK1*
2010	Gao et al. (The Plant Journal)	Heritable targeted mutagenesis in maize using a designed endonuclease	An engineered meganuclease I-*Cre*I were introduced to maize to disrupt *liguleless1 (lg1)* gene via indel mutation	*Agrobacterium*-mediated (immature embryos)	Meganuclease	*lg1*
2013	Djukanovic et al. (The Plant Journal)	Male-sterile maize plants produced by targeted mutagenesis of the cytochrome P450-like gene (*MS26*) using a re-designed I-*Cre*I homing endonuclease	An engineered meganuclease I-*Cre*I wre introduced to maize to disrupt cytochrome P450-like gene (*MS26*)	*Agrobacterium*-mediated (immature embryos)	Meganucleases	*MS26*
2013	Ainley et al. (Plant Biotechnology Journal)	Trait stacking via targeted genome editing	Complex trait stacking was achieved by pre-integrated trait landing pads and corresponding ZFN	Particle gun bombardment (immature embryos)	ZFN	*pat* and *aad1*

2014	Liang et al. (Journal of Genetics and Genomics)	Targeted mutagenesis in *Zea mays* using TALENs and the CRISPR/Cas system	TALENS were used to disrupt PA synthesis in maize through the creation of INDELS	PEG transfection, *Agrobacterium*-mediated (protoplast, immature embryo)	TALENS	*ZmIPK1A*, *ZmIPK*, and *ZmMRP4*
			CRISPRs were used to disrupt PA synthesis in maize through the creation of INDELS	PEG transfection (protoplast)	CRISPR/Cas9	*ZmIPK1A*, *ZmIPK*, and *ZmMRP4*
2014	Xing et al. (BMC Plant Biology)	A CRISPR/Cas9 toolkit for multiplex genome editing in plants	Maize-codon-optimized Cas9 demonstrated efficient targeted mutagenesis in maize protoplasts and transgenic T0 lines	PEG, *Agrobacterium*-mediated (protoplast, immature embryo)	CRISPR/Cas9	*ZmHKT1*
2015	Svitashev et al. (Plant Physiology)	Targeted mutagenesis, precise gene editing, and site-specific gene insertion in maize using Cas9 and guide RNA	Demonstrated multiple gene knock outs, targeted nucleotide modification, and site specific integration by homology direct repair	Particle gun bombardment (immature embryos)	CRISPR/Cas9	*ALS*, *LIG3*, *Ms26*, *Ms45*
2015	Char et al. (Plant Biotechnology Journal)	Heritable site-specific mutagenesis using TALENs in maize	TALENs were employed to generate stable, heritable mutations at the maize glossy2 (gl2) locus	*Agrobacterium*-mediated (immature embryos)	TALENS	*gl2*
2016	Svitashev et al. (Nature Communications)	Genome editing in maize directed by CRISPR-Cas9 ribonucleoprotein complexes	Delivered gRNA and CRISPR endonuclease directly into the immature embryos to create precision genome edits	Particle gun bombardment (immature embryos)	CRISPR/Cas9 (RNPs)	*ALS2*, *LIG*, *MS26*, *MS45*

(*Continued*)

Table 1 (*Continued*)

Time	First author (journal)	Title	Innovation	Transformation methods (type of explant)	Type of Genome editing	Genes
2016	Qi et al. (BMC Biotechnology)	High-efficiency CRISPR/Cas9 multiplex gene editing using the glycine tRNA-processing system-based strategy in maize	Maize glycine tRNA-processing system-based multiplex gene editing system	*Agrobacterium*-mediated (immature embryos)	CRISPR/Cas9	*MADS, MYBR, AP2, RPL, PPR, LncRNA*
2017	Zong et al. (Nature Biotechnology)	Precise base editing in rice, wheat and maize with a Cas9-cytidine deaminase fusion	Demonstrated base editing by nickase Cas9 protein in maize protoplast and stable transformation	PEG transfection, *Agrobacterium*-mediated (protoplast, immature embryo)	CRISPR/Cas9	*ZmCENH3*
2017	Li et al. (Journal of Genetics and Genomics)	Generation of thermosensitive male-sterile maize by targeted knockout of the *ZmTMS5* gene	Achieving thermo-sensitive male sterile maize plants by knocking-out target gene using CRISPR	Particle gun bombardment (Immature embryos)	CRISPR/Cas9	*ZmTMS5*
2017	Shi et al. (Plant Biotechnology Journal)	ARGOS8 variants generated by CRISPR-Cas9 improve maize grain yield under field drought stress conditions	Obtaining drought-tolerant trait by replacing or targeted inserting of promoter on endogenous ethylene regulator ARGOS	Particle gun bombardment (immature embryos)	CRISPR/Cas9	*ARGOS8, GOS2*
2017	Char et al. (Plant Biotechnology Journal)	An *Agrobacterium* -delivered CRISPR / Cas9 system for high-frequency targeted mutagenesis in maize	Demonstrated the ISU maize CRISPR platform, high-frequency mutagenesis by *Agrobacterium* delivery of CRISPR/Cas	*Agrobacterium*-mediated (immature embryos)	CRISPR/Cas9	*ZmAgo18a, ZmAgo18b, a1, a4*

2017	Feng et al. (Plant Biotechnology Journal)	High-efficiency genome editing using a *dmc1* promoter-controlled CRISPR/Cas9 system in maize	Demonstrated high-efficiency dmc1 promoter controlled CRISPR/Cas9 system in maize	*Agrobacterium*-mediated (immature embryos)	CRISPR/Cas9	*zb7*
2018	Wu et al. (PLoS Genetics)	Role of heterotrimeric Gα proteins in maize development and enhancement of agronomic traits	Achieving multiplex genome editing by knocking out three ortholog of *ZmXLG* genes	*Agrobacterium*-mediated (immature embryos)	CRISRP/Cas9	*ZmXLG1, ZmXLG3a, ZmXLG3b*
2019	Kelliher et al. (Nature Biotechnology)	One-step genome editing of elite crop germplasm during haploid induction	Used pollen to edit the plant genome without a stable insertion of DNA cassette	Haploid inducer pollen-mediated (nascent zygote embryos)	CRISPR/Cas9	*matl, GW2, VLHP1*
2019	Wang et al. (Molecular Plant)	Development of haploid-inducer mediated genome editing system for accelerating maize breeding	Pairing a haploid inducer with CRISPR to create homozygous lines of edited maize	Haploid inducer pollen-mediated (nascent zygote embryos)	CRISPR/Cas9	*ZmLG1*
2019	Young et al. (Scientific Reports)	CRISPR-Cas9 editing in maize: systematic evaluation of off-target activity and its relevance in crop improvement	Evaluation of off-target activity of CRISPR/Cas9 in maize genome	Particle gun bombardment (immature embryos)	CRISPR/Cas9 (RNPs)	*Ms26, Ms45, Lig1*
2019	Lee et al. (Plant Biotechnology Journal)	Activities and specificities of CRISPR/Cas9 and Cas12a nuclease for targeted mutagenesis in maize	Comparison of Cas9 and Cas12a activities by targeting gl2 coding region recognized by both nucleases	*Agrobacterium*-mediated (immature embryos)	CRISPR Cas9 and CRISPR Cas12a	*gl2*

(*Continued*)

Table 1 (*Continued*)

Time	First author (journal)	Title	Innovation	Transformation methods (type of explant)	Type of Genome editing	Genes
2020	Jaqueth et al. (The Plant Journal)	Fertility restoration of maize CMS-C altered by a single amino acid substitution within the *Rf4* bHLH transcription factor	Used CRISPR-gRNA RNP directly into immature embryos and created single amino acid substitution at the target site to validate putative gene	Particle gun bombardment (immature embryos)	CRISPR/Cas9 (RNPs)	*Rf4*
2020	Liu et al. (Plant Cell)	High-throughput CRISPR/Cas9 mutagenesis streamlines trait gene identification in maize	Established high-throughput CRISPR-Cas9 genome editing pipeline in maize	*Agrobacterium*-mediated (immature embryos)	CRISPR/Cas9	1244 candidate genes related to agronomic tratis of maize
2020	Wu et al. (Proc Natl Acad Sci USA)	The maize heterotrimeric G protein β subunit controls shoot meristem development and immune responses	Reported the results of *Gβ* (*ZB1*) gene knockout using CRISPR/Cas9 in maize	*Agrobacterium*-mediated (immature embryos)	CRISPR/Cas9	*Gβ (ZB1)*

expression and thus conferring herbicide resistance. Successful HDR events were obtained in 30% to 40% of recovered events (D'Halluin et al., 2008).

While the application of natural meganucleases for genome editing is feasible, it was largely hindered by limited targetability, due to their extremely rare recognition sites in a given genome (e.g. one per 6.87×10^{10} bp for I-*Sce*I). Therefore, the engineering of meganucleases with new specificities was pursued, and it was demonstrated that the I-*Cre*I meganuclease could be modified to target new sites (Rosen et al., 2006; Arnould et al., 2007). In maize, an engineered I-*Cre*I endonuclease was used to disrupt the *liguleless1* (*lg1*) gene (Gao et al., 2010). A ~3% mutagenesis frequency in T_0 transgenic plants was achieved via *Agrobacterium*-mediated transformation of immature embryos, resulting in deletions ranging between 2 bp and 220 bp (Gao et al., 2010). Similarly, another engineered I-*Cre*I endonuclease was used to create male-sterile plants by inducing DSBs in the cytochrome P450-like gene *MS26* (Djukanovic et al., 2013). The I-*Cre*I endonuclease was engineered to recognize a 22 bp target sequence in the 5th exon of *MS26*. *Agrobacterium*-mediated transformation with the modified I-*Cre*I gene resulted in targeted mutations in 5.8% of transgenic plants recovered (Djukanovic et al., 2013).

Although both natural and engineered meganucleases have been used with success in genome editing, the systems are cumbersome to design and the extent to which recognition sequences can be modified is limited. Therefore, meganucleases have not been widely used for targeted genome engineering (Silva et al., 2011; Dunn and Pinkert, 2014).

2.2 Zinc finger nucleases (ZFNs)

Zinc finger nucleases (ZFNs) have two domains: a zinc-finger-based DNA recognition domain, which can be assembled from modules to target specific sequences, and a non-specific *Fok* I endonuclease domain to cleave the targeted DNA (Kim et al., 1996). Shukla et al. (2009) were the first group that used ZFNs to create targeted mutations in maize, both indel mutations, and targeted DNA integrations. Four pairs of ZFNs were selected from 66 designed to knockout the maize *IPK1* gene (inositol-pentakisphosphate 2-kinase). *IPK* is involved in phytate biosynthesis, which accounts for a majority of the seeds phosphorus but is indigestible by non-ruminant animals. Knocking out *IPK* increases the nutritional value of animal feed by reducing the amount of phytate (Shukla et al., 2009). The ZFN reagents were introduced into type II calli of the maize Hi II genotype using the silica carbide whiskers (WHISKERS™)-based transformation method. The reagents included a donor repair template, which had an herbicide resistance *PAT* (phosphinothricin acetyltransferase) gene with or without a promoter (autonomous and non-autonomous *PAT*, respectively) and sequences homologous to the *IPK1* target site (Shukla et al., 2009). Successful integration of the repair template would inactivate the *IPK1* gene and confer herbicide resistance. All four pairs of ZFNs demonstrated precise targeted insertion via HDR. Insertion frequencies with autonomous *PAT* ranged from 3.4% to 22.1%, averaging 19.6%, while promoter-less constructs (non-autonomous *PAT*) showed enrichment for targeted integration, ranging from 16.7% to 100% and averaging 40% (Shukla et al., 2009). The higher frequency of HDR success of the promoter-less compared to autonomous constructs may be attributed to the need for HDR to occur in order for the promoter-less individuals to survive herbicide selection. Many of the autonomous constructs likely resulted in insertion events that were not a result of HDR (Shukla et al., 2009).

ZFNs were also used by the same group to perform complex trait stacking in the maize genome by using HDR and pre-integrated trait landing pads (TLPs) (Ainley et al., 2013). In this work, trait stacking was conducted in two steps. First, a transgene containing the *pat* gene, which confers resistance to the herbicide glufosinate, and a TLP with a ZFN target sequence were introduced into the maize genome by WHISKER™ transformation. Next, a second transgene, *aad1* (aryloxylkanoate dioxygenase1), which provides tolerance to aryloxyphenoxypropionate herbicides, and a ZFN expression construct were co-transferred to immature embryos of maize by particle bombardment. The *aad1* transgene was successfully integrated into the TLP adjacent to the *pat* gene in 5% of the transformed embryos. Molecular and segregation analyses of the T_0 events and their progenies showed that the two herbicide tolerance traits co-segregated, confirming the ability for targeted genome editing to create stacked transgenes (Ainley et al., 2013). Advantages of this technology are that DNA segments too large to be efficiently introduced by conventional transformation methods can be assembled sequentially, and that the position effect on transgene expression can be eliminated.

Although these two studies demonstrated successful maize genome editing by ZFNs, engineering ZFNs for a specific target site remains time- and resource-intensive. Another possible reason for the limited distribution of ZFN technology is its restricted intellectual property options (Cloney, 2016). Therefore, its utilization in maize community has been limited.

2.3 Transcription activator-like effector nucleases (TALENs)

Compared to ZFNs, the TALEN system offers a clear advantage for designing cleavage of desired target sequences. Similar to ZFNs, TALENs utilize the *Fok*I endonuclease to cut target DNAs. The TALEN DNA-binding structure is made up of repeats of 34 amino acids, with amino acids at positions 12 and 13 determining the binding of a specific nucleotide. DNA encoding these repeats can be assembled to target a particular sequence (Joung and Sander, 2013). TALENs were first shown to edit the maize genome by using a maize protoplast transient expression system (Liang et al., 2014). TALEN constructs were created to target the *ZmPDS* (phytoene desaturase) gene, as well as genes involved in the phytic acid biosynthetic pathway (*ZmIPK1A*, *ZmIPK*, and *ZmMRP4*). This transient system demonstrated that indel mutations could be successfully induced by TALENs at rates between 9.1% and 23.1% in protoplasts. The same TALEN constructs were then introduced into Hi-II immature embryos via *Agrobacterium*-mediated transformation. Sequencing of T_0 plants showed that different mutation patterns at the target sites were created at a rate of about 39% for each of the five constructs tested (Liang et al., 2014). Char et al. (2015) were the first to demonstrate that TALEN-induced mutations could be inherited to the next generation. TALENs, targeting the endogenous *glossy2* (*gl2*) gene in maize, were introduced into immature embryos using *Agrobacterium*-mediated transformation. Induced *gl2* mutations were found at frequencies of 10% and 3.7% in Hi-II and B104 T_0 plants, respectively, suggesting genotypic effects on maize genome editing (Char et al., 2015).

2.4 Clustered regularly interspaced short palindromic repeat (CRISPR)/CRISPR-associated proteins (Cas)

Currently, CRISPR/Cas9 is the most widely used genome engineering tool. Unlike ZFNs and TALENs, which rely on relatively complex protein engineering to target a particular DNA sequence (Kim et al., 1996; Joung and Sander, 2013), CRISPR/Cas9 uses an approximately 20-nt guide sequence in a single-guide RNA (sgRNA) to target a corresponding site in the genome (Jinek et al., 2012). This target site must be next to protospacer adjacent motifs (PAM) with the sequence NGG, where N is any base, to be cut by the Cas9 nuclease to create a DSB. Given this simplicity and its high efficiency in genome editing applications, it is not surprising that there was only one more report testing the TALENs (Char et al., 2015) while there have been more than 10 CRISPR/Cas9 publications in maize genome editing since Liang et al. (2014) published the first use of TALENs and CRISPR/Cas9 in maize genome editing (Table 1).

Liang et al. (2014) compared the efficiency of CRISPR/Cas9 and TALENs in maize by targeting genes associated with the phytic acid (PA) biosynthesis pathway. CRISPR/Cas9 and two gRNAs targeting the *ZmIPK* gene were delivered into maize protoplasts via polyethylene glycol (PEG)-mediated transformation. Both gRNA1 and gRNA2 successfully introduced indel mutations at the target site with frequencies of 16.4% and 19.1%, respectively (Liang et al., 2014). Overall this early version of CRISPR/Cas9 editing demonstrated similar efficiencies (13.1%) compared to TALEN constructs (9.1%) when targeting similar genetic regions (Liang et al., 2014).

In other work, Xing et al. (2014) demonstrated CRISPR-directed targeted mutagenesis in the *ZmHKT1* (maize high-affinity potassium transporter) gene in protoplasts and in stably transformed T_0 lines via *Agrobacterium*-mediated transformation. About 60% of the T_0 lines showed biallelic mutations, indicating efficient targeted mutagenesis. Li et al. (2017) generated a thermosensitive male-sterile maize line by knocking out the *ZmTMS5* gene. Shi et al. (2017) improved maize grain yield under drought conditions by precisely replacing the promoter of the endogenous ethylene regulator *ARGOS8* with a strong GOS2 promoter.

Svitashev et al. (2015) were the first to demonstrate the deployment of CRISPR technology in three aspects of maize genome engineering: targeted mutagenesis at multiple sites, precise gene editing via HDR, and site-specific gene insertion. First, they showed that Cas9-gRNAs could be used to generate multiple gene knockouts simultaneously. Twelve different sites at five genomic locations were simultaneously targeted for CRISPR/Cas9-mediated cleavage via particle gun bombardment. Second, they reported precise endogenous gene editing via HDR using CRISPR/Cas9. The *Acetolactate synthase2* (*ALS2*) gene was chosen to demonstrate HDR because a single amino acid change (proline at amino acid position 165 to serine, P165S) confers resistance to sulfonylurea herbicides. CRISPR/Cas9 reagents and repair templates were introduced into immature maize embryos via particle bombardment. The donor repair templates for *ALS2* were either one of two 127-nt single-stranded oligos (ssOligos) or a 794 bp double-stranded template DNA cloned into a plasmid vector (Svitashev et al., 2015). The group was able to recover a total of nine chlorosulfuron-resistant events from 3000 bombarded embryos using the three different donor DNA templates. Although the frequencies of HDR in the *ALS* gene (0.2% to 0.4%) were not high, all herbicide-resistant events had DNA base replacements leading to P165S amino acid change that were inherited in the progenies. Third, they compared the efficiency of site-specific gene insertion via HDR using biolistic delivery of DNA constructs carrying either meganuclease or CRISPR/Cas9 to create *PAT* gene insertions at the same *LIG1* target sites. The meganuclease achieved 0.7% targeted insertion, whereas CRISPR/Cas9 delivered by bombardment showed 2.5% to 4.0% efficiency. Interestingly, the same CRISPR/Cas9 and repair templates delivered by *Agrobacterium* strain, LBA4404, did not produce any targeted insertion events, suggesting that bombardment might be a more efficient delivery method for targeted gene insertion via HDR in maize (Svitashev et al., 2015).

Char et al. (2017) demonstrated efficient genome editing by *Agrobacterium*-mediated CRISPR/Cas9 delivery in maize Hi-II and B104 genotypes. Two single gRNA constructs targeting *Argonaute* (*Ago*) genes (*ZmAgo18a* or *b*) showed similar mutation frequencies of 70% to 74% in Hi-II. In addition, when two gRNAs transcription cassettes were cloned into a single plasmid vector targeting *ZmAgo18a/b* or *anthocyaninless* genes (*a1* and *a4*), mutant lines carrying biallelic mutations at both target genes were recovered, indicating highly efficient multiplex genome editing in maize via *Agrobacterium*-mediated transformation.

Wu et al. (2018) used multiplex CRISPR genome editing to edit three extra-large GTP binding protein (XLG) genes (*ZmXLG1*, *ZmXLG3a* and *ZmXLG3b)* simultaneously to examine the G protein α subunit (Gα) function in maize meristem regulation. The EHA101 strain of *Agrobacterium* was used to deliver DNA constructs containing multiple gRNAs and Cas9 into Hi-II immature embryos. One event was recovered with three null alleles that was later bred to other genotypes. The double mutants with weak Gα phenotype showed reduction in plant height without change in shoot apical meristem size (Wu et al., 2018). Multiplex genome editing was also achieved by adapting the tRNA-processing system to deliver multiple gRNAs processed from a single transcript driven by the maize U6 promoter (Qi et al., 2016). These tRNA-gRNA achieved 57.1% to 100% on-target mutation frequencies at six target loci in the T_0 lines (Qi et al., 2016).

Optimizing promoters for CRISPR/Cas9 reagents can further improve genome-editing efficiency. Feng et al. (2018) reported that a Cas9 gene driven by a maize *dmc1* promoter and gRNA driven by a maize U3 promoter achieved highly efficient targeted mutagenesis: 100% of the T_0 callus lines carried on-target mutations, and importantly, around 66% of T_0 lines (156/237) had the same (homozygous) or different (biallelic) mutation patterns on both alleles, without detectable off-target effects.

An important method of maize genome editing is transient expression of the CRISPR/Cas reagents in cells to reduce off-target effects and avoid regulatory concerns due to transgene integration. Svitashev et al. (2016) demonstrated that preassembled CRISPR/Cas9-gRNA ribonucleoprotein complexes (RNPs) can be introduced by bombardment into maize immature embryos to achieve targeted indel mutations. In addition, co-bombardment of ssOligo repair donor templates and CRISPR/Cas9-gRNA RNPs demonstrated HDR-mediated sequence replacement in the target gene *ALS2*, conferring resistance to sulfonylurea herbicides (Svitashev et al., 2016).

Using the same approach, Jaqueth et al. (2020) achieved a single amino acid substitution to validate the identity of a candidate gene as a restorer of fertility 4 (*Rf4*), which is the major fertility-restorer gene of type C cytoplasmic male sterility (CMS-C). One heterozygous event from CMS-C restoring line (*Rf4*-F187) containing a single amino acid substitution (F187Y), a synonymous mutation at the PAM site, and without plasmid DNA integration, was achieved by co-bombardment of RNPs and donor template. Testcross to a CMS-C non-restoring line (*Rf4*-Y187) showed a 1:1 segregation ratio for male fertility, validating this amino acid change is responsible for the *Rf4* phenotype (Jaqueth et al., 2020).

Another method used to edit the maize genome without stable integration of genome editing reagents takes advantage of the aberrant reproductive process of haploid induction, which is ubiquitous in maize breeding programs (Kelliher et al., 2019; Wang et al., 2019). When maize is pollinated, one sperm fertilizes the egg to form a zygote while the other is fused with the two polar nuclei to form the endosperm. During haploid induction, fertilization occurs, but the male genome is lost, producing haploid embryos of solely maternal origin (Zhang et al., 2008; Zhao et al., 2013). By having CRISPR/Cas9 constructs in the male inducer, post-fertilization loss of the male genome gives temporal expression of CRISPR/Cas and gRNA cassettes to create DSBs in the maternally derived genome of the haploid zygote. Edited maternal haploid plants can be recovered in the progeny and can be treated with microtubule disrupting agents such as colchicine to restore fertility and produce a completely homozygous diploid line when self-pollinated. Kelliher et al. (2019) were able to achieve a rate of more than 3% for five of six constructs tested, with one construct not working at all (Kelliher et al., 2019). Wang et al. (2019) reported that 245 haploid plants were recovered from 339 haploid candidate seeds, 10 of which contained mutations at the *Lg1* locus, demonstrating a 4.1% editing efficiency (Wang et al., 2019). This approach can be used to produce homozygous mutants in multiple transformation recalcitrant lines in just two generations once a transgenic inducer is produced. Because it essentially uses the same method as current haploid breeding programs, this method is simple to integrate and scale-up (Kelliher et al., 2019; Wang et al., 2019).

One of the concerns with targeted genome editing is the potential for off-target activity of the CRISPR/Cas9 nuclease, which has been addressed in two maize studies (Lee et al., 2019; Young et al., 2019). These two studies utilized *in silico* predictions (Cas-Offinder: Bae et al., 2014) and in vitro cleavage assays, CIRCLE-seq (Tsai et al., 2017) and CLEAVE-seq (Young et al., 2019), to identify genome-wide potential off-targets. Selected potential off-targets were monitored for *in planta* mutagenesis, but no significant off-target activities were observed when there were more than two mismatches in the protospacer, indicating that unnecessary off-target effects can be avoided by designing highly specific gRNAs (Lee et al., 2019; Young et al., 2019).

Maize genome editing toolkits have been further expanded in recent years. Base editing, which can introduce single nucleotide changes without making DSBs or needing a donor template, was first demonstrated in maize by Zong et al. (2017). They used a cytosine base editor (CBE), which utilizes a CRISPR-Cas9 nickase, which creates single strand breaks, fused with a cytosine deaminase and a uracil glycosylase inhibitor. CBE converts C·G base pairs to T·A within a narrow editing window (nucleotides 3–8 in the protospacer). Using PEG-mediated transformation of maize protoplasts, Zong et al. (2017) first showed that CBE can convert the first base pair in codon 66 of the blue fluorescent protein **C**AC (histidine) to **T**AC (tyrosine) resulting in a color shift in 0.22% of the protoplasts. To test the feasibility of base editing of endogenous genes in maize, *ZmCENH3* (centromere-specific histone H3) was selected as a target (Zong et al., 2017). CENH3 is a kinetochore protein that facilitates chromosome movement in actively dividing cells. CENH3 is also active in recruiting and co-localizing with other proteins responsible for chromosome movement (Maheshwari et al., 2015). Their work showed that CBE successfully introduced C-to-T conversions in both protoplasts (up to 9%) and transgenic maize plants (10%) via *Agrobacterium*-mediated transformation of immature embryos (Zong et al., 2017). A key advantage of using CBE compared to other CRISPR technologies was that it produced base pair changes without producing many indels.

More recently, Lee et al. (2019) tested CRISPR/Cas9 and Cas12a (also known as Cpf1) activities and specificities in the maize inbred line B104. LbCas12a, from *Lachnospiraceae bacterium*, requires 'TTTV' PAM sequences (V=A, C, or G), thus can expand targetable genomic regions where 'NGG' PAM for Cas9 is not available (Zetsche et al., 2015). Lee et al. (2019) showed that although LbCas12a can introduce targeted mutations in *gl2*, the mutagenesis efficiency was lower than that of Cas9 at the two target sites (0–60% vs. 90–100%) suggesting that further optimization is required for efficient Cas12a-mediated maize genome editing. The low mutagenesis efficiency of Cas12a could have been due to the low temperatures (<28°C) used during the maize transformation procedures (Malzahn et al., 2019). They tested this hypothesis in rice protoplasts, using three different Cas12a nucleases (AsCas12a (from *Acidaminococcus bacterium)*, FnCas12a (from *Francisella novicida)*, and LbCas12a) and four different temperatures (22°C, 28°C, 32°C, and 37°C). All three Cas12a nucleases introduced indel mutations at the two target sites under the four temperature conditions, but the highest mutation frequencies were observed at elevated temperatures 28°C and 32°C compared to 22°C and 37°C (Malzahn et al., 2019). In *Arabidopsis*, editing efficiency was dramatically higher at 29°C compared to 22°C. Moreover, in maize 51.4% to 100% of the T_1 plants obtained from two independent monoallelic mutant T_0 lines grown at 28°C and pollinated by wild-type B104 (Lee et al., 2019) exhibited biallelic or monoallelic mutations at the target sites indicating high LbCas12a activities at 28°C (Malzahn et al., 2019), further suggesting that a high-temperature treatment during transformation could enhance Cas12a-mediated genome editing in maize.

3 Current challenges in maize genome editing

As with all new and emerging technologies, problems and bottlenecks exist that limit its effectiveness. Limiting factors may include time, money, regulations, availability of specialized laboratory equipment, and germplasm. Establishing proofs of concept through transient assays and reporter gene experiments are often the first steps in overcoming these challenges. Once a technology has initially been shown as effective it will draw the attention of other laboratories who can continue to optimize and expand it.

3.1 The tissue culture and transformation bottleneck

Precise genome engineering inevitably requires an efficient delivery of the editing reagents into maize cells; the most popular methods are *Agrobacterium*- and biolistic-mediated transformation (Altpeter et al., 2016). Current methods often require using antibiotic or herbicide-resistant genes for selection through tissue culture in order to increase the efficiency of recovering a gene-edited event. The presence of this introduced DNA could be a regulatory concern. In addition, regulatory hurdles are a concern for any commercial product produced through genetic transformation, especially via *Agrobacterium*-mediated transformation. The use of biolistics for delivering CRISPR reagents as RNP avoids regulatory hurdles associated with *Agrobacterium* being classified as a plant pathogen. In addition, RNP delivery avoids the integration of CRISPR reagents carrying transgenes into the genome, reducing undesired off-target effects. *Rhizobium* and *Ensifer adhaerens* are also capable of producing transgenic events in plants, and their use, or a future non-pathogenic bacterium may side-step these regulations (Broothaerts et al., 2005; Altpeter et al., 2016).

Whether through *Agrobacterium*-mediated or biolistic delivery, current protocols for transformation and gene editing in maize require a tissue culture step. Traditional selection in tissue culture is time- and labor-intensive, and requires skill in aseptic technique, delicate handling, and proper evaluation of materials throughout the process. The ability to form regenerable callus for tissue culture is also limited to a few lines that are of limited agronomic importance. It also requires the production of immature zygotic embryos that are costly and time-consuming. Additionally, the media and methods of callus selection and regeneration are not uniform between these lines (Raji et al., 2018; Lee and Zhang, 2016). Automation of tissue culture could improve throughput and make transformation scalable and affordable to independent laboratories as a service, as well as making large-scale knockout screens feasible (Altpeter et al., 2016; Liu et al., 2020). Reports of success in automation of maize tissue culture are currently lacking, and even well-funded biotech companies rely on teams of specialty technicians. This presents a bottleneck for any projects involving gene editing, because researchers either have to learn to perform plant transformation themselves, or order from one of the increasingly busy facilities that provide plant transformation as a service. Arguably the greatest improvement to gene editing in maize would be a fast, high-throughput transformation system that is easy to perform like the floral dip method of *Arabidopsis* (Clough and Bent, 1998).

Lowe et al. (2016) developed a method using co-delivery of the morphogenic genes, Baby boom (*Bbm*) and Wuschel2 (*Wus2*), to produce transgenic callus from maize embryos, as well as from explants such as leaves and mature seed, that were previously not transformable. Constitutive expression of *Bbm* and *Wus2* is detrimental to regeneration of plants, thus inducible *Cre/lox* excision of the developmental genes was required (Lowe et al., 2016). Subsequent versions skipped the callus step through the use of non-excised scutellum-specific and auxin-inducible promoters for driving *Bbm* and *Wus*, respectively (Lowe et al., 2018), and later a heat-shock induced promoter for *Cre/lox* excision of the *Bbm/Wus* construct was used (Masters et al., 2020). The haploid induction editing method described in Section 2.4 is another way to overcome recalcitrance to tissue culture-based transformation (Kelliher et al., 2019). While both of these methods allow more lines to be edited without lengthy backcrossing schemes, both still require a tissue culture step.

3.2 Fidelity and efficiency

The potential for off-target mutations is an important issue for precise genome editing in maize and other crops. Screening for off-target mutations by targeted sequencing of predicted off-target sites or by whole genome sequencing is feasible but resource intensive, especially for academic laboratories. Off-target mutations can have undesirable consequences, and may need to be removed by crossing, which adds substantially to the time required to get the desired product. Recent work has carefully examined CRISPR/Cas9 off-target activity in maize and improved enzymes are being developed. As described above in Section 2.4, two groups recently extensively examined off-target activity in maize (Lee et al., 2019; Young et al., 2019). Future work using similar approaches would test specificities and off-target activities of other versions of nucleases, including base editors and Cas12a, before their extensive applications for genome editing.

An increasing number of engineered Cas9 and other nucleases with higher specificities and noncanonical PAM sequences are becoming available. The latter is especially important for base editing and HDR, which require the DNA be cut very close to the site to be modified. A modified SpCas9, xCas9, with higher specificity and an expanded PAM recognition (including NG, GAA, and GAT) (Hu et al., 2018), has been used successfully in rice (Li et al., 2019; Zhong et al., 2019). Similarly, a highly specific SaCas9 (PAM sequence NNGRRT) nuclease has been developed (Tan et al., 2019) and SaCas9/ortholog chimeras have been created that can recognize up to one-fourth of possible target sites (Ma et al., 2019). More recently, Walton et al. (2020) created near–PAM-less SpCas9 variants with higher specificities and reduced off-target activities. These and other improved nucleases for genome editing will surely be applied to maize, thus editing specificity and targeting limitations are rapidly becoming a minor issue.

3.3 Chromosome rearrangements in biolistic transformation

Biolistic transformation is a method commonly used to introduce DNA constructs into plant tissues, including constructs for editing DNA. It is a method that can lead to high copy-number insertions, but by reducing the amount of DNA bombarded, single or low copy-number events are more common (Lowe et al., 2009). Editing events are often analyzed by PCR amplification and/or sequencing of target sites and likely off-target sites, and it has generally been assumed that there are not larger-scale events from the biolistic transformation process. Liu et al. (2019) examined genome-scale effects of biolistic transformation with lambda and plasmid DNA which would not be observed by analysis of only the edited sites. They found that transformation events could range from simple insertions to complex fragmentation and rejoining of small DNA fragments as well as megabase-scale genome rearrangements. Though genome rearrangements can also occur with *Agrobacterium*-mediated transformation, it is thought to occur at a lower level (Anderson et al., 2016), and for this reason *Agrobacterium*-mediated transformation may be the preferred method for introducing editing reagents as DNA into maize. Supporting evidence was recently reported by Banakar et al. (2019) who found that random DNA fragments originating from the bombarded plasmid and genomic DNAs were inserted at targeted DSBs. CRISPR/Cas9 reagents delivered by particle bombardment both as RNP and plasmid DNA induced high-frequency random insertions (over 14%), whereas *Agrobacterium*-mediated transformation did not produce such events (Banakar et al., 2019). Biolistics is the most widely used method for delivering editing reagents as RNPs to create fertile regenerants in maize (Svitashev et al., 2016); however, it is not yet clear how to reduce genomic damage caused by the microprojectiles (Krysiak et al., 1999).

3.4 Increasing HDR frequency for gene modification or insertion

In work involving maize and other plants, genome editing is most commonly used to create small indels caused by NHEJ at DSBs. Larger deletions can also be created by using two nuclease target sites flanking the region to be deleted. On the other hand, genomic DNA modification by HDR using a DNA repair template is reported less often, which is likely due to the low efficiency of HDR in plants. Though not yet reported in maize, promising methods to increase HDR have been developed.

One approach of enhancing HDR is to increase the template DNA concentration. The geminivirus vector system has been used for this purpose.

Geminiviruses have small circular ssDNA genomes that replicate to high levels from a dsDNA replicative form during infection. Baltes et al. (2014) used a vector based on bean yellow dwarf geminivirus as a repair template in *Arabidopsis*, and Wang et al. (2017) used a wheat dwarf geminivirus vector to achieve an almost 20% gene knock-in efficiency in rice. The limitation of this method is that geminiviruses, like other plant viruses, have a limited capacity for added DNA fragments in their genomes, and thus these experiments were done in transgenic tissue already expressing Cas9. Wheat dwarf geminivirus can replicate to high levels in maize cells (Timmermans et al., 1992) and maize streak geminivirus has been developed as a vector for maize (Palmer and Rybicki, 2001), so there is potential for using this method in maize.

Another approach is to increase the local effective concentration of the repair template by tethering it to the Cas nuclease. This makes the template immediately available once the DNA is cut, thus allowing HDR to compete more effectively with the NHEJ pathway. Sharon et al. (2018) used a modified bacterial retron to synthesize in vivo gRNA-ssDNA HDR template fusions that were used with Cas9 for high-throughput genome editing in yeast. Anzalone et al. (2019) developed 'prime editing' using a Cas9 nickase-reverse transcriptase fusion, which can directly introduce small edits to the target DNA via reverse transcription. Nicked non-target strand DNA serves as a primer and the 3′ portion of the extended gRNA (pegRNA) serves as an RT template to directly introduce specific sequence modifications at the target site. For RNP delivery, Aird et al. (2018) developed a Cas9-HUH endonuclease fusion that can be assembled in vitro with an unmodified ssDNA template that spontaneously forms a covalent attachment with the endonuclease. These methods have great potential for CRISPR-mediated HDR in maize.

3.5 DNA-free editing-RNPs

Delivery of genome editing reagents as RNPs has several advantages. First, the half-life of the RNP is limited because the nucleases are delivered as protein, thus reducing the potential for off-target editing. Second, there is no concern about having to remove DNA encoding the editing reagents, or other DNA fragments by making crosses with progeny plants. Though genome editing in maize can be done without selection (Svitashev et al., 2016), DNA carrying a selectable marker is often co-transfected with the RNP to enrich for edited cells, with the selectable marker being outcrossed in the progeny. RNPs are generally introduced into plant tissue by biolistic delivery after coating onto gold microcarriers using TransIT®-2020, a transfection reagent from Mirus Bio (Madison, WI) that is commonly used for animal cell transfection (Svitashev et al., 2016). RNPs can also be delivered into plant protoplasts via transfection, but regeneration of transgenic maize plants from protoplasts is problematic

(Morocz et al., 1990). Increasing the delivery efficiency of RNPs, particularly in the absence of a selectable marker, will be advantageous.

3.6 Regulatory landscape

Not all challenges in maize genome editing are technical: regulation and proper stewardship are also important challenges. Maize is an important and valuable food crop, so proper stewardship of transgenic plants is important for both safety and public trust. Because it is wind-pollinated, maize can theoretically pollinate plants many kilometers away, though realistic distances are likely limited to a few hundred meters (Luna et al., 2001; Bannert and Stamp, 2007). Proper isolation, handling, storage, disposal, and labeling of transgenic plants are imperative. Recently policy changes have been made in Australia and the United States, which ease the regulatory burden on gene-edited organisms (Parrott et al., 2020). Provided the edit could have been produced by random mutation, and the organism is not transgenic, the path to approval for commercial use is within reasonable reach for academic labs and smaller companies (Waltz, 2018; Mallapaty, 2019; USDA, 2019, 2020).

4 Case study: gene editing of a recalcitrant line by co-bombardment of morphogenic genes (*Babyboom/Wuschel*) and CRISPR/Cas9 reagents

To overcome the genotype-dependent tissue regeneration challenge, plant morphogenic genes (*Bbm/Wus*) described in Section 3.1 have been used in maize. Lowe et al. (2016; 2018) reported the successful production of transgenic lines from previously non-transformable maize genotypes using *Bbm/Wus*. Svitashev et al. (2016) used the co-bombardment of *Bbm/Wus* genes and CRISPR/Cas9 RNP complex for more efficient recovery of targeted mutations in the transformable Hi II genotype. Here we report an attempt to generate targeted mutation in a recalcitrant maize genotype using the combination of *Bbm/Wus* and CRISPR/Cas9 via biolistic gun-mediated transformation.

Fast-Flowering Mini-Maize A (FFMM-A) is a rapid-cycling, miniature line of maize capable of going from seed-to-seed in 60 days (McCaw et al., 2016). FFMM-A is recalcitrant to transformation using a conventional maize transformation protocol (Raji et al., 2018) due to its inability to form regenerable callus. However, using a *Bbm/Wus* containing construct, Lowe et al. (2018) achieved a success of ~90% transformation frequency with this genotype.

In an attempt to establish *Bbm/Wus*-assisted transformation in our laboratory, we first compared the tissue culture responses and transformation frequencies among three LBA4404 *Agrobacterium* strains that carry: (1) pTF3, a superbinary vector (Komari et al., 2006) that contains the selectable marker *bar* gene and visual marker *gus* gene, (2) PHP79066, a *Bbm/Wus* construct containing the maize *Bbm* gene under a scutellum-specific promoter $Pltp_{pro}$ and the maize *Wus2* gene under an auxin-inducible promoter $Axig1_{pro}$ (Fig. 1c in Lowe et al., 2018), or (3) PHP79066/PHP71539, the *Bbm/Wus* construct PHP79066 and a 'helper' plasmid PHP71539 carrying an extra copy of *vir* genes for increasing *Agrobacterium*-mediated T-DNA delivery and integration (Fig. 1c in Anand et al., 2018, GenBank accession# MF788073).

Transformation of FFMM-A immature embryos was carried out using the protocol described in Lowe et al. (2018). Immature embryos infected with the PHP79066/PHP71539 construct responded most vigorously and developed a high frequency (~90%) of germinating transgenic plantlets (Fig. 1a). This is consistent with previously reported results (Lowe et al., 2018). In comparison, immature embryos infected with the PHP79066-only strain showed delayed and localized responses (Fig. 1c) and a lower transformation rate (~6%). Additionally, the transformed embryos formed callus that was friable, maintainable, herbicide resistant, and proliferated rapidly on shoot formation media (Lowe et al., 2018), but was difficult to regenerate into plants (Fig. 1e and f). Control infection with the pTF3 strain that does not carry the *Bbm/Wus* genes did not produce callus, but some protrusions did form (Fig. 1d). These protrusions, which can be seen in Fig. 1b, are clearly distinguished from callus by their watery, grainy-textured appearance. They are not suitable material for producing transgenic plants, and their formation has been observed without *Agrobacterium* infection. These results confirmed what was reported by Lowe et al. (2018). They also suggested that the combination of the *Bbm/Wus* construct and the *Agrobacterium vir* enhancing 'helper' plasmid is important to effectively transform FFMM-A using this *Agrobacterium*-mediated method using LBA4404.

We then performed gene editing experiments by co-bombarding immature embryos of FFMM-A using plasmid DNAs carrying *Bbm/Wus* genes and CRISPR/Cas9 reagents. The CRISPR/Cas9 construct, A844B (Lee et al., 2019), contains Cas9-gRNA cassettes targeting maize *Gl2* and an herbicide-resistance *bar* gene for plant selection. The target locus *Gl2* is responsible for the formation of an epicuticular wax layer in juvenile leaf tissues, and loss of function mutants (*gl2*) exhibit a screenable phenotype characterized by water droplets adhering to the surface of seedling and juvenile leaves when misted. For co-bombardments, 50 mg each of plasmids PHP79066 (*Bbm/Wus*) and A844B (Cas9-gRNA) were co-precipitated onto 0.15 mg of gold particles for each shot. They were bombarded into 77 FFMM-A embryos as described in Wang et al. (2020) with the following modifications: embryos were dissected onto 605J media (Jones et al., 2019) and moved onto osmotic media composed of 605J media with an additional 36.4 g/L each of sorbitol and mannitol. After bombardment, embryos were treated as per Jones et al. (2019), but with 12-day resting on 605J and two, two-week periods on 289O media with 1.5 mg/L then 3.0 mg/L bialaphos replacing imazapyr for selection.

(i)

```
B104 Reference    ATTTATCGTTTCTTTGGTCACAGATCACAAACTTCAA-ATGCGGTGGGCTGGCGCTGGGG (WT)
FFMM-A Plant 1    ATTTATCGTTTCTTTGGTCACAGATCACAAACTTCAAAATGCGGTGGGCTGGCGCTGGGG (+1)
FFMM-A Plant 2    ATTTATCGTTTCTTTGGTCACAGATCACAAACTTCAAAATGCGGTGGGCTGGCGCTGGGG (+1)
Control FFMM-A    ATTTATCGTTTCTTTGGTCACAGATCACAAACTTCAA-ATGCGGTGGGCTGGCGCTGGGG (WT)
                  ************************************* **********************
```

Figure 1 Tissue development of FFMM-A embryos infected by an *Agrobacterium* (LBA4404) strain containing a *Bbm/Wus* construct and a 'helper' plasmid (PHP79066/PHP71539) (a) and by a strain without *Bbm/Wus* construct (pTF3). (b) Representative callus morphology observed on embryos infected by an Agro strain carrying a *Bbm/Wus* construct (PHP79066 only) (c) and a strain without *Bbm/Wus* construct (pTF3). (d) Morphology of the rapidly growing callus that was recalcitrant to forming shoots. (e) and (f) Misted leaf of a T_0 plant (PHP79066/PHP71539) with wild-type *Gl2* (g) and a T_0 plant (co-bombarded PHP79066/A844B) with loss of function *gl2* (h) phenotypes. Sequenced mutations aligned to the B104 and FFMM-A wild-type alleles shows 1 bp (a) insertion (i). gRNA target sequence is highlighted in blue, PAM is in orange, and 1 bp insertion is labeled in red.

The response of the bombarded embryos was similar to *Agrobacterium*-mediated transformation using the LBA4404 strain with PHP79066-only, described above, which does not carry the helper plasmid PHP71539. Only 3/77 embryos produced friable, bialaphos-resistant callus after a long delay.

The callus was able to be maintained for a couple of months, but only one callus event was able to be regenerated into twin plantlets. Those plantlets tested PCR positive for both PHP79066 and A844B. In addition, these plants displayed a phenotype of water droplet adhesion to seedling leaves, the typical phenotype for a *gl2* loss-of-function mutant (Fig. 1h), while plants regenerated from an *Agrobacterium* infection experiment using only PHP79066/PHP71539 did not (Fig. 1g). Sanger sequencing analysis confirmed a homozygous 1 bp (A) insertion at the target site (Fig. 1i). These results indicated that the biolistic delivery of *Bbm/Wus* and Cas9-gRNA plasmid DNAs can generate targeted mutations in a recalcitrant maize genotype.

These plants were grown to maturity in a growth chamber, but failed to produce pollen and seed, so T_1 analysis could not be performed. It is important to note that *Gl2* wax deposition occurs primarily in the first six seedling leaves (Bianchi et al., 1975; Avato et al., 1985), so its phenotype can only be readily observed in those leaves. Water can adhere to adult leaves next to the tassel in relatively small FFMM-A PHP79066 *Gl2* regenerants and may also adhere immediately after removal from tissue culture or on damaged tissue. Plants were acclimatized to air and soil for at least a week before testing, and only lower seedling leaves were tested for phenotype.

These positive results show that co-bombardment with *Bbm/Wus* is likely a viable way to produce gene-edited plants in traditionally non-transformable lines, though changes to medium composition or protocol to improve callus regeneration into plantlets may be needed. The delayed response and production of callus rather than somatic embryos after bombardment, or when infected with *Agrobacterium* without a helper plasmid is interesting. Lowe et al. (2016) noted that Wuschel can move to neighboring cells to cause nearby, untransformed cells to divide. Most recently, Hoerster et al. (2020) reported the use of non-integrating *Zm-Wus2* vectors to enhance maize transformation, a method designated as 'altruistic transformation'. The strong but transient expression of *Wus2* stimulated cell growth in neighboring cells transformed by a separate construct with selectable markers, and led to enhanced regeneration frequency of T_0 plants containing the selectable marker construct without *Wus2*.

With the PHP79066 plasmid used in the above experiments, expression of *Bbm/Wus* is meant to be transient; *Bbm* is driven by a tissue-specific promotor, and *Wus2* is driven by an auxin-inducible promoter induced by synthetic auxins supplied in the resting medium. It is possible that a diffusion of Wuschel from other nearby infected/transformed cells provides a feedback that is necessary for early rapid growth and differentiation into somatic embryos using PHP79066. Without high transformation density within the local tissue, transformed events may not receive this feedback, and make callus with undesirable morphology. This protocol may be improved by increasing the density of transformed cells on the scutellum, or by using a different version of *Bbm/Wus* that expresses constitutively but must be excised.

5 Conclusion and future trends

This chapter provides a summary of the efforts and progress in maize genome editing in recent years. Maize is one of the cereal crops that has received the most attention in research and development owing to its global economic importance, as well as its well-established genetic techniques and knowledge. As discussed in Section II, all SSN technologies developed in the past two decades, such as ZFN, engineered meganucleases, TALENs and CRISPR/Cas, have been demonstrated in maize. While all SSN technologies can work in maize, the CRISPR/Cas system is undoubtedly the dominant technology due to its simplicity and effectiveness. An increasing number of genes and pathways have been perturbed by the CRISPR/Cas technology for functional analysis, and many discoveries resulting from its use are sure to come. CRISPR/Cas reagents can be introduced into maize cells using either *Agrobacterium*- or biolistic-mediated delivery methods. Edited plants that carry either the CRISPR reagents or selectable marker genes can be backcrossed to non-transgenic plants to segregate the transgene away from the edit. CRISPR/Cas9 reagents can also be delivered as an in vitro-assembled RNP complex for DNA-free editing, and improved methods to recover edited events from this method are a goal for academics and industry alike. Because transgene-free single base-pair editing may be exempt from USDA regulation, this technology may be democratized compared to traditional transgenic approaches, and novel edited maize lines may begin appearing on the market from collaborations between academic labs and small companies as well as the usual large seed companies.

The following scientific and technological areas are likely to draw great attention in the coming years. (1) Advanced bioinformatics tools that allow researchers to create more efficient gRNAs that limit the number of off-target cut sites (Young et al., 2019). (2) Future advances in CRISPR/Cas-mediated gene editing tools will probably include engineered Cas endonucleases and new Cas orthologs with a wider range of base-pair editing and HDR capabilities and novel PAM recognition sequences to provide a more flexibility in target location (Luo et al., 2016). (3) Gene function studies by knockout are generally limited to non-vital genes, and hemizygous individuals (Wu et al., 2020); inducible expression of gene editing systems could enable the study of vital genes as well. (4) Because maize transformation technology remains a major challenge for the research community, greatly improved cell/tissue transformation/regenerability and reagent delivery will be needed to make genome editing accessible for most public research groups. A simple method that by-passes tissue culture entirely would change the field of maize precision genome editing overnight. With fewer regulations (USDA, 2019, 2020; Parrott et al., 2020), increased computing power, better engineered reagents, and enhanced delivery methods, precision genome editing in maize is primed for future improvements (Svitashev et al., 2015; Tomlinson, 2018; Lee et al., 2019; Palermo et al., 2019).

6 Acknowledgements

This project was partially supported by the National Science Foundation Plant Genome Research Program Grant 1725122 and 1917138, by the Agriculture and Food Research Initiative Competitive Grant # 2019-67013-29016 from the USDA National Institute of Food and Agriculture (NIFA) to K.W., by the Predictive Plant Phenomics Research Traineeship Program (National Science Foundation Grant DGE-1545453) to J.Z., by the USDA NIFA Hatch project #IOW04714 and by the State of Iowa funds and by the Crop Bioengineering Center of Iowa State University.

7 Where to look for further information

CRISPR/Cas technologies are rapidly advancing and the Addgene CRISPR blog is a good place to share newly developed approaches and guidance. The blog posts introduced recently identified CRISPR/Cas effectors and their applications for precise genome engineering or other purposes. The Addgene's eBook 'CRISPR 101: A Desk Resource' provides a thorough introduction to CRISPR/Cas technologies and their practical applications.

- Addgene CRISPR blog: https://blog.addgene.org/topic/crispr.
- Addgene's eBook: https://info.addgene.org/download-addgenes-ebook-crispr-101-2nd-edition.

Many web-based gRNA design tools are available in public. The following links are for gRNA design tools which suggest potential off-target sites for maize based on the reference genome sequence.

- CGAT: http://cbc.gdcb.iastate.edu/cgat/.
- Cas-Designer: http://www.rgenome.net/cas-designer/.
- CHOPCHOP: http://chopchop.cbu.uib.no/.
- CRISPOR: http://crispor.tefor.net/.
- E-CRISP: http://www.e-crisp.org/E-CRISP/designcrispr.html.
- CRISPR Design tool (Horizon discovery): https://horizondiscovery.com/en/products/tools/CRISPR-Design-Tool.
- CRISPR Design Tool (Synthego): https://www.synthego.com/products/bioinformatics/crispr-design-tool.

Genetic Engineering and Biotechnology News has articles about genome editing in plants: https://www.genengnews.com/.

Scientific journals provide peer reviewed research on current and relevant data. Journals with high impact rating like *Nature* are excellent sources for finding news on GMO production from researchers who are currently active in the field.

- https://www.nature.com/scitable/topicpage/genetically-modified-organisms-gmos-transgenic-crops-and-732/.
- https://www.nature.com/articles/s41598-018-21284-2: Impact of genetically engineered maize on agronomic, environmental and toxicological traits: a meta-analysis of 21 years of field data.
- https://www.ncbi.nlm.nih.gov/pmc/articles/PMC212689/: Genetically modified corn environmental benefits and risks, Virginia Gewin.
- https://www.frontiersin.org/journals/genome-editing/sections/genome-editing-in-plants: Frontiers Genome Editing in plants.

In the United States three government agencies, the Food and Drug Administration (FDA), the United States Department of Agriculture (USDA) and the Environmental Protection Agency (EPA) oversee and regulate GMO products. The United States of America Food and Drug Administration (FDA) website gives a good overview of the state of GMOs, including corn, and their prevalence in the United States. The information available discusses the benefits to consuming and growing GMO products. The FDA regulates food safety and the process by how they are shipped and stored. The EPA regulates GMOs and the possible impact they can have on the environment. The USDA regulates GMOs as they relate to other plants.

- https://www.fda.gov/food/agricultural-biotechnology/gmo-crops-animal-food-and-beyond.
- https://www.fda.gov/food/agricultural-biotechnology/how-gmo-crops-impact-our-world.
- https://www.fda.gov/food/agricultural-biotechnology/how-gmos-are-regulated-food-and-plant-safety-united-states.
- https://www.epa.gov/regulation-biotechnology-under-tsca-and-fifra/overview-plant-incorporated-protectants.
- https://www.aphis.usda.gov/aphis/home.

The National Corn Growers Association, which was founded in 1957 and is currently representing the interest of more than 300 000 corn growers in the United States, is an excellent organization to learn about and become involved with growing GMO corn. The National Corn Growers Association hosts interactive events and discussions and publishes newsletters related to GMO corn.

- National Corn Growers Association (Social Organization): https://www.ncga.com/.

8 References

Ainley, W. M., Sastry-Dent, L., Welter, M. E., Murray, M. G., Zeitler, B., Amora, R., Corbin, D. R., Miles, R. R., Arnold, N. L., Strange, T. L., Simpson, M. A., Cao, Z., Carroll, C., Pawelczak, K. S., Blue, R., West, K., Rowland, L. M., Perkins, D., Samuel, P., Dewes, C. M., Shen, L., Sriram, S., Evans, S. L., Rebar, E. J., Zhang, L., Gregory, P. D., Urnov, F. D., Webb, S. R. and Petolino, J. F. (2013). Trait stacking via targeted genome editing, *Plant Biotechnol. J.* 11(9): 1126–1134.

Aird, E. J., Lovendahl, K. N., St Martin, A., Harris, R. S. and Gordon, W. R. (2018). Increasing Cas9-mediated homology-directed repair efficiency through covalent tethering of DNA repair template, *Commun. Biol.* 1: 54.

Altpeter, F., Springer, N. M., Bartley, L. E., Blechl, A. E., Brutnell, T. P., Citovsky, V., Conrad, L. J., Gelvin, S. B., Jackson, D. P., Kausch, A. P., Lemaux, P. G., Medford, J. I., Orozco-Cárdenas, M. L., Tricoli, D. M., Van Eck, J., Voytas, D. F., Walbot, V., Wang, K., Zhang, Z. J. and Stewart, C. N., Jr. (2016). Advancing crop transformation in the era of genome editing, *Plant Cell* 28(7): 1510–1520.

Anand, A., Bass, S. H., Wu, E., Wang, N., McBride, K. E., Annaluru, N., Miller, M., Hua, M. and Jones, T. J. (2018). An improved ternary vector system for *Agrobacterium*-mediated rapid maize transformation, *Plant Mol. Biol.* 97(1–2): 187–200.

Anand, A., Wu, E., Li, Z., TeRonde, S., Arling, M., Lenderts, B., Mutti, J. S., Gordon-Kamm, W., Jones, T. J. and Chilcoat, N. D. (2019). High efficiency *Agrobacterium*-mediated site-specific gene integration in maize utilizing the FLP-FRT recombination system, *Plant Biotechnol. J.* 17(8): 1636–1645.

Anderson, J. E., Michno, J. M., Kono, T. J. Y., Stec, A. O., Campbell, B. W., Curtin, S. J. and Stupar, R. M. (2016). Genomic variation and DNA repair associated with soybean transgenesis: a comparison to cultivars and mutagenized plants, *BMC Biotechnol.* 16(1): 41.

Andorf, C., Beavis, W. D., Hufford, M., Smith, S., Suza, W. P., Wang, K., Woodhouse, M., Yu, J. and Lubberstedt, T. (2019). Technological advances in maize breeding: past, present and future, *Theor. Appl. Genet.* 132(3): 817–849.

Anzalone, A. V., Randolph, P. B., Davis, J. R., Sousa, A. A., Koblan, L. W., Levy, J. M., Chen, P. J., Wilson, C., Newby, G. A., Raguram, A. and Liu, D. R. (2019). Search-and-replace genome editing without double-strand breaks or donor DNA, *Nature* 576(7785): 149–157.

Arnould, S., Perez, C., Cabaniols, J. P., Smith, J., Gouble, A., Grizot, S., Epinat, J. C., Duclert, A., Duchateau, P. and Pâques, F. (2007). Engineered I-CreI derivatives cleaving sequences from the human XPC gene can induce highly efficient gene correction in mammalian cells, *J. Mol. Biol.* 371(1): 49–65.

Avato, P., Bianchi, G. and Salamini, F. (1985). Absence of long chain aldehydes in the wax of the Glossy II mutant of maize, *Phytochemistry* 24(9): 1995–1997.

Bae, S., Park, J. and Kim, J. S. (2014). Cas-OFFinder: a fast and versatile algorithm that searches for potential off-target sites of Cas9 and RNA-guided endonucleases, *Bioinformatics* 30(10): 1473–1475.

Baltes, N. J., Gil-Humanes, J., Cermak, T., Atkins, P. A. and Voytas, D. F. (2014). DNA replicons for plant genome engineering, *Plant Cell* 26(1): 151–163.

Banakar, R., Eggenberger, A. L., Lee, K., Wright, D. A., Murugan, K., Zarecor, S., Lawrence-Dill, C. J., Sashital, D. G. and Wang, K. (2019). High-frequency random DNA insertions upon co-delivery of CRISPR-Cas9 ribonucleoprotein and selectable marker plasmid in rice, *Sci. Rep.* 9(1): 19902.

Bannert, M. and Stamp, P. (2007). Cross-pollination of maize at long distance, *Eur. J. Agron.* 27(1): 44–51.

Belhaj, K., Chaparro-Garcia, A., Kamoun, S. and Nekrasov, V. (2013). Plant genome editing made easy: targeted mutagenesis in model and crop plants using the CRISPR/Cas system, *Plant Methods* 9(1): 39.

Bianchi, G., Avato, P. and Salamini, F. (1975). Glossy mutants of maize. VI. Chemical constituents of *glossy-2* epicuticular waxes, *Maydica* 20: 165–173.

Brandle, J. E., McHugh, S. G., James, L., Labbé, H. and Miki, B. L. (1995). Instability of transgene expression in field grown tobacco carrying the *csr1-1* gene for sulfonylurea herbicide resistance, *Nat. Biotechnol.* 13(9): 994–998.

Broothaerts, W., Mitchell, H. J., Weir, B., Kaines, S., Smith, L. M., Yang, W., Mayer, J. E., Roa-Rodríguez, C. and Jefferson, R. A. (2005). Gene transfer to plants by diverse species of bacteria, *Nature* 433(7026): 629–633.

Char, S. N., Neelakandan, A. K., Nahampun, H., Frame, B., Main, M., Spalding, M. H., Becraft, P. W., Meyers, B. C., Walbot, V., Wang, K. and Yang, B. (2017). An *Agrobacterium*-delivered CRISPR/Cas9 system for high-frequency targeted mutagenesis in maize, *Plant Biotechnol. J.* 15(2): 257–268.

Char, S. N., Unger-Wallace, E., Frame, B., Briggs, S. A., Main, M., Spalding, M. H., Vollbrecht, E., Wang, K. and Yang, B. (2015). Heritable site-specific mutagenesis using TALENs in maize, *Plant Biotechnol. J.* 13(7): 1002–1010.

Chen, K., Wang, Y., Zhang, R., Zhang, H. and Gao, C. (2019). CRISPR/Cas genome editing and precision plant breeding in agriculture, *Annu. Rev. Plant Biol.* 70: 667–697.

Cloney, R. (2016). Patent law and genome engineering: a short guide to a rapidly changing landscape, *Mol. Ther.* 24(3): 419–421.

Clough, S. J. and Bent, A. F. (1998). Floral dip: a simplified method for *Agrobacterium*-mediated transformation of *Arabidopsis thaliana*, *Plant J.* 16(6): 735–743.

D'Halluin, K., Vanderstraeten, C., Stals, E., Cornelissen, M. and Ruiter, R. (2008). Homologous recombination: a basis for targeted genome optimization in crop species such as maize, *Plant Biotechnol. J.* 6(1): 93–102.

Djukanovic, V., Smith, J., Lowe, K., Yang, M., Gao, H., Jones, S., Nicholson, M. G., West, A., Lape, J., Bidney, D., Carl Falco, S., Jantz, D. and Lyznik, L. A. (2013). Male-sterile maize plants produced by targeted mutagenesis of the cytochrome P450-like gene (*MS26*) using a re-designed I-*Cre*I homing endonuclease, *Plant J.* 76(5): 888–899.

Doebley, J. (2004). The genetics of maize evolution, *Annu. Rev. Genet.* 38(1): 37–59.

Dunn, D. A. and Pinkert, C. A. (2014). Chapter 8. Gene editing. In: *Transgenic Animal Technology: A Laboratory Handbook* (3rd edn.) Pinkert, C. (Ed.), Elsevier Inc, pp 229–248.

Edgell, D. R. (2009). Selfish DNA: homing endonucleases find a home, *Curr. Biol.* 19(3): R115–R117.

Eichten, S. R., Foerster, J. M., de Leon, N., Kai, Y., Yeh, C. T., Liu, S., Jeddeloh, J. A., Schnable, P. S., Kaeppler, S. M. and Springer, N. M. (2011). B73-Mo17 near-isogenic lines demonstrate dispersed structural variation in maize, *Plant Physiol.* 156(4): 1679–1690.

Elmayan, T. and Vaucheret, H. (1996). Expression of single copies of a strongly expressed 35S transgene can be silenced post-transcriptionally, *Plant J.* 9(6): 787–797.

Feng, C., Su, H., Bai, H., Wang, R., Liu, Y., Guo, X., Liu, C., Zhang, J., Yuan, J., Birchler, J. A. and Han, F. (2018). High-efficiency genome editing using a dmc1 promoter-controlled CRISPR/Cas9 system in maize, *Plant Biotechnol. J.* 16(11): 1848–1857.

Gaj, T., Gersbach, C. A. and Barbas, C. F. (2013). ZFN, TALEN, and CRISPR/Cas-based methods for genome engineering, *Trends Biotechnol.* 31(7): 397-405.

Gao, H., Smith, J., Yang, M., Jones, S., Djukanovic, V., Nicholson, M. G., West, A., Bidney, D., Falco, S. C., Jantz, D. and Lyznik, L. A. (2010). Heritable targeted mutagenesis in maize using a designed endonuclease, *Plant J.* 61(1): 176–187.

Gewin, V. (2003). Genetically modified corn - environmental benefits and risks, *PLoS Biol.* 1(1): E8.

Gorbunova, V. and Levy, A. A. (1999). How plants make ends meet: DNA double-strand break repair, *Trends Plant Sci.* 4(7): 263–269.

Gordon-Kamm, W. J., Spencer, T. M., Mangano, M. L., Adams, T. R., Daines, R. J., Start, W. G., O'Brien, J. V., Chambers, S. A., Adams, W. R., Willetts, N. G., Rice, T. B., Mackey, C. J., Krueger, R. W., Kausch, A. P. and Lemaux, P. G. (1990). Transformation of maize cells and regeneration of fertile transgenic plants, *Plant Cell* 2(7): 603–618.

Hoerster, G., Wang, N., Ryan, L., Wu, E., Anand, A., McBride, K., Lowe, K., Jones, T. and Gordon-Kamm, B. (2020). Use of non-integrating Zm-Wus2 vectors to enhance maize transformation, *In Vitro Cell. Dev. Biol.-Plant* 56(3): 265–279.

Hu, J. H., Miller, S. M., Geurts, M. H., Tang, W., Chen, L., Sun, N., Zeina, C. M., Gao, X., Rees, H. A., Lin, Z. and Liu, D. R. (2018). Evolved Cas9 variants with broad PAM compatibility and high DNA specificity, *Nature* 556(7699): 57–63.

ISAAA (2017). Global status of commercialized Biotech/GM crops in 2017: biotech crop adoption surges as economic benefits accumulate in 22 years. ISAAA No. 53 (vol. 2017).

Jaganathan, D., Ramasamy, K., Sellamuthu, G., Jayabalan, S. and Venkataraman, G. (2018). CRISPR for crop improvement: an update review, *Front. Plant Sci.* 9: 985.

Jaqueth, J. S., Hou, Z., Zheng, P., Ren, R., Nagel, B. A., Cutter, G., Niu, X., Vollbrecht, E., Greene, T. W. and Kumpatla, S. P. (2020). Fertility restoration of maize CMS-C altered by a single amino acid substitution within the Rf4 bHLH transcription factor, *Plant J.* 101(1): 101–111.

Ji, Q., Xu, X. and Wang, K. (2013). Genetic transformation of major cereal crops, *Int. J. Dev. Biol.* 57(6-8): 495–508.

Jinek, M., Chylinski, K., Fonfara, I., Hauer, M., Doudna, J. A. and Charpentier, E. (2012). A programmable dual RNA-guided DNA endonuclease in adaptive bacterial immunity, *Science* 337(6096): 816–821.

Jones, T., Lowe, K., Hoerster, G., Anand, A., Wu, E., Wang, N., Arling, M., Lenderts, B. and Gordon-Kamm, W. (2019). Maize transformation using the morphogenic genes *Baby Boom* and *Wuschel2*. In: *Transgenic Plants*, Kumar, S., Barone, P. and Smith, M. (Eds), Humana Press, New York, NY, pp 81–93.

Joung, J. K. and Sander, J. D. (2013). TALENs: a widely applicable technology for targeted genome editing, *Nat. Rev. Mol. Cell. Biol.* 14(1): 49–55.

Kelliher, T., Starr, D., Su, X., Tang, G., Chen, Z., Carter, J., Wittich, P. E., Dong, S., Green, J., Burch, E., McCuiston, J., Gu, W., Sun, Y., Strebe, T., Roberts, J., Bate, N. J. and Que, Q. (2019). One-step genome editing of elite crop germplasm during haploid induction, *Nat. Biotechnol.* 37(3): 287–292.

Kim, Y. G., Cha, J. and Chandrasegaran, S. (1996). Hybrid restriction enzymes: zinc finger fusions to FokI cleavage domain, *Proc. Natl. Acad. Sci. U.S.A.* 93(3): 1156–1160.

Komari, T., Takakura, Y., Ueki, J., Kato, N., Ishida, Y. and Hiei, Y. (2006). Binary vectors and super-binary vectors. In: *Agrobacterium Protocols*, Wang, K. (Ed)., Humana Press, Totowa, NJ, pp 15–41.

Krysiak, C., Mazuś, B. and Buchowicz, J. (1999). Generation of DNA double-strand breaks and inhibition of somatic embryogenesis by tungsten microparticles in wheat, *Plant Cell Tissue Organ Cult.* 58(3): 163–170.

Lee, H. and Zhang, Z. J. (2016). Maize (*Zea mays*) Hi-II transformation via *Agrobacterium*-mediated T-DNA transfer, Curr. Protoc, *Plant Biol.* 1(1): 121–137.

Lee, K., Zhang, Y., Kleinstiver, B. P., Guo, J. A., Aryee, M. J., Miller, J., Malzahn, A., Zarecor, S., Lawrence-Dill, C. J., Joung, J. K., Qi, Y. and Wang, K. (2019). Activities and specificities of CRISPR-Cas9 and Cas12a nucleases for targeted mutagenesis in maize, *Plant Biotechnol. J.* 17(2): 362–372.

Liang, Z., Zhang, K., Chen, K. and Gao, C. (2014). Targeted mutagenesis in *Zea mays* using TALENs and the CRISPR/Cas system, *J. Genet. Genomics* 41(2): 63–68.

Li, J., Zhang, H., Si, X., Tian, Y., Chen, K., Liu, J., Chen, H. and Gao, C. (2017). Generation of thermosensitive male-sterile maize by targeted knockout of the *ZmTMS5* gene, *J. Genet. Genomics* 44(9): 465–468.

Li, J., Luo, J., Xu, M., Li, S., Zhang, J., Li, H., Yan, L., Zhao, Y. and Xia, L. (2019). Plant genome editing using xCas9 with expanded PAM compatibility, *J. Genet. Genomics* 46(5): 277–280.

Liu, H. J., Jian, L., Xu, J., Zhang, Q., Zhang, M., Jin, M., Peng, Y., Yan, J., Han, B., Liu, J., Gao, F., Liu, X., Huang, L., Wei, W., Ding, Y., Yang, X., Li, Z., Zhang, M., Sun, J., Bai, M., Song, W., Chen, H., Sun, X., Li, W., Lu, Y., Liu, Y., Zhao, J., Qian, Y., Jackson, D., Fernie, A. R. and Yan, J. (2020). High-throughput CRISPR/Cas9 mutagenesis streamlines trait gene identification in maize, *Plant Cell* 32(5): 1397–1413.

Liu, J., Nannas, N. J., Fu, F. F., Shi, J., Aspinwall, B., Parrott, W. A. and Dawe, R. K. (2019). Genome-scale sequence disruption following biolistic transformation in rice and maize, *Plant Cell* 31(2): 368–383.

Liu, S., Yeh, C. T., Ji, T., Ying, K., Wu, H., Tang, H. M., Fu, Y., Nettleton, D. and Schnable, P. S. (2009). *Mu* transposon insertion sites and meiotic recombination events co-localize with epigenetic marks for open chromatin across the maize genome, *PLoS Genet.* 5(11): e1000733.

Lowe, B. A., Shiva Prakash, N., Way, M., Mann, M. T., Spencer, T. M. and Boddupalli, R. S. (2009). Enhanced single copy integration events in corn via particle bombardment using low quantities of DNA, *Transgen. Res.* 18(6): 831–840.

Lowe, K., La Rota, M., Hoerster, G., Hastings, C., Wang, N., Chamberlin, M., Wu, E., Jones, T. and Gordon-Kamm, W. (2018). Rapid genotype "independent" *Zea mays* L. (maize) transformation via direct somatic embryogenesis, *In Vitro Cell. Dev. Biol. Plant* 54(3): 240–252.

Lowe, K., Wu, E., Wang, N., Hoerster, G., Hastings, C., Cho, M. J., Scelonge, C., Lenderts, B., Chamberlin, M., Cushatt, J., Wang, L., Ryan, L., Khan, T., Chow-Yiu, J., Hua, W., Yu, M.,

Banh, J., Bao, Z., Brink, K., Igo, E., Rudrappa, B., Shamseer, P. M., Bruce, W., Newman, L., Shen, B., Zheng, P., Bidney, D., Falco, C., Register, J., Zhao, Z. Y., Xu, D., Jones, T. and Gordon-Kamma, W. (2016). Morphogenic regulators Baby Boom and Wuschel improve monocot transformation, *Plant Cell* 28(9): 1998–2015.
Luna, V. S., Figueroa, M. J., Baltazar, M. B., Gomez, L. R., Townsend, R. and Schoper, J. B. (2001). Maize pollen longevity and distance isolation requirements for effective pollen control, *Crop Sci.* 41(5): 1551–1557.
Luo, M., Gilbert, B. and Ayliffe, M. (2016). Applications of CRISPR/Cas9 technology for targeted mutagenesis, gene replacement and stacking of genes in higher plants, *Plant Cell Rep.* 35(7): 1439–1450.
Ma, D., Xu, Z., Zhang, Z., Chen, X., Zeng, X., Zhang, Y., Deng, T., Ren, M., Sun, Z., Jiang, R. and Xie, Z. (2019). Engineer chimeric Cas9 to expand PAM recognition based on evolutionary information, *Nat. Commun.* 10(1): 560.
Maheshwari, S., Tan, E. H., West, A., Franklin, F. C., Comai, L. and Chan, S. W. (2015). Naturally occurring differences in CENH3 affect chromosome segregation in zygotic mitosis of hybrids, *PLOS Genet.* 11(1): e1004970.
Mallapaty, S. (2019). Australian gene-editing rules adopt 'middle ground', *Nature*. https://doi.org/10.1038/d41586-019-01282-8.
Malzahn, A. A., Tang, X., Lee, K., Ren, Q., Sretenovic, S., Zhang, Y., Chen, H., Kang, M., Bao, Y., Zheng, X., Deng, K., Zhang, T., Salcedo, V., Wang, K., Zhang, Y. and Qi, Y. (2019). Applications of CRISPR-Cas12a temperature sensitivity for improved genome editing in rice, maize, and Arabidopsis, *BMC Biol.* 17(1): 9.
Manova, V. and Gruszka, D. (2015). DNA damage and repair in plants-from models to crops, *Front. Plant Sci.* 6: 885.
Martin-Ortigosa, S., Peterson, D. J., Valenstein, J. S., Lin, V. S.-Y., Trewyn, B. G., Lyznik, L. A. and Wang, K. (2014). Mesoporous silica nanoparticle mediated intracellular Cre protein delivery for maize genome editing via loxP sites excision, *Plant Physiol.* 164(2): 537–547.
Masters, A., Kang, M., McCaw, M., Zobrist, J. D., Gordon-Kamm, W., Jones, T. and Wang, K. (2020). *Agrobacterium*-mediated immature embryo transformation of recalcitrant maize inbred lines using morphogenic genes, *J. Vis. Exp.* 156(156): e60782.
McCaw, M. E., Wallace, J. G., Albert, P. S., Buckler, E. S. and Birchler, J. A. (2016). Fast-flowering mini-maize: seed to seed in 60 days, *Genetics* 204(1): 35–42.
Mlynárová, L., Keizer, L. C. P., Stiekema, W. J. and Nap, J. P. (1996). Approaching the lower limits of transgene variability, *Plant Cell* 8(9): 1589–1599.
Morocz, S., Donn, G., Nemeth, J. and Dudits, D. (1990). An improved system to obtain fertile regenerants via maize protoplasts isolated from a highly embryogenic suspension culture, *Theor. Appl. Genet.* 80(6): 721–726.
Nagaya, S., Kato, K., Ninomiya, Y., Horie, R., Sekine, M., Yoshida, K. and Shinmyo, A. (2005). Expression of randomly integrated single complete copy transgenes does not vary in *Arabidopsis thaliana*, *Plant Cell Physiol.* 46(3): 438–444.
Najera, V. A., Twyman, R. M., Christou, P. and Zhu, C. (2019). Applications of multiplex genome editing in higher plants, *Curr. Opin. Biotech.* 59: 93–102.
Nannas, N. J. and Dawe, R. K. (2015). Genetic and genomic toolbox of *Zea mays*, *Genetics* 199(3): 655–669. https://doi.org/10.1534/genetics.114.165183.
Neuffer, M. G., Coe, E. H. and Wessler, S. R. (1997). *Mutants of Maize*, Cold Spring Harbor Laboratory Press, New York, NY.

Ni, M., Cui, D., Einstein, J., Narasimhulu, S., Vergara, C. E. and Gelvin, S. B. (1995). Strength and tissue specificity of chimeric promoters derived from the octopine and mannopine synthase Genes, *Plant J.* 7(4): 661–676.

Ow, D. W. (2005). 2004 SIVB Congress Symposium Proceeding: transgene management via multiple site-specific recombination systems, *In Vitro Cell. Dev. Biol. - Plant* 41(3): 213–219.

Ow, D. W. (2016). The long road to recombinase-mediated plant transformation, *Plant Biotechnol. J.* 14(2): 441–447.

Palermo, G., Ricci, C. G. and McCammon, J. A. (2019). The invisible dance of CRISPR-Cas9. Simulations unveil the molecular side of the gene-editing revolution, *Phys. Today* 72(4): 30–36.

Palmer, K. E. and Rybicki, E. P. (2001). Investigation of the potential of maize streak virus to act as an infectious gene vector in maize plants, *Arch. Virol.* 146(6): 1089–1104.

Parrott, W. A., Harbell, J., Kaeppler, H., Jones, T., Tomes, D., Van Eck, J., Wang, K. and Wenck, A. (2020). The proposed APHIS regulation modernization could enhance agriculture biotechnology research and development in the USA, *In Vitro Cell. Dev. Biol.-Plant* 56(1): 1–7.

Qi, W., Zhu, T., Tian, Z., Li, C., Zhang, W. and Song, R. (2016). High-efficiency CRISPR/Cas9 multiplex gene editing using the glycine tRNA-processing system-based strategy in maize, *BMC Biotechnol.* 16(1): 58.

Raji, J. A., Frame, B., Little, D., Santoso, T. and Wang, K. (2018). *Agrobacterium*- and biolistic- mediated transformation of maize B104 inbred. In: *Maize: Methods and Protocols*, Lagrimini, L. M. (Ed.), Springer, USA, pp. 15–40.

Rosen, L. E., Morrison, H. A., Masri, S., Brown, M. J., Springstubb, B., Sussman, D., Stoddard, B. L. and Seligman, L. M. (2006). Homing endonuclease I-CreI derivatives with novel DNA target specificities, *Nucleic Acids Res.* 34(17): 4791–4800.

Sauer, B. and Henderson, N. (1988). Site-specific DNA recombination in mammalian cells by the Cre recombinase of bacteriophage P1, *Proc. Natl. Acad. Sci. U.S.A.* 85(14): 5166–5170.

Schnable, P. S., Ware, D., Fulton, R. S., Stein, J. C., Wei, F., Pasternak, S., Liang, C., Zhang, J., Fulton, L., Graves, T. A., Minx, P., Reily, A. D., Courtney, L., Kruchowski, S. S., Tomlinson, C., Strong, C., Delehaunty, K., Fronick, C., Courtney, B., Rock, S. M., Belter, E., Du, F., Kim, K., Abbott, R. M., Cotton, M., Levy, A., Marchetto, P., Ochoa, K., Jackson, S. M., Gillam, B., Chen, W., Yan, L., Higginbotham, J., Cardenas, M., Waligorski, J., Applebaum, E., Phelps, L., Falcone, J., Kanchi, K., Thane, T., Scimone, A., Thane, N., Henke, J., Wang, T., Ruppert, J., Shah, N., Rotter, K., Hodges, J., Ingenthron, E., Cordes, M., Kohlberg, S., Sgro, J., Delgado, B., Mead, K., Chinwalla, A., Leonard, S., Crouse, K., Collura, K., Kudrna, D., Currie, J., He, R., Angelova, A., Rajasekar, S., Mueller, T., Lomeli, R., Scara, G., Ko, A., Delaney, K., Wissotski, M., Lopez, G., Campos, D., Braidotti, M., Ashley, E., Golser, W., Kim, H., Lee, S., Lin, J., Dujmic, Z., Kim, W., Talag, J., Zuccolo, A., Fan, C., Sebastian, A., Kramer, M., Spiegel, L., Nascimento, L., Zutavern, T., Miller, B., Ambroise, C., Muller, S., Spooner, W., Narechania, A., Ren, L., Wei, S., Kumari, S., Faga, B., Levy, M. J., McMahan, L., Van Buren, P., Vaughn, M. W., Ying, K., Yeh, C. T., Emrich, S. J., Jia, Y., Kalyanaraman, A., Hsia, A. P., Barbazuk, W. B., Baucom, R. S., Brutnell, T. P., Carpita, N. C., Chaparro, C., Chia, J. M., Deragon, J. M., Estill, J. C., Fu, Y., Jeddeloh, J. A., Han, Y., Lee, H., Li, P., Lisch, D. R., Liu, S., Liu, Z., Nagel, D. H., McCann, M. C., SanMiguel, P., Myers, A. M., Nettleton, D., Nguyen, J., Penning, B. W., Ponnala, L., Schneider, K. L., Schwartz, D. C., Sharma, A., Soderlund,

C., Springer, N. M., Sun, Q., Wang, H., Waterman, M., Westerman, R., Wolfgruber, T. K., Yang, L., Yu, Y., Zhang, L., Zhou, S., Zhu, Q., Bennetzen, J. L., Dawe, R. K., Jiang, J., Jiang, N., Presting, G. G., Wessler, S. R., Aluru, S., Martienssen, R. A., Clifton, S. W., McCombie, W. R., Wing, R. A. and Wilson, R. K. (2009). The B73 maize genome: complexity, diversity, and dynamics, *Science* 326(5956): 1112-1115.

Settles, A. M., Holding, D. R., Tan, B. C., Latshaw, S. P., Liu, J., Suzuki, M., Li, L., O'Brien, B. A., Fajardo, D. S., Wroclawska, E., Tseung, C. W., Lai, J., Hunter, C. T., Avigne, W. T., Baier, J., Messing, J., Hannah, L. C., Koch, K. E., Becraft, P. W., Larkins, B. A. and McCarty, D. R. (2007). Sequence-indexed mutations in maize using the UniformMu transposon-tagging population, *BMC Genomics* 8(116): 116.

Sharon, E., Chen, S.-A. A., Khosla, N. M., Smith, J. D., Pritchard, J. K. and Fraser, H. B. (2018). Functional genetic variants revealed by massively parallel precise genome editing, *Cell* 175(2): 544-557.e16.

Shi, J., Gao, H., Wang, H., Lafitte, H. R., Archibald, R. L., Yang, M., Hakimi, S. M., Mo, H. and Habben, J. E. (2017). ARGOS8 variants generated by CRISPR-Cas9 improve maize grain yield under field drought stress conditions, *Plant Biotechnol. J.* 15(2): 207-216.

Shukla, V. K., Doyon, Y., Miller, J. C., DeKelver, R. C., Moehle, E. A., Worden, S. E., Mitchell, J. C., Arnold, N. L., Gopalan, S., Meng, X., Choi, V. M., Rock, J. M., Wu, Y. Y., Katibah, G. E., Zhifang, G., McCaskill, D., Simpson, M. A., Blakeslee, B., Greenwalt, S. A., Butler, H. J., Hinkley, S. J., Zhang, L., Rebar, E. J., Gregory, P. D. and Urnov, F. D. (2009). Precise genome modification in the crop species *Zea mays* using zinc-finger nucleases, *Nature* 459(7245): 437-441.

Silva, G., Poirot, L., Galetto, R., Smith, J., Montoya, G., Duchateau, P. and Pâques, F. (2011). Meganucleases and other tools for targeted genome engineering: perspectives and challenges for gene therapy, *Curr. Gene Ther.* 11(1): 11-27.

Srivastava, V. and Thomson, J. (2016). Gene stacking by recombinases, *Plant Biotechnol. J.* 14(2): 471-482.

Stewart, C. N., Jr., Halfhill, M. D. and Warwick, S. I. (2003). Transgene introgression from genetically modified crops to their wild relatives, *Nat. Rev. Genet.* 4(10): 806-817.

Svitashev, S., Schwartz, C., Lenderts, B., Young, J. K. and Mark Cigan, A. (2016). Genome editing in maize directed by CRISPR-Cas9 ribonucleoprotein complexes, *Nat. Commun.* 7: 13274.

Svitashev, S., Young, J. K., Schwartz, C., Gao, H., Falco, S. C. and Cigan, A. M. (2015). Targeted mutagenesis, precise gene editing, and site-specific gene insertion in maize using Cas9 and guide RNA, *Plant Physiol.* 169(2): 931-945.

Tan, Y., Chu, A. H. Y., Bao, S., Hoang, D. A., Kebede, F. T., Xiong, W., Ji, M., Shi, J. and Zheng, Z. (2019). Rationally engineered Staphylococcus aureus Cas9 nucleases with high genome-wide specificity, *Proc. Natl. Acad. Sci. U.S.A.* 116(42): 20969-20976.

Timmermans, M. C., Das, O. P. and Messing, J. (1992). Trans replication and high copy numbers of wheat dwarf virus vectors in maize cells, *Nucleic Acids Res.* 20(15): 4047-4054.

Tomlinson, T. (2018). A CRISPR future for gene-editing regulation: a proposal for an updated biotechnology regulatory system in an era of human genomic editing, *Fordham Law Rev.* 87(1): 437-483.

Tsai, S. Q., Nguyen, N. T., Malagon-Lopez, J., Topkar, V. V., Aryee, M. J. and Joung, J. K. (2017). CIRCLE-seq: a highly sensitive in vitro screen for genome-wide CRISPR-Cas9 nuclease off-targets, *Nat. Methods* 14(6): 607-614.

Tuxill, J., Reyes, L. A., Moreno, L. L., Uicab, V. C. and Jarvis, D. I. (2010). All maize is not equal: maize variety choices and Mayan foodways in rural Yucatan, Mexico. In: *Pre-Columbian Foodways*, Staller, J. and Carrasco, M. (Eds), Springer, New York, NY. https://doi.org/10.1007/978-1-4419-0471-3_19.

USDA, 84 Fed. Reg 26514 (2019) (to be codified at 7. C.F.R. pts. 340, & 372). Available at: https://www.federalregister.gov/documents/2019/06/06/2019-11704/movement-of-certain-genetically-engineered-organisms.

USDA (2020). Secretary Perdue issues USDA statement on plant breeding innovation. Retrieved July 26, 2020, Available at: https://www.aphis.usda.gov/aphis/ourfocus/biotechnology/brs-news-and-information/2018_brs_news/plant_breeding.

Voytas, D. F. and Gao, C. (2014). Precision genome engineering and agriculture: opportunities and regulatory challenges, *PLoS Biol.* 12(6): e1001877.

Walton, R. T., Christie, K. A., Whittaker, M. N. and Kleinstiver, B. P. (2020). Unconstrained genome targeting with near-PAMless engineered CRISPR-Cas9 variants, *Science* 368(6488): 290–296.

Waltz, E. (2018). With a free pass, CRISPR-edited plants reach market in record time, *Nat. Biotechnol.* 36(1): 6–7.

Wang, B., Zhu, L., Zhao, B., Zhao, Y., Xie, Y., Zheng, Z., Li, Y., Sun, J. and Wang, H. (2019). Development of a haploid-inducer mediated genome editing system for accelerating maize breeding, *Mol. Plant* 12(4): 597–602.

Wang, K., Zhu, H. and McCaw, M. (2020). Biolistic DNA delivery in maize immature embryos. *Methods Mol. Biol.* 2124: 177–195.

Wang, M., Lu, Y., Botella, J. R., Mao, Y., Hua, K. and Zhu, J. K. (2017). Gene targeting by homology-directed repair in rice using a geminivirus-based CRISPR/Cas9 system, *Mol. Plant* 10(7): 1007–1010.

Wu, Q., Regan, M., Furukawa, H. and Jackson, D. (2018). Role of heterotrimeric Gα proteins in maize development and enhancement of agronomic traits, *PLoS Genet.* 14(4): e1007374.

Wu, Q., Xu, F., Liu, L., Char, S. N., Ding, Y., Je, B. Il., Schmelz, E., Yang, B. and Jackson, D. (2020). The maize heterotrimeric G protein β subunit controls shoot meristem development and immune responses, *Proc. Natl. Acad. Sci. U.S.A.* 117(3): 1799–1805.

Xing, H. L., Dong, L., Wang, Z. P., Zhang, H. Y., Han, C. Y., Liu, B., Wang, X. C. and Chen, Q. J. (2014). A CRISPR/Cas9 toolkit for multiplex genome editing in plants, *BMC Plant Biol.* 14: 327.

Young, J., Zastrow-Hayes, G., Deschamps, S., Svitashev, S., Zaremba, M., Acharya, A., Paulraj, S., Peterson-Burch, B., Schwartz, C., Djukanovic, V., Lenderts, B., Feigenbutz, L., Wang, L., Alarcon, C., Siksnys, V., May, G., Chilcoat, N. D. and Kumar, S. (2019). CRISPR-Cas9 editing in maize: systematic evaluation of off-target activity and its relevance in crop improvement, *Sci. Rep.* 9(1): 6729.

Zetsche, B., Gootenberg, J. S., Abudayyeh, O. O., Slaymaker, I. M., Makarova, K. S., Essletzbichler, P., Volz, S. E., Joung, J., van der Oost, J., Regev, A., Koonin, E. V. and Zhang, F. (2015). Cpf1 Is a single RNA-guided endonuclease of a class 2 CRISPR-Cas system, *Cell* 163(3): 759–771.

Zhang, Z., Qiu, F., Liu, Y., Ma, K., Li, Z. and Xu, S. (2008). Chromosome elimination and in vivo haploid production induced by stock 6-derived inducer line in maize (*Zea mays* L.), *Plant Cell Rep.* 27(12): 1851–1860.

Zhao, X., Xu, X., Xie, H., Chen, S. and Jin, W. (2013). Fertilization and uniparental chromosome elimination during crosses with maize haploid inducers, *Plant Physiol.* 163(2): 721–731.

Zhong, Z., Sretenovic, S., Ren, Q., Yang, L., Bao, Y., Qi, C., Yuan, M., He, Y., Liu, S., Liu, X., Wang, J., Huang, L., Wang, Y., Baby, D., Wang, D., Zhang, T., Qi, Y. and Zhang, Y. (2019). Improving plant genome editing with high-fidelity xCas9 and non-canonical PAM-targeting Cas9-NG, *Mol. Plant* 12(7): 1027–1036.

Zong, Y., Wang, Y., Li, C., Zhang, R., Chen, K., Ran, Y., Qiu, J. L., Wang, D. and Gao, C. (2017). Precise base editing in rice, wheat and maize with a Cas9-cytidine deaminase fusion, *Nat. Biotechnol.* 35(5): 438–440.

Chapter 14

Genome editing in sorghum

Aixia Li, Shandong University, China; and David R. Holding, University of Nebraska-Lincoln, USA

1 Introduction

Sorghum (Sorghum bicolor (L.) Moench) is a drought-tolerant C4 plant, which can be grown in arid and semi-arid regions (Hariprasanna and Patil, 2015). It is one of the most prominent cereal grain crops and ranks fifth in the world in terms of production. Sorghum grain is a major staple food of humans in Africa and a significant source for livestock feed in the US sorghum has great potential for expanded cultivation as a gluten-free source of calories, protein, fiber and phytochemicals for human nutrition. Sorghum is also used for alcoholic beverage production in some Asian countries such as China, the Philippines and India. However, sorghum confers poor characteristics such as lysine-deficient and low digestibility proteins, and high tannin content (Taylor and Schussler, 1986; Aboubacar et al., 2001; Wu et al., 2012). Although sorghum is drought-tolerant, it is often susceptible to various disease and pests which cause the loss of yield (Chandrashekar and Satyanarayana, 2006). These unfavorable factors often contribute to food shortage and malnutrition especially in Africa. Genetic improvement through breeding and biotechnology is needed to create new sorghum varieties with better agronomic traits.

Conventional genetic improvement in crops is based on the utilization of random natural variations and chemical/irradiation mutagenesis and long-term breeding procedures all of which are time-consuming. With the rise in next-generation sequencing techniques and availability of complete genome sequences, genetic engineering has become feasible for modifying specific

http://dx.doi.org/10.19103/AS.2020.0082.21

genes and this is having great impact for plant functional genomics research and genetic improvement. However, conventional genetic engineering is achieved via RNAi, antisense or over-expression which require permanent transgene systems (Tang et al., 2007) and may result in unpredictable and adverse effects. In addition, the introduction of permanent foreign DNA in organisms is still causing major issues with consumer acceptance (Dong and Ronald, 2019). Genetically modified organism (GMO) crops are an unacceptable food resource to the public in many countries, which stifles scientific innovation, GM food production, marketing and consumption. Rapid and high-efficiency site-specific genetic modification using a removable transgene offers a major move toward socially acceptable genetic improvement and breeding programs in sorghum and other crops.

Several gene-editing technologies based on specific DNA recognition and binding properties of specialized proteins such as customized homing nuclease (meganuclease), zinc-finger nucleases (ZFNs) and transcription activator-like effector nucleases (TALENs) have been developed and applied to create mutations in target site in many organisms (Puchta and Fauser, 2014; Voytas and Gao, 2014). However, these methods have the disadvantage of complicated design principles and require sophisticated protein assembly to establish specific target site recognition (Smith et al., 2006; Arnould et al., 2006; Maeder et al., 2008; Boch et al., 2009; Weber et al., 2011). In recent years, a simple RNA-guided clustered regularly interspersed short palindromic repeat (CRISPR)/CRISPR-associated protein 9 (Cas9) tool has been developed from a type II bacterial immune system (Jinek et al., 2012), and is faster, cheaper and easier to use in targeted gene editing than ZFNs and TALENs (Reis et al., 2014).

In the CRISPR/Cas9 system, the Cas9 endonuclease is directed by a synthesized guide RNA (gRNA, the spacer region is about 20 bp), and can specifically bind double DNA strands by base pairing of the gRNA spacer region and the targeted DNA sequence that is followed by a 5′-NGG protospacer-adjacent motif (PAM) on the non-complementary strand. The Cas9 then cuts DNA double strands at 3 bp upstream of the PAM, forming a double-strand break (DSB) site (Jinek et al., 2012). The resulting DSB site is then repaired by the cell's endogenous non-homologous end joining or homology-directed repair mechanisms (Wyman and Kanaar, 2006; Chapman et al., 2012), which often induces small indels or specific base replacements at the targeted genome site. gRNAs can be designed manually or using online tools, and can be easily inserted into the Cas9/gRNA expression vectors by BsaI-based Gibson assembly or Golden gate ligation (Xing et al., 2014; Ma et al., 2015). This Cas9/gRNA technique is relatively straightforward for researchers. Moreover, the targeted site edits are produced by first making transgenic plants of genes encoding Cas9 and one or more gRNAs. The transgenes can be removed by subsequent self-pollination or outcrossing because the heritability of the edits is not dependent on the continued presence of the CRISPR machinery. Thus, only the permanent edits are left behind for further plant functional research and crop breeding. Since CRISPR events are non-permanent, the USDA currently exempts many gene-edited crops without carrying the transgene from the stringent and prohibitively expensive regulations governing conventional GMO crops (Belhaj et al., 2013; Agarwal et al., 2018). Therefore, CRISPR/Cas9 offers tremendous potential in genetic improvement and has been wildly applied to improve yield traits, herbicide resistance, disease resistance, abiotic tolerance or nutrient quality in major cereal crops such as rice, wheat, maize, as summarized by many review articles on plant gene editing (Song et al., 2016; Liu et al., 2017a,b; Soda et al., 2018; Hussain et al., 2018; Agarwal et al., 2018; Abdelrahman et al., 2018; Zafar et al., 2020). These successful cases demonstrate that precise CRISPR/Cas9-mediated gene or gene family editing can be used efficiently and rapidly for crop improvement.

Sorghum is the first C4 crop to have its genome sequenced, and has a smaller size genome (about 730 Mb) than maize (Paterson et al., 2009; Schnable et al., 2009). The availability of the genome sequences facilitates genetic engineering and functional genomics research in sorghum and the emergence and development of CRISPR/Cas9 technique dramatically increases this potential. This chapter describes the advance of gene editing adopted in sorghum by highlighting the few studies completed. A study of the successful application of gene editing in sorghum for protein improvement is described in detail, and the future prospects for CRISPR/Cas9-mediated gene editing in sorghum genetic improvement and agricultural productivity are addressed.

2 Advances of gene editing in sorghum

In sorghum, only a few examples of gene editing have been reported thus far (Table 1). Jiang et al. (2013) performed the first gene editing attempt in sorghum callus cells derived from immature embryos by targeting a red fluorescence protein (*DsRED2*) gene located in T-DNA region (Jiang et al., 2013). In this experiment, they constructed an *A. tumefaciens* binary vector (Y158) carrying four independent genes, an out-of-frame red fluorescence protein gene (*DsRED2*) as the Cas9/sgRNA targeting gene, a monocot codon optimized synthetic *Cas9* gene, a *U6* promoter-driven sgRNA gene and a *CaMV 35S* promoter-driven GFP-NptII fusion visual marker gene, and were delivered into immature sorghum embryos by *Agrobacterium*-mediated transformation. Approximately 1/3 (5 of 18) of the stably transformed GFP positive groups of cells expressed DsRED2 fluorescent protein indicating a high frequency of restoring the *DsRED2* open-reading frame during NHEJ DNA repair by Cas9/sgRNA targeting. Notably, this work only involved editing of the T-DNA itself, rather than endogenous sorghum genes.

Table 1 Application of CRISPR/Cas9 genome editing in sorghum

Target genes	Gene function	Cas9 version	Cas9 promoter	sgRNA promoter	Delivery method	Mutation frequency (%)	Reference
DsRED2	Red fluorescence protein	Monocot codon optimized synthetic Cas9	Rice Actin 1	OsU6	Agrobacterium-mediated transformation	33	Jiang et al. (2013)
SbCENH3	Centromere-specific histone H3	spCas9	ZmUBI 1	ZmU6	Agrobacterium-mediated transformation	0.7–29	Che et al. (2018)
k1C gene family	α-kafirin	ZmCas9	ZmUBI	TaU3P	Agrobacterium-mediated transformation	92.4	Li et al. (2018)
CAD	Cinnamy1 alcohol dehydrogenase	-	Zm UBI	OsU3	Biolistic bombardment	25	Liu et al. (2019)
SbFT; SbGA2ox5	Flowering time; GA2 oxidase5	pCas9	Zm UBI	OsU6	Agrobacterium-mediated transformation	33.3–83.3	Char et al. (2020)

DuPont Pioneer's scientists reported a ternary vector system that facilitated *Agrobacterium* transformation and successfully applied in the popular variety T×430 and three recalcitrant African varieties including Macia, Malisor 84-7 and Tegemeo (Che et al., 2018). They further assembled the *spCas9 and sgRNA* cassette into a T-DNA expressing vector of this ternary system, and developed the first stable CRISPR/Cas9-mediated gene knockouts in T×430 sorghum by targeting a sorghum *centromere-specific histone H3* (*Sb-CENH3*) (Che et al., 2018). The *Sb-CENH3* knockouts may have the potential of utilization in the development of a sorghum haploid inducer because of the important gene function in the regulation of chromosome segregation. Moreover, this CRISPR/Cas9 system exhibited a targeted editing efficiency of 37–40%, and a knockout efficiency of 20–37% (Che et al., 2018). Sander (2019) further described Che's strategy of CRISPR/Cas9 gene editing including vector construction, molecular analyses of T0 plants, and the segregation and propagation of stably edited plants without T-DNA and vector backbone (Sander, 2019). Here, the key technique of Che's gene editing work was to apply Cas9/gRNA in a ternary vector system with broadly applicable sorghum transformation.

Because of the existence of several compatible wild sorghum relatives in the United States, conventional transgenic sorghum is not approved for release and is unlikely to be. The CRISPR/Cas9 strategy may therefore present great potential in gene editing of various sorghum ecotypes for sorghum breeding. Once the transgene has been demonstrably removed, leaving behind only the targeted gene edits, the sorghum lines are no different from the products of sorghum breeding programs. Thus, depending on the nature of the edited gene itself, the effects of rare hybridization events with wild species do not pose any more danger than with conventionally bred sorghum.

Sorghum gene editing engineered by Jiang et al. (2013) and Che et al. (2018) utilized the *Agrobacterium*-mediated transformation system to deliver CRISPR/Cas9 into sorghum immature embryo cells. Recently, Liu et al. (2019) reported a detailed protocol for the editing of cinnamy1 alcohol dehydrogenase (*CAD*) gene by CRISPR/Cas9 in sorghum genotype T×430 through biolistic bombardment (Liu et al., 2019). This report highlighted three crucial factors for an efficient CRISPR/Cas9 system for gene editing in sorghum, which include the efficient transformation system, gRNA design for targeted sequence, and the effective expression of Cas9 and gRNA elements. In the CRISPR/Cas9 system designed by Liu et al. (2019), Cas9 gene was driven by maize *ubiquitin* promoter and the gRNA was driven by rice *U3* promoter, which proved to be effective in sorghum. The homozygous and heterozygous editing of *CAD* has been successfully achieved in T_0 primary transgenic lines.

Generally, one single gene is targeted and edited by CRISPR/Cas9 system using one gRNA. Alternatively, two gRNAs can also be assembled into a CRISPR/Cas9 vector to simultaneously target two sites in one gene or more than one gene. Char et al. (2020) developed two Cas9/gRNA1-gRNA2 constructs to engineer targeted site modification in two endogenous genes, *SbFT* and *SbGA2ox5*, that may influence flowering time and plant height in sorghum (Char et al., 2020). The *Cas9* gene was directed by the maize *ubiquitin 1* gene promoter, and two gRNAs in one construct were individually driven by two different rice *U6* promoters. Two Cas9/gRNA1-gRNA2 constructs for targeting *SbFT* and *SbGA2ox5* were delivered into sorghum genotype P898012 and resulted in the editing efficiency of 33.3% and 83.3% in T0, individually.

3 Application of CRISPR/Cas9 gene editing for sorghum protein improvement

The above CRISPR/Cas9 strategies in sorghum focus on editing genes individually. In recent years, gene family editing was also used for crop breeding. Sanchez-Leon et al. (2018) reported a successful example of α-gliadin gene family editing using a consensus gRNA-mediated CRISPR/Cas9 system for the reduction of gluten in wheat. Here, we also applied a single gRNA CRISPR/Cas9 system to target a *k1C* family of 20 gene copies encoding the α-kafirins products in sorghum with the goal of protein quality improvement (Li et al., 2018). Kafirins (the sorghum prolamins), conferring low lysine content and indigestibility to the grain, account for 70% of the total protein in sorghum seeds and have a dominant influence on protein quality and digestibility (Oria et al., 2000; Aboubacar et al., 2001). Kafirins are classified as α-, β-, γ-, and δ-subunits that deposited in the ER-localized protein bodies. The α-kafirin protein moiety alone comprises about 80% of the total kafirins and is encoded by two *k1A* genes, one *k1D* gene and 20 highly similar *k1C* gene copies which are located in one cluster on chromosome 5. A floury kernel sorghum mutant, P721Q, generated by EMS mutagenesis, contains a point mutation resulting in a single amino acid substitution of the conserved signal peptide site of a *k1C* gene encoding 22-kD α-kafirin. This mutation reduces the Lys-devoid kafirin proteins and increases Lys-rich non-kafirin proteins, thereby enhancing the overall protein quality of the sorghum grain. Moreover, the reduction of kafirin accumulation alters the morphology of protein bodies and enhances protein digestibility, probably through enhanced accessibility to gastric proteases (Oria et al., 2000). In an earlier study, Kumar et al. (2012) used an RNA interference (RNAi) approach to suppress the α-kafirins and, in doing so, improved sorghum protein quality by enhancing lysine content and protein digestibility (Kumar et al., 2012). These works indicate protein quality improvement can be achieved by specific gene manipulation. It is feasible to perform α-kafirin gene editing by CRISPR/Cas9 system for sorghum protein improvement.

Thus, we developed a workflow of creating different α-kafirin genes edits by a single guide RNA-mediated CRISPR/Cas9 system targeting the *k1C* family of 20 highly similar genes for sorghum protein improvement (Fig. 1). It included gRNA design for gene family and vector construction, the delivery of cas9/gRNA system to sorghum, T7E1 assay for mutation detection, and Illumina MiSeq analysis for determining target site mutations. We also summarized the screening of stably edited lines, the measurement of protein digestibility and lysine content, and the introgression of high lysine/high digestibility into elite varieties for sorghum breeding.

3.1 sgRNA design of k1C *gene family and Cas9/gRNA vector construction*

For single gene editing, a gRNA can be designed by using the online software CRISPOR (http://crispor.tefor.net/) or CRISPR-P2.0 (http://crispr.hzau.edu.cn/CRISPR2/). Previous studies indicate that Cas9 can tolerate several mismatches between the gRNA and target region (Kim et al., 2015). Therefore, one consensus gRNA can be designed to target homologous genes within a gene family with some degree of divergence at the target site. Sequence alignment is first performed using bioinformatics software such as MegAlign of DNASTAR.

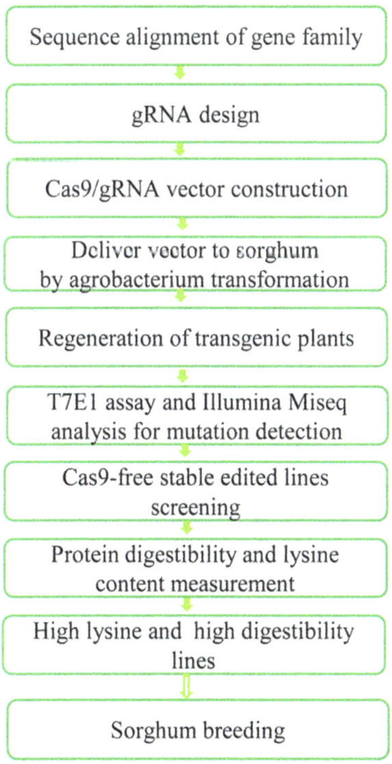

Figure 1 Workflow of CRISPR/Cas9-mediated gene family editing for sorghum protein improvement.

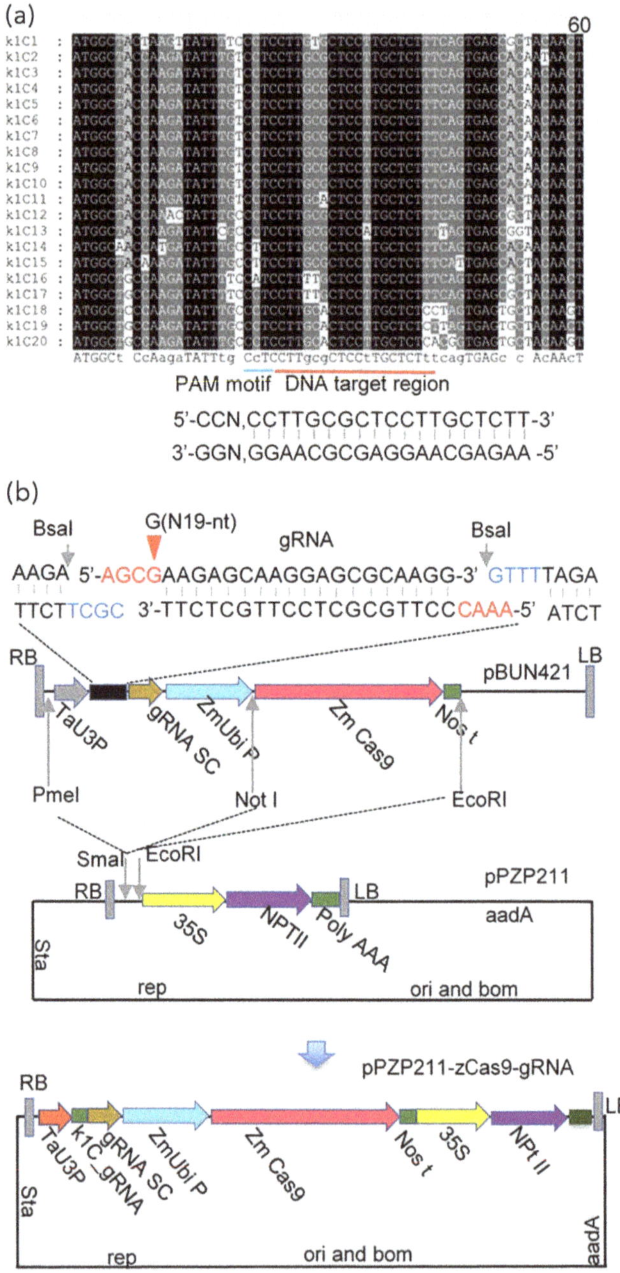

Figure 2 Construction of sgRNA/Cas9 system. (a) Partial sequence alignment of the *k1C* gene family from the reference and sgRNA design. (b) Schematic diagram of the construction of the sgRNA/Cas9 system. Gray arrows indicate the cohesive end produced by BsaI digestion in the vector. The synthetic double-stranded sgRNA with

The next step is to select a consensus sequence with PAM motif (5′-NGG-3′), and then detect sequence specificity by blasting against NCBI (the National Center for Biotechnology Information) or Phytozome to evaluate potential off-target binding sites. The gRNA spacer sequence is about 20 bp nucleotides and can be designed according to targeted DNA sequence that is followed by a PAM motif on the non-complementary strand. Moreover, when using one gRNA for gene editing, knock-out mutations are best generated by designing the gRNA sequence in 5′-end of coding region so as to create the best chance of frameshift mutations that result in truncated and/or non-sense mutations.

In this experiment, to effectively target most alpha kafirin genes (20 *k1C* gene family members) using one gRNA, a consensus sequence of 19 nucleotides following a 5′NGG PAM motif on the non-complementary strand was selected in the most highly conserved region encoding the z1C N-terminal ER signal peptide (Fig. 2a). The complementary sequence of the consensus 19 nucleotides in the coding strand was assembled into the CRISPR/Cas9 vector. An extra G base from the BsaI-digested CRISPR/Cas9 vector was fused with the 19-nt sequence in the complementary strand of *k1C* genes, which formed a final 20-bp gRNA with the 'G' at the 5′-end. Details of the CRISPR/Cas9 constructs can be found in the paper published by Li et al. (2018).

3.2 Delivery of Cas9/gRNA system and regeneration of transgenic sorghum plants

Three genetic transformation methods including protoplast-based transformation, particle bombardment, and agrobacterium-mediated transformation have been reported in sorghum (Battraw and Hall, 1991; Casas et al., 1993; Zhao et al., 2000). For sorghum transformation, the protoplast method has some limitations. For example, it is a labor-intensive method requiring considerable skills for adequate regeneration and that its success varies by genotype. Particle bombardment and agrobacterium-mediated transformation are two successful methods for generating stable transgenic sorghum. Agrobacterium-mediated

Figure 2 (*Continued*)
four-nucleotide 5′ overhangs at both ends was inserted into the sgRNA/Cas9 expression system. The red triangle represents a G base in the TaU3 promoter followed by the 5′-end of the 19-nucleotide sgRNA that resulted in a final 20-bp sgRNA with G (N19-nt) in the guide RNA/Cas9 system. RB/LB represent the left and right borders of the vector; TaU3P is the wheat U3 gene promoter; the gRNA site refers to the guide RNA clone site; gRNA SC is the gRNA scaffold; Ubi P is the ubiquitin gene promoter; Zm Cas9 is the maize codon-optimized Cas9; Nos t is the Nos terminator; 35S P is the 35S promoter; Npt II is the neomycin phosphotransferase II gene. Figure modified from Li et al. (2018), copyright American Society of Plant Biologists.

transformation is the preferred method for most researches because of its advantages in producing transgenic plants with single or low copies inserts.

Protoplast transformation has been commonly used to test gRNA efficiency in crops, and needs high-quality protoplasts for gene editing. However, the gRNA efficiency predicted in protoplasts may not correlate with the efficiency observed in intact plant tissues. Recently, an agrobacterium-mediated transient transformation system with intact sorghum leaves was established, which can be used to test gRNA expression construct efficiency (Sharma et al., 2020). For gene editing in plants, particle bombardment and agrobacterium-mediated transformation are two conventional methods to deliver Cas9/gRNA system into plant tissues, and both have been successfully applied in sorghum (Che et al., 2018; Liu et al., 2019; Char et al., 2020).

For generating single or low copies transgenic plants, the current anthors used agrobacterium-mediated transformation to deliver the cas9/gRNA system to sorghum (Li et al., 2018). The binary vector pPZP211_zCas9_Kaf. gRNA was mobilized into *Agrobacterium tumefaciens* via triparental mating, and the derived transconjugants were used for the transformation of sorghum (Sorghum bicolor) genotype T×430, as previously described (Howe et al., 2006). In this work, a total of 26 T0 events were generated and propagated in the greenhouse. After self-pollination, T1 generation seeds were collected.

3.3 T7E1 assay and Illumina MiSeq analysis for detection and determination of mutations

Target mutations can be preliminarily detected by several methods such as RNA-guided engineered nuclease-restriction fragment length polymorphism (Kim et al., 2014), PCR-restriction enzyme assay (Shan et al., 2014), T7 endonuclease I (T7E1) assay (Vouillot et al., 2015), and high-throughput qPCR-based method (Peng et al., 2018). The T7E1 assay is a popular method to detect target site edits as it is a relatively simple and quick way to determine if more in-depth sequencing is warranted. The T7E1 enzyme can recognize base mismatches and cleaved heteroduplexes formed by annealing of wild-type (WT) DNA strand and mutated DNA strand (Vouillot et al., 2015). In a T7E1 assay, specific primers are designed to individually amplify DNA fragments spanning target region of WT, mutant, and the mixture of WT and mutant. The amplicons are reannealed and digested by T7E1 enzyme. T7E1 will digest all heteroduplexes which reveals mutations.

Li et al. (2018) utilized the T7E1 assay to preliminarily evaluate targeted mutagenesis of the *k1C* gene family in transgenic plants. First, a set of primers with degenerate specificity to the gene family was designed to amplify the target region in the T×430 control, T1 transgenic plants, and the mixture of T×430 and T1 plants. Amplified fragments were subjected to a complete denaturation-slow annealing program to form heteroduplexes, and then digested with T7E1 enzyme. Digested PCR products were separated and visualized using polyacrylamide gel electrophoresis (PAGE) and stained with ethidium bromide. Use of PAGE rather than agarose allows the detection of small quantities of fragments from the cleavage of heteroduplexes in gene family amplicons. If mutations exist in the T1 plant, the T1 plant and the mixture of T×430 and T1 plant will exhibit an additional band compared to the T×430 control.

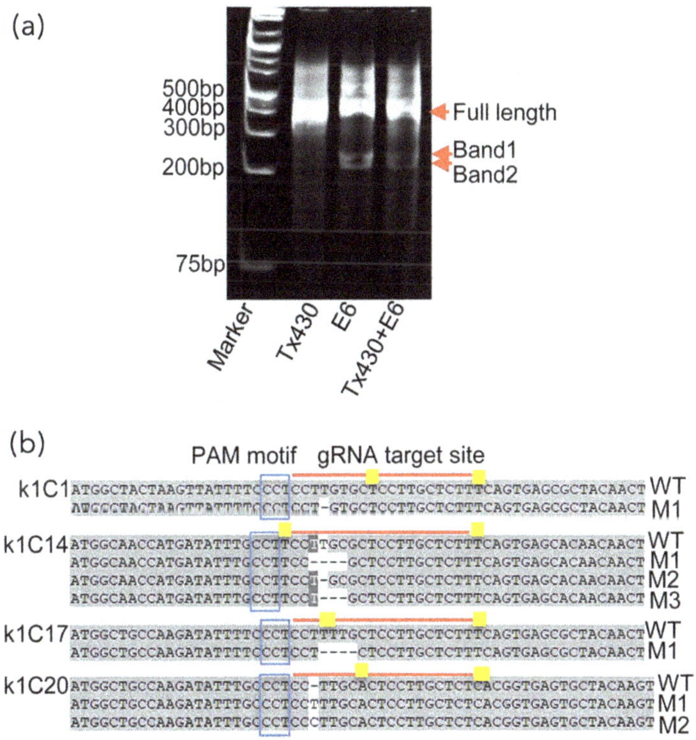

Figure 3 Mutation detection of a T1 edited plant derived from E6 event. (a) T7E1 assay. The amplicons spanning target site of *k1C* family were obtained in T×430, E6, and the mixture of T×430 and E6 using a set of degenerate primers, and were reannealing and digested by T7E1 enzyme. Digested products were separated and visualized using an 8% PAGE gel–two lanes including E6, and the mixture of T×430 and E6 shows the additional bands compared to the lane T×430. (b) Target site mutations in multiple *k1C* genes identified by the amplicons sequencing. Red lines indicate target sites. Green rectangles present PAM motif. Blue stars show mismatch between gRNA and target site. Yellow boxes indicate mismatches of gRNA to target site of *k1C* genes. WT is the sequence from T×430 as reference. M presents the sequence from mutant alleles. Figure modified from Li et al. (2018), copyright American Society of Plant Biologists.

Although the T7E1 assay demonstrates the occurrence of gene editing, subsequent direct sequencing is needed to accurately determine target site mutations. To survey target site mutations in the *k1C* gene family, PCR products spanning the k1C target site were obtained using a set of degenerate primers in T1 plants. PCR products were sequenced using highly multiplexed Illumina Miseq. Figure 3 shows mutation detection in a T1-edited plant of event E6 by T7E1 assay combined with the amplicon sequencing.

Extensive edits in 25 of 26 events occurred in one or multiple *k1C* family members, and almost all *k1C* genes members, including 13 *k1C* members/ group (*k1C1*, *k1C2*, *k1C3-9*, *k1C10*, *k1C11*, *k1C12*, *k1C13*, *k1C14*, *k1C16*, *k1C17*, *k1C18*, *k1C19*, and *k1C20*), were edited in a varying number of events in the T1 generation (Li et al., 2018). A large repertoire of mutations was determined (Table 2), which included small deletions of 1–33 bp and small insertions of 1–16 bp. The predominant mutation type was a single base T insertion or deletion, which occurred at 3 bp upstream of PAM motif that was considered as the Cas9 cleavage site (Jinek et al., 2012). These *k1C* site mutations may influence gene function and *α* kafirin accumulation because of frameshifts or in-frame mutations that lead to gene knockout or the substitution or deletion of essential conserved amino acids.

This study indicated, for the first time, that a CRISPR/Cas9 system, based on one degenerate single gRNA, can effectively induce extensive editing of all members of a gene family and reduction of their encoded proteins. It was also the first example applying CRISPR to improve a major limiting output trait in sorghum.

3.4 Screening of stably edited, transgene-free lines

CRISPR/Cas9 constructs are heritable and can keep Cas9 editing activities in the next generation if the transgenes encoding the CRISPR/Cas9 machinery are maintained in the edited plants. To obtain stable mutation lines, it is essential to eliminate CRISPR/Cas9 constructs in edited plants. Generally, Cas9 is removed by self-pollination or backcrossing. Cas9-free edited lines can be screened by simple PCR assays. Removal of the transgene is necessary so as to prevent additional edits at target or off-target sites. Furthermore, transgene-free edited lines can then be propagated for large-scale yield trials, bulking and eventual production in unregulated field plots.

Transgenes can also be removed from edited plants using molecular tools. For example, a transgene killer CRISPR (TKC) system was developed in rice (He et al., 2018). This technology enables self-elimination of the transgene after the targeted gene is edited. In detail, the TKC technology combines Cas9/gRNA cassette with two suicide genes (*CMS2* and *BARNASE*) that can

Table 2 The mutation types occurred in T1 edited plants

The mutation types[a]	Distance (bp) to start codon "ATG"[b]	Insertion/deletion base number	Frequency[c]
−TGGCTACCAAGATATTT(/C)GT(/C)CCTCCT	2	−25	2/26
−TGGCTACCAAGATATTCGCCCTCC	2	−24	1/26
−GCTGCCAAGATATTTTCCCTCCTTTT	3	−26	1/26
−GCTACCAAGATATTT(/C)GT(/C)CCTCCTT	4	−24	5/26
−GCTACCAAGATATTT(/C)GT(/C)CCTCCTTGC	4	−26	4/26
−GCTACCAAGATATTTGTCCTCCTTGCGCTCCTT	4	−33	1/26
−CCAAGATATTT(/C)GT(/C)CCTCCTTG	8	−21	1/26
+ACAATGTCTACCC	11	−13	1/26
−ATATTTGCCCTCCTTGCA	13	−18	1/26
−ATATTTGCCCTCCTTGCACTCCTTGCTCTC	13	−30	1/26
−TATTTGCCCTCCTTGCAC	14	−18	1/26
−TTTGTCCTCCTT	16	−12	1/26
−TTTGTCCTCCTTGC	16	−14	1/26
−TTT(/C)GT(/C)CCTC	16	−9	3/26
−TTTGTCCTCC	16	−10	1/26
−TTT(/C)GT(/C)CCTCCT	16	−11	2/26
−TTTGTCCTCCTTG	16	−13	1/26
−TTTGTCCTCCTTGCG	16	−15	1/26
−TTTGTCCTCCTTGCGCTCCT	16	−20	1/26
−TTT(/C)GT(/C)CCTCCTTGCGCTCCTTGCTC	16	−26	2/26
−TT(/C)GT(/C)CCTCC	17	−9	4/26
−TTGTCCTCCTTGC	17	−13	1/26

(*Continued*)

Table 2 *(Continued)*

The mutation types[a]	Distance (bp) to start codon "ATG"[b]	Insertion/deletion base number	Frequency[c]
−TTGTCCTCCTTGCGCTCC	17	−18	1/26
−TCCTCCTTGCGC	20	−12	2/26
−TCCTCCTTGCGCTCCTTGCTCT	20	−22	1/26
+AA	21	+2	1/26
+AAGTGAAA	22	+8	1/26
−CTCCTTG<u>C(/T) G(/A)</u>	22	−6	12/26
+T	23	+1	5/26
−T	23	−1	9/26
−TCCT	23	−4	7/26
+C	24	+1	13/26
+CACAA	24	+5	1/26
+CCGCACTGCTC	24	+11	1/26
−C	24	−1	12/26
−CCTTG	24	−5	8/26
+CGCAGTGCTCA	25	+11	1/26
−CTTG	25	−4	13/26
−CTTG<u>C(/T) G(/A)</u>	25	−6	13/26
+T	26	+1	23/26
+TGAATA	26	+6	1/26
−T	26	−1	23/26
−TT,26	26	−2	4/26
+A	27	+1	5/26

+G	27	+1	3/26
+C	27	+1	7/26
+GAAACAGCAAGG	27	+12	1/26
−TG	27	−2	10/26
−TGC	27	−3	10/26
−TGCG	27	−4	3/26
−TGCGCTCCTTG	27	−11	2/26
+C	28	+1	4/26
−G	28	−1	13/26
−GC	28	−2	13/26
−GCG(/A)	28	−3	10/26
−GCGCTC	28	−6	1/26
−C	29	−1	3/26
+AAA	29	+3	1/26
+ACAACGTTCTCA	29	+12	1/26
+GGACTAAATTCTATGT	30	−16	1/26
−TCCTTGCTCTCCT	32	−13	1/26
−C	33	−1	4/26

[a]"+" &"−" indicate insert and deletion; (/) presents base polymorphism located in different *k1C* gene with edits.
[b]Mutation locations are presented in the distance (bp) to start codon "ATG"; PAM motif lies in the 21-23 bp distance to start codon "ATG" of *k1C* genes.
[c]Frequency presents the occurrence rate of mutations in 26 events. Yellow color highlights the most common mutations in T1 edited plants.

kill transgene-containing pollen and embryos, respectively. The Cas9/gRNA element is first expressed during transformation and vegetative growth phase, and the suicide genes are expressed to kill all transgene-containing pollen and embryo in the reproductive phase, which produces transgene-free seeds. Thus, stably edited plants without CRISPR/Cas9 transgene can be achieved in T1 generation. The TKC technology was further improved by replacing 35S promoter with rice promoters to drive suicide gene expression that triggers self-elimination of transgenes without reducing gene editing efficiency (He et al., 2019). This technology can greatly accelerate the isolation of Cas9-free edited lines and has potential in functional genomics research and crop improvement of sorghum and other crops.

3.5 Sorghum digestibility and protein quality assessment

Previously, low kafirin and high non-kafirin sorghum lines with high digestibility and high lysine (HDHL) traits showed completely opaque seeds (Weaver et al., 1998; Oria et al., 2000). This is a potentially undesirable phenotype as it can lead to fungal susceptibility, mechanical damage during harvest and impaired germination. T2 HDHL populations lines created using CRISPR derived from multiple independent event lineages display a heritable semi-vitreous phenotype (slightly reduced thickness of the vitreous endosperm layer). This results from irregular protein body structure and reduced protein body size (Figs 4 and 5). These lines were selected for the measurement of protein digestibility and lysine content. Protein digestibility was measured using an *in vitro* pepsin digestibility assay described by Mertz (Mertz et al., 1984) with minor modifications. Free and protein-bound amino acids were extracted and measured using an ultra-performance liquid chromatography-tandem mass spectrometry (UPLC-MS/MS) system (Angelovici et al., 2013). Thus, high lysine/high digestibility lines with low kafirins and semi-vitreous kernel phenotype were identified by analyzing protein digestibility and lysine content (Table 3).

3.6 High lysine/high digestibility line selection and breeding strategies

For breeding and release, three steps need to be performed. The first is selection of high lysine/high digestibility edited lines. The second is transferring high lysine/high digestibility traits into the desired sorghum varieties and transgene removal. The third is verification of mutations in target genes and reassessment of protein quality index in the new varieties. The breeding strategy in its simplest form is to remove the transgene with a simple cross to WT T×430

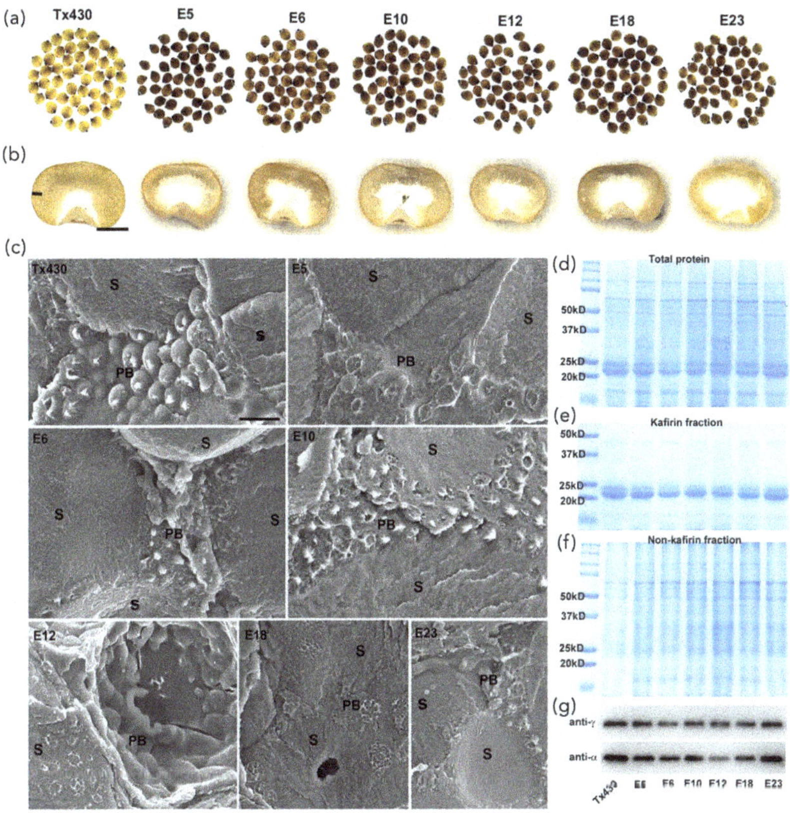

Figure 4 Mature kernel phenotypes and protein profiles of selected T2 events. (a) T2 kernels with partial opacity (reduction in vitreous endosperm) of six independent lines compared with the WT T×430 in the light box. (b) Dissecting microscope images of representative basal halves of T2 kernels showing reductions in the thickness of the vitreous endosperm layer. Bar = 1 mm and applies to all images. The short bar in T×430 represents the distance from the pericarp (300 μm) examined in all images in (c). (c) Scanning electron micrographs of representative basal kernel halves, as shown in (b). S represents starch grains, and PB points to protein bodies. Bar = 2 μm and applies to all images. (d-f) SDS-PAGE analysis of total kernel proteins, kafirin protein fractions, and nonkafirin protein fractions, respectively. Coomassie Blue-stained gels show loading from 250 μg of mature kernel flour in all lanes. (g) Western-blot analysis of α-kafirin accumulation in six selected T2 events and WT T×430. α-Kafirins were detected by maize α-zein antiserum (1:5,000 dilution) using a 1:100 dilution of the total protein loaded on the Coomassie Blue-stained gel in (d). Sample identities in (d–g) are as labeled in (g). Figure reproduced from Li et al. (2018), copyright American Society of Plant Biologists.

and maintain the edits in this background since T×430 is a well-performing and widely used sorghum inbred in the United States. Alternatively, selected events can be out-crossed to a line carrying another desirable trait such as high

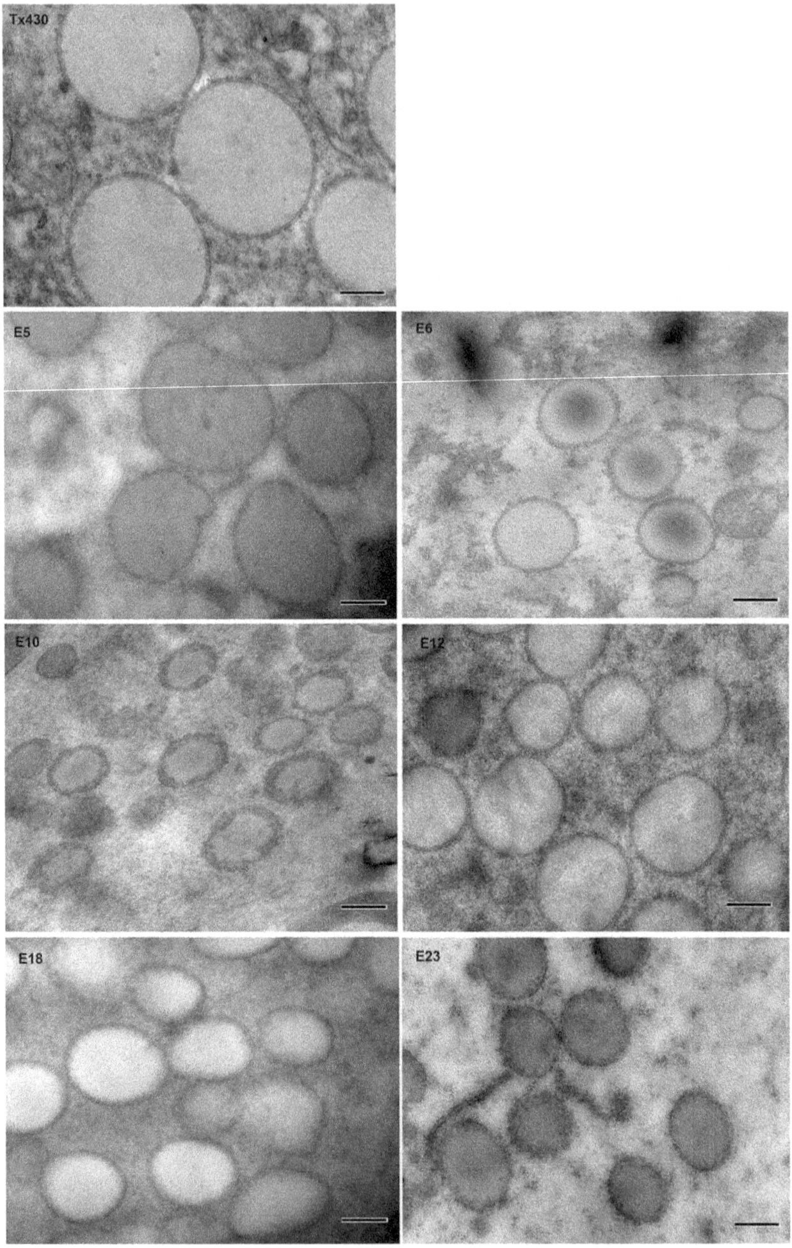

Figure 5 Kafirin protein body phenotypes in developing kernels. Transmission electron micrographs of 20 DAP (±2 DAP) developing sorghum endosperm from T3 kernels showing 90 nm Spur's resin sections through kafirin protein bodies in the 3rd–4th sub-aleurone layer at medial distal end of the kernel. T×430 is the untransformed control. Scale bar in all panels represents 400 nm.

Table 3 Protein digestibility and lysine content of raw flour derived from T2 events

Events	Protein content (g/100 g flour)	Protein digestibility (g/100 g flour)	Protein bound lysine (g/100 g flour)	Free lysine (g/100 g flour)
T×430	12.19 ± 0.14	48.05 ± 5.43	0.17 ± 0.01	1.52E-03 ± 1.12E-04
E5	15.47 ± 0.35	64.30 ± 0.43	0.31 ± 0.03	1.15E-02 ± 2.65E-04
E6	12.55 ± 0.13	72.03 ± 3.10	0.29 ± 0.03	1.03E-02 ± 6.81E-04
E10	14.45 ± 0.23	69.18 ± 0.89	0.35 ± 0.02	1.28E-02 ± 1.86E-03
E12	15.19 ± 0.10	69.81 ± 2.34	0.40 ± 0.01	1.61E-02 ± 2.57E-03
E18	13.89 ± 0.26	71.35 ± 1.46	0.35 ± 0.01	8.98E-03 ± 2.29E-04
E23	13.96 ± 0.11	60.38 ± 2.46	0.31 ± 0.03	3.31E-03 ± 2.81E-04

Note: Percent protein refers to mean protein percentage ± standard deviation in raw flour (n = 3); percent digestibility indicates the mean percentage of digested protein standard deviation (n = 3); protein bound and free lysine content refers to mean ± standard deviation (n = 3).

nitrogen-use efficiency. The use of a different recurrent parent would require multiple generations of backcrossing with the removal of the transgene by PCR selection at the F2 stage, and visual, biochemical and genetic selection of the target site edits at each generation. Hybrid production in sorghum is more challenging than in corn since it must be achieved through a male sterile system. To introduce the HDHL edits into such a system would be possible but would need to be done through introgression of edits into both parents in the hybrid system of choice. The sorghum transformation approach used is optimized from the T×430 genotype while other elite backgrounds are invariably more recalcitrant toward transformation. In future, the development of more universal methods for sorghum transformation will be essential for accelerating sorghum gene editing and realizing its potential.

4 Future trends

Sorghum improvement mainly focuses on grain yield and its component traits, resistance to insect pests, diseases, striga, drought, salinity, cold, and nutritional quality. Since the 1930s, sorghum improvement has consisted of several long-term processes: (1) pure line selection of local landraces (1930s–1960s); (2) variety development with desired traits by hybridization and selection (1960s–1970s); and (3) marker-assisted selection breeding (MAS) using DNA markers linked to desired traits/QTL, EMS technology creating point mutations, and transgenic method (1980s to present).

With the availability of the sorghum whole-genome sequence in 2009, it became possible to target specific genes and employ genomic tools to tailor sorghum for various needs. A user-friendly CRISPR/Cas9 gene-editing technique, developed in 2013, offers tremendous potential to accelerate crop improvement. This technique has been employed for gene editing in sorghum (Char et al., 2020; Che et al., 2018; Li et al., 2018; Liu et al., 2019). The study described in detail here to improve sorghum digestibility and lysine content by CRISPR/Cas9-mediated *k1C* gene family editing both highlights the technological advances of the technique but shows its immediate applicability to crop improvement (Li et al., 2018). More recently, using CRISPR/Cas9-edited sorghum, it was demonstrated that the benefit of LGS1-mediated resistance to the parasitic weed Striga hermonthica (Delile) Benth strongly depends on parasite genotype and abiotic environment (Bellis et al., 2020). These examples indicate that breeders can engineer the editing of genes or gene families related to key traits using CRISPR/Cas9 system to pursue specific genetic improvement in sorghum.

Currently, grain quality and yield are two critical traits for sorghum breeding and productivity. High digestibility, high lysine, vitamin A, and low tannin content are the aims of nutritional quality improvement for human food and animal fodder. Moreover, low lignin, increased sugar, and high cellulose content are also important desired traits in sorghum biomass and fodder improvement (Venkatesh Bhat et al., 2015). Grain yield was closely correlated with head weight, weight of a standard length of peduncle perimeter, head length and breadth (Vance, 1976). The cloning of key related genes in sorghum could offer opportunities for CRISPR/Cas9-mediated specific gene editing to develop new elite cultivars. For example, it was found that a *Tan1* gene, encoding a WD40 protein, controls tannin biosynthesis in sorghum (Wu et al., 2012). As a result breeders can design editing strategies of the *Tan1* gene to reduce tannin content in sorghum. Jiao et al. (2018) identified the multi-seeded gene, which increases grain number per panicle, and has the potential to boost grain yield (Jiao et al., 2018). Such genes can be targeted to improve sorghum production.

Sorghum is often affected by different biotic stresses (such as fungal, bacterial and viral diseases, and insect pests) or abiotic stresses (such as drought, salinity and low temperatures). Such stresses often result in the reduction of sorghum yield and quality. Generally, biotic stress is regulated by different signaling pathways controlled by phytohormones such as jasmonic acid (JA), salicylic acid (SA) and ethylene, and abiotic stress is largely controlled by the abscisic acid signaling pathway (Atkinson and Urwin, 2012). CRISPR/Cas9 could be used to modify key genes involved in plant hormone signaling pathways to enhance resistance to various diseases/insect pests or improve drought/low-temperature tolerance.

The non-permanent transgene system CRISPR/Cas9 can efficiently create various mutations including gene knockouts which mimic the presence/absence variations present in natural populations. These can not only be used to elucidate gene function in various metabolic and regulatory pathways, but also have great potential in further genetic improvement related to key agronomic traits in sorghum and other crops. The CRISPR/Cas9 technique will, thus, likely open new avenues in future sorghum breeding.

5 Conclusion

Sorghum genetic engineering is a rapid means to improve the crop with the addition of better traits and removal of undesirable features. However, the release of commercial GM crops has multiple regulatory and societal barriers. The novel CRISPR/Cas9 system is a powerful gene-editing tool and induces precise mutations in targeted regions of a plant genome. CRISPR/Cas9 is a non-permanent transgene system because the constructs can be segregated out by backcrossing or eliminated by self-killer gene techniques (He et al., 2018, 2019), which offer great potential in accelerating genetic engineering and crop improvement. The successful application of CRISPR/

Cas9-mediated gene editing for several sorghum genes, especially the *k1C* gene family editing for sorghum protein quality improvement, gave a model for genetic modification programs in other important genes/traits of sorghum. In the future, the CRISPR/Cas9 technique will accelerate the generation of new sorghum cultivars with desired traits and can exert great impacts in sorghum production for addressing the problem of food deficiency and nutrient malnutrition in the developing world. Improving input and output traits for reinvigorating and expanding sorghum cultivation, and realizing its superior sustainability characteristics in a changing and uncertain climate, can also be accelerated through CRISPR.

6 References

Abdelrahman, M., Al-Sadi, A. M., Pour-Aboughadareh, A., Burritt, D. J. and Tran, L. P. 2018. Genome editing using CRISPR/Cas9-targeted mutagenesis: an opportunity for yield improvements of crop plants grown under environmental stresses. *Plant Physiology and Biochemistry: PPB* 131, 31–6.

Aboubacar, A., Axtell, J. D., Huang, C. P. and Hamaker, B. R. 2001. A rapid protein digestibility assay for identifying highly digestible sorghum lines. *Cereal Chemistry Journal* 78(2), 160–5.

Agarwal, A., Yadava, P., Kumar, K., Singh, I., Kaul, T., Pattanayak, A. and Agrawal, P. K. 2018. Insights into maize genome editing via CRISPR/Cas9. *Physiology and Molecular Biology of Plants: An International Journal of Functional Plant Biology* 24(2), 175–83.

Angelovici, R., Lipka, A. E., Deason, N., Gonzalez-Jorge, S., Lin, H., Cepela, J., Buell, R., Gore, M. A. and DellaPenna, D. 2013. Genome-wide analysis of branchedchain amino acid levels in Arabidopsis seeds. *Plant and Cell Physiology* 25, 4827–43.

Arnould, S., Chames, P., Perez, C., Lacroix, E., Duclert, A., Epinat, J. C., Stricher, F., Petit, A. S., Patin, A., Guillier, S., Rolland, S., Prieto, J., Blanco, F. J., Bravo, J., Montoya, G., Serrano, L., Duchateau, P. and Pâques, F. 2006. Engineering of large numbers of highly specific homing endonucleases that induce recombination on novel DNA targets. *Journal of Molecular Biology* 355(3), 443–58.

Atkinson, N. J. and Urwin, P. E. 2012. The interaction of plant biotic and abiotic stresses: from genes to the field. *Journal of Experimental Botany* 63(10), 3523–43.

Battraw, M. and Hall, T. C. 1991. Stable transformation of sorghum-bicolor protoplasts with chimeric neomycin phosphotransferase-Ii and beta-glucuronidase genes. *TAG. Theoretical and Applied Genetics. Theoretische und Angewandte Genetik* 82(2), 161–8.

Belhaj, K., Chaparro-Garcia, A., Kamoun, S. and Nekrasov, V. 2013. Plant genome editing made easy: targeted mutagenesis in model and crop plants using the CRISPR/Cas system. *Plant Methods* 9(1), 39.

Bellis, E. S., Kelly, E. A., Lorts, C. M., Gao, H., DeLeo, V. L., Rouhan, G., Budden, A., Bhaskara, G. B., Hu, Z., Muscarella, R., Timko, M. P., Nebie, B., Runo, S. M., Chilcoat, N. D., Juenger, T. E., Morris, G. P., dePamphilis, C. W. and Lasky, J. R. 2020. Genomics of sorghum local adaptation to a parasitic plant. *Proceedings of the National Academy of Sciences of the United States of America* 117(8), 4243–51.

Boch, J., Scholze, H., Schornack, S., Landgraf, A., Hahn, S., Kay, S., Lahaye, T., Nickstadt, A. and Bonas, U. 2009. Breaking the code of DNA binding specificity of TAL-type III effectors. *Science* 326(5959), 1509-12.

Casas, A. M., Kononowicz, A. K., Zehr, U. B., Tomes, D. T., Axtell, J. D., Butler, L. G., Bressan, R. A. and Hasegawa, P. M. 1993. Transgenic sorghum plants via microprojectile bombardment. *Proceedings of the National Academy of Sciences of the United States of America* 90(23), 11212-6.

Chandrashekar, A. and Satyanarayana, K. V. 2006. Disease and pest resistance in grains of sorghum and millets. *Journal of Cereal Science* 44(3), 287-304.

Chapman, J. R., Taylor, M. R. G. and Boulton, S. J. 2012. Playing the end game: DNA double-strand break repair pathway choice. *Molecular Cell* 47(4), 497-510.

Char, S. N., Wei, J., Mu, Q., Li, X., Zhang, Z. J., Yu, J. and Yang, B. 2020. An Agrobacterium-delivered CRISPR/Cas9 system for targeted mutagenesis in sorghum. *Plant Biotechnology Journal* 18(2), 319-21.

Che, P., Anand, A., Wu, E., Sander, J. D., Simon, M. K., Zhu, W., Sigmund, A. L., Zastrow-Hayes, G., Miller, M., Liu, D., Lawit, S. J., Zhao, Z. Y., Albertsen, M. C. and Jones, T. J. 2018. Developing a flexible, high-efficiency Agrobacterium-mediated sorghum transformation system with broad application. *Plant Biotechnology Journal* 16(7), 1388-95.

Dong, O. X. and Ronald, P. C. 2019. Genetic engineering for disease resistance in plants: recent progress and future perspectives. *Plant Physiology* 180(1), 26-38.

Hariprasanna, K. and Patil, J. V. 2015. Sorghum: origin, classification, biology and improvement, in R. Madhusudhana et al. (Eds), *Sorghum Molecular Breeding*, pp. 3-18. India: Springer.

He, Y. B., Zhu, M., Wang, L. H., Wu, J., Wang, Q., Wang, R. and Zhao, Y. 2018. Programmed self-elimination of the CRISPR/Cas9 construct greatly accelerates the isolation of edited and transgene-free rice plants. *Molecular Plant* 11(9), 1210-3.

He, Y. B., Zhu, M., Wang, L., Wu, J., Wang, Q., Wang, R. and Zhao, Y. 2019. Improvements of TKC technology accelerate isolation of transgene-free CRISPR/Cas9-edited rice plants. *Rice Science* 26(2), 109-17.

Howe, A., Sato, S., Dweikat, I., Fromm, M. and Clemente, T. 2006. Rapid and reproducible Agrobacterium-mediated transformation of sorghum. *Plant Cell Reports* 25(8), 784-91.

Hussain, B., Lucas, S. J. and Budak, H. 2018. CRISPR/Cas9 in plants: at play in the genome and at work for crop improvement. *Briefings in Functional Genomics* 17(5), 319-28.

Jiang, W., Zhou, H., Bi, H., Fromm, M., Yang, B. and Weeks, D. P. 2013. Demonstration of CRISPR/Cas9/sgRNA-mediated targeted gene modification in Arabidopsis, tobacco, sorghum and rice. *Nucleic Acids Research* 41(20), e188.

Jiao, Y. P., Lee, Y. K., Gladman, N., Chopra, R., Christensen, S. A., Regulski, M., Burow, G., Hayes, C., Burke, J., Ware, D. and Xin, Z. 2018. MSD1 regulates pedicellate spikelet fertility in sorghum through the jasmonic acid pathway. *Nature Communications* 9(1), 822.

Jinek, M., Chylinski, K., Fonfara, I., Hauer, M., Doudna, J. A. and Charpentier, E. 2012. A programmable dual-RNA-guided DNA endonuclease in adaptive bacterial immunity. *Science* 337(6096), 816-21.

Kim, D., Bae, S., Park, J., Kim, E., Kim, S., Yu, H. R., Hwang, J., Kim, J. I. and Kim, J. S. 2015. Digenome-seq: genome-wide profiling of CRISPR-Cas9 off-target effects in human cells. *Nature Methods* 12(3), 237-43.

Kim, J. M., Kim, D., Kim, S. and Kim, J. S. 2014. Genotyping with CRISPR-Cas-derived RNA-guided endonucleases. *Nature Communications* 5, 3157.

Kumar, T., Dweikat, I., Sato, S., Ge, Z., Nersesian, N., Chen, H., Elthon, T., Bean, S., Ioerger, B. P., Tilley, M. and Clemente, T. 2012. Modulation of kernel storage proteins in grain sorghum (Sorghum bicolor (L.) Moench). *Plant Biotechnology Journal* 10(5), 533-44.

Li, A., Jia, S., Yobi, A., Ge, Z., Sato, S. J., Zhang, C., Angelovici, R., Clemente, T. E. and Holding, D. R. 2018. Editing of an alpha-kafirin gene family increases, digestibility and protein quality in sorghum. *Plant Physiology* 177(4), 1425-38.

Liu, G. Q., Li, G. Q. and Godwin, I. D. 2019. Genome editing by CRISPR/Cas9 in sorghum through biolistic bombardment. *Methods in Molecular Biology* 1931, 169-83.

Liu, X., Wu, S. R., Xu, J., Sui, C. and Wei, J. H. 2017a. Application of CRISPR/Cas9 in plant biology. *Acta Pharmaceutica Sinica. B* 7(3), 292-302.

Liu, X. J., Xie, C. X., Si, H. J. and Yang, J. X. 2017b. CRISPR/Cas9-mediated genome editing in plants. *Methods* 121-122, 94-102.

Ma, X. L., Zhang, Q. Y., Zhu, Q. L., Liu, W., Chen, Y., Qiu, R., Wang, B., Yang, Z., Li, H., Lin, Y., Xie, Y., Shen, R., Chen, S., Wang, Z., Chen, Y., Guo, J., Chen, L., Zhao, X., Dong, Z. and Liu, Y. G. 2015. A robust CRISPR/Cas9 system for convenient, high-efficiency multiplex genome editing in monocot and dicot plants. *Molecular Plant* 8(8), 1274-84.

Maeder, M. L., Thibodeau-Beganny, S., Osiak, A., Wright, D. A., Anthony, R. M., Eichtinger, M., Jiang, T., Foley, J. E., Winfrey, R. J., Townsend, J. A., Unger-Wallace, E., Sander, J. D., Müller-Lerch, F., Fu, F., Pearlberg, J., Göbel, C., Dassie, J. P., Pruett-Miller, S. M., Porteus, M. H., Sgroi, D. C., Iafrate, A. J., Dobbs, D., McCray, P. B., Cathomen, T., Voytas, D. F. and Joung, J. K. 2008. Rapid "Open-Source" Engineering of customized zinc-finger nucleases for highly efficient gene. *Molecular Cell* 31(2), 294-301.

Mertz, E. T., Hassen, M. M., Cairnswhittern, C., Kirleis, A. W., Tu, L. and Axtell, J. D. 1984. Pepsin digestibility of proteins in sorghum and other major cereals. *Proceedings of the National Academy of Sciences of the United States of America* 81(1), 1-2.

Oria, M. P., Hamaker, B. R., Axtell, J. D. and Huang, C. P. 2000. A highly digestible sorghum mutant cultivar exhibits a unique folded structure of endosperm protein bodies. *Proceedings of the National Academy of Sciences of the United States of America* 97(10), 5065-70.

Paterson, A. H., Bowers, J. E., Bruggmann, R., Dubchak, I., Grimwood, J., Gundlach, H., Haberer, G., Hellsten, U., Mitros, T., Poliakov, A., Schmutz, J., Spannagl, M., Tang, H., Wang, X., Wicker, T., Bharti, A. K., Chapman, J., Feltus, F. A., Gowik, U., Grigoriev, I. V., Lyons, E., Maher, C. A., Martis, M., Narechania, A., Otillar, R. P., Penning, B. W., Salamov, A. A., Wang, Y., Zhang, L., Carpita, N. C., Freeling, M., Gingle, A. R., Hash, C. T., Keller, B., Klein, P., Kresovich, S., McCann, M. C., Ming, R., Peterson, D. G., Mehboob-ur-Rahman, Ware, D., Westhoff, P., Mayer, K. F., Messing, J. and Rokhsar, D. S. 2009. The Sorghum bicolor genome and the diversification of grasses. *Nature* 457(7229), 551-6.

Peng, C., Wang, H., Xu, X. L., Wang, X., Chen, X., Wei, W., Lai, Y., Liu, G., Godwin, I. D., Li, J., Zhang, L. and Xu, J. 2018. High-throughput detection and screening of plants modified by gene editing using quantitative real-time polymerase chain reaction. *Plant Journal: For Cell and Molecular Biology* 95(3), 557-67.

Puchta, H. and Fauser, F. 2014. Synthetic nucleases for genome engineering in plants: prospects for a bright future. *Plant Journal: For Cell and Molecular Biology* 78(5), 727-41.

Reis, A., Hornblower, B., Robb, B. and Tzertzinis, G. 2014. CRISPR/Cas9 and targeted genome editing: a new era in molecular biology. *New England BioLabs* Expressions 1, 3-6.

Sander, J. D. 2019. Gene editing in sorghum through Agrobacterium. *Methods in Molecular Biology* 1931, 155-68.

Schnable, P. S., Ware, D., Fulton, R. S., Stein, J. C., Wei, F., Pasternak, S., Liang, C., Zhang, J., Fulton, L., Graves, T. A., Minx, P., Reily, A. D., Courtney, L., Kruchowski, S. S., Tomlinson, C., Strong, C., Delehaunty, K., Fronick, C., Courtney, B., Rock, S. M., Belter, E., Du, F., Kim, K., Abbott, R. M., Cotton, M., Levy, A., Marchetto, P., Ochoa, K., Jackson, S. M., Gillam, B., Chen, W., Yan, L., Higginbotham, J., Cardenas, M., Waligorski, J., Applebaum, E., Phelps, L., Falcone, J., Kanchi, K., Thane, T., Scimone, A., Thane, N., Henke, J., Wang, T., Ruppert, J., Shah, N., Rotter, K., Hodges, J., Ingenthron, E., Cordes, M., Kohlberg, S., Sgro, J., Delgado, B., Mead, K., Chinwalla, A., Leonard, S., Crouse, K., Collura, K., Kudrna, D., Currie, J., He, R., Angelova, A., Rajasekar, S., Mueller, T., Lomeli, R., Scara, G., Ko, A., Delaney, K., Wissotski, M., Lopez, G., Campos, D., Braidotti, M., Ashley, E., Golser, W., Kim, H., Lee, S., Lin, J., Dujmic, Z., Kim, W., Talag, J., Zuccolo, A., Fan, C., Sebastian, A., Kramer, M., Spiegel, L., Nascimento, L., Zutavern, T., Miller, B., Ambroise, C., Muller, S., Spooner, W., Narechania, A., Ren, L., Wei, S., Kumari, S., Faga, B., Levy, M. J., McMahan, L., Van Buren, P., Vaughn, M. W., Ying, K., Yeh, C. T., Emrich, S. J., Jia, Y., Kalyanaraman, A., Hsia, A. P., Barbazuk, W. B., Baucom, R. S., Brutnell, T. P., Carpita, N. C., Chaparro, C., Chia, J. M., Deragon, J. M., Estill, J. C., Fu, Y., Jeddeloh, J. A., Han, Y., Lee, H., Li, P., Lisch, D. R., Liu, S., Liu, Z., Nagel, D. H., McCann, M. C., SanMiguel, P., Myers, A. M., Nettleton, D., Nguyen, J., Penning, B. W., Ponnala, L., Schneider, K. L., Schwartz, D. C., Sharma, A., Soderlund, C., Springer, N. M., Sun, Q., Wang, H., Waterman, M., Westerman, R., Wolfgruber, T. K., Yang, L., Yu, Y., Zhang, L., Zhou, S., Zhu, Q., Bennetzen, J. L., Dawe, R. K., Jiang, J., Jiang, N., Presting, G. G., Wessler, S. R., Aluru, S., Martienssen, R. A., Clifton, S. W., McCombie, W. R., Wing, R. A. and Wilson, R. K. 2009. The B73 maize genome: complexity, diversity, and dynamics. *Science* 326(5956), 1112-5.

Shan, Q. W., Wang, Y. P., Li, J. and Gao, C. X. 2014. Genome editing in rice and wheat using the CRISPR/Cas system. *Nature Protocols* 9(10), 2395-410.

Sharma, R., Liang, Y., Lee, M. Y., Pidatala, V. R., Mortimer, J. C. and Scheller, H. V. 2020. Agrobacterium-mediated transient transformation of sorghum leaves for accelerating functional genomics and genome editing studies. *BMC Research Notes* 13(1), 116.

Smith, J., Grizot, S., Arnould, S., Duclert, A., Epinat, J. C., Chames, P., Prieto, J., Redondo, P., Blanco, F. J., Bravo, J., Montoya, G., Pâques, F. and Duchateau, P. 2006. A combinatorial approach to create artificial homing endonucleases cleaving chosen sequences. *Nucleic Acids Research* 34(22), e149.

Soda, N., Verma, L. and Giri, J. 2018. CRISPR-Cas9 based plant genome editing: significance, opportunities and recent advances. *Plant Physiology and Biochemistry: PPB* 131, 2-11.

Song, G. Y., Jia, M. L., Chen, K., Kong, X., Khattak, B., Xie, C., Li, A. and Mao, L. 2016. CRISPR/Cas9: a powerful tool for crop genome editing. *The Crop Journal* 4(2), 75-82.

Tang, G. L., Galili, G. and Zhuang, X. 2007. RNAi and microRNA: breakthrough technologies for the improvement of plant nutritional value and metabolic engineering. *Metabolomics* 3(3), 357-69.

Taylor, J. R. N. and Schussler, L. 1986. The protein compositions of the different anatomical parts of sorghum grain. *Journal of Cereal Science* 4(4), 361–9.

Vance, P. N. 1976. The relationship of several plant characters with grain yield in sorghum and their use in estimating grain loss through pest activity. *Australian Journal of Experimental Agriculture* 16(78), 129–34.

Venkatesh Bhat, B., Balakrishna, D., Madhu, P. and Patil, J. V. 2015. Chapter10: Genetic engineering for novel traits, in R. Madhusudhana et al. (Eds), *Sorghum Molecular Breeding*, pp. 217–26. Cambridge, UK: Cambridge University Press.

Vouillot, L., Thelie, A. and Pollet, N. 2015. Comparison of T7E1 and surveyor mismatch cleavage assays to detect mutations triggered by engineered nucleases. *G3* 5(3), 407–15.

Voytas, D. F. and Gao, C. X. 2014. Precision genome engineering and agriculture: opportunities and regulatory challenges. *PLoS Biology* 12(6), e1001877.

Weaver, C. A., Hamaker, B. R. and Axtell, J. D. 1998. Discovery of grain sorghum germ plasm with high uncooked and cooked in vitro protein digestibilities. *Cereal Chemistry Journal* 75(5), 665–70.

Weber, E., Engler, C., Gruetzner, R., Werner, S. and Marillonnet, S. 2011. A modular cloning system for standardized assembly of multigene constructs. *PLoS ONE* 6(2), e16765.

Wu, Y., Li, X., Xiang, W., Zhu, C., Lin, Z., Wu, Y., Li, J., Pandravada, S., Ridder, D. D., Bai, G., Wang, M. L., Trick, H. N., Bean, S. R., Tuinstra, M. R., Tesso, T. T. and Yu, J. 2012. Presence of tannins in sorghum grains is conditioned by different natural alleles of Tannin1. *Proceedings of the National Academy of Sciences of the United States of America* 109(26), 10281–6.

Wyman, C. and Kanaar, R. 2006. DNA double-strand break repair: all's well that ends well. *Annual Review of Genetics* 40, 363–83.

Xing, H. L., Dong, L., Wang, Z. P., Zhang, H. Y., Han, C. Y., Liu, B., Wang, X. C. and Chen, Q. J. 2014. A CRISPR/Cas9 toolkit for multiplex genome editing in plants. *BMC Plant Biology* 14, 327.

Zafar, S. A., Zaidi, S. S., Gaba, Y., Singla-Pareek, S. L., Dhankher, O. P., Li, X., Mansoor, S. and Pareek, A. 2020. Engineering abiotic stress tolerance via CRISPR-Cas mediated genome editing. *Journal of Experimental Botany* 71(2), 470–9.

Zhao, Z. Y., Cai, T., Tagliani, L., Miller, M., Wang, N., Pang, H., Rudert, M., Schroeder, S., Hondred, D., Seltzer, J. and Pierce, D. 2000. Agrobacterium-mediated sorghum transformation. *Plant Molecular Biology* 44(6), 789–98.

Chapter 15

CRISPR/Cas9-mediated genome editing in *Brassica*

Cheng Dai, Xia Tian and Chaozhi Ma, Huazhong Agricultural University, China

1 Introduction

Brassica plants are composed of many species, and varieties, which are widely cultivated for food, edible oil, animal feed, and biodiesel, such as cabbage, cauliflower, broccoli, Chinese cabbage, mustard, radish, rapeseed, and so on. Most agronomic traits controlled by breeding are quantitatively inherited, being controlled by many genes, and strongly environmentally influenced. Crop improvement by breeding has been largely based on genetic diversity accumulated through a long history of natural mutations.

In the past few decades, mutagenesis was introduced by physical, chemical, and specific DNA sequence insertion (γ radiation, ethyl methanesulfonate (EMS), T-DNA, and transposon insertion) (Hua et al., 2019). All these random mutagenesis approaches generate large numbers of mutations throughout the plant genome (https://mvd.iaea.org). Mutagenesis-based breeding has produced more than 3000 commercial varieties of food crops (Hua et al., 2019). The process of screening is time-consuming and labor-intensive, because the initial mutagenesis must be followed by the screening of large populations to identify mutants with desirable properties, especially for polyploid crops (Parry et al., 2009). Additionally, health and environmental concerns are associated with genetic transformation. Development of new cultivars, such as oilseed, using traditional cross-breeding strategies, is time-consuming and complicated because they are allotetraploids (Mason and Snowdon, 2016; Smith et al., 2008).

http://dx.doi.org/10.19103/AS.2020.0082.23

Therefore, new breeding technologies that can introduce one or a few traits into an elite background would appear useful for developing new cultivars.

Compared to the imprecise random mutagenesis methods, targeted genome modification in plants is first reported in the 1980s (Paszkowski et al., 1988). In this study, the foreign DNA is recombined within overlapping parts of the protein-coding region, resulting in the formation of an active gene in the host chromosome after transformation (Paszkowski et al., 1988). However, this technology remained far from practical breeding applications at the beginning, waiting for improvement in efficiency. Thus far, three genome-editing tools have been well developed: Zinc-Finger Nucleases (ZFNs), Transcription Activator-Like Effector Nucleases (TALENs) and Clustered Regularly Interspaced Palindromic Repeat (CRISPR)-associated protein 9 system (CRISPR/Cas9) (Chen and Gao, 2014). Some of the principles for these tools are similar, such as generating site-specific double-strand breaks (DSBs) in the genome followed by error-prone DNA repair. There are two primary endogenous DNA repair pathways, namely homology-dependent repair (HDR) and non-homologous end-joining (NHEJ) (Bortesi et al., 2016). NHEJ is the predominant mechanism in somatic plant cells, which frequently causes short insertions or deletions (indels) around the DSBs (Bortesi et al., 2016). If the targeted site is in the coding region, the introduced indels frequently generate frameshifts, resulting in gene disruption. The ZFNs, and TALENs recognize genomic target sites using protein motifs (zinc finger domain, or transcription activator), which can be combined with numerous effector domains to impact genomic structure and function, including nucleases, transcriptional activators and repressors, recombinases, transposases, DNA histone methyltransferases, and histone acetyltransferases (Gaj et al., 2013). However, the target design of ZFNs and TALENS relied on ribonucleotide complex formation, and they are challenging design and manipulate, especially for multiple edits (Gaj et al., 2013). The CRISPR/Cas system contained two important parts: the sgRNA part, using the sgRNA to target DNA sequence by base complementarity for specific target recognition and the Cas protein part, to induce double-strand break. Besides gene deletions, CRISPR/Cas9 is useful for inserting specific DNA fragment into target sites and specifically altering the transcriptional activity of genes by fusing transcriptional activation or repression domains to an inactivated Cas9 (Li et al., 2013). Recently, other Cas proteins were isolated from diverse bacteria, and could recognize different PAM region (Hua et al., 2019). Because of its simplicity and diversity, the CRISPR/Cas system is the preferred choice for plant genome editing.

The genome-editing technology, combined with traditional breeding, can accelerate the introduction of desired traits and greatly reduce costs (Hua et al., 2019). In addition, the genetic elements required for genome editing, such as Cas9 and sgRNAs, can be removed from the genome through genetic crosses or following segregation in the progeny, which makes genome-edited plants free of exogenous DNA (Yang et al., 2017), and the mutations were stable inheritable. To date, genome editing has been mostly applied to improve agronomy traits such as crop yield, quality, and stress resistance (Hua et al., 2019).

2 Gene disruption in *Brassica* species

Rapeseed (*Brassica napus*, AACC, 2n = 38) originated from a spontaneous hybridization between *Brassica rapa* (AA, 2n = 20) and *Brassica oleracea* (CC, 2n = 18). In *B. napus*, a majority of genes have multiple paralogues with high similarities, hindering gene-functional studies. The gene disruption mutants of *Brassica napus* have been generated by CRISPR/Cas9 toolbox (Table 1). For example, rapeseed architecture traits such as plant height and branch numbers can be efficiently affected by editing the *BnaRGAs* and *BnaMAX1s* genes, respectively (Yang et al., 2017; Zheng et al., 2020). Conventional breeding methods have been used to manipulate these traits because natural allelic variants exist for *BnaRGAs* and *BnaMAX1s*, but the CRISPR-based approach is much faster and far less labor-intensive. Yellow-seeded rapeseed produces higher oil, higher protein, and lower fiber contents than brown-seeded types, which is a desirable trait with great potential for improving seed quality in *Brassica* crops. Unfortunately, no natural or induced yellow seed germplasms have been found in *Brassica napus*. CRISPR/Cas9 has been recently used to knock out basic helix-loop-helix (bHLH) genes BnTT8s, which dramatically decreases the anthocyanin concentration in seed coat without greatly affecting yield (Zhai et al., 2020). Editing four paralogues of *BnaARF2*s can produce large seeds and increase the yield per plants (Tang et al., 2018).

Besides rapeseed, there are few reports about genome editing in *B. rapa* and *B. oleracea* (Table 1). FLC and FRI are the two major flowering repressors in plant species; knocking out BrFLC and BrFRI in Chinese cabbage by CRISPR/Cas9 promotes flowering (Murovec et al., 2018; Jeong et al., 2019). Same results are also reported in *B. oleracea* (Murovec et al., 2018). *BolC.GA4.a* is an orthologue of *Arabidopsis GA4* which encodes AtGA3OX1, the last enzyme in the biosynthesis of bioactive gibberellins. Genome editing *BolC.GA4.a* affects plant architecture and the pod valve margin (Lawrenson et al., 2015). Cabbage (*Brassica oleracea var. capitata*) is a biennial plant with strong self-incompatibility. Using the S-receptor kinase gene *BoSRK3* and the male-sterility-associated gene *BoMS1* as the target genes, multisite and multiple gene mutations were achieved using a construct with tandemly arrayed tRNA-sgRNA architecture to express multiple sgRNAs. The *BoSRK3* gene mutation suppressed self-incompatibility completely, converting the self-incompatible line into a self-compatible line. In addition, the *BoMS1* gene mutation produced

Table 1 A list of publications on genome editing in *Brassica* species

Species	Genome editing tool	Target gene	Gene function or phenotype	Reference
Brassica napus	CRISPR/Cas9	*BnaRGA*	Versatile phenotypes due to enhanced GA signaling, such as longer stems, etc.	Yang et al. (2017)
Brassica napus	CRISPR/Cas9	*BnaDA1*	Increase in organ size	Yang et al. (2017)
Brassica napus	CRISPR/Cas9	*BnaFUL*	Failure of dehiscence, premature rupture of the carpel valves	Yang et al. (2017)
Brassica napus	CRISPR/Cas9	*BnaALC*	Increase shatter resistance to avoid seed loss	Braatz et al. (2017)
Brassica napus	CRISPR/Cas9	*BnaCRT1a*	Reduce plant susceptibility to *Verticillium longisporum*	Probsting et al. (2020)
Brassica napus	CRISPR/Cas9	*BnaAP2*	Homeotic transition of sepals to carpels	Zhang et al. (2018)
Brassica napus	CRISPR/Cas9	*BnaCLV3*	Produce multilocular siliques with a higher number of seeds/silique	Yang et al. (2018)
Brassica napus	CRISPR/Cas9	*BnaTT8*	Yellow-seeded	Zhai et al. (2020)
Brassica napus	CRISPR/Cas9	*BnaFAD2*	Increase the content of oleic acid	Okuzaki et al. (2018)
Brassica napus	CRISPR/Cas9	*BnaIND*	Increase shatter resistance to avoid seed loss	Zhai et al. (2019)
Brassica napus	CRISPR/Cas9	*BnaJAG*	Produce cylindrical hollow callus, pseudoseeds, and false septum	Zaman et al. (2019)
Brassica napus	CRISPR/Cas9	*BnaLPAT*	Decrease the oil content	Zhang et al. (2019)
Brassica napus	CRISPR/Cas9	*BnaMAX1*	Improve plant architecture and increase the yield	Zheng et al. (2020)
Brassica napus	CRISPR/Cas9	*BnaSFAR*	Elevate seed oil content	Karunarathna et al. (2020)
Brassica napus	CRISPR/Cas9	*BnaWRKY11*	Enhance resistance to Sclerotinia	Sun et al. (2018)
Brassica napus	CRISPR/Cas9	*BnaWRKY70*	Enhance resistance to Sclerotinia	Sun et al. (2018)
Brassica napus	CRISPR/Cas9-CBE	*BnaALS*	Enhanced herbicide resistance	Wu et al. (2020)
Brassica napus	ZFN	*KASII*	Reduce total saturated fatty acid contents	Gupta et al. (2012)

Brassica rapa	CRISPR/Cas9	*BrFLC*	Promote flowering	Jeong et al. (2019)
Brassica rapa	CRISPR/Cas9	*BrFRI*	Promote flowering	Murovec et al. (2018)
Brassica rapa	CRISPR/Cas9	*BrPDS*	Exhibit albino phenotype	Murovec et al. (2018)
Brassica oleracea	TALEN and CRISPR/Cas9	*BoFRI*	Promote flowering	Murovec et al. (2018)
Brassica oleracea	CRISPR/Cas9	*BoPDS*	Exhibit albino phenotype	Ma et al. (2019), Murovec et al. (2018)
Brassica oleracea	CRISPR/Cas9	*BolC.GA4.a*	Reduce GA level	Lawrenson et al. (2015)
Brassica oleracea	CRISPR/Cas9	*BoSRK*	Suppressed self-incompatibility	Ma et al. (2019)
Brassica oleracea	CRISPR/Cas9	*BoMS1*	Produce male-sterile	Ma et al. (2019)

a completely male-sterile mutant, which was highly cross-compatible with its nonmutant isoline at the flowering stage as a result of a simultaneous *BoSRK3* gene mutation, enabling the economic propagation of the male-sterile line through bee-mediated cross-pollination (Ma et al., 2019).

3 Homologous recombination-based editing in *Brassica* species

Precise genome modification such as knock-in and gene replacement via homologous recombination (HR) is a powerful tool and widely used in many organisms. The gene targeting (GT) technology in mice won the Nobel Prize in Physiology & Medicine in 2007. However, GT is still a very challenging task in higher plant species, because of low frequency of homologous recombination. Some experiments reported that high concentrations of donor DNA can significantly improve gene-targeting efficiency via geminiviral replicons (Baltes et al., 2014; Wang et al., 2017). By adjusting the ratio of CRISPR construct and donor DNA, the ratio of HR events could also be increased (Li et al., 2018; Sun et al., 2016). In the company 'Cibus', researchers successfully applied the combination of single-stranded oligonucleotides (ssODN) and CRISPR/Cas9 to develop a herbicide-tolerant plant (Sauer et al., 2016). Using this technology, they have generated non-transgenic sulfonylurea (SU) herbicide-tolerant canola, providing an alternative weed control option for canola farmers globally (https://cibus.com/crops.php).

4 Case study: generating mutants by CRISPR/Cas9 toolbox

Agrobacterium-mediated transformation is the main and most successful method for the production of genetically modified plants. Here we provide a case study of generating rapeseed mutants by CRISPR/Cas9 toolbox (All data are from our previous study, Yang et al., 2017). To apply CRISPR/Cas9 in *B. napus*, the sgRNA-Cas9 vectors from a previous report were used (Xing et al., 2014). The sgRNAs were predicted by the online tool CRISPR-P (http://cbi.hzau.edu.cn/cgi-bin/CRISPR) (Lei et al., 2014). In one construct, two sgRNAs with a high score were selected and driven by *Arabidopsis* U6-26 and U6-29 promoter, whereas the expression of *Cas9* was driven by the *CaMV 35S* promoter. We planned to test a total of four *B. napus* genes including *BnaA9.RGA*, *BnaC9.RGA*, *BnaA6.RGA*, and *BnaC7.RGA* (Table 2). *BnaA9.RGA*, *BnaC9.RGA*, *BnaA6.RGA*, and *BnaC7.RGA* are paralogous genes of the *BnaRGA* family, orthologues of *Arabidopsis RGA* (*REPRESSOR OF GA1-3*) gene, acting as a master repressor in gibberellic signaling, and the loss-of-function mutation will cause phenotypes mimicking a high GA level (Silverstone et al., 1997; Sun and Gubler, 2004). To target one gene, two sgRNAs at gene-specific regions were designed to make

Table 2 Selected target genes in *B. napus* used to assess CRISPR/Cas9

Target gene family	Gene identifier	Expected knock-out phenotype
BnaRGA	BnaA09g18700D	Versatile phenotypes due to enhanced GA signaling, such as longer stems, etc.
	BnaC09g52270D	
	BnaA06g34810D	
	BnaC07g20900D	

the constructs. The goal was to delete the entire gene of multiple members of the gene family by using one pair of sgRNA at each end of the gene, and study single gene function. To target the paralogous genes of one gene family, two sgRNAs derived from conserved regions were designed. The goal was to delete the entire gene family by using one pair of sgRNA. Together we made five constructs, four of which target one single gene (Table 2) and one of which target the paralogues in one gene family (Table 2). Then all the constructs were transformed into oilseed callus.

First, we evaluated the mutation efficiency of CRISPR/Cas9 for single-gene targeted sgRNAs in *B. napus* T0 generation. Because the paralogous genes have high sequence similarity in the allotetraploid *B. napus*, we confirmed the specificity of the primers for genotyping by direct sequencing. A total of 108 Cas9-positive T0 transgenic lines for the four constructs were identified (Table 3). To accurately calculate the mutation rate, each target site was directly sequenced from the transgenic plants. We manually checked the sequencing chromatograms for each line. We concluded that mutations had been successfully introduced when the sequencing chromatograms indicated a nucleotide change. Among the sgRNAs, the mutation rate varied from 5.3% to 100.0% (Table 3). The average mutation rate was 66.3% (Table 3).

Table 3 Percentage of mutated plants in the T0 generation induced by targeted sgRNAs

Target gene	sgRNA	No. of plants examined	No. of plants with mutations	Mutation rate (%)
BnaA9.RGA	sgRNA1	30	24	80.0
	sgRNA2		21	70.0
BnaC9.RGA	sgRNA1	29	29	100.0
	sgRNA2		29	100.0
BnaA6.RGA	sgRNA1	30	19	63.3
	sgRNA2		21	70.0
BnaC7.RGA	sgRNA1	19	1	5.3
	sgRNA2		8	42.1
Total (average)		108		(66.3)

Simultaneously mutating paralogous genes in one gene family is important in allotetraploids, such as *B. napus*. To test the efficiency of simultaneous gene mutations in *B. napus*, two sgRNAs targeting the conserved sequences among the four paralogues of *BnaRGAs*, and a total of 29 Cas9-positive transgenic lines were created (Table 4). The mutation frequency at each target site for sgRNA-mediate multiple-gene targeted mutagenesis ranged from 27.6% to 96.6% (Table 2), which was similar to sgRNA-mediated single-gene targeted mutagenesis. The double, triple and quadruple mutants were identified in T0 generation, and the relative rate was about 3.4%, 6.9% and 86.2%, respectively.

Next, the sequencing data from all the target sites was combined and analyzed to determine the mutation types and frequencies induced by CRISPR/Cas9 in *B. napus*. In total, the sequencing results of 238 PCR amplicons were analyzed by the decoding website DSDecode (http://dsdecode.scgene.com/). Part of the results was further confirmed by TA cloning and sequencing. Several types of mutations were observed: deletions, insertions, substitutions, and combined mutations (i.e., more than one mutation type in one allele). Among all the types of mutations, 58.4% were deletions, 36.1% were insertions, 3.4% were combined mutations, and 2.1% were substitutions (Fig. 1a and b). The mutations were predominantly short nucleotide changes (≤3 bp) (48.3%) (Fig. 1c), a majority of which (37.8%) were one-nucleotide insertions (Fig. 1b). The length of deletions ranged from one bp to hundreds of bp, and 9.2% of mutations exhibited a >100-bp deletion. The longest deletion was 152 bp. These long fragment deletions are caused by the simultaneous repairing of two DSBs generated by two sgRNAs in one construct in our system.

There are two alleles for each gene, both of which might be mutated by CRISPR/Cas9. Thus, CRISPR/Cas9 could produce five genotypes: homozygote (the two alleles have the same mutation) (L31, *BnaA6.RGA* and *BnaC7.RGA* sgRNA2 target site), bi-allele (the two alleles have different mutations) (L31, *BnaA6.RGA* sgRNA1 target site, and *BnaC9.RGA* sgRNA2 target site), heterozygote (only one allele is mutated) (L31, *BnaA9.RGA* sgRNA2 target site), chimera (more than two different mutations exist) (L31, *BnaA9.RGA* and *BnaC9.RGA* sgRNA1 target site), and WT-type (no mutation) (L43) (Fig. 2). Based on the genotyping results, most of them contained at least one mutation site. However, no mutations were found in the transgenic lines, even when the Cas9 and sgRNAs existed.

When 3 bp or multiples of 3 bp are deleted from an exon without affecting the reading frame, a few amino acids may be deleted from the middle of a protein, which may alter its biochemical properties. *BnaA6.RGA* encodes a DELLA protein that negatively regulates GA signaling in many plant species. It is reported that deletion of the conserved TVHYNP domain resulted in a dwarf phenotype and insensitivity to exogenous GA (Silverstone et al., 1997; Sun and Gubler, 2004). Among the transgenic lines of *BnaA6.RGA-sgRNA*, two T0 lines

Table 4 Percentage of mutated sites in the T0 generation nduced by multiple-gene-targeted sgRNAs

Gene family	Target gene	sgRNA	No. of plants examined	No. of plants with mutations	Mutation rate (%)	Homozygous mutations	% of homozygous mutations	% of plants mutated at different target genes
BnaRGA	*BnA9.RGA*	sgRNA1	29	28	96.6	3	10.3	One gene is mutated: 0
		sgRNA2		8	27.6	0	0	Two genes are mutated: 3.4
	BnC9.RGA	sgRNA1	29	25	86.2	0	0	Three genes are mutated: 6.9
		sgRNA2		9	31.0	0	0	Four genes are mutated: 86.2
	BnA6.RGA	sgRNA1	29	28	96.6	6	20.7	
		sgRNA2		10	34.5	0	0	
	BnC7.RGA	sgRNA1	29	27	93.1	5	17.2	
		sgRNA2		12	41.4	0	0	
Total			29					

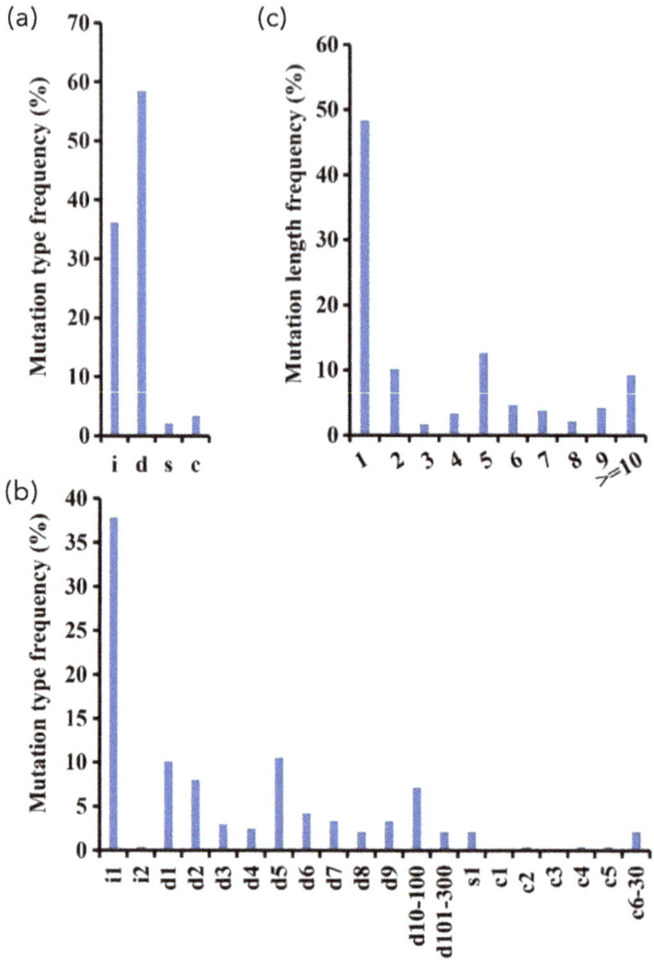

Figure 1 Frequency of CRISPR/Cas9-induced mutation types. (a, b) Frequency of each mutation type for all the mutations induced by the five constructs in the T0 generation. i: insertion; d: deletion; s: substitution; c: combined mutation. d#, number of base pairs (bp) deleted from the target site; i#, number of bp inserted at the target site, c#, combined mutations. (c) Frequency of different mutation lengths regardless of the mutation types using the data from (a).

(T0-L4 and T0-L6) were dwarf (Fig. 3a), and the phenotype was inherited in the T1 generation (Fig. 3b). According to the sequencing results, the T0-L4 and T0-L6 plants have a 6- and 12-nt deletion at the target sites, respectively (Fig. 3c), that causes a 2- and 4-amino acid deletion in the TVHYNP domain (Fig. 3c).

Because a majority of genes have multiple paralogues with functional redundancy in *B. napus* (allotetraploid), single gene knockouts probably

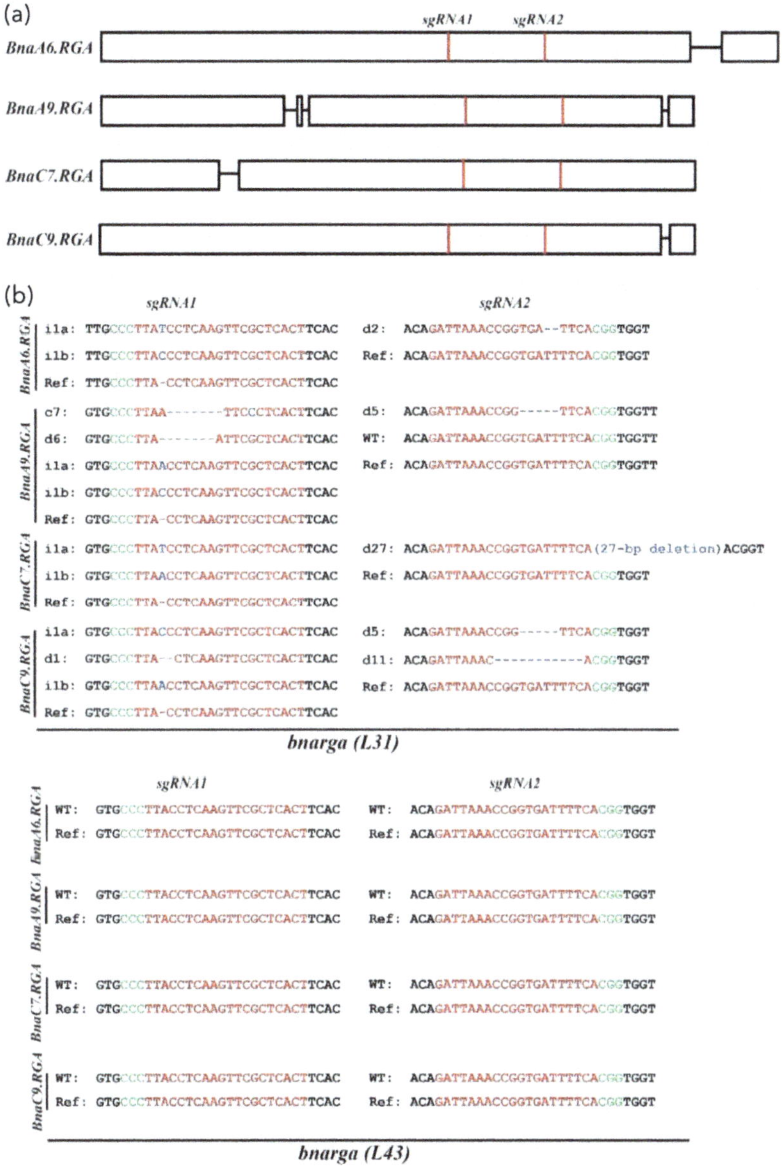

Figure 2 The four knockout paralogues of BnaRGA. (a) The four paralogues of BnaRGA gene model includes exons (box) separated by introns (represented by the solid line). The red vertical line in the gene model indicates the sgRNA target sites. (b) Genotype of the bnarga mutant in the T0 generation. The protospacer adjacent motif (PAM) region is marked in green. The sgRNA is denoted by red. d#, # of bp deleted from a target site; i#, # of bp inserted at target site; i#a, same number of insertion at one site; i#b, same number of insertion of different nucleotides at the same site; c#, combined mutation.

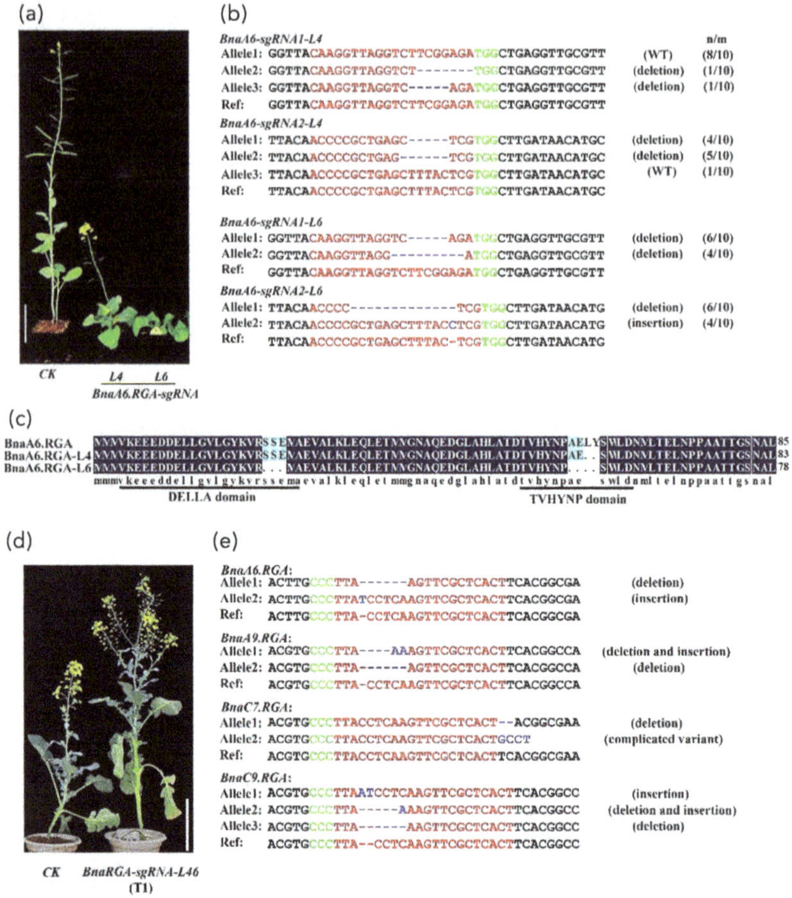

Figure 3 Phenotype and genotype of a BnaA6.RGA-sgRNA single mutant (T0) and a BnaRGA-sgRNA quadruple mutant (T1). (a) Morphology of BnaA6.RGA-sgRNA transgenic plants at the same age. L4 and L6: two individual T0 transgenic lines. CK: BnaA6.DA1-sgRNA (L10). Bar = 15 cm. (b) Genotype of BnaA6.RGA-sgRNA lines (L4 and L6) in the T0 generation. (c) Amino acid sequence alignment of the target sites derived from WT and two BnaA6.RGA-sgRNA transgenic lines (L4 and L6). The DELLA and TVHYNP domains are underlined. The amino acid deletion in the DELLA domain is caused by sgRNA1. The amino acid deletion in the TVHYNP domain is caused by sgRNA2. (d) Image showing the morphology of the quadruple mutant induced by BnaRGA-sgRNA (L46-T1) at the same age as the control. CK: WT plant. Bar = 15 cm. (e) Genotype of the quadruple mutant isolated from BnaRGA-sgRNA-L46 in the T1 generation. The PAM sequence is indicated with green. The sgRNA is indicated with red. The mutation sites are indicated with blue. For the term n/m, m indicates the number of clones examined, and n indicates the number of clones showing the indicated genotype.

Table 5 Percentage of T1 plants with or without the Cas9 transgene

Target gene	Line	No. of plants examined	Cas9 positive plants	Cas9 negative rate (%)	Cas9 negative plants
BnaRGA	L27	30	26	13.30	4
BnaRGA	L33	30	29	3.30	1
BnaRGA	L38	28	25	10.70	3
BnaRGA	L40	38	37	2.60	1
BnaRGA	L44	30	26	13.33	4
BnaRGA	L46	30	28	6.70	2
BnaA6.RGA	L4	24	11	54.20	13
BnaA6.RGA	L5	30	19	36.70	11
BnaA6.RGA	L6	10	8	20.00	2
Total		250	209		41 (16.4%)

will not show obvious phenotypes. In this study, the T1 quadruple mutant of *BnaRGA* grew longer stems than control plants (Fig. 3d). Genotyping results indicated that the four paralogues of *BnaRGA* could be efficiently knocked out with homozygous or bi-allelic mutations in the T0 generation (Fig. 3e). Together, these data make the case that CRISPR/Cas9 has great advantages for gene function studies in *B. napus*.

Genetically manipulated materials without T-DNA insertions are largely favored for crop improvement because they are classified as non-GMO plants in the United States, Germany, and Sweden (https://geneticliteracyproject.org/2017/04/20/crispr-conundrum-line-gmos-natural-crops-genes-edited/). Indeed, the T-DNA insertions were lost in T1 generation (Table 5).

5 Conclusion and future trends

In summary, we demonstrated that the CRISPR/Cas9 is a highly efficient tool for genome editing in *B. napus*. We found that sgRNAs derived from conserved sequences could simultaneously induce homozygous or bi-allelic mutations at multiple loci. Therefore, CRISPR/Cas9 provides a powerful tool for studying gene function in oilseed and the fastest method for the breeding of polyploid crops. Thus, CRISPR/Cas9 is useful for both basic and applied research in *B. napus*.

6 Where to look for further information

Genome-editing toolkits have been used to effect precise modifications in many crops, which have had a great influence on basic research as well as crop improvement. One of the advantages of these technologies is that the transgenes cassettes used to induce genetic modification can be easily removed from the genome by genetic segregation, which make the resulting plants typically indistinguishable from naturally occurring genetic variants. Scientists, policymakers, and regulatory authorities have extensively debated the regulation of genome-edited plants. The European Court of Justice ruled on the interpretation of the definition of the term 'genetically modified organism' in the GMO Directive 2001/18/EC on 25 July 2018, ruling that all organisms produced by genome editing are subject to the legal framework applicable to release, placing on the market, labeling, and traceability of GMOs.

However, individual countries also provide their own guidance. For example, German science academies and the German Research Foundation published statement 'Toward a scientifically justified, differentiated regulation of genome edited plants in the EU':

- https://www.leopoldina.org/uploads/tx_leopublication/2019_Stellungnahme_Genomeditierte_Pflanzen_web_02.pdf.

This statement concludes that, 'due to the mounting divergence between scientific progress and legal standardization, the primarily process-based European regulatory approach is no longer justifiable' and that 'potential risks can only emanate from the modified traits of the organism as a product of the breeding process, and not from the process itself'. Though there is still a long way to let the public accept the genome-editing crops, it is a first step for ruling genome-editing crops are non-GMO. The genome-editing technologies offer a great chance for improving crops and ensuring global food security, and we should grasp this opportunity to increase *Brassica* crop productivity.

7 References

Baltes, N. J., Gil-Humanes, J., Cermak, T., Atkins, P. A. and Voytas, D. F. (2014). DNA replicons for plant genome engineering. *Plant Cell* 26(1), 151–163.

Bortesi, L., Zhu, C., Zischewski, J., Perez, L., Bassie, L., Nadi, R., Forni, G., Lade, S. B., Soto, E., Jin, X., Medina, V., Villorbina, G., Munoz, P., Farre, G., Fischer, R., Twyman, R. M., Capell, T., Christou, P. and Schillberg, S. (2016). Patterns of CRISPR/Cas9 activity in plants, animals and microbes. *Plant Biotechnol. J.* 14(12), 2203–2216.

Braatz, J., Harloff, H. J., Mascher, M., Stein, N., Himmelbach, A. and Jung, C. (2017). CRISPR-Cas9 targeted mutagenesis leads to simultaneous modification of different homoeologous gene copies in polyploid oilseed rape (Brassica napus). *Plant Physiol.* 174(2), 935-942.

Chen, K. and Gao, C. (2014). Targeted genome modification technologies and their applications in crop improvements. *Plant Cell Rep.* 33(4), 575–583.

Gaj, T., Gersbach, C. A. and Barbas, C. F., 3rd. (2013). ZFN, TALEN, and CRISPR/Cas-based methods for genome engineering. *Trends Biotechnol.* 31, 397–405, doi:10.1016/j.tibtech.2013.04.004.

Gupta, M., DeKelver, R. C., Palta, A., Clifford, C., Gopalan, S., Miller, J. C., Novak, S., Desloover, D., Gachotte, D., Connell, J., Flook, J., Patterson, T., Robbins, K., Rebar, E. J., Gregory, P. D., Urnov, F. D. and Petolino, J. F. (2012). Transcriptional activation of Brassica napus beta-ketoacyl-ACP synthase II with an engineered zinc finger protein transcription factor. *Plant Biotechnol. J.* 10(7), 783–791.

Hua, K., Zhang, J., Botella, J. R., Ma, C., Kong, F., Liu, B. and Zhu, J. K. (2019). Perspectives on the application of genome-editing technologies in crop breeding. *Mol. Plant* 12(8), 1047–1059.

Jeong, S. Y., Ahn, H., Ryu, J., Oh, Y., Sivanandhan, G., Won, K. H., Park, Y. D., Kim, J. S., Kim, H., Lim, Y. P. and Kim, S. G. (2019). Generation of earlyflowering Chinese cabbage (Brassica rapa spp.pekinensis) through CRISPR/Cas9-mediated genome editing. *Plant Biotechnol. Rep.* 13, 491–499.

Karunarathna, N. L., Wang, H., Harloff, H. J., Jiang, L. and Jung, C. (2020). Elevating seed oil content in a polyploid crop by induced mutations in seed fatty acid reducer genes. *Plant Biotechnol. J.* https://doi.org/10.1111/pbi.13381.

Lawrenson, T., Shorinola, O., Stacey, N., Li, C., Ostergaard, L., Patron, N., Uauy, C. and Harwood, W. (2015). Induction of targeted, heritable mutations in barley and Brassica oleracea using RNA-guided Cas9 nuclease. *Genome Biol.* 16, 258.

Lei, Y., Lu, L., Liu, H. Y., Li, S., Xing, F. and Chen, L. L. (2014). CRISPR-P: a web tool for synthetic single-guide RNA design of CRISPR-system in plants. *Mol. Plant* 7(9), 1494–1496.

Li, J., Zhang, X., Sun, Y., Zhang, J., Du, W., Guo, X., Li, S., Zhao, Y. and Xia, L. (2018). Efficient allelic replacement in rice by gene editing: A case study of the NRT1.1B gene. *J. Integr. Plant Biol.* 60(7), 536–540.

Li, J. F., Norville, J. E., Aach, J., McCormack, M., Zhang, D., Bush, J., Church, G. M. and Sheen, J. (2013) Multiplex and homologous recombination-mediated genome editing in Arabidopsis and Nicotiana benthamiana using guide RNA and Cas9. *Nat. Biotechnol.* 31, 688–691. doi:10.1038/nbt.2654.

Ma, C., Zhu, C., Zheng, M., Liu, M., Zhang, D., Liu, B., Li, Q., Si, J., Ren, X. and Song, H. (2019). CRISPR/Cas9-mediated multiple gene editing in Brassica oleracea var. capitata using the endogenous tRNA-processing system. *Hortic. Res.* 6, 20.

Mason, A. S. and Snowdon, R. J. (2016). Oilseed rape: learning about ancient and recent polyploid evolution from a recent crop species. *Plant Biol.* 18(6), 883–892.

Murovec, J., Gucek, K., Bohanec, B., Avbelj, M. and Jerala, R. (2018). DNA-free genome editing of Brassica oleracea and B. rapa protoplasts using CRISPR-Cas9 ribonucleoprotein complexes. *Front. Plant Sci.* 9, 1594.

Okuzaki, A., Ogawa, T., Koizuka, C., Kaneko, K., Inaba, M., Imamura, J. and Koizuka, N. (2018). CRISPR/Cas9-mediated genome editing of the fatty acid desaturase 2 gene in Brassica napus. *Plant Physiol. Biochem.* 131, 63–69.

Parry, M. A., Madgwick, P. J., Bayon, C., Tearall, K., Hernandez-Lopez, A., Baudo, M., Rakszegi, M., Hamada, W., Al-Yassin, A., Ouabbou, H., Labhilili, M. and Phillips, A. L. (2009). Mutation discovery for crop improvement. *J. Exp. Bot.* 60(10), 2817–2825.

Paszkowski, J., Baur, M., Bogucki, A. and Potrykus, I. (1988). Gene targeting in plants. *EMBO J.* 7(13), 4021–4026.

Probsting, M., Schenke, D., Hossain, R., Hader, C., Thurau, T., Wighardt, L., Schuster, A., Zhou, Z., Ye, W., Rietz, S., Leckband, G. and Cai, D. (2020). Loss-of-function of CRT1a (calreticulin) reduces plant susceptibility to Verticillium longisporum in both Arabidopsis thaliana and oilseed rape (Brassica napus). *Plant Biotechnol. J.* 10.1111/pbi.13394

Sauer, N. J., Narvaez-Vasquez, J., Mozoruk, J., Miller, R. B., Warburg, Z. J., Woodward, M. J., Mihiret, Y. A., Lincoln, T. A., Segami, R. E., Sanders, S. L., Walker, K. A., Beetham, P. R., Schopke, C. R. and Gocal, G. F. (2016). Oligonucleotide-mediated genome editing provides precision and function to engineered nucleases and antibiotics in plants. *Plant Physiol.* 170(4), 1917–1928.

Silverstone, A. L., Mak, P. Y., Martinez, E. C. and Sun, T. P. (1997). The new RGA locus encodes a negative regulator of gibberellin response in Arabidopsis thaliana. *Genetics* 146(3), 1087–1099.

Smith, J. S. C., Hussain, T., Jones, E. S., Graham, G., Podlich, D., Wall, S. and Williams, M. (2008). Use of doubled haploids in maize breeding: implications for intellectual property protection and genetic diversity in hybrid crops. *Mol. Breed.* 22(1), 51–59.

Sun, Q., Lin, L., Liu, D., Wu, D., Fang, Y., Wu, J. and Wang, Y. (2018). CRISPR/Cas9-mediated multiplex genome editing of the BnWRKY11 and BnWRKY70 genes in Brassica napus L. *Int. J. Mol. Sci.* 19(9), 2716.

Sun, T. P. and Gubler, F. (2004). Molecular mechanism of gibberellin signaling in plants. *Annu. Rev. Plant Biol.* 55, 197–223.

Sun, Y., Zhang, X., Wu, C., He, Y., Ma, Y., Hou, H., Guo, X., Du, W., Zhao, Y. and Xia, L. (2016). Engineering herbicide-resistant rice plants through CRISPR/Cas9-mediated homologous recombination of acetolactate synthase. *Mol. Plant* 9(4), 628–631.

Tang, T., Yu, X., Yang, H., Gao, Q., Ji, H., Wang, Y., Yan, G., Peng, Y., Luo, H., Liu, K., Li, X., Ma, C., Kang, C. and Dai, C. (2018). Development and validation of an effective CRISPR/Cas9 vector for efficiently isolating positive transformants and transgene-free mutants in a Wide Range of plant species. *Front. Plant Sci.* 9, 1533.

Wang, M., Lu, Y., Botella, J. R., Mao, Y., Hua, K. and Zhu, J. K. (2017). Gene targeting by homology-directed repair in rice using a geminivirus-based CRISPR/Cas9 system. *Mol. Plant* 10(7), 1007–1010.

Wu, J., Chen, C., Xian, G., Liu, D., Lin, L., Yin, S., Sun, Q., Fang, Y., Zhang, H. and Wang, Y. (2020). Engineering herbicide-resistant oilseed rape by CRISPR/Cas9-mediated cytosine base-editing. *Plant Biotechnol. J.* 18(9), 1857–1859.

Xing, H. L., Dong, L., Wang, Z. P., Zhang, H. Y., Han, C. Y., Liu, B., Wang, X. C. and Chen, Q. J. (2014). A CRISPR/Cas9 toolkit for multiplex genome editing in plants. *BMC Plant Biol.* 14, 327.

Yang, H., Wu, J. J., Tang, T., Liu, K. D. and Dai, C. (2017). CRISPR/Cas9-mediated genome editing efficiently creates specific mutations at multiple loci using one sgRNA in Brassica napus. *Sci. Rep.* 7(1), 7489.

Yang, Y., Zhu, K., Li, H., Han, S., Meng, Q., Khan, S. U., Fan, C., Xie, K. and Zhou, Y. (2018). Precise editing of CLAVATA genes in Brassica napus L. regulates multilocular silique development. *Plant Biotechnol. J.* 16(7), 1322–1335.

Zaman, Q. U., Chu, W., Hao, M., Shi, Y., Sun, M., Sang, S. F., Mei, D., Cheng, H., Liu, J., Li, C. and Hu, Q. (2019). CRISPR/Cas9-mediated multiplex genome editing of JAGGED gene in *Brassica napus* L. *Biomolecules* 9(11), 725.

Zhai, Y., Cai, S., Hu, L., Yang, Y., Amoo, O., Fan, C. and Zhou, Y. (2019). CRISPR/Cas9-mediated genome editing reveals differences in the contribution of INDEHISCENT homologues to pod shatter resistance in Brassica napus L. *Theor. Appl. Genet.* 132(7), 2111–2123.

Zhai, Y., Yu, K., Cai, S., Hu, L., Amoo, O., Xu, L., Yang, Y., Ma, B., Jiao, Y., Zhang, C., Khan, M. H. U., Khan, S. U., Fan, C. and Zhou, Y. (2020). Targeted mutagenesis of BnTT8 homologs controls yellow seed coat development for effective oil production in Brassica napus L. *Plant Biotechnol. J.* 18(5), 1153–1168.

Zhang, K., Nie, L., Cheng, Q., Yin, Y., Chen, K., Qi, F., Zou, D., Liu, H., Zhao, W., Wang, B. and Li, M. (2019). Effective editing for lysophosphatidic acid acyltransferase 2/5 in allotetraploid rapeseed (Brassica napus L.) using CRISPR-Cas9 system. *Biotechnol. Biofuels* 12, 225.

Zhang, Y., Huang, S., Wang, X., Liu, J., Guo, X., Mu, J., Tian, J. and Wang, X. (2018). Defective APETALA2 genes lead to sepal modification in Brassica Crops. *Front. Plant Sci.* 9, 367.

Zheng, M., Zhang, L., Tang, M., Liu, J., Liu, H., Yang, H., Fan, S., Terzaghi, W., Wang, H. and Hua, W. (2020). Knockout of two BnaMAX1 homologs by CRISPR/Cas9-targeted mutagenesis improves plant architecture and increases yield in rapeseed (Brassica napus L.). *Plant Biotechnol. J.* 18(3), 644–654.

Chapter 16

Genome editing of tomatoes and other Solanaceae

Joyce Van Eck, The Boyce Thompson Institute and Cornell University, USA

1 Introduction

The *Solanaceae* family is a remarkably diverse group that contains approximately 3000–4000 plant species that comprise nearly 90 genera (Knapp et al. 2004). In addition to their inherent diversity, their uses are also greatly diverse including utility as food crops, medicinal applications, and ornamentals. The family members *Solanum lycopersicum* (tomato), *S. tuberosum* (potato), *S. melongena* (eggplant), and *Capsicum annuum* (pepper) are some of the most important food crops with potato being the fourth most important worldwide. In addition to the reliance on tomato, potato, eggplant, and pepper for culinary uses, they also have significant economic value. Using tomato as an example of economic value, there are two classifications of the fruit, genotypes that are used as a source for the fresh fruit market and those used for processing products such as pasta sauce, ketchup, and salsa. In 2017, the combined US dollar value for fresh market and processed tomatoes was 1.7 billion (https://www.agmrc.org/commodities-products/vegetables/tomatoes).

http://dx.doi.org/10.19103/AS.2020.0082.25

As for medicinal purposes, the *Solanaceae* family contains a number of species that have played a vital role as sources of extracts for treating various afflictions through traditional medicine for centuries. These medicinally important species belongs to genera that include: *Atropa*, *Capsicum*, *Datura*, *Hyoscymus*, *Physalis*, and *Withania* (Shah et al. 2013; Dar et al. 2015; Zhang and Tong 2016). Extracts from different plant parts have been used to treat inflammatory conditions, headaches, and asthma. Steroidal compounds, specifically withanolides, from species in the *Withania* and *Physalis* genera, have been reported to be effective for their anti-inflammatory properties and growth inhibition of cancer cell lines (Sang-Ngern et al. 2016; Wang et al. 2019b).

Being a diverse family, it is not surprising that there is a vast array of floral morphologies, flower sizes and colors exhibited by members of the *Solanaceae*. This assortment of flower characteristics provided the foundation for developing several species for ornamental purposes through selection and breeding. The *Solanaceae* utilized as ornamentals include *Petunia hybrida* (petunia), *Salpiglosis sinuata* (velvet tongue), *Schizanthus pinnatus* (butterfly flower), and *Calibrachoa* spp (Knapp et al. 2004; Facciuto et al. 2006). Of these species, petunia is the most widely grown in the United States.

In addition to their food, medicinal, and ornamental value, members of the *Solanaceae* family have also been utilized as model plant species for genetic, biological, and gene function studies. *Nicotiana tabacum*, *Nicotiana benthamiana*, tomato, and petunia have all be exploited as models for investigations on various plant mechanisms related to plant development, pest resistance, fruit biology, and secondary metabolites. *N. benthamiana*, in particular, has played a large role in plant virology studies because of its susceptibility to a wide array of viruses (Goodin et al. 2008). In part, this susceptibility has also made it a valuable tool for the application of Virus-Induced Gene-Silencing (VIGS) for gene function studies (Senthil-Kumar and Mysore 2014).

With the *Solanaceae* representing economically important food crops and species used as models for research, it is not unexpected that several of the earliest reports of gene editing in plants involved this family. Resources that allowed gene editing of these species to be realized include publicly available genome sequences and methodologies for genetic engineering, both important prerequisites for application of editing (Tomato Genome Consortium 2012; Van Eck et al. 2019). Applications of editing approaches for solanaceous plant species include reports related to Transcription Activator-Like Effector Nucleases (TALENs) and Clustered Regularly Interspaced Short Palindromic Repeats (CRISPR)/CRISPR-Associated Proteins (CRISPR/Cas). The majority of the reports for solanaceous species represent the creation of indels to disrupt gene expression to understand function; however, there are reports of CRISPR/Cas-based technologies such as base editing and DNA-free gene editing through delivery of ribonucleoprotein complexes, which will be discussed in this chapter. One of the earliest reports that demonstrated CRISPR/Cas editing in plants was by Nekrasov et al. (2013). They utilized a transient expression assay by agroinfiltration into *N. benthamiana* leaves with *Agrobacterium tumefaciens* and generated stable transgenic lines where the *PHYTOENE DESATURASE* (*PDS*) gene was targeted. Resultant plant material that contained edits in *PDS* exhibited the expected phenotype of a white, bleached appearance.

While there have been reports of modification through TALENs, the focus here will be on CRISPR/Cas in solanaceous species. For some of these reports, namely the earliest ones, they were simply proof-of-concept experiments to demonstrate that editing was feasible in *Solanaceae* utilized as model species. Whereas in the reports that followed, the work presented demonstrated that application of gene editing was effective at modification of traits such as plant architecture, disease resistance, fruit and potato tuber characteristics. The ability to precisely modify traits to create germplasm variability where needed will help advance breeding programs to play a role in addressing the most critical deficiencies that hinder progress in agricultural production and food security.

2 Gene editing proof-of-concept investigations

The greatest number of reports by far of gene editing in solanaceous species is through the CRISPR/Cas9 approach to generate indels. The earliest reports could be classified as proof-of-concept experiments because they simply demonstrate the feasibility of the editing technology for the *Solanaceae* (Table 1). The first reports of successful gene editing in tomato were both in 2014 (Ron et al. 2014; Brooks et al. 2014). One of these reports demonstrated successful editing in hairy roots recovered after transformation by *Agrobacterium rhizogenes* (Ron et al. 2014). Whereas the report by Brooks et al. showed that a plant phenotype could be affected by recreation of the tomato wiry mutant in stably edited lines generated by *Agrobacterium tumefaciens*-mediated transformation. Since those first reports, gene editing to create indels to disrupt gene expression in tomato has become routine and utilized for biological studies (Reem and Van Eck 2019). In addition to tomato, proof-of-concept of editing in *Nicotiana tabacum* was demonstrated by targeting the *PHYTOENE DESATURASE* (*NtPDS*) gene and a PDR-type transporter, *NtPDR6*, which is involved in strigolactone transport (Gao et al. 2015). Results revealed a mutation frequency of 81.8% with a gRNA designed to *NtPDS* and 87.5% for an *NtPDR6* gRNA in the primary transgenic lines (T0).

To demonstrate CRISPR/Cas9 editing in potato, *S. tuberosum*, Butler et al. (2015) designed a gRNA to mutate the potato *ACETOLACTATE SYNTHASE 1* (*StALS1*) gene in diploid and tetraploid lines. Recovery of edited lines for

Table 1 CRISPR/Cas editing approaches used for solanaceous species

Editing approach	Intended Purpose	Species	Reference
Indels	Proof of concept	Tomato	Brooks et al. (2014)
			Ron et al. (2014)
		Potato	Butler et al. (2015)
		Tobacco	Gao et al. (2015)
	Disease tolerance	Tomato	Nekrasov et al. (2017)
			Ortigosa et al. (2019)
			Tashkandi et al. (2018)
			Zhang et al. (2020)
		Nicotiana benthamiana	Ali et al. (2019)
	Enhanced lycopene	Tomato	Li et al. (2018c)
	Pink-colored fruit		Deng et al. (2018)
	Carotenoid genes		D'Ambrosio et al. (2018)
	Ripening-related traits		Ito et al. (2015, 2017)
			Li et al. (2018a)
			Wang et al. (2019a)
	Fruit size		Xu et al. (2015)
			Rodriguez-Leal et al. (2017)
	Tuber starch content	Potato	Tuncel et al. (2019)
	Plant architecture	Tomato	Soyk et al. (2017)
	Inflorescence development	Tomato	Soyk et al. (2019)
	de novo domestication	Tomato wild species	Li et al. (2018b)
			Zsogon et al. (2017)
		Groundcherry	Lemmon et al. (2018)
Geminivirus replicon	Proof of concept	Tomato	Cermak et al. (2017)
Allele swapping	Extended shelf life	Tomato	Yu et al. (2017)
DNA insertion	Proof of concept	Tomato	Danilo et al. (2019)
		Nicotiana tabacum	Hirohata et al. (2019)
Multiplex editing	Proof of concept	Tomato	Hashimoto et al. (2018)
CRISPR/Cas RNP	Proof of concept	*Nicotiana attenuata*	Woo et al. (2015)
		Potato	Andersson et al. (2018)
		Petunia	Subburaj et al. (2019)
Base Editing	Proof of concept	Tomato	Shimatani et al. (2017)
		Potato	Veillet et al. (2019)

each *A. tumefaciens* transformation experiment ranged from 3–60%. Mutations observed in the T0 generations persisted through to clonal and Cas9-free germline populations.

Following reports of successful gene editing by CRISPR/Cas9 in various organisms, technology was designed for simultaneously targeting multiple genes all within one gene editing construct, including designs developed for use in plants. For instance, there is a multiplex toolbox available for plants that provides a protocol and reagents to assemble genome editing constructs for dicots and monocots (Lowder et al. 2015). To assess mutation efficiencies of the multiplex editing approach in tomato, Hashimoto et al. (2018) used a multiplex CRISPR/Cas9 vector system that contained various promoters driving Cas9 expression along with different gRNA expression combinations. They found that the multiplex system was an efficient approach, specifically when Cas9 expression was driven by the promoter from the tomato *ELONGATION FACTOR-1α* (*SLEF1 α*) gene, which resulted in various types of mutations at a high efficiency. Based on their results, they recommended optimization of the CRISPR/Cas9 vector system for different plant species, primarily the selection of the promoter used for Cas9 expression.

3 DNA sequence insertion

Compared with the number of reports on modification of gene expression through the creation of indels with the CRISPR/Cas system, there are significantly fewer reports regarding the insertion of DNA segments or allele swapping in plants. These types of modifications are often referred to as knock-ins and they can range from a single nucleotide to larger amounts of DNA. The paucity of reports is due to the challenge of accomplishing knock-ins because they rely on homology directed repair (HDR) through homologous recombination (HR), which is less prevalent, less efficient in plants than the other mode of DNA repair, non-homologous end joining (NHEJ) (Steinert et al. 2016). HR efficiencies in plant cells are low ranging from 10^{-4} to 10^{-6} (Voytas 2013) making it difficult to obtain a high level of DNA sequence insertion by HDR. While *A. tumefaciens*-mediated transformation was very effective for the creation of indels to affect gene expression, this method proved to be less efficient for knock-ins or allele swapping. Therefore, gene delivery methods based on polyethylene glycol (PEG)-mediated transformation of protoplasts and biolistics (gene gun) have been exploited for this gene editing approach.

While it is difficult to accomplish knock-ins, there are reports of HDR by exploiting HR in tomato and tobacco (Danilo et al. 2019; Hirohata et al. 2019) (Table 1). Danilo et al. (2019) demonstrated HDR by editing the tomato *ACETOLACTATE SYNTHASE 1* (*SlALS1*) gene. They utilized a gRNA to target *SlALS1* in combination with a synthetic DNA donor fragment that contained several amino acid modifications. The binary vector contained the gRNA, Cas9, and the synthetic donor DNA and was introduced by *A. tumefaciens*-mediated transformation into cotyledons. T0 plants were analyzed by PCR and sequencing to determine the level of editing in *SlALS*. Of the thirty-seven plants evaluated, 84% had edits in the gene and all HDR-related edits were transmitted to the next generation (T1). A similar approach exploiting *ALS* and a second gene was used to demonstrate the feasibility of HDR in *N. tabacum* (Hirohata et al. 2019). Cas9/gRNA expression and HDR donor cassettes contained in two different constructs were used to modify the tobacco *ALS* gene, *SuRB*, and the *ANTHOCYANIN 2* gene, *An2*, which is a MYB transcription factor involved in anthocyanin accumulation. HDR donors included a version of *SuRB* that contained the W568L mutation known to confer herbicide resistance (Townsend et al. 2009) and the *An2* driven by the strong, constitutive CaMV35S promoter. While the frequency of HR in *SuRB* was low in the T1 and T2 populations, no HR modifications of the endogenous *An2* were observed.

4 Application of CRISPR/Cas ribonucleoprotein complexes

The ability to accomplish transgene-free, often referred to as DNA-free, gene editing has the advantage that there is no integration of any DNA vector-related sequences. This in turn decreases the incidence of any unintended phenotypic effects, simplifies the regulatory approval process, and has the potential to improve consumer acceptance. For DNA-free editing, the mutation of a gene of interest is facilitated by delivery of a CRISPR/Cas-Ribonucleoprotein (RNP) complex that contains a gRNA complementary to that gene without the integration of any sequences (DeWitt et al. 2017). It is believed that this form of editing, especially in plants, will assuage consumer concerns and lessen regulatory requirements regarding genetic engineering of food crops being that there is no introduction of any foreign DNA (Wolter and Puchta 2017).

One obstacle that prevents this editing approach from being widely adopted in plants at this time is that *Agrobacterium*-mediated delivery of CRISPR/Cas-RNP components into plant cells has not been achieved. What has been shown to be the best delivery methods in plants, including members of the *Solanaceae*, is delivery of CRISPR/Cas-RNP complexes through transformation of protoplasts and biolistics, which has been used for crops such as maize (Svitashev et al. 2016; Andersson et al. 2018; Park et al. 2019). Both protoplast and biolistics pose some obstacles as compared to delivery by *Agrobacterium*-mediated methods. The isolation of protoplasts that can sustain division and result in plants is technically challenging and has not been reported for a wide number of species as have *Agrobacterium*-mediated methods. For biolistics, the main obstacle is having the appropriate target plant material where delivery of the DNA-coated particles can reach a sufficient number of cells and that those are cells with plant regenerability potential.

One of the earliest reports in plants described the delivery of a CRISPR/Cas-RNP complex by polyethylene glycol-(PEG) mediated transformation into *Nicotiana attenuata* protoplasts to target the *Allene Oxide Cyclase* (*AOC*) gene (Woo et al. 2015) (Table 1). Edits were detected in the expected region of the gene in plants regenerated from the protoplast and persisted into the next generations. PEG-mediated transformation was also used for CRISPR/Cas-RNP complex delivery into protoplasts of *Petunia x hybrida* and potato (Subburaj et al. 2016; Andersson et al. 2018). For petunia, six gRNAs were designed to target the *Petunia nitrate reductase* (*NR*) gene. Plants were not regenerated from the protoplasts; however, genomic DNA was isolated and assessed for mutations. The frequency of mutations ranged from 2.4% to 21% depending on the gRNA design. To test the efficacy in potato, Andersson et al. (2018) used gRNAs designed to mutate the *Granule Bound Starch Synthase* (*GBSS*) gene in the tetraploid potato cultivar, Kuras. Unlike the petunia report, plants were regenerated from the potato protoplasts. All four *GBSS* alleles were found to be mutated in 2–3% of the recovered plants, which resulted in a complete knockout of *GBSS* expression.

5 Base editing

The ability to precisely modify genes at the base pair level was first demonstrated in mammalian cell lines (Komor et al. 2016) and yeast (Nishida et al. 2016). Through the development of complexes that contained various deaminases and other design features such as the inclusion of a uracil DNA glycosylase inhibitor, methodologies became available for an irreversible conversion of C to T and A to G. Base editing allows the creation of single polymorphisms (SNPS) that not only result in precise modifications at the gene level, but can also affect phenotypes resulting in the ability to achieve desirable traits. Until recently, there were no reports of technologies available for the conversion of A to T. However, in late 2019 a report by Anzalone et al. (2019) described an approach they termed prime editing that broadens the options to achieve any of the possible twelve base conversions. Prime editing builds on the foundation of CRISPR; however, it uses a CAS9 variant called a nickase that cuts only one DNA strand, which is fused to a reverse transcriptase. To date, it has only been applied to human cell lines, but application in plants may soon be reported.

The first report of base editing of a solanaceous food crop was in, not surprisingly, tomato, and it centered on C to T conversion through the incorporation of a cytidine deaminase in the constructs (Shimatani et al. 2017) (Table 1). Using MicroTom as the test model, Shimatani et al. (2017) used gRNAs that were designed to target *DELLA* and *ETR1*, which are hormone-signaling regulators. The constructs were delivered into cells through *A. tumefaciens*-mediated transformation. Assessment of DNA from the T0 and T1 generations revealed an assortment of mutations that included conversions of C to T and C to G in addition to DNA substitutions and indels. Recently, they published a protocol chapter that outlines steps for achieving base editing in tomato (Shimatani et al. 2019).

A second study on C to T base conversion in tomato also included another important *Solanaceae* food crop, potato (Veillet et al. 2019). Their approach was to target *ALS* in both species with the intent of generating edited plant lines that were chlorsulfuron-resistant. The plants were assessed for conversions at various cytosine locations within the tomato *ALS* locus. They found that greater than 80% of the edits at the C_{-14} location were C to T conversions. As for potato, the editing was expected to be more challenging because it is a tetraploid. Twenty plants were recovered of which fifteen contained indels. Analysis of base conversion in the five remaining plants revealed that base conversion was mainly C to G and C to T; however, C to A was observed, but at a lesser frequency.

6 Modification of gene function for improved disease tolerance

Strengthening the resilience of crops to withstand infection by plant pathogens has been the focus of many plant breeding programs both in the public and private sectors. Various approaches for gene editing add to the toolbox of options for building tolerance or resistance to invasion by pathogens. Inactivation of susceptibility genes is one viable approach for application of CRISPR/Cas by creating indels in such genes and rendering them inactive thus shutting down access by pathogens (Zaidi et al. 2018). The strategy of susceptibility gene inactivation was applied to build resistance to a fungal pathogen of tomato (Nekrasov et al. 2017) (Table 1). The tomato (*S. lycopersicum*) *MILDEW RESISTANT LOCUS* (*SlMlo*) confers susceptibility to *Oidium neolycopersici*, which causes powdery mildew disease. Through gene targeting and disruption of *SlMlo* expression, Nekrasov et al. (2017) generated non-transgenic T1 progeny that were found to be fully resistant to *O. neolycopersici*. This material, designated Tomelo, maintained characteristics similar to wild type plants as related to plant growth and fruit size.

Two recent publications reported investigations of the feasibility of using CRISPR/Cas9-mediated editing for studies related to *Pseudomonas syringae* pv. *tomato* (*Pto*) DC3000, the causal agent of tomato bacterial speck disease (Ortigosa et al. 2019; Zhang et al. 2020). Ortigosa et al. (2019) took the approach of blocking the control of stomata opening by coronatine (COR), which is produced by the bacteria, because this opening promotes colonization on the leaves. Being that stomata response to COR in *Arabidopsis* requires a co-receptor, they targeted the tomato ortholog (*SLJAZ2*) of the *Arabidopsis* COR co-receptor, *AtJAZ2*. As a result, they generated edited tomato lines that were resistant to *Pto* DC3000.

For their studies related to *P. syringae* pv. *tomato*, Zhang et al. (2020) developed an approach for using CRISPR/Cas9 to confirm the functional relevance amongst a large number of candidate immunity-associated genes. To facilitate the efficient screening of 63 immunity-associated genes, they first evaluated gRNA efficiency by transient expression in tomato leaves followed by the generation of stable transgenic lines with '*Agrobacterium* pools' of 2–4 CRISPR/Cas9 constructs. This approach allowed them to systematically generate mutant lines as compared to a gene-by-gene approach for targeting. Currently, they are advancing the lines of interest to the next generation to develop homozygous lines that are Cas9 free. The progeny will be used to determine if mutations in any of the immunity-associated genes confer an immune response to *P. syringae* pv. *tomato*.

In addition to bacterial and fungal pathogens, *Solanaceae* family members are also susceptible to viruses that can be equally devastating and for some viruses genetic control can be difficult to obtain. Based on two reports related to the Tomato yellow leaf curl virus, application of CRISPR/Cas shows promise as an approach for conferring resistance. Using *N. benthamiana* as a model for engineering resistance to geminiviruses, Ali et al. (2019) provided an outline of a strategy and a protocol that can be used as a foundation for using CRISPR/Cas9 to directly degrade the viral genome. In a separate study on *N. benthamiana* and tomato, immunity to Tomato yellow leaf curl virus (TYLCV), also a geminivirus, was accomplished through gRNAs designed to target either the coat protein or replicase (Tashkandi et al. 2018).

7 Editing to affect fruit and tuber quality

Availability of genomic resources, editing protocols, and *A. tumefaciens*-mediated transformation methodologies for tomato and potato made application of editing readily achievable for gene function studies (Xu et al. 2011; Tomato Genome Consortium 2012; Nadakuduti et al. 2019; Reem and Van Eck 2019; Van Eck et al. 2019). Many of the reports in tomato have focused on fruit characteristics such as size, ripening, and color. Color can also affect nutritional quality of fruits and vegetables; therefore, modification of color can also enhance health benefits from these foods.

Carotenoids are one class of compounds in plants that can impart color to flowers, fruits, and vegetables. To demonstrate the efficacy of manipulating the carotenoid biosynthesis pathway in tomato flowers and fruit with CRISPR/Cas9, D'Ambrosio et al. (2018) designed gRNAs for two genes in the pathway, *PHYTOENE SYNTHASE 1* (*PSY1*) and *BETA CAROTENE HYDROXYLASE 2* (*CrtR-b2*). The majority of the mutations observed were a single deletion or insertion near the cleavage site. Significant effects were observed on the carotenoid profiles based on HPLC analysis. Mutated lines with significant reductions in carotenoids exhibited white flowers and yellow fruit, thus demonstrating the feasibility of using CRISPR to manipulate the carotenoid pathway.

Lycopene is one of the major carotenoids produced in tomato contributing to its red characteristic color and imparts health benefits with reports related to decreased incidence of cancer, cardiovascular disease, and eye disorders (Perveen et al. 2015). Efforts to increase lycopene content have been accomplished through breeding, overexpression and gene editing of key carotenoid biosynthesis genes. A study where multi-gene targeting was applied to key parts of the carotenoid pathway resulted in a significant increase, approximately fivefold, in lycopene content (Li et al. 2018c). Mutations that resulted were primarily small indels less than 10 bp; however, there were two additional types of mutations, a large DNA deletion and an inversion. Further characterization showed that there was stable inheritance and no off-target sites were detected.

Additional approaches that resulted in color modification did not involve manipulation of the carotenoid pathway. Pink-fruited tomatoes were covered within one year by targeting the transcription factor *SlMYB12*, a key player in flavonoid accumulation (Deng et al. 2018). Whereas, purple-fruited MicroTom tomatoes were recovered by a different approach than disrupting gene expression. In a proof-of-concept investigation, a geminivirus replicon was used to demonstrate the feasibility of accomplishing homologous recombination through gene editing technology at a high efficiency in plants. Cermak et al. (2015) took advantage of an endogenous anthocyanin gene, *Anthocyanin Mutant 1* (*ANT1*) by swapping in the strong, constitutive Cauliflower Mosaic Virus 35S promoter upstream to drive expression. This resulted in anthocyanin accumulation in plant tissues including the fruit that persisted into the next generation.

Understanding the ripening process is essential for providing consistently good-quality fruits to consumers. Tomato has been extensively used as a model for elucidating the interplay of pathways and related gene networks involved in fruit ripening. CRISPR/Cas9 was used to further the understanding of the role of a fleshy fruit ripening-associated gene, *RIN*, and its mutant counterpart, *rin* (Ito et al. 2015, 2017). Based on earlier reports, *rin* was believed to be a null mutation. However, through their studies, Ito et al. (2015, 2017) concluded that *rin* is actually a gain-of-function mutation and that while *RIN* is indeed involved in ripening it does not play a role in the very early stages. In a different approach related to control of ripening, Li et al. (2018a) focused on long non-coding RNAs (lncRNAs), which are non-protein-coding transcripts that are longer than 200 nucleotides. They identified and cloned an lncRNA from tomato, designated *lncRNA1459*. To verify its biological function, they generated loss-of-function mutants using CRISPR/Cas9 mutagenesis. Fruit ripening in these mutants was significantly inhibited.

Fruit softening is also a part of the ripening process. Characteristics that can prevent damage of softened fruit as a result of transportation and storage are important considerations for maintaining good-quality fruits with a long shelf life. As with ripening, the softening process has been well studied and numerous factors have been identified including the role of pectins, which are cell wall polysaccharides. To add to the knowledge on factors that control fruit softening, CRISPR/Cas9 was used to induce mutations in three genes that encode for pectin degrading enzymes in tomato (Wang et al. 2019a). gRNAs were designed to target the following genes: *Pectate Lyase* (*PL*), *polygalacturonase 2a* (*PG2a*), and *Beta-galactanase* (*TBG4*). Reverse Transcription Quantitative PCR (RTqPCR) was used to verify that gene expression of the target genes was indeed reduced in CRISPR-mutated lines. Firmer fruits were only recovered from lines where expression of *PL* was disrupted.

Extended shelf life of fruits and vegetables is a very desirable characteristic; however, retaining high quality during storage can be a challenge. If good quality could be maintained for longer periods of time, there is the potential to reduce the amount of waste generated from discarded fruits and vegetables. In their investigation regarding the feasibility of using CRISPR approaches to confer long-shelf life, Yu et al. (2017) focused on replacement of the tomato ALCOBACA (*ALC)* gene with its recessive counterpart *alc* via HDR. While the replacement efficiency was low at approximately 8% in the T0, they did observe improvement of storage time in the T1 generation.

In addition to the fruit characteristics already described, size is also an important trait when it comes to quality. The mature fruit of the closest wild progenitor (*Solanum pimpinellifolium*), to what consumers know as a tomato today, is the size of a large pea. Through spontaneous mutations, selection, and breeding there is an incredible array of tomato shapes and sizes. Numerous reports describe the genetic control over fruit size, including control by the tomato *CLAVATA 3* (*SlClV3*) gene. To investigate the mechanism that underlies the control by *SlCLV3* on meristem and fruit size, Xu et al. (2015) designed two gRNAs to target the coding sequence. Analysis revealed that that disruption of *SlCLV3* expression caused a lack of arabinose residues on the CLV protein as a result of associated arabinosyltransferases being affected. Targeting various regions of the *SlCLV3* promoter has also resulted in a range of fruit trait variation (Rodriguez-Leal et al. 2017).

For potato tubers, one of the main quality traits of interest is the amount and type of starch. In addition to cooking and processing quality, assessment of starch content also takes into consideration its digestibility in the human gut. With recent reports of increases in obesity and diabetes, namely type II, the focus of potato tuber quality improvement has moved toward modifications that would increase the content of less digestible resistant starch, which is primarily composed of amylose and long-unbranched amylopectin. To determine the feasibility of applying gene editing to enhance resistant starch content, Tuncel et al. (2019) used CRISPR/Cas9 to induce mutations in two starch-branching enzymes genes, *SBE1* and *SBE2*, in the potato cultivar, Desiree. Three different constructs were designed with gRNAs that individually targeted each gene in two of the constructs and in the third the gRNAs were combined for multiplex editing. The level of decreased expression of these enzymes resulted in an array of starch phenotypes and effects on starch granule development with large decreases being required for reduced amylopectin branching.

8 Plant growth and development modifications

Reports on utilization of CRISPR/Cas gene editing to decipher the intricacies of genes and networks involved in growth and development in tomato far outnumber any other member of the *Solanaceae*. These reports include but are not limited to elucidation of control of plant and inflorescence architecture, inflorescence development, and stem cell proliferation (Soyk et al. 2017, 2019; Rodriguez-Leal et al. 2019). One of the first reports that exploited gene editing to affect plant architecture in tomato was by Soyk et al. (2017) where they targeted both the *SELF-PRUNING* (*SP*) gene, which is an anti-florigen, and its homolog *SP5G*, which is responsible for daylength sensitivity. Plant architecture and flowering time were dramatically affected resulting in compact plants that had fruit ready for harvest two weeks earlier than the wild-type counterpart.

CRISPR/Cas editing proved to be a valuable tool to tease apart how cryptic or unforeseen variation that naturally exists has affected inflorescence development and yield in tomato (Soyk et al. 2019). This was revealed by focusing efforts on the interplay of mutations in *JOINTLESS2* (*J2*) and *ENHANCER OF J2* (*EJ2*). Results demonstrated how selection and breeding over time for jointless tomatoes and inflorescence phenotypes affected harvestability and yield.

The homeostasis of stem cell proliferation in plant meristems underlies continued growth and development. The intricacies of the genetic control over stem cell homeostasis has been well-studied in various plant species, including tomato. Two key genes involved in this process in tomato are *SlCLV3* and *CLV3/EMBRYO-SURROUNDING REGION* (*SlCLE*). To tease apart the mechanisms of the interplay, Rodriguez-Leal et al. (2019) focused on *SlCLE9*, which in their earlier study was shown to be involved in a compensatory type interaction with *SlCLV3* (Xu et al. 2015). They were able to demonstrate that there is indeed a compensation relationship between *SlCLV3* and *SlCLE9*. This compensation was identified by analysis of a double mutant designated *slclv3* slcle9 in parallel with plant lines that were a combination of overexpression and mutants generated by CRISPR/Cas9 editing of additional meristem-related tomato genes.

9 Case study: application of gene editing for de novo domestication

When we think of domesticated crop species, we envision wild progenitor species that in some instances hardly resemble their modern cultivated counterparts. Transformation of these wild species into viable food crops spanned hundreds to thousands of years and occurred through natural, chance mutations combined with human intervention to select for ideal characteristics that were further improved by plant breeding. 'Domestication syndrome' has been used to denote what is considered a collection of classic traits that were modified in wild species (Hammer 1984; Doebley et al. 2006). Characteristics that distinguish modern cultivars from their wild progenitors include increased fruit and grain size, a more moderate plant architecture, higher productivity, and daylength neutrality.

While domestication of wild and underutilized plant species can take hundreds to thousands of years, three recent reports describe the potential utilization of CRISPR/Cas editing for what is referred to as *de novo* domestication to target some of the classic domestication traits to accelerate improvement. These simultaneously reported studies centered on the *Solanaceae* family, two wild species of tomato and a member of the *Physalis* genus. One report presented a case study for *de novo* domestication of *Solanum galapagense*, a wild species of tomato that is very salt tolerant, but has pea-sized fruit and a large, sprawling growth habit (Zsogon et al. 2017). A list of target traits was presented that when modified through various approaches of gene editing beyond the creation of mutations through indels had the potential to transform this wild species into what would be considered a food crop ideotype. The second report utilizing a wild species centered on *S. pimpinellifolium*, the closest wild ancestor to the cultivated tomato (Li et al. 2018b). Li et al. (2018b) edited four tomato domestication-related genes within one CRISPR/Cas9 construct through the use of the Csy4 multi-gRNA approach (Cermak et al. 2017). An array of mutations was observed in the target genes that affected characteristics such as plant architecture and fruit size.

The third solanaceous species where CRISPR/Cas9 was used to affect domestication-related traits was *Physalis pruinosa* (groundcherry), which has been referred to as an orphan species because it has wild characteristics and has not be subjected to any trait improvement (Lemmon et al. 2018). Groundcherry has an unmanageable growth habit with fruit that is small (marble-sized), mildly sweet, and surrounded by a papery husk (Wolff et al. 2013). The fruit has uneven ripening and drops to the ground even before fully ripe. To demonstrate the feasibility of applying CRSIPR/Cas9 to rapidly improve characteristics that would be viewed as undesirable from an agricultural production standpoint, Lemmon et al. (2018) focused on modification of three genes. To increase the marble-sized fruit, two gRNAs were designed to target the *CLAVATA1* (*CLV1*) gene. Fruit harvested from the modified lines had a 24% increase in size as compared to the unmodified wild type. As for plant growth habit, based on knowledge about *SP* and *SP5G* in tomato as described above, these were obvious choices for modifying the plant growth habit. Plants recovered from editing *SP* were significantly more compact, whereas, *SP5G* targeting resulted somewhat more compact; however, there was an increase in the number of fruits along each groundcherry stem.

10 Conclusion and future trends

With the human population increasing along with decreasing land availability for expanding crop production, crop improvement plays a vital role in food security now and in the future. In addition to needing robust crops, reliable food productions systems require a multifaceted coordination of aspects related to farming practices, politics, land management, social considerations, economics, and others. As for crop improvement, availability of gene editing approaches has provided the ability to fine-tune changes precisely at the DNA sequence level, which broadens the possibilities for achieving desirable plant traits. Like an artist with a palette of colors, plant scientists now have a larger set of options for trait improvement, such as strengthening resilience to abiotic and biotic stresses, improving yield, enhancing nutritional quality, and transforming undesirable characteristics of underutilized crops to allow their adoption into agricultural production possible.

Plant biodiversity is also an important part of securing an ample food supply. The *Solanaceae* family is comprised of an amazing array of plant species that are the source of not only some of the most relied upon foods worldwide, but also a source of incredible diversity. This wealth of diversity can be tapped for crop improvement programs and provides a source of species whose potential as food crops is yet to be realized. As described in this report, gene editing has already had an impact on the precise modification of traits such as those related to disease resistance, fruit and tuber quality, and plant growth habit of several solanaceous species. Equally exciting are the reports on *de novo* domestication that lay the groundwork and inspiration for turning underutilized, orphan species into ones that could contribute to diversifying our food crops beyond those that we rely on so heavily. While these new editing technologies have sparked considerable excitement and expand the options for trait modification, it is unrealistic to believe that any one tool alone is going help reach the desired goals. More likely, it is going to be a combination of various crop improvement approaches that will be necessary to obtain the ideal plant characteristics needed for producing a reliable and healthy food supply.

11 Where to look for further information

11.1 Further reading

For information on the mechanism and application of CRISPR/Cas gene editing approaches, there are various online resources and several excellent videos that illustrate the process.

- https://www.livescience.com/58790-crispr-explained.html.
- http://theconversation.com/what-is-crispr-gene-editing-and-how-does-it-work-84591.
- https://www.google.com/search?ei=lZWMXsOQEMKxytMPwPOyoAU&q=how+to+crispr+a+gene+youtube&oq=how+to+crispr+a+gene+you+tu&gs_lcp=CgZwc3ktYWIQARgBMgUIIRCgATIECCEQCjoECAAQRzoGCAAQFhAeOgUIABDNAjoICCEQFhAdEB5KCggXEgYxMi0xMDFKCAgYEgQxMi05UM1GWJxYYLxpaABwAngAgAG6AYgB0QaSAQMwLjeYAQCgAQGqAQdnd3Mtd2l6&sclient=psy-ab#kpvalbx=_ipaMXp37NMGlytMPqoOeSA11.
- https://www.youtube.com/watch?v=MnYppmstxls.
- https://www.youtube.com/watch?v=4YKFw2KZA5o.

In addition to the online resources, there are a number of publications including book chapters that provide information on the CRISPR/Cas system and application in plants.

- An early publication on CRISPR/Cas9 outlines how it works and future applications by the pioneers of its use for editing. The new frontier of genome engineering with CRISPR-Cas9, J. Doudna and E. Charpentier, 2014, Science 346:1-9.
- Provides background on gene editing approaches and application in plants including tomato and potato, Plant Genome Editing with CRISPR Systems, 2019, edited by Y. Qi.
- Protocols for transformation methodologies of various plant species including tomato and potato, 2019, Transgenic Plants, edited by S. Kumar, P. Barone, M. Smith.
- Review article that provides background on different CRISPR/Cas approaches and application in plants. The CRISPR toolkit for genome editing and beyond, M. Adli, 2019, Nature Communications, 9:1-13.
- A very comprehensive review article regarding application of CRISPR/Cas approaches to affect traits in plants. Applications of CRISPR-Cas in agriculture and plant biotechnology, Zhu et al., 2020, Nature Reviews doi: 10.1038/s41580-020-00288-9.

11.2 Key journals and conferences

- SIVB (Society for In Vitro Biology) is well attended by members of the private and public research sectors.
- Solanaceae conferences are held annually in different countries around the world.Information can be found on the Solanaceae Genomics Network (solgenomics.net).

- PAG (Plant and Animal Genome) is well attended by members of the Solanaceae community and there is a Solanaceae workshop held each year as part of the conference program.

11.3 Major international research projects

- Information on Solanaceae-related international projects can be found at the Solanaceae Genomics Network, solgenomics.net.
- The SolCode Project links genes to traits in the Solanaceae, https://solgenomics.net/projects/solcode/.
- G2P-Sol connects genetic resources, genomes, and phenotypes of solanaceous crops, http://www.g2p-sol.eu/.

12 References

Ali, Z., Zaidi, S. S., Tashkandi, M. and Mahfouz, M. M. (2019). A simplified method to engineer CRISPR/Cas9-mediated geminivirus resistance in plants. In: Kobayashi, K. and Nishiguchi, M. (Eds) *Antiviral Resistance in Plants. Methods in Molecular Biology*. Humana Press Inc., New York, 2028:167–183.

Andersson, M., Turesson, H., Olsson, N., Fält, A. S., Ohlsson, P., Gonzalez, M. N., Samuelsson, M. and Hofvander, P. (2018). Genome editing in potato via CRISPR-Cas9 ribonucleoprotein delivery. *Physiol. Plant.* 164(4):378–384.

Anzalone, A. V., Randolph, P. B., Davis, J. R., Sousa, A. A., Koblan, L. W., Levy, J. M., Chen, P. J., Wilson, C., Newby, G. A., Raguram, A. and Liu, D. R. (2019). Search-and-replace genome editing without double-strand breaks or donor DNA. *Nature* 576(7785):149–157.

Brooks, C., Nekrasov, V., Lipppman, Z. B. and Van Eck, J. (2014). Efficient gene editing in tomato in the first generation using the clustered regularly interspaced short palindromic repeats/CRISPR-associated9 system. *Plant Physiol.* 166(3):1292–1297.

Butler, N. M., Atkins, P. A., Voytas, D. F. and Douches, D. S. (2015). Generation and inheritance of targeted mutations in potato (*Solanum tuberosum* L.) using the CRISPR/Cas system. *PLoS ONE* 10(12):e0144591.

Cermak, T., Baltes, N. J., Cegan, R., Zhang, Y. and Voytas, D. F. (2015). High-frequency, precise modification of the tomato genome. *Genome Biol.* 16:232.

Cermak, T., Curtin, S. J., Gil-Humanes, J., Čegan, R., Kono, T. J. Y., Konečná, E., Belanto, J. J., Starker, C. G., Mathre, J. W., Greenstein, R. L. and Voytas, D. F. (2017). A multipurpose toolkit to enable advanced genome engineering in plants. *Plant Cell* 29(6):1196–1217.

D'Ambrosio, C., Stigliani, A. L. and Giorio, G. (2018). CRISPR/Cas9 editing of carotenoid genes in tomato. *Transgenic Res.* 27(4):367–378.

Danilo, B., Perrot, L., Mara, K., Botton, E., Nogué, F. and Mazier, M. (2019). Efficient and transgene-free gene targeting using *Agrobacterium*-mediated delivery of the CRISPR/Cas9 system in tomato. *Plant Cell Rep.* 38(4):459–462.

Dar, N. J., Hamid, A. and Ahmad, M. (2015). Pharmacologic overview of *Withania somnifera*, the Indian ginseng. *Cell. Mol. Life Sci.* 72(23):4445–4460.

Deng, L., Wang, H., Sun, C., Li, Q., Jiang, H., Du, M., Li, C. B. and Li, C. (2018). Efficient generation of pink-fruited tomatoes using CRISPR/Cas9 system. *J. Genet. Genomics* 45(1):51–54.

DeWitt, M. A., Corn, J. E. and Carroll, D. (2017). Genome editing via delivery of Cas9 ribonucleoprotein. *Methods* 121-122:9-15.

Doebley, J. F., Gaut, B. S. and Smith, B. D. (2006). The molecular genetics of crop domestication. *Cell* 127(7):1309-1321.

Facciuto, G., Pannunzio, M. J., Coviella, M. A., Soto, S., Hagiwara, J. C. and Borja, M. (2006). Characterization of the ornamental value of *Calibrachoa* spp. native to Argentina. *Acta Hortic.* 714:37-42.

Gao, J., Wang, G., Ma, S., Xie, X., Wu, X., Zhang, X., Wu, Y., Zhao, P. and Xia, Q. (2015). CRISPR/Cas9-mediated targeted mutagenesis in *Nicotiana tabacum*. *Plant Mol. Biol.* 87(1-2):99-110.

Goodin, M. M., Zaitlin, D., Naidu, R. A. and Lommel, S. A. (2008). *Nicotiana benthamiana*: its history and future as a model for plant-pathogen interactions. *Mol. Plant Microbe Interact.* 21(8):1015-1026.

Hammer, K. (1984). Das Domestikationssyndrom. *Die Kult.* 32(1):11-34.

Hashimoto, R., Ueta, R., Abe, C., Osakabe, Y. and Osakabe, K. (2018). Efficient multiplex genome editing induces precise, and self-ligated type mutations in tomato plants. *Front. Plant Sci.* 9:916.

Hirohata, A., Sato, I., Kaino, K., Iwata, Y., Koizumi, N. and Mishiba, K. I. (2019). CRISPR/Cas9-mediated homologous recombination in tobacco. *Plant Cell Rep.* 38(4):463-473.

Ito, Y., Nishizawa-Yokoi, A., Endo, M., Mikami, M., Shima, Y., Nakamura, N., Kotake-Nara, E., Kawasaki, S. and Toki, S. (2017). Re-evaluation of the *rin* mutation and the role of *RIN* in the induction of tomato ripening. *Nat. Plants* 3(11):866-874.

Ito, Y., Nishizawa-Yokoi, A., Endo, M., Mikami, M. and Toki, S. (2015). CRISPR/Cas9-mediated mutagenesis of the *RIN* locus that regulates tomato fruit ripening. *Biochem. Biophys. Res. Commun.* 467(1):76-82.

Knapp, S., Bohs, L., Nee, M. and Spooner, D. M. (2004). Solanaceae - a model for linking genomics with biodiversity. *Comp. Funct. Genomics* 5(3):285-291.

Komor, A. C., Kim, Y. B., Packer, M. S., Zuris, J. A. and Liu, D. R. (2016). Programmable editing of a target base in genomic DNA without double-stranded DNA cleavage. *Nature* 533(7603):420-424.

Lemmon, Z. H., Reem, N. T., Dalrymple, J., Soyk, S., Swartwood, K. E., Rodriguez-Leal, D., Van Eck, J. and Lippman, Z. B. (2018). Rapid improvement of domestication traits in an orphan crop by genome editing. *Nat. Plants* 4(10):766-770.

Li, R., Fu, D., Zhu, B., Luo, Y. and Zhu, H. (2018a). CRISPR/Cas9-mediated mutagenesis of lncRNA1459 alters tomato fruit ripening. *Plant J.* 94(3):513-524.

Li, T., Yang, X., Yu, Y., Si, X., Zhai, X., Zhang, H., Dong, W., Gao, C. and Xu, C. (2018b). Domestication of wild tomato is accelerated by genome editing. *Nat. Biotechnol.* 36:1160-1163.

Li, X., Wang, Y., Chen, S., Tian, H., Fu, D., Zhu, B., Luo, Y. and Zhu, H. (2018c). Lycopene is enriched in tomato fruit by CRISPR/Cas9-mediated multiplex genome editing. *Front. Plant Sci.* 9:559.

Lowder, L. G., Zhang, D., Baltes, N. J., Paul, J. W. III, Tang, X., Zheng, X., Voytas, D. F., Hsieh, T.-F., Zhang, Y., and Qi, Y. (2015). A CRISPR/Cas9 toolbox for multiplexed plant genome editing and transcriptional regulation. *Plant Physiol.* 169:971-985.

Nadakuduti, S. S., Starker, C. G., Voytas, D. F., Buell, C. R. and Douches, D. S. (2019). Genome editing in potato with CRISPR/Cas9. In: Qi, Y. (Ed.) *Plant Genome Editing With CRISPR Systems: Methods and Protocols*. Humana Press, New York, NY 1917:183-201.

Nekrasov, V., Staskawicz, B., Weigel, D., Jones, J. D. and Kamoun, S. (2013). Targeted mutagenesis in the model plant *Nicotiana benthamiana* using Cas9 RNA-guided endonuclease. *Nat. Biotechnol.* 31(8):691-693.

Nekrasov, V., Wang, C., Win, J., Lanz, C., Weigel, D. and Kamoun, S. (2017). Rapid generation of a transgene-free powdery mildew resistant tomato by genome deletion. *Sci. Rep.* 7(1):482.

Nishida, K., Arazoe, T., Yachie, N., Banno, S., Kakimoto, M., Tabata, M., Mochizuki, M., Miyabe, A., Araki, M., Hara, K. Y., Shimatani, Z. and Kondo, A. (2016). Targeted nucleotide editing using hybrid prokaryotic and vertebrate adaptive immune systems. *Science* 353(6305):1-8.

Ortigosa, A., Gimenez-Ibanez, S., Leonhardt, N. and Solano, R. (2019). Design of a bacterial speck resistant tomato by CRISPR/Cas9-mediated editing of SlJAZ2. *Plant Biotechnol. J.* 17(3):665-673.

Park, J., Choi, S., Park, S., Yoon, J., Park, A. Y. and Choe, S. (2019). DNA-free genome editing via ribonucleoprotein (RNP) delivery of CRISPR/Cas in lettuce. In: Qi, Y. (Ed.) *Plant Genome Editing with CRISPR Systems: Methods and Protocols*. Humana Press, New York, NY 1917:337-354.

Perveen, R., Suleria, H. A., Anjum, F. M., Butt, M. S., Pasha, I. and Ahmad, S. (2015). Tomato (*Solanum lycopersicum*) carotenoids and lycopenes chemistry; metabolism, absorption, nutrition, and allied health claims–a comprehensive review. *Crit. Rev. Food Sci. Nutr.* 55(7):919-929.

Reem, N. T. and Van Eck, J. (2019). Application of CRISPR/Cas9-mediated gene editing in tomato. In: Qi, Y. (Ed.) *Plant Genome Editing with CRISPR Systems: Methods and Protocols*. Humana Press, New York, 1917:171-182.@@

Rodriguez-Leal, D., Lemmon, Z. H., Man, J., Bartlett, M. E. and Lippman, Z. B. (2017). Engineering quantitative trait variation for crop improvement by genome editing. *Cell* 171(2):470-480.e8.

Rodriguez-Leal, D., Xu, C., Kwon, C. T., Soyars, C., Demesa-Arevalo, E., Man, J., Liu, L., Lemmon, Z. H., Jones, D. S., Van Eck, J., Jackson, D. P., Bartlett, M. E., Nimchuk, Z. L. and Lippman, Z. B. (2019). Evolution of buffering in a genetic circuit controlling plant stem cell proliferation. *Nat. Genet.* 51(5):786-792.

Ron, M., Kajala, K., Pauluzzi, G., Wang, D., Reynoso, M. A., Zumstein, K., Garcha, J., Winte, S., Masson, H., Inagaki, S., Federici, F., Sinha, N., Deal, R. B., Bailey-Serres, J. and Brady, S. M. (2014). Hairy root transformation using *Agrobacterium rhizogenes* as a tool for exploring cell type-specific gene expression and function using tomato as a model. *Plant Physiol.* 166(2):455-469.

Sang-Ngern, M., Youn, U. J., Park, E. J., Kondratyuk, T. P., Simmons, C. J., Wall, M. M., Ruf, M., Lorch, S. E., Leong, E., Pezzuto, J. M. and Chang, L. C. (2016). Withanolides derived from *Physalis peruviana* (Poha) with potential anti-inflammatory activity. *Bioorg. Med. Chem. Lett.* 26(12):2755-2759.

Senthil-Kumar, M. and Mysore, K. S. (2014). Tobacco rattle virus-based virus-induced gene silencing in *Nicotiana benthamiana*. *Nat. Protoc.* 9(7):1549-1562.

Shah, V. V., Shah, N. D. and Patrekar, P. V. (2013). Medicinal plants from solanaceae family. *Res. J. Pharm. Tech.* 6:143-151.

Shimatani, Z., Ariizumi, T., Fujikura, U., Kondo, A. H. and Nishida, K. (2019). Targeted base editing with CRISPR-deaminase in tomato. In: Qi, Y. (Ed.) *Plant Genome Editing With CRISPR Systems: Methods and Protocols*. Humana Press, New York, 1917:297-307.

Shimatani, Z., Kashojiya, S., Takayama, M., Terada, R., Arazoe, T., Ishii, H., Teramura, H., Yamamoto, T., Komatsu, H., Miura, K., Ezura, H., Nishida, K., Ariizumi, T. and Kondo, A. (2017). Targeted base editing in rice and tomato using a CRISPR-Cas9 cytidine deaminase fusion. *Nat. Biotechnol.* 35(5):441–443.

Soyk, S., Lemmon, Z. H., Sedlazeck, F. J., Jiménez-Gómez, J. M., Alonge, M., Hutton, S. F., Van Eck, J., Schatz, M. C. and Lippman, Z. B. (2019). Duplication of a domestication locus neutralized a cryptic variant that caused a breeding barrier in tomato. *Nat. Plants* 5(5):471–479.

Soyk, S., Muller, N. A., Park, S. J., Schmalenbach, I., Jiang, K., Hayama, R., Zhang, L., Van Eck, J., Jiménez-Gómez, J. M. and Lippman, Z. B. (2017). Variation in the flowering gene *SELF PRUNING 5G* promotes day-neutrality and early yield in tomato. *Nat. Genet.* 49(1):162–168.

Steinert, J., Schiml, S. and Puchta, H. (2016). Homology-based double-stranded break-induced genome engineering in plants. *Plant Cell Rep.* 35(7):1429–1438.

Subburaj, S., Chung, S. J., Lee, C., Ryu, S. M., Kim, D. H., Kim, J. S., Bae, S. and Lee, G. J. (2016). Site-directed mutagenesis in *Petunia* × *hybrida* protoplast system using direct delivery of purified recombinant Cas9 ribonucleoproteins. *Plant Cell Rep.* 35(7):1535–1544.

Svitashev, S., Schwartz, C., Lenderts, B., Young, J. K. and Mark Cigan, A. (2016). Genome editing in maize directed by CRISPR-Cas9 ribonucleoprotein complexes. *Nat. Commun.* 7:13274.

Tashkandi, M., Ali, Z., Aljedaani, F., Shami, A. and Mahfouz, M. M. (2018). Engineering resistance against *Tomato yellow leaf curl virus* via the CRISPR/Cas9 system in tomato. *Plant Signal. Behav.* 13(10):e1525996.

Tomato Genome Consortium (2012). The tomato genome sequence provides insights into fleshy fruit evolution. *Nature* 485(7400):635–641.

Townsend, J. A., Wright, D. A., Winfrey, R. J., Fu, F., Maeder, M. L., Joung, J. K. and Voytas, D. F. (2009). High-frequency modification of plant genes using engineered zinc-finger nucleases. *Nature* 459(7245):442–445.

Tuncel, A., Corbin, K. R., Ahn-Jarvis, J., Harris, S., Hawkins, E., Smedley, M. A., Harwood, W., Warren, F. J., Patron, N. J. and Smith, A. M. (2019). Cas9-mediated mutagenesis of potato starch-branching enzymes generates a range of tuber starch phenotypes. *Plant Biotechnol. J.* 17(12):2259–2271.

Van Eck, J., Keen, P. and Tjahjadi, M. (2019). *Agrobacterium tumefaciens*-mediated transformation of tomato. In: Kumar, S., Barone, P. and Smith, M. (Eds) *Transgenic Plants: Methods and Protocols*. Humana Press, New York, 1864:225–234.

Veillet, F., Perrot, L., Chauvin, L., Kermarrec, M. P., Guyon-Debast, A., Chauvin, J. E., Nogué, F. and Mazier, M. (2019). Transgene-free genome editing in tomato and potato plants using *Agrobacterium*-mediated delivery of a CRISPR/Cas9 cytidine base editor. *Int. J. Mol. Sci.* 20(2):402.

Voytas, D. F. (2013). Plant genome engineering with sequence-specific nucleases. *Annu. Rev. Plant Biol.* 64:327–350.

Wang, D., Samsulrizal, N. H., Yan, C., Allcock, N. S., Craigon, J., Blanco-Ulate, B., Ortega-Salazar, I., Marcus, S. E., Bagheri, H. M., Perez Fons, L., Fraser, P. D., Foster, T., Fray, R., Knox, J. P. and Seymour, G. B. (2019a). Characterization of CRISPR mutants targeting genes modulating pectin degradation in ripening tomato. *Plant Physiol.* 179(2):544–557.

Wang, H. C., Hu, H. H., Chang, F. R., Tsai, J. Y., Kuo, C. Y., Wu, Y. C. and Wu, C. C. (2019b). Different effects of 4β-hydroxywithanolide E and withaferin A, two withanolides from Solanaceae plants, on the Akt signaling pathway in human breast cancer cells. *Phytomedicine* 53:213–222.

Wolff, S. A., Coelho, L. H., Zabrodina, M., Brinckmann, E. and Kittang, A.-I. (2013). Plant mineral nutrition, gas exchange and photosynthesis in space: a review. *Adv. Space Res.* 51(3):465–475.

Wolter, F. and Puchta, H. (2017). Knocking out consumer concerns and regulator's rules: efficient use of CRISPR/Cas ribonucleoprotein complexes for genome editing in cereals. *Genome Biol.* 18(1):43.

Woo, J. W., Kim, J., Kwon, S. I., Corvalán, C., Cho, S. W., Kim, H., Kim, S. G., Kim, S. T., Choe, S. and Kim, J. S. (2015). DNA-free genome editing in plants with preassembled CRISPR-Cas9 ribonucleoproteins. *Nat. Biotechnol.* 33(11):1162–1164.

Xu, C., Liberatore, K. L., Macalister, C. A., Huang, Z., Chu, Y. H., Jiang, K., Brooks, C., Ogawa-Ohnishi, M., Xiong, G., Pauly, M., Van Eck, J., Matsubayashi, Y., van der Knaap, E. and Lippman, Z. B. (2015). A cascade of arabinosyltransferases controls shoot meristem size in tomato. *Nat. Genet.* 47(7):784–792.

Xu, X., Pan, S., Cheng, S., Zhang, B., Mu, D., Ni, P., Zhang, G., Yang, S., Li, R., Wang, J., Orjeda, G., Guzman, F., Torres, M., Lozano, R., Ponce, O., Martinez, D., De la Cruz, G., Chakrabarti, S. K., Patil, V. U., Skryabin, K. G., Kuznetsov, B. B., Ravin, N. V., Kolganova, T. V., Beletsky, A. V., Mardanov, A. V., Di Genova, A., Bolser, D. M., Martin, D. M., Li, G., Yang, Y., Kuang, H., Hu, Q., Xiong, X., Bishop, G. J., Sagredo, B., Mejía, N., Zagorski, W., Gromadka, R., Gawor, J., Szczesny, P., Huang, S., Zhang, Z., Liang, C., He, J., Li, Y., He, Y., Xu, J., Zhang, Y., Xie, B., Du, Y., Qu, D., Bonierbale, M., Ghislain, M., Herrera, Mdel R., Giuliano, G., Pietrella, M., Perrotta, G., Facella, P., O'Brien, K., Feingold, S. E., Barreiro, L. E., Massa, G. A., Diambra, L., Whitty, B. R., Vaillancourt, B., Lin, H., Massa, A. N., Geoffroy, M., Lundback, S., DellaPenna, D., Buell, C. R., Sharma, S. K., Marshall, D. F., Waugh, R., Bryan, G. J., Destefanis, M., Nagy, I., Milbourne, D., Thomson, S. J., Fiers, M., Jacobs, J. M., Nielsen, K. L., Sønderkær, M., Iovene, M., Torres, G. A., Jiang, J., Veilleux, R. E., Bachem, C. W., de Boer, J., Borm, T., Kloosterman, B., van Eck, H., Datema, E., Hekkert, Bt, Goverse, A., van Ham, R. C. and Visser, R. G. (2011). Genome sequence and analysis of the tuber crop potato. *Nature* 475(7355):189–195.

Yu, Q. H., Wang, B., Li, N., Tang, Y., Yang, S., Yang, T., Xu, J., Guo, C., Yan, P., Wang, Q. and Asmutola, P. (2017). CRISPR/Cas9-induced targeted mutagenesis and gene replacement to generate long-shelf life tomato lines. *Sci. Rep.* 7(1):11874.

Zaidi, S. S., Mukhtar, M. S. and Mansoor, S. (2018). Genome editing: targeting susceptibility genes for plant disease resistance. *Trends Biotechnol.* 36(9):898–906.

Zhang, N., Roberts, H. M., Van Eck, J. and Martin, G. B. (2020). Generation and molecular characterization of CRISPR/Cas9-induced mutations in 63 immunity-associated genes in tomato reveals specificity and a range of gene modifications. *Front. Plant Sci.* 11:10.

Zhang, W. N. and Tong, W. Y. (2016). Chemical constituents and biological activities of plants from the genus *Physalis*. *Chem. Biodivers.* 13(1):48–65.

Zsogon, A., Cermak, T., Voytas, D. and Peres, L. E. (2017). Genome editing as a tool to achieve the crop ideotype and *de novo* domestication of wild relatives: case study in tomato. *Plant Sci.* 256:120–130.

Chapter 17

Genome editing of woody perennial trees

Chung-Jui Tsai, University of Georgia, USA

1 Introduction

The CRISPR (clustered regularly interspaced short palindromic repeats) technology is revolutionizing biology and paving the way for the future of agriculture and forestry. Its ease of adoption has democratized the field, making gene editing accessible to non-model species or traditionally less-tractable systems, including woody perennials. While many experimental challenges remain for long-lived perennials, CRISPR now makes it practical to generate null mutants in primary transformants of many tree species (Osakabe et al., 2016; Tsai and Xue, 2015; Bewg et al., 2018). This enabling technology is poised to greatly accelerate not only the discovery of gene functions and genotype-phenotype associations, but also perennial tree improvement. With the ever-growing array of new and diverse CRISPR systems and the constantly evolving technical platforms to accommodate designer modifications, the CRISPR era may still be in its infancy.

Like other revolutionary technologies, the power of CRISPR has drawn attention to limitations that prevent realization of its full potential. For many crops including perennial trees, those limitations include recalcitrance to tissue culture regeneration and ineffective genetic transformation methods. The

http://dx.doi.org/10.19103/AS.2020.0082.27

advent of CRISPR has already motivated research and development efforts to improve, streamline or create new methods of delivering gene-editing reagents into plant cells for the production of edited plants, as featured in several other chapters of this book.

This chapter will discuss additional challenges facing woody perennial trees, namely the heterozygosity of outcrossing species and limited genomic resources. Gene knockouts (KO) represent the predominant applications of the CRISPR technology in woody perennial crops to date and will be the focus of this chapter. The issues considered for gene KO are generally applicable to other CRISPR applications targeting regulatory sequences or non-coding genes. Case studies from the *Populus* 4-coumarate:CoA ligase (4CL) family are presented to demonstrate the power of CRISPR in elucidating functional redundancy as well as specificity of duplicated genes in a perennial woody species.

2 The heterozygous genome challenge in gene editing

The programmable feature of CRISPR-based gene editing relies on a short guide RNA (gRNA), 17-20 nucleotides in length, that base-pairs with the target region(s) in the genome for cleavage. As such, any sequence polymorphisms in the gRNA target loci can impede gene-editing specificity and efficiency. This is of concern for outcrossing species, including woody perennial crops, with heterozygous genomes (Zhou et al., 2015).

Take the first sequenced tree genome–that of *Populus trichocarpa*–as an example; more than 1.2 million (M) single nucleotide polymorphisms (SNPs) and small insertion/deletion polymorphisms (indels) were detected in a single individual 'Nisqually-1' (Tuskan et al., 2006). The number of SNPs increased considerably when comparing multiple individuals by genome re-sequencing: Wang et al. (2016) reported ~5.9 M and ~6.3 M SNPs from 24 *P. tremula* and 22 *P. tremuloides* genotypes, respectively, whereas a population-level study of 544 unrelated *P. trichocarpa* individuals identified nearly 18 M SNPs (Evans et al., 2014).

Genome heterozygosity is exacerbated in hybrids which are critical for breeding and domestication of forest trees, fruit/nut trees and ornamental trees (Wu et al., 2018; Miller and Gross, 2011; Smith and Nichols, 1941). Hybrids often possess desirable agronomic traits, and hybrid genotypes are frequently used in tree transformation studies (Donmez et al., 2013; Busov et al., 2010), including those with gene editing (Nishitani et al., 2016; Zhou et al., 2015).

The finding of more than 10 M SNPs, equivalent to one SNP per ~40 bp sequence (Xue et al., 2015) illustrates the hybrid genome complexity in a re-sequenced hybrid aspen genotype *P. tremula* × *P. alba* INRA 717-1B4 (717) widely used in poplar transformation. Although high, this level of heterozygosity is likely an underestimate as other forms of polymorphisms and small-scale structural rearrangements common in hybrid genomes were not considered. High levels of genome heterozygosity have also been reported for other woody perennials, such as grapevine (one SNP per ~100 bp; Velasco et al., 2007), apple (one SNP per ~210 bp; Velasco et al., 2010) and sweet orange (one SNP per ~280 bp; Xu et al., 2013).

The frequent occurrence of sequence polymorphisms in outcrossing species can result in mismatches between gRNA and target sequences and ruin gene-editing experiments. This was demonstrated unequivocally in a poplar study where one gRNA design based on the *P. trichocarpa* Nisqually-1 reference genome failed to edit the target gene in the transformation genotype 717 due to the presence of SNPs, while 100% biallelic mutations were achieved by other gRNAs at perfectly matched loci (see case studies in Section 6) (Zhou et al., 2015).

Numerous computational programs and web services have been developed, along with various on-target and off-target prediction and scoring criteria to aid in the design of gRNAs (Doench et al., 2016; Bae et al., 2014; Liu et al., 2017; Hsu et al., 2013; Stemmer et al., 2015). However, a major limitation of those programs is the universal reliance on haplotype genomes as the reference, without considering allelic variation in heterozygous species of both animals and plants.

Nearly all publicly available reference genomes represent consensus sequences of sequenced alleles, and thus are pseudo-haplotypes in practice for heterozygous species. As such, gRNAs designed by popular algorithms for outcrossing species may be of unknown reliability and require additional scrutiny to avoid sequence polymorphisms at the target loci. This issue deserves attention especially when transformation genotypes or cell lines differ from those that have been genome-sequenced, because additional sequence polymorphisms are to be expected.

The potential impact of sequence polymorphisms on gRNA specificity was illustrated by a series of *in silico* experiments we previously reported (Zhou et al., 2015; Xue et al., 2015). We first used the >10 M SNPs and indels identified between the *P. trichocarpa* Nisqually-1 reference (pseudo-haplotype) genome and the *P. tremula × alba* '717' re-sequencing data to generate a variant-substituted custom genome for 717 and associated variant database (http://aspendb.uga.edu/s717). We designed gRNAs and calculated specificity scores using publicly available pipelines (Hsu et al., 2013; Lei et al., 2014) against either the *P. trichocarpa* or the variant-substituted 717 genomes. We then screened the gRNA target sites for SNPs or indels in the 717 dataset.

The results showed that there is more than 50% chance that a gRNA design based on the closely related *P. trichocarpa* reference genome will encounter sequence polymorphisms in the corresponding 717 target site. Using the variant-substituted consensus 717 genome for gRNA design reduced the gRNA-target mismatch frequency somewhat, but a very substantial fraction (>40%) of the designed gRNAs remains questionable with heterozygous target sequences (Zhou et al., 2015; Xue et al., 2015). The analysis underscores the challenge of working with outcrossing species such as woody perennials.

Heterozygous genome complexity may explain some of the poor-performing gRNAs reported in the literature. Fortunately, the issue of sequence polymorphisms can be proactively addressed by incorporating additional considerations during gRNA design as discussed in the next section. Increasing awareness of the heterozygous genome challenge with non-model species should motivate the development of new computational programs or functionalities, such as AlleleAnalyzer (Keough et al., 2019), for allele-specific gRNA design and/or design with individualized genomes.

It should be noted that performance of any variant-sensitive gRNA design pipeline, whether manual or computational, is ultimately linked to the quality of the input reference genome or gene sequences (e.g., whether variants are included). A related challenge is that most published gRNA design tools are based on data derived from animal systems, with unknown applicability to plants. In our hands, for instance, 100% biallelic KO was achieved in stably transformed poplar (Zhou et al., 2015; Tsai et al., 2020) using gRNAs with a medium or low score by currently available algorithms (Haeussler et al., 2016; Stemmer et al., 2015; Housden et al., 2015; Hsu et al., 2013). A recent study also found no correlation between observed and predicted gRNA performance, using eight online gRNA design programs, in both transient and stable transformation experiments (Naim et al., 2020). This highlights the need for improved algorithms to support plant genome editing research.

3 Genome resource limitation of woody perennial crops

A clear understanding of the target gene(s) to be modified is essential for successful design and execution of gene-editing experiments. This includes sequence confirmation of the target gene(s) from the transformation genotype to identify sequence polymorphisms for reasons discussed in the previous section. For gene KO experiments, the most straightforward strategy is to target exonic (coding) sequences so that small indels created by CRISPR/Cas9 are more likely to result in frameshift mutations and disrupt protein functions. Thus, information on gene structure annotation, including exon-intron junctions and 5′ or 3′ untranslated regions (UTRs), is necessary to avoid erroneously targeting introns or exon-exon junctions.

Unlike the previous generation of antisense-, RNAi- and microRNA-based gene silencing methods that require prior knowledge of target gene expression for appropriate construct design (such as the choice of promoter), CRISPR gene editing occurs at the DNA level which means the basic CRISPR construct design is applicable to any gene of interest. Nevertheless, information about target gene expression in the transformation genotype is important for several reasons. It helps to ascertain that the target gene is indeed expressed and is not a computational annotation artifact or a pseudogene.

Gene expression information is especially important if the transformation genotype is different from that of the reference genome, or the target gene was identified from a genome-wide association or a genomic selection study. In the latter cases, the underlying genetic variants may occur in part of the population but not the transformation genotype. For example, a gene-trait association that is linked to unusual expression in a limited number of genotypes will not be verified by KO experiments in a transformation genotype that does not express the target gene.

When reference genome and/or genome re-sequencing resources are not available for the transformation genotype, RNA-seq is the most cost-effective means to obtain both sequence and expression information in an unbiased manner. RNA-seq reads can be mapped to a closely related reference genome or *de novo* assembled for transcript abundance estimation and variant discovery using well-established pipelines (Xue et al., 2015; Weber, 2015). With the continuing growth of RNA-seq data in open repositories (such as Sequence Read Archive at NCBI), publicly available RNA-seq datasets from the genotype of interest can be used for this purpose.

For many species and genotypes, expressed sequence tags (ESTs) represent another data source to mine for gene expression and variant information (Boguski et al., 1993), although the technology has become obsolete. In the absence of large scale or high-throughput sequencing data, PCR cloning and Sanger sequencing remain a tested, low-cost method to confirm target gene sequence from the transformation genotype. Full-length gene cloning is not necessary as long as the amplicon spans potential gRNA target sites. Multiple primer pairs should be used to ensure that no allelic biases are introduced inadvertently by primers that harbor SNPs. Specific PCR products can then be Sanger-sequenced directly for rapid SNP detection by decoding heterozygous chromatograms either manually or using various software tools (Xue and Tsai, 2015; Crowe, 2005).

4 Gene-editing reagents

Essentially all published gene-editing studies of perennial trees to date used the classical CRISPR/Cas9 endonuclease from *Streptococcus pyogenes* (SpCas9). Various forms of SpCas9 codon-optimized for human (Mali et al., 2013) or plant expression (Ma et al., 2015; Li et al., 2013) have been used with success. The gRNA is usually expressed by an RNA Polymerase III (Pol III) promoter, such as U6 and U3, and fused to an artificial tracer RNA referred to as scaffold.

Heterozygous U6 and U3 from *Arabidopsis* and soybean are effective in a wide range of species, including poplars, grapevine, orange (Elorriaga et al., 2018; Fan et al., 2015; Zhou et al., 2015; Ren et al., 2016; Peng et al., 2017), alleviating the need to identify and clone endogenous U6/U3 promoters. The largely species-independent nature of the CRISPR reagents is one key characteristic rendering CRISPR gene editing a powerful tool with broad application.

Most published tree studies have employed two or more gRNAs per gene for KO (Fan et al., 2015; Elorriaga et al., 2018; Muhr et al., 2018), although single gRNA has also been found effective (Zhou et al., 2015; Tsai et al., 2020; Peng et al., 2017; Fellenberg et al., 2020). The inclusion of multiple gRNAs targeting several sites per gene provides insurance against potential sequence mismatches and can lead to deletions of intervening fragments, thereby improving gene KO efficiency (Li et al., 2013). However, constructs with multiple U6/U3-gRNA-scaffold cassettes can be technically challenging to build due to repetitive promoter and scaffold sequences in the vector.

Some researchers use distinct U6/U3 promoters to avoid direct promoter repeats (Muhr et al., 2018; Ma et al., 2015), though scaffold repeats remain. Others employ a polycistronic design by incorporating self-processing tRNAs in-between gRNAs as one transcriptional unit under control of a single Pol III (Xie et al., 2015). Although the design alleviates the need for multiple Pol III promoters, it nevertheless introduces repeating units of tRNAs, which can be difficult to assemble.

Given the sequence heterozygosity in woody perennials, the use of multiple gRNAs may increase the likelihood of unintended off-target cleavages. For this reason, a single variant-free gRNA per gene following the design considerations outlined in Section 2 is preferred in our lab. The single-gRNA design has been used to knock out individual, duplicated and tandemly repeated genes at high (near 100%) efficiencies in our lab (Zhou et al., 2015; Tsai et al., 2020).

5 Stability

Stable inheritance of genetic edits is essential for the application of the CRISPR technology to crop species. This has been unequivocally demonstrated in herbaceous annuals, including *Arabidopsis*, rice, oilseed rape, and maize (Feng et al., 2014; Jiang et al., 2014; Pan et al., 2016; Braatz et al., 2017; Xu et al., 2015; Zhu et al., 2016). For long-lived trees, elite germplasms are usually propagated by vegetative cuttings for faithful maintenance of desired traits. Stable transmission of CRISPR/Cas9-mediated mutations in asexual propagules is therefore of topmost importance for woody perennials.

In theory, mutations at the target site by CRISPR/Cas9 will render the edited sequence unrecognizable by the same Cas9-gRNA complex. A related issue is the long-term effects of potential off-target activities in the presence of Cas9-gRNA. We have addressed those concerns by amplicon deep sequencing of both on-target and off-target sites in a suite of transgenic poplars after multiple years of vegetative propagation or repeated cycles of cutback-and-sprouting (Bewg et al., 2018). We found on-target edits to be stable following vegetative propagation by tissue culture or rooted cuttings. Similar observations have also been reported in several other tree studies (Elorriaga et al., 2018; Muhr et al., 2018).

At one off-target site with two SNPs (one per allele at two different positions), we observed no editing in several independent transgenic events after multiple cycles of cutback and regrowth over a 4-year period in the greenhouse (Bewg et al., 2018). This suggests that Cas9-gRNA specificity is maintained during clonal propagation. Obviously, segregants that are free of T-DNA from *Agrobacterium*-mediated transformation will be desired to avoid any unintended effects. However, this is not always possible for outcrossing woody perennials due to their long generation times and, in many cases, dioecious reproduction (meaning that compatible male and female parent genotypes need to be edited before crossing). Nonetheless, our data offer experimental evidence to support long-term stability and specificity of CRISPR/Cas9-editing outcomes in vegetatively propagated woody perennials even in the presence of T-DNA.

6 CRISPR/Cas9-empowered characterization of duplicated genes

Most land plants have experienced recurring episodes of gene duplications ranging in scale from whole genome, to segmental, to transposon-mediated, to tandem duplications. Gene duplications are sources of evolutionary novelty, as duplicated genes can retain the same function to provide redundancy or undergo pseudo-functionalization, sub-functionalization or neo-functionalization during evolution (Flagel and Wendel, 2009; Freeling, 2009; Panchy et al., 2016). Discerning the multiple fates of duplicated genes can be difficult in non-model systems that lack large collections of KO mutants.

Previous approaches for gene silencing, such as antisense or RNAi, often target highly homologous sequences (Frizzi and Huang, 2010) and are therefore of limited use for functional dissection of closely related gene duplicates. Even in model systems like *Arabidopsis* with a comprehensive catalog of mutants generated by chemical (e.g. EMS), physical (e.g. γ-ray), or biological (e.g. T-DNA) mutagenesis, multiple rounds of crosses are required to obtain higher-order KO mutants to investigate functions of duplicated or otherwise redundant genes. In addition, higher-order mutants are not readily obtained by crossing when the distance between target gene loci is small and limits recombination (Roy Choudhury et al., 2020).

The CRISPR technology is well-suited to address this challenge. By targeting regions with and without sequence polymorphisms among duplicates, gRNAs can be designed to knock out individual as well as multiple copies of gene family members to unravel their functional relatedness. A case study concerns the *4CL* family in *Populus*. *4CL* is a key enzyme of phenylpropanoid metabolism which accounts for ~40% of photosynthetically fixed carbon in the forms of lignin, flavonoids, chlorogenic acids and a host of other phenolic compounds with structural as well as protective functions (Croteau et al., 2000).

In all sequenced plant genomes, *4CL* is encoded by multiple genes in two phylogenetically distinct groups (Chen et al., 2014). Group I *4CLs* include all known lignin biosynthesis-associated isoforms in diverse species (Hu et al., 1999; Lee et al., 1997; Kajita et al., 1996; Xu et al., 2011; Gui et al., 2011). Group II *4CLs*, in comparison, have been under-investigated and are thought to participate in biosynthesis of nonstructural phenolics based on gene expression inference (Harding et al., 2002; Hu et al., 1998; Chen et al., 2014).

Like most angiosperm species, poplar contains a single Group II gene, *4CL2*. CRISPR/Cas9-KO of *4CL2* in poplar results in reduced accrual of flavonoid-derived proanthocyanidins (Zhou et al., 2015), also known as condensed tannins, an abundant pool of nonstructural phenylpropanoids in leaves, barks and roots (Tsai et al., 2006). This provides direct evidence for the involvement of Group II *4CL* in flavonoid biosynthesis.

Group I *4CL* is expanded in angiosperms, with two subgroups in most sequenced genomes (Chen et al., 2014). Poplar contains four Group I members in two pairs, *4CL1-4CL5* and *4CL3-4CL4*, derived from the Salicoid genome duplication event (Tuskan et al., 2006). Among them, *4CL1* has been experimentally characterized for its involvement in lignin biosynthesis based on molecular, biochemical and reverse genetics evidence (Voelker et al., 2010; Hu et al., 1999). However, the roles of the other Group I *4CLs* remain unclear.

The *4CL1* and *4CL5* paralogs are the only *4CL* genes expressed in lignifying tissues, with *4CL1* accounting for 80-90% of the *4CL* transcripts detected in developing xylem of multiple poplar species (Tsai et al., 2020) (Fig. 1a and b). *4CL1* and *4CL5* differ in their enzyme kinetics and substrate preferences. *4CL1* has higher catalytic activities than *4CL5* and exhibits similar affinities for 4-coumaric acid, caffeic acid, and ferulic acid, whereas the low-affinity *4CL5* has a strong preference for caffeic acid (Chen et al., 2013).

CRISPR/Cas9 editing of *4CL1* in poplar confirmed its involvement in lignin biosynthesis as expected, with ~20% reduction in lignin content (Zhou et al., 2015). *4CL5*-KO at the homologous site (Fig. 1a) did not affect stem lignification under normal growth conditions (Tsai et al., 2020); however, double KO mutants are not

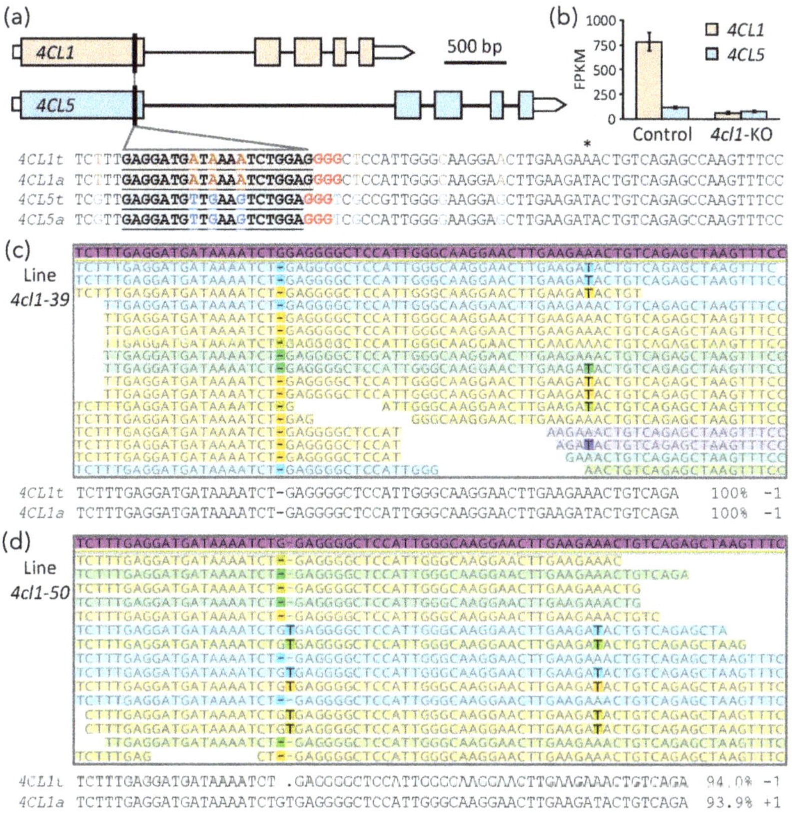

Figure 1 CRISPR-knockout of *4CL1* or *4CL5* gene in poplar. (a) Schematics of *4CL1* and *4CL5* gene structures. Colored boxes are coding exons, white boxes are UTRs, and the black vertical lines denote the gRNA target sites, with the allelic sequences shown below (a, *P. alba* allele; t, *P. tremula* allele). The protospacer adjacent motif (PAM) is shown in red. Orange and blue fonts denote *4CL1* and *4CL5*-specific sequences, respectively. Asterisk denotes a SNP between the two *4CL1* alleles. (b) Transcript levels of *4CL1* and *4CL5* in control (wild type and *Cas9*-only vector control) and *4cl1*-KO lines. Data are mean±SE (n=6). FPKM, fragments per kilobase of transcript per million of mapped reads as determined by RNA-seq. (c-d) Representative screenshots of RNA-seq reads mapped to the *4CL1* target site (reference sequence in purple) from two KO lines *4cl1-39* (c) and *4cl1-50* (d). The editing patterns detected by amplicon deep sequencing are shown below each panel. The percentage values are the fractions of mapped reads (>8800 for *4cl1-39* and >12 000 for *4cl1-50*) with the indicated indels.

viable. This suggests that *4CL5* is dispensable in the presence of *4CL1* and plays a minor or conditional role in poplar lignification. The *4cl1* KO mutants provide an ideal study system for an analysis of *4CL5* function in lignin biosynthesis.

RNA-seq analysis of *4cl1* mutants revealed greatly (92%) reduced *4CL1* transcript levels in developing xylem relative to wild type (WT) and Cas9 vector controls (Fig. 1b) (Tsai et al., 2020). The reduction can be ascribed to nonsense-mediated mRNA decay (NMD) triggered by premature stop codons resulting from CRISPR/Cas9-induced indels in the first exon of *4CL1* (Fig. 1a). Indeed, a close examination of the mutant-derived RNA-seq reads at the *4CL1* target site confirms that the residual *4CL1* transcripts are aberrant, harboring frameshifting indels that match those identified by amplicon-sequencing (Fig. 1c and d).

4CL5 transcript levels were not significantly altered in the *4cl1* mutants relative to the controls (Tsai et al., 2020). In fact, *4CL5* transcript levels trended lower in the mutants (Fig. 1b), perhaps associated with nonspecific silencing by *4CL1*-NMD. Regardless, *4CL5* became the predominant functional *4CL* transcripts in the mutant xylem. We observed upregulation of lignin pathway genes encoding enzymes involved in biosynthesis of caffeic acid, the preferred substrate of 4CL5: a caffeoyl shikimate esterase and a recently discovered bifunctional ascorbate peroxidase/4-coumarate 3-hydroxylase (Barros et al., 2019). This is consistent with increased levels of caffeic acid and its various conjugates in the mutant xylem extracts (Tsai et al., 2020).

Those changes are expected to enable caffeic acid catalysis by the low-affinity 4CL5. The results are consistent with previously undemonstrated plasticity in the lignin pathway. Given the abundance of duplicated genes in plant genomes, compensatory redundancies may be a common mechanism for taxon-specific phenotypic plasticity. CRISPR/Cas9-mediated gene editing now affords the precision and specificity for functional dissection of redundant genes.

7 Conclusion and future trends

Woody perennial horticultural and forest trees provide numerous services and benefits to our society in the form of food, pulp and paper, timber, renewable energy, and atmosphere carbon sequestration. As experimental systems, however, trees have been under-utilized and under-investigated because of their long generation times and generally less-tractable genetics. CRISPR genome editing is rapidly leveling the playing field, providing those traditionally non-model species with a powerful platform to accelerate genetic improvement. The ability to generate null mutants in the first generation (i.e. primary transformants) enables experimental characterization of gene functions at a rapid pace and with less ambiguity. As demonstrated for the *4CL* gene family, functional redundancy can now be dissected with gene-specific precision.

Challenges remain for researchers and practitioners working with woody perennials. Chief among them are generally high levels of genome heterozygosity and low efficiencies of genetic transformation and plant regeneration systems. Thanks to advances in new sequencing technologies and decreasing costs, sequence polymorphisms can be readily identified in any species to aid specific gRNA design.

Genetic transformation is a bottleneck in many crop species and genotypes, including trees. Recent studies have demonstrated the power of morphogenic regulators, such as Baby Boom (BBM) and Wuschel (WUS), which can reprogram somatic cells into stem cells to regenerate embryos for whole plant recovery (Lowe et al., 2016). This approach appears widely applicable to monocots and dicots, including trees such as poplar and coffee (Arroyo-Herrera et al., 2008; Deng et al., 2009). However, these transcription factors need to be under strict spatiotemporal regulation in order to recover morphologically normal transgenic plants without persistent and recurring embryogenesis. Thus, how to implement morphogenic regulator-based regeneration systems in trees remains a research priority (Nagle et al., 2018).

Another challenge concerns the persistence of transgenes such as *Cas9* and selectable marker genes that may hinder regulatory approval of genome-edited trees for field trials or eventual release for commercial, environmental, or conservation applications. Given the demonstrated high rates of biallelic edits in transgenic trees, it is possible to obtain transgene-free mutants in one cross. This approach when used in conjunction with early-flowering technologies (Hoenicka et al., 2014) would drastically shorten the tree improvement cycle. The transgene-free, gene-edited progenies could potentially be exempt from regulatory requirements in some countries, thereby promoting field evaluations and real-world application.

However, for dioicous species like poplar, biallelic mutants in genetically compatible male and female backgrounds are necessary for this approach to work. For many species including poplars, precocious flowering induction by chemical and/or transgenic means remains inefficient and future research into this bottleneck is needed in order to realize the full potential of genome editing.

8 Where to look for further information

8.1 Further reading

- A review on the history and current and future development of genome editing in woody plants by Osakabe et al. (2016).
- A review on tree genome editing patterns and long-term stability by Bewg et al. (2018).
- A review on genome editing of fruit trees by Wang et al. (2018).
- A collection of review articles on 'Forest Genomics and Biotechnology', with various current and future applications that can benefit from genome editing technology, published in Frontiers in Plant Science, edited by Allona et al. (2019).

9 Acknowledgements

Research on CRISPR genome editing and associated genomic resources in the Tsai lab was supported in part by the National Institute of Food and Agriculture in the Department of Agriculture (2015-67013-22812), the National Science Foundation (IOS-1546867), and The Center for Bioenergy Innovation, a Department of Energy's Research Center funded by the Office of Biological and Environmental Research in the DOE Office of Science.

10 References

Allona, I., Kirst, M., Boerjan, W., Strauss, S. and Sederoff, R. 2019. Editorial: forest genomics and biotechnology. *Frontiers in Plant Science* 10, 1187.

Arroyo-Herrera, A., Ku Gonzalez, A., Canche Moo, R., Quiroz-Figueroa, F. R., Loyola-Vargas, V. M., Rodriguez-Zapata, L. C., Burgeff D'Hondt, C., Suárez-Solís, V. M. and Castaño, E. 2008. Expression of WUSCHEL in Coffea canephora causes ectopic morphogenesis and increases somatic embryogenesis. *Plant Cell Tissue and Organ Culture* 94(2), 171–180.

Bae, S., Park, J. and Kim, J. S. 2014. Cas-OFFinder: a fast and versatile algorithm that searches for potential off-target sites of Cas9 RNA-guided endonucleases. *Bioinformatics* 30(10), 1473–1475.

Barros, J., Escamilla-Trevino, L., Song, L., Rao, X., Serrani-Yarce, J. C., Palacios, M. D., Engle, N., Choudhury, F. K., Tschaplinski, T. J., Venables, B. J., Mittler, R. and Dixon, R. A. 2019. 4-Coumarate 3-hydroxylase in the lignin biosynthesis pathway is a cytosolic ascorbate peroxidase. *Nature Communications* 10(1), 1994.

Bewg, W. P., Ci, D. and Tsai, C. J. 2018. Genome editing in trees: from multiple repair pathways to long-term stability. *Frontiers in Plant Science* 9, 1732.

Boguski, M. S., Lowe, T. M. J. and Tolstoshev, C. M. 1993. dbEST–database for "expressed sequence tags". *Nature Genetics* 4(4), 332–333.

Braatz, J., Harloff, H. J., Mascher, M., Stein, N., Himmelbach, A. and Jung, C. 2017. CRISPR-Cas9 targeted mutagenesis leads to simultaneous modification of different homoeologous gene copies in polyploid oilseed rape (*Brassica napus*). *Plant Physiology* 174(2), 935–942.

Busov, V. B., Strauss, S. H. and Pilate, G. 2010. Transformation as a tool for genetic analysis in *Populus*. In: Jansson, S., Bhalerao, R. P. and Groover, A. T. (Eds), *Genetics and Genomics of Populus*. New York: Springer. pp. 113–133.

Chen, H. C., Song, J., Williams, C. M., Shuford, C. M., Liu, J., Wang, J. P., Li, Q., Shi, R., Gokce, E., Ducoste, J., Muddiman, D. C., Sederoff, R. R. and Chiang, V. L. 2013. Monolignol pathway 4-coumaric acid:coenzyme A ligases in Populus trichocarpa: novel specificity, metabolic regulation, and simulation of coenzyme A ligation fluxes. *Plant Physiology* 161(3), 1501–1516.

Chen, H.-Y., Babst, B. A., Nyamdari, B., Hu, H., Sykes, R., Davis, M. F., Harding, S. A. and Tsai, C.-J. 2014. Ectopic expression of a loblolly pine Class II 4-coumarate:CoA ligase alters soluble phenylpropanoid metabolism but not lignin biosynthesis in Populus. *Plane Cell Physiology*, 55, 1669–1678.

Croteau, R., Kutchan, T. M. and Lewis, N. G. 2000. Chapter 24: Natural products (secondary metabolites). In: Buchanan, B., Gruissem, W. and Jones, R. (Eds), *Biochemistry and Molecular Biology of Plants*. Rockville, MR: American Society of Plant Physiologists. pp. 1250–1318.

Crowe, M. L. 2005. SeqDoC: rapid SNP and mutation detection by direct comparison of DNA sequence chromatograms. *BMC Bioinformatics* 6, 133.

Deng, W., Luo, K., Li, Z. and Yang, Y. 2009. A novel method for induction of plant regeneration via somatic embryogenesis. *Plant Science* 177(1), 43–48.

Doench, J. G., Fusi, N., Sullender, M., Hegde, M., Vaimberg, E. W., Donovan, K. F., Smith, I., Tothova, Z., Wilen, C., Orchard, R., Virgin, H. W., Listgarten, J. and Root, D. E. 2016. Optimized sgRNA design to maximize activity and minimize off-target effects of CRISPR-Cas9. *Nature Biotechnology* 34(2), 184–191.

Donmez, D., Simsek, O., Izgu, T., Kacar, Y. A. and Mendi, Y. Y. 2013. Genetic transformation in *Citrus*. *The Scientific World Journal* 2013, 491207.

Elorriaga, E., Klocko, A. L., Ma, C. and Strauss, S. H. 2018. Variation in mutation spectra among CRISPR/Cas9 mutagenized poplars. *Frontiers in Plant Science* 9, 594.

Evans, L. M., Slavov, G. T., Rodgers-Melnick, E., Martin, J., Ranjan, P., Muchero, W., Brunner, A. M., Schackwitz, W., Gunter, L., Chen, J. G., Tuskan, G. A. and Difazio, S. P. 2014. Population genomics of *Populus trichocarpa* identifies signatures of selection and adaptive trait associations. *Nature Genetics* 46(10), 1089–1096.

Fan, D., Liu, T., Li, C., Jiao, B., Li, S., Hou, Y. and Luo, K. 2015. Efficient CRISPR/Cas9-mediated targeted mutagenesis in *Populus* in the first generation. *Scientific Reports* 5, 12217–12217.

Fellenberg, C., Corea, O., Yan, L. H., Archinuk, F., Piirtola, E. M., Gordon, H., Reichelt, M., Brandt, W., Wulff, J., Ehlting, J. and Peter Constabel, C. 2020. Discovery of salicyl benzoate UDP-glycosyltransferase, a central enzyme in poplar salicinoid phenolic glycoside biosynthesis. *Plant Journal. For Cell and Molecular Biology* 102(1), 99–115.

Feng, Z., Mao, Y., Xu, N., Zhang, B., Wei, P., Yang, D.-L., Wang, Z., Zhang, Z., Zheng, R., Yang, L., Zeng, L., Liu, X. and Zhu, J.-K. 2014. Multigeneration analysis reveals the inheritance, specificity, and patterns of CRISPR/Cas-induced gene modifications in Arabidopsis. *Proceedings of the National Academy of Sciences of the United States of America* 111, 4632–4637.

Flagel, L. E. and Wendel, J. F. 2009. Gene duplication and evolutionary novelty in plants. *New Phytologist* 183(3), 557–564.

Freeling, M. 2009. Bias in plant gene content following different sorts of duplication: tandem, whole-genome, segmental, or by transposition. *Annual Review of Plant Biology* 60, 433–453.

Frizzi, A. and Huang, S. 2010. Tapping RNA silencing pathways for plant biotechnology. *Plant Biotechnology Journal* 8(6), 655–677.

Gui, J., Shen, J. and Li, L. 2011. Functional characterization of evolutionarily divergent 4-coumarate:coenzyme A ligases in rice. *Plant Physiology* 157(2), 574–586.

Haeussler, M., Schönig, K., Eckert, H., Eschstruth, A., Mianné, J., Renaud, J. B., Schneider-Maunoury, S., Shkumatava, A., Teboul, L., Kent, J., Joly, J. S. and Concordet, J. P. 2016. Evaluation of off-target and on-target scoring algorithms and integration into the guide RNA selection tool CRISPOR. *Genome Biology* 17(1), 148.

Harding, S. A., Leshkevich, J., Chiang, V. L. and Tsai, C. J. 2002. Differential substrate inhibition couples kinetically distinct 4-coumarate: coenzyme A ligases with spatially distinct metabolic roles in quaking aspen. *Plant Physiology* 128(2), 428–438.

Hoenicka, H., Lehnhardt, D., Nilsson, O., Hanelt, D. and Fladung, M. 2014. Successful crossings with early flowering transgenic poplar: interspecific crossings, but not transgenesis, promoted aberrant phenotypes in offspring. *Plant Biotechnology Journal* 12(8), 1066-1074.

Housden, B. E., Valvezan, A. J., Kelley, C., Sopko, R., Hu, Y., Roesel, C., Lin, S., Buckner, M., Tao, R., Yilmazel, B., Mohr, S. E., Manning, B. D. and Perrimon, N. 2015. Identification of potential drug targets for tuberous sclerosis complex by synthetic screens combining CRISPR-based knockouts with RNAi. *Science Signaling* 8(393), rs9.

Hsu, P. D., Scott, D. A., Weinstein, J. A., Ran, F. A., Konermann, S., Agarwala, V., Li, Y., Fine, E. J., Wu, X., Shalem, O., Cradick, T. J., Marraffini, L. A., Bao, G. and Zhang, F. 2013. DNA targeting specificity of RNA-guided Cas9 nucleases. *Nature Biotechnology* 31(9), 827-832.

Hu, W. J., Kawaoka, A., Tsai, C. J., Lung, J., Osakabe, K., Ebinuma, H. and Chiang, V. L. 1998. Compartmentalized expression of two structurally and functionally distinct 4-coumarate:CoA ligase genes in aspen (Populus tremuloides). *Proceedings of the National Academy of Sciences of the United States of America* 95(9), 5407-5412.

Hu, W. J., Harding, S. A., Lung, J., Popko, J. L., Ralph, J., Stokke, D. D., Tsai, C. J. and Chiang, V. L. 1999. Repression of lignin biosynthesis promotes cellulose accumulation and growth in transgenic trees. *Nature Biotechnology* 17(8), 808-812.

Jiang, W., Yang, B. and Weeks, D. P. 2014. Efficient CRISPR/Cas9-mediated gene editing in *Arabidopsis thaliana* and inheritance of modified genes in the T2 and T3 generations. *PLoS ONE* 9(6), e99225.

Kajita, S., Katayama, Y. and Omori, S. 1996. Alterations in the biosynthesis of lignin in transgenic plants with chimeric genes for 4-coumarate: coenzyme A ligase. *Plant and Cell Physiology* 37(7), 957-965.

Keough, K. C., Lyalina, S., Olvera, M. P., Whalen, S., Conklin, B. R. and Pollard, K. S. 2019. AlleleAnalyzer: a tool for personalized and allele-specific sgRNA design. *Genome Biology* 20(1), 167.

Lee, D., Meyer, K., Chapple, C. and Douglas, C. J. 1997. Antisense suppression of 4-coumarate: coenzyme A ligase activity in Arabidopsis leads to altered lignin subunit composition. *Plant Cell* 9(11), 1985-1998.

Lei, Y., Lu, L., Liu, H. Y., Li, S., Xing, F. and Chen, L. L. 2014. CRISPR-P: a web tool for synthetic single-guide RNA design of CRISPR-system in plants. *Molecular Plant* 7(9), 1494-1496.

Li, J. F., Norville, J. E., Aach, J., Mccormack, M., Zhang, D., Bush, J., Church, G. M. and Sheen, J. 2013. Multiplex and homologous recombination-mediated genome editing in Arabidopsis and Nicotiana benthamiana using guide RNA and Cas9. *Nature Biotechnology* 31(8), 688-691.

Liu, H., Ding, Y., Zhou, Y., Jin, W., Xie, K. and Chen, L. L. 2017. CRISPR-P 2.0: an improved CRISPR-Cas9 tool for genome editing in plants. *Molecular Plant* 10(3), 530-532.

Lowe, K., Wu, E., Wang, N., Hoerster, G., Hastings, C., Cho, M. J., Scelonge, C., Lenderts, B., Chamberlin, M., Cushatt, J., Wang, L., Ryan, L., Khan, T., Chow-Yiu, J., Hua, W., Yu, M., Banh, J., Bao, Z., Brink, K., Igo, E., Rudrappa, B., Shamseer, P. M., Bruce, W., Newman, L., Shen, B., Zheng, P., Bidney, D., Falco, C., Register, J., Zhao, Z. Y., Xu, D., Jones, T. and Gordon-Kamm, W. 2016. Morphogenic regulators Baby Boom and Wuschel improve monocot transformation. *Plant Cell* 28(9), 1998-2015.

Ma, X., Zhang, Q., Zhu, Q., Liu, W., Chen, Y., Qiu, R., Wang, B., Yang, Z., Li, H., Lin, Y., Xie, Y., Shen, R., Chen, S., Wang, Z., Chen, Y., Guo, J., Chen, L., Zhao, X., Dong, Z. and Liu,

Y. G. 2015. A robust CRISPR/Cas9 system for convenient, high-efficiency multiplex genome editing in monocot and dicot plants. *Molecular Plant* 8(8), 1274-1284.

Mali, P., Yang, L., Esvelt, K. M., Aach, J., Guell, M., Dicarlo, J. E., Norville, J. E. and Church, G. M. 2013. RNA-guided human genome engineering via Cas9. *Science* 339(6121), 823-826.

Miller, A. J. and Gross, B. L. 2011. From forest to field: perennial fruit crop domestication. *American Journal of Botany* 98(9), 1389-1414.

Muhr, M., Paulat, M., Awwanah, M., Brinkkötter, M. and Teichmann, T. 2018. CRISPR/Cas9-mediated knockout of Populus BRANCHED1 and BRANCHED2 orthologs reveals a major function in bud outgrowth control. *Tree Physiology* 38(10), tpy088-tpy088.

Nagle, M., Déjardin, A., Pilate, G. and Strauss, S. H. 2018. Opportunities for innovation in genetic transformation of forest trees. *Frontiers in Plant Science* 9, 1443.

Naim, F., Shand, K., Hayashi, S., O'Brien, M., Mcgree, J., Johnson, A. A. T., Dugdale, B. and Waterhouse, P. M. 2020. Are the current gRNA ranking prediction algorithms useful for genome editing in plants? *PLoS ONE* 15(1), e0227994.

Nishitani, C., Hirai, N., Komori, S., Wada, M., Okada, K., Osakabe, K., Yamamoto, T. and Osakabe, Y. 2016. Efficient genome editing in apple using a CRISPR/Cas9 system. *Scientific Reports* 6, 31481.

Osakabe, Y., Sugano, S. S. and Osakabe, K. 2016. Genome engineering of woody plants: past, present and future. *Journal of Wood Science* 62(3), 217-225.

Pan, C., Ye, L., Qin, L., Liu, X., He, Y., Wang, J., Chen, L. and Lu, G. 2016. CRISPR/Cas9-mediated efficient and heritable targeted mutagenesis in tomato plants in the first and later generations. *Scientific Reports* 6, 24765.

Panchy, N., Lehti-Shiu, M. and Shiu, S. H. 2016. Evolution of gene duplication in plants. *Plant Physiology* 171(4), 2294-2316.

Peng, A., Chen, S., Lei, T., Xu, L., He, Y., Wu, L., Yao, L. and Zou, X. 2017. Engineering canker-resistant plants through CRISPR/Cas9 targeted editing of the susceptibility gene CsLOB1 promoter in citrus. *Plant Biotechnology Journal* 15(12), 1509-1519.

Ren, C., Liu, X., Zhang, Z., Wang, Y., Duan, W., Li, S. and Liang, Z. 2016. CRISPR/Cas9-mediated efficient targeted mutagenesis in Chardonnay (Vitis vinifera L.). *Scientific Reports* 6, 32289-32289.

Roy Choudhury, S., Li, M., Lee, V., Nandety, R. S., Mysore, K. S. and Pandey, S. 2020. Flexible functional interactions between G-protein subunits contribute to the specificity of plant responses. *Plant Journal: For Cell and Molecular Biology* 102(2), 207-221, 10.1111/tpj.14714.

Smith, E. C. and Nichols, C. 1941. Species hybrids in forest trees. *Journal of the Arnold Arboretum* 22, 443-454.

Stemmer, M., Thumberger, T., Del Sol Keyer, M., Wittbrodt, J. and Mateo, J. L. 2015. CCTop: an intuitive, flexible and reliable CRISPR/Cas9 target prediction tool. *PLoS ONE* 10(4), e0124633.

Tsai, C. J., Harding, S. A., Tschaplinski, T. J., Lindroth, R. L. and Yuan, Y. 2006. Genome-wide analysis of the structural genes regulating defense phenylpropanoid metabolism in *Populus*. *New Phytologist* 172(1), 47-62.

Tsai, C. J. and Xue, L. J. 2015. CRISPRing into the woods. *GM Crops and Food* 6(4), 206-215.

Tsai, C. J., Xu, P., Xue, L.-J., Hu, H., Nyamdari, B., Naran, R., Zhou, X., Goeminne, G., Gao, R., Gjersing, E., Dahlen, J., Pattathil, S., Hahn, M. G., Davis, M., Ralph, J., Boerjan, W. A. and Harding, S. A. 2020. Compensatory guaiacyl lignin biosynthesis at the expense of syringyl lignin in 4CL1-knockout poplar. *Plant Physiology* 183, 123-136.

Tuskan, G. A., Difazio, S., Jansson, S., Bohlmann, J., Grigoriev, I., Hellsten, U., Putnam, N., Ralph, S., Rombauts, S., Salamov, A., Schein, J., Sterck, L., Aerts, A., Bhalerao, R. R., Bhalerao, R. P., Blaudez, D., Boerjan, W., Brun, A., Brunner, A., Busov, V., Campbell, M., Carlson, J., Chalot, M., Chapman, J., Chen, G. L., Cooper, D., Coutinho, P. M., Couturier, J., Covert, S., Cronk, Q., Cunningham, R., Davis, J., Degroeve, S., Déjardin, A., Depamphilis, C., Detter, J., Dirks, B., Dubchak, I., Duplessis, S., Ehlting, J., Ellis, B., Gendler, K., Goodstein, D., Gribskov, M., Grimwood, J., Groover, A., Gunter, L., Hamberger, B., Heinze, B., Helariutta, Y., Henrissat, B., Holligan, D., Holt, R., Huang, W., Islam-Faridi, N., Jones, S., Jones-Rhoades, M., Jorgensen, R., Joshi, C., Kangasjärvi, J., Karlsson, J., Kelleher, C., Kirkpatrick, R., Kirst, M., Kohler, A., Kalluri, U., Larimer, F., Leebens-Mack, J., Leplé, J. C., Locascio, P., Lou, Y., Lucas, S., Martin, F., Montanini, B., Napoli, C., Nelson, D. R., Nelson, C., Nieminen, K., Nilsson, O., Pereda, V., Peter, G., Philippe, R., Pilate, G., Poliakov, A., Razumovskaya, J., Richardson, P., Rinaldi, C., Ritland, K., Rouzé, P., Ryaboy, D., Schmutz, J., Schrader, J., Segerman, B., Shin, H., Siddiqui, A., Sterky, F., Terry, A., Tsai, C. J., Uberbacher, E., Unneberg, P., Vahala, J., Wall, K., Wessler, S., Yang, G., Yin, T., Douglas, C., Marra, M., Sandberg, G., Van De Peer, Y. and Rokhsar, D. 2006. The genome of black cottonwood, *Populus trichocarpa* (Torr. & Gray). *Science* 313(5793), 1596–1604.

Velasco, R., Zharkikh, A., Troggio, M., Cartwright, D. A., Cestaro, A., Pruss, D., Pindo, M., Fitzgerald, L. M., Vezzulli, S., Reid, J., Malacarne, G., Iliev, D., Coppola, G., Wardell, B., Micheletti, D., Macalma, T., Facci, M., Mitchell, J. T., Perazzolli, M., Eldredge, G., Gatto, P., Oyzerski, R., Moretto, M., Gutin, N., Stefanini, M., Chen, Y., Segala, C., Davenport, C., Demattè, L., Mraz, A., Battilana, J., Stormo, K., Costa, F., Tao, Q., Si-Ammour, A., Harkins, T., Lackey, A., Perbost, C., Taillon, B., Stella, A., Solovyev, V., Fawcett, J. A., Sterck, L., Vandepoele, K., Grando, S. M., Toppo, S., Moser, C., Lanchbury, J., Bogden, R., Skolnick, M., Sgaramella, V., Bhatnagar, S. K., Fontana, P., Gutin, A., Van De Peer, Y., Salamini, F. and Viola, R. 2007. A high quality draft consensus sequence of the genome of a heterozygous grapevine variety. *PLoS ONE* 2(12), e1326.

Velasco, R., Zharkikh, A., Affourtit, J., Dhingra, A., Cestaro, A., Kalyanaraman, A., Fontana, P., Bhatnagar, S. K., Troggio, M., Pruss, D., Salvi, S., Pindo, M., Baldi, P., Castelletti, S., Cavaiuolo, M., Coppola, G., Costa, F., Cova, V., Dal Ri, A., Goremykin, V., Komjanc, M., Longhi, S., Magnago, P., Malacarne, G., Malnoy, M., Micheletti, D., Moretto, M., Perazzolli, M., Si-Ammour, A., Vezzulli, S., Zini, E., Eldredge, G., Fitzgerald, L. M., Gutin, N., Lanchbury, J., Macalma, T., Mitchell, J. T., Reid, J., Wardell, B., Kodira, C., Chen, Z., Desany, B., Niazi, F., Palmer, M., Koepke, T., Jiwan, D., Schaeffer, S., Krishnan, V., Wu, C., Chu, V. T., King, S. T., Vick, J., Tao, Q., Mraz, A., Stormo, A., Stormo, K., Bogden, R., Ederle, D., Stella, A., Vecchietti, A., Kater, M. M., Masiero, S., Lasserre, P., Lespinasse, Y., Allan, A. C., Bus, V., Chagné, D., Crowhurst, R. N., Gleave, A. P., Lavezzo, E., Fawcett, J. A., Proost, S., Rouzé, P., Sterck, L., Toppo, S., Lazzari, B., Hellens, R. P., Durel, C. E., Gutin, A., Bumgarner, R. E., Gardiner, S. E., Skolnick, M., Egholm, M., Van De Peer, Y., Salamini, F. and Viola, R. 2010. The genome of the domesticated apple (*Malus* × *domestica* Borkh.). *Nature Genetics* 42(10), 833–839.

Voelker, S. L., Lachenbruch, B., Meinzer, F. C., Jourdes, M., Ki, C., Patten, A. M., Davin, L. B., Lewis, N. G., Tuskan, G. A., Gunter, L., Decker, S. R., Selig, M. J., Sykes, R., Himmel, M. E., Kitin, P., Shevchenko, O. and Strauss, S. H. 2010. Antisense down-regulation of 4CL expression alters lignification, tree growth, and saccharification potential of field-grown poplar. *Plant Physiology* 154(2), 874–886.

Wang, J., Street, N. R., Scofield, D. G. and Ingvarsson, P. K. 2016. Variation in linked selection and recombination drive genomic divergence during allopatric speciation of European and American Aspens. *Molecular Biology and Evolution* 33(7), 1754–1767.

Wang, X., Tu, M., Li, Z., Wang, Y. and Wang, X. 2018. Current progress and future prospects for the clustered regularly interspaced short palindromic repeats (CRISPR) genome editing technology in fruit tree breeding. *Critical Reviews in Plant Sciences* 37(4), 233–258.

Weber, A. P. M. 2015. Discovering new biology through sequencing of RNA. *Plant Physiology* 169(3), 1524–1531.

Wu, G. A., Terol, J., Ibanez, V., López-García, A., Pérez-Román, E., Borredá, C., Domingo, C., Tadeo, F. R., Carbonell-Caballero, J., Alonso, R., Curk, F., Du, D., Ollitrault, P., Roose, M. L., Dopazo, J., Gmitter, F. G., Rokhsar, D. S. and Talon, M. 2018. Genomics of the origin and evolution of *Citrus*. *Nature* 554(7692), 311–316.

Xie, K., Minkenberg, B. and Yang, Y. 2015. Boosting CRISPR/Cas9 multiplex editing capability with the endogenous tRNA-processing system. *Proceedings of the National Academy of Sciences of the United States of America* 112(11), 3570–3575.

Xu, B., Escamilla-Treviño, L. L., Sathitsuksanoh, N., Shen, Z., Shen, H., Percival Zhang, Y. H., Dixon, R. A. and Zhao, B. 2011. Silencing of 4-coumarate:coenzyme A ligase in switchgrass leads to reduced lignin content and improved fermentable sugar yields for biofuel production. *New Phytologist* 192(3), 611–625.

Xu, Q., Chen, L. L., Ruan, X., Chen, D., Zhu, A., Chen, C., Bertrand, D., Jiao, W. B., Hao, B. H., Lyon, M. P., Chen, J., Gao, S., Xing, F., Lan, H., Chang, J. W., Ge, X., Lei, Y., Hu, Q., Miao, Y., Wang, L., Xiao, S., Biswas, M. K., Zeng, W., Guo, F., Cao, H., Yang, X., Xu, X. W., Cheng, Y. J., Xu, J., Liu, J. H., Luo, O. J., Tang, Z., Guo, W. W., Kuang, H., Zhang, H. Y., Roose, M. L., Nagarajan, N., Deng, X. X. and Ruan, Y. 2013. The draft genome of sweet orange (*Citrus sinensis*). *Nature Genetics* 45(1), 59–66.

Xu, R. F., Li, H., Qin, R. Y., Li, J., Qiu, C. H., Yang, Y. C., Ma, H., Li, L., Wei, P. C. and Yang, J. B. 2015. Generation of inheritable and "transgene clean" targeted genome-modified rice in later generations using the CRISPR/Cas9 system. *Scientific Reports* 5, 11491.

Xue, L. J. and Tsai, C. J. 2015. AGEseq: analysis of genome editing by sequencing. *Molecular Plant* 8(9), 1428–1430, 10.1016/j.molp.2015.06.001.

Xue, L.-J., Alabady, M. S., Mohebbi, M. and Tsai, C.-J. 2015. Exploiting genome variation to improve next-generation sequencing data analysis and genome editing efficiency in *Populus tremula* x *alba* 717-1B4. *Tree Genetics & Genomics* 11, 82.

Zhou, X., Jacobs, T. B., Xue, L. J., Harding, S. A. and Tsai, C. J. 2015. Exploiting SNPs for biallelic CRISPR mutations in the outcrossing woody perennial Populus reveals 4-coumarate:CoA ligase specificity and redundancy. *New Phytologist* 208(2), 298–301.

Zhu, J., Song, N., Sun, S., Yang, W., Zhao, H., Song, W. and Lai, J. 2016. Efficiency and inheritance of targeted mutagenesis in maize using CRISPR-Cas9. *Journal of Genetics and Genomics = Yi Chuan Xue Bao* 43(1), 25–36.

Index

www.ingramcontent.com/pod-product-compliance
Ingram Content Group UK Ltd.
Pitfield, Milton Keynes, MK11 3LW, UK
UKHW021824100926
8556IPUK00003B/147

9 781786 76447